P9-AQF-642

Optical Information Processing

Optical Information Processing

FRANCIS T. S. YU

Electrical Engineering Department
The Pennsylvania State University

A Wiley-Interscience Publication

JOHN WILEY & SONS

New York Chichester Brisbane Toronto Singapore

Library of Congress Cataloging in Publication Data

Yu, Francis T. S., 1934–
Optical information processing.

Rev. ed. of: Introduction to diffraction, information processing, and holography. [1973]
Includes bibliographies and index.
1. Optical data processing. I. Title.
TA1630.Y8 1982 621.36 82-11057
ISBN 0-471-09780-2

Printed in the United States of America

10 9 8 7 6 5 4 3 2 1

To my students

Preface

The technique of wave front reconstruction was conceived by D. Gabor about three decades ago, which ultimately led to his receiving a Nobel prize in 1970. It was, however, the spatial frequency carrier concept of E. N. Leith and J. Upatnieks in 1962 that brought about holographic imaging of high quality for the first time. A few years before this rediscovery of holography by Leith and Upatnieks, L. J. Cutrona et al. brought optical data processing to light. Then in 1963, A. Vander Lugt's concept of complex spatial filtering brought optical information processing into wider use. Since then these applications have stimulated a profound relationship between optical processing and electrical engineering. This vigorous trend has continued, not only because optical systems are capable of performing certain complex operations, but also because the underlying optical theory is analogous to the input-output system concept. Because of the vast demand for engineering applications, about a decade ago I wrote a book entitled *Introduction to Diffraction, Information Processing, and Holography*. It was based mostly on lecture notes that I had written for my classes. Because of the rapid advances in this area, the book quickly fell behind the current trend of research. In order to close this gap, I have revised it for a more up-to-date text. Several sections that I now consider irrelevant have been deleted and I have added about 50 percent new material. This revised version includes current research in white-light optical processing, rainbow holography, and new techniques and applications of optical information processing.

The aim of this new text is to provide the student with a basic but comprehensive background in modern optical information processing and to make available an up-to-date text that is suitable for the current trend of interest. Furthermore, this book can also serve interested physicists and members of technical staffs. It contains three main topics—diffraction, information processing, and holography. The first part is intended to serve as a foundation, particularly for readers who are not familiar with the basic concepts of diffraction. This part covers essentially the same materials as are found in some of the well-known texts cited at the ends of the chapters. The second and third parts of the book are the main body of the text. Much of the material therein was derived from recently published articles.

In writing this book, I have for the most part approached the analysis by means of an elementary point concept (impulse response) and linear system theory. There are two major reasons for using this approach:

1 It simplifies the analysis, so that the solution of the problem can be actually calculated.
2 Electrical engineers are more familiar with the basic concepts of impulse response and linear system theory.

The material in this book, together with a few additional topics, was used in a series of two-quarter courses in optical information processing at The Pennsylvania State University. The students were mostly senior and first year graduate students. I have found that it is possible to teach the whole book without any significant omissions in a two-quarter period. With a few additional materials, the text can also be used for a two-semester course. It may be emphasized that the book is not intended to cover the vast area of optical information processing, but rather it is restricted to a region that I consider to be of practical importance.

In view of the great number of contributors in this area, I apologize for possible omissions of appropriate references in various parts of the book. In this connection, the excellent book, *Principles of Optics* by M. Born and E. Wolf deserves a special mention.

I am indebted to the late Dr. H. K. Dunn for his encouragement, criticism, and enormous assistance during the preparation of my first manuscript. I am grateful for the excellent research and publications of the workers in this field; special mention must be made of Professors L. J. Cutrona and E. N. Leith. I would also like to express my appreciation to the following: T. H. Chao, M. S. Dymek, P. H. Ruterbusch, and S. L. Zhuang for their excellent work and proofreading of most parts of the manuscript; X. X. Chen, who spent much of his valuable time making the drawings; and Miss Kim Williams and Miss Karen Penland, who typed most of the manuscript. As for my students, I appreciate their constant interest, enthusiasm, and motivation. Without their encouragement this book could not have been brought to completion. Finally, thanks go to my wife, Lucy, and our children, Peter, Ann, and Edward, for their unbounded love, patience, and understanding.

FRANCIS T. S. YU

October 1982
University Park, Pennsylvania

Contents

1

Linear System Theory and Fourier Transformation

It is well known that optical and quadrupolar systems exhibit similarities. For instance, any optical lens can be conveniently represented by a quadrupolar model. Thus the concepts of linear system theory are rather important in the analysis of modern coherent optical systems, for at least two reasons: (1) a great number of applications in modern optics may be assumed linear, at least within some specified ranges; and (2) an exact solution in the analysis of linear optical problems may be obtained by means of standard techniques.

Except for a very few special cases, there is no general procedure for analyzing nonlinear optical systems. Of course, there are practical ways of solving nonlinear optical problems that may involve some graphical or experimental approaches. Approximations are often necessary in solving nonlinear problems, and each situation may require special techniques. Fortunately, a great number of optical problems are linear, and these are generally solvable. However, we would like to emphasize that no optical system is strictly linear; linearization always involves the imposition of restrictions.

Since linearity leads to a simplification in mathematical analysis, it is our aim in this chapter to review some of the mathematical tools of linear analysis. However, it may be admitted that the development of the mathematics in this chapter is by no means rigorous, but it is rather along introductory lines. It may also be emphasized that optical information processing is generally two-dimensional in nature. Therefore in the development of the mathematical tools, we restrict our presentation to the two-dimensional case.

1.1 LINEAR SYSTEMS FROM A PHYSICAL VIEWPOINT

It is a common practice to describe the behavior of a physical system by relating the input excitation to the output response (fig. 1.1). Both excitation and re-

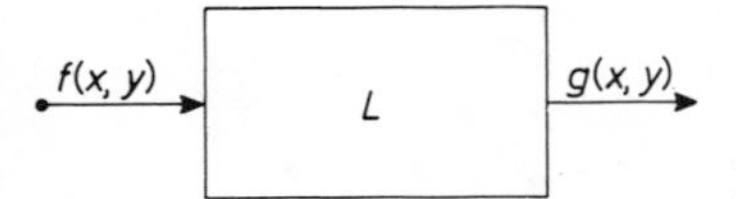

Fig. 1.1 A block diagram of a linear system.

sponse may be physically measurable quantities. Suppose an input excitation $f_1(x, y)$ produces an output response $g_1(x, y)$, and a second excitation $f_2(x, y)$ produces a second response $g_2(x, y)$. We may write

$$f_1(x, y) \to g_1(x, y) \tag{1.1}$$

and

$$f_2(x, y) \to g_2(x, y), \tag{1.2}$$

respectively. Then for a linear system we have

$$f_1(x, y) + f_2(x, y) \to g_1(x, y) + g_2(x, y). \tag{1.3}$$

Equation (1.3), in conjunction with eqs. (1.1) and (1.2), represents the *additivity* property of the linear system. Thus a necessary condition for a system to be linear is that the *principle of superposition* hold. The principle of superposition implies that the presence of one excitation does not affect the response due to the other excitations.

Now, if the input excitation of a physical system is $Cf_1(x, y)$, where C is an arbitrary constant, then the output response is $Cg_1(x, y)$, ie.,

$$Cf_1(x, y) \to Cg_1(x, y). \tag{1.4}$$

Equation (1.4) represents the *homogeneity* characteristic of the linear system. Thus a property of a linear system is the preserving of the magnitude scale factor. Therefore, a physical system is linear if and only if eqs. (1.3) and (1.4) are satisfied. In other words, if a system possesses the additivity and homogeneity properties, then it is a linear system.

There is, however, another important physical aspect that characterizes a linear system with constant parameters. That is, if the input excitation $f(x, y)$ applied to such a system is an alternating function of x and y with spatial frequencies p and q, and results in an alternating output response $g(x, y)$ with the same spatial frequencies p and q, then the system is said to have a *spatial invariance*. In other words, a spatially invariant system will not generate new spatial frequencies. The qualification of spatial invariance implies that, if

$$f(x, y) \to g(x, y),$$

then

$$f(x - x_1, y - y_1) \to g(x - x_1, y - y_1), \tag{1.5}$$

where x_1 and y_1 are arbitrary spatial constants. A linear system that possesses the spatial invariance property of eq. (1.5) is called a *linear spatially invariant system*.

The physical characteristics of linear systems will be much clearer from the viewpoint of Fourier analysis, the development of which is given in the following few sections.

1.2 FOURIER TRANSFORMATION AND THE SPATIAL FREQUENCY SPECTRUM

Fourier transforms are particularly important in the analysis of modern optical information processing and in wave front reconstruction. In this section we consider first a class of complex functions $f(x, y)$ that satisfy the following sufficient conditions:

1 $f(x, y)$ must be sectionally continuous in every finite region over the (x, y) plane, i.e., there must be only a finite number of discontinuities.

2 $f(x, y)$ must be absolutely integrable over the (x, y) plane, i.e.,

$$\iint_{-\infty}^{\infty} |f(x, y)| \, dx \, dy < \infty. \tag{1.6}$$

Then these functions can be represented by the equation

$$f(x, y) = \frac{1}{4\pi^2} \iint_{-\infty}^{\infty} F(p, q) \exp\left[i(px + py)\right] dp \, dq, \tag{1.7}$$

where

$$F(p, q) = \iint_{-\infty}^{\infty} f(x, y) \exp\left[-i(px + qy)\right] dx \, dy, \tag{1.8}$$

and p and q are the corresponding spatial frequency variables.

Equations (1.7) and (1.8) are known as the two-dimensional *Fourier transform pair*. Equation (1.8) is often called the *Fourier transform*, and eq. (1.7) is known as the *inverse Fourier transform*. For brevity, eqs. (1.7) and (1.8) can be written as

$$f(x, y) = \mathscr{F}^{-1}[F(p, q)] \tag{1.9}$$

and

$$F(p, q) = \mathscr{F}[f(x, y)]. \tag{1.10}$$

where $\mathscr{F}^{-1}$ and $\mathscr{F}$ denote the inverse and direct Fourier transformations, respectively.

It may be noted that $F(p, q)$ is, in general, a complex function:

$$F(p, q) = |F(p, q)| \exp [i\phi(p, q)], \tag{1.11}$$

where $|F(p, q)|$ and $\phi(p, q)$ are frequently referred to as the *amplitude* and *phase* spectra, and $F(p, q)$ is known as the *Fourier spectrum* or *spatial frequency spectrum*.

A complex function of two independent variables is called a *separable* function if and only if it can be written as a product of two functions, each of which is a single-variable function:

$$f(x, y) = f(x) f(y). \tag{1.12}$$

From eq. (1.12) we may conclude that the Fourier transform of a two-dimensional separable function is the product of the Fourier transform of each of the one-dimensional functions, i.e.,

$$F(p, q) = F(p) F(q). \tag{1.13}$$

By way of illustration, we next consider the Fourier analyses of two elementary functions that are of considerable importance in optical information processes.

Fourier Transformation of a Circular Symmetric Function

Given the circular symmetric function

$$f(x, y) = f(r) = \begin{cases} 1, & r \leq 1, \\ 0, & \text{otherwise}, \end{cases} \tag{1.14}$$

where $r^2 = x^2 + y^2$. To determine the Fourier transformation of eq. (1.14), we first apply the coordinate transformations:

$$\begin{aligned} x &= r \cos \theta, & r^2 &= x^2 + y^2, \\ y &= r \sin \theta, & \theta &= \tan^{-1} (y/x), \\ p &= \rho \cos \phi, & \rho^2 &= p^2 + q^2, \\ q &= \rho \sin \phi, & \phi &= \tan^{-1} (q/p), \\ dx\, dy &= r\, dr\, d\theta. \end{aligned} \tag{1.15}$$

By proper substitution of eqs. (1.14) and (1.15) in eq. (1.8), we have

$$F(\rho) = F(p, q) = \int_0^1 r\, dr \int_0^{2\pi} \exp [-ir \rho \cos (\theta - \phi)]\, d\theta. \tag{1.16}$$

By the use of the identity

$$J_0(z) = \frac{1}{2\pi} \int_0^{2\pi} \exp[-iz \cos(\theta - \phi)]\, d\theta, \tag{1.17}$$

where J_0 is the zero-order Bessel function of the first kind, eq. (1.16) can be reduced to

$$F(\rho) = 2\pi \int_0^1 J_0(r\rho)\, r\, dr. \tag{1.18}$$

By applying the well-known integral

$$z J_1(z) = \int_0^z \alpha J_0(\alpha)\, d\alpha, \tag{1.19}$$

eq. (1.18) becomes

$$F(\rho) = \frac{2\pi}{\rho} J_1(\rho), \tag{1.20}$$

where J_1 is the first-order Bessel function of the first kind. Equation (1.20) is the required Fourier transform.

By the way, it can be shown that the Fourier transform of any circular symmetric function is

$$F(\rho) = 2\pi \int_0^\infty f(r) J_0(\rho r)\, r\, dr. \tag{1.21}$$

and its inverse Fourier transform is

$$f(r) = \frac{1}{2\pi} \int_0^\infty F(\rho) J_0(\rho r)\, \rho\, d\rho. \tag{1.22}$$

Equation (1.21) is frequently referred to as the *Hankel transform* (of order zero) of $f(r)$. It may be emphasized that the Hankel transform is no more than a special case of the two-dimensional Fourier transform. Thus any property of the Fourier transform can equivalently be applied to the Hankel transform.

Spectrum of the Two-Dimensional Dirac Delta Function

The two-dimensional Dirac delta function, $\delta(x - x_0, y - y_0)$, is defined on the real spatial domain of the xy plane, as existing at (x_0, y_0), but having zero value elsewhere on the xy spatial plane. Also by definition, it has the property

$$\iint_{-\infty}^{\infty} \delta(x - x_0, y - y_0)\, dx\, dy = 1. \tag{1.23}$$

Thus the delta function is a pulse of zero spatial duration but of infinite amplitude. The Fourier transform of the delta function is

$$F(p, q) = \mathscr{F}[\delta(x - x_0, y - y_0)] = \exp[-i(px_0 + qy_0)]. \quad (1.24)$$

Therefore, the corresponding amplitude and phase components may be given, respectively, as

$$|F(p, q)| = 1 \quad (1.25)$$

and

$$\phi(p, q) = -(px_0 + qy_0). \quad (1.26)$$

Equation (1.25) shows that the amplitude spectrum of the delta function is a contiuous spatial frequency function of the unit height, which extends over the whole spatial frequency domain.

If the delta function occurs at the origin (0, 0), of the spatial domain, then the phase spectrum of eq. (1.26) vanishes, and eq. (1.24) becomes

$$F(p, q) = 1. \quad (1.27)$$

It should also be noted that, for any other value of (x_0, y_0), the unit spectral vector rotates continuously with the phase angle of $-(px_0 + qy_0)$.

1.3 PROPERTIES OF THE FOURIER TRANSFORM

We now consider some of the basic properties of the Fourier transform that we find useful in optical information processing and wave front reconstruction. These basic properties are presented here as mathematical theorems. Since the proofs are similar to those for the one-dimensional case, the proofs are simply sketched in, with no pretense at rigor.

1 Linearity

Given that C_1 and C_2 are two arbitrary complex constants. If $f_1(x, y)$ and $f_2(x, y)$ are Fourier transformable, i.e., if

$$\mathscr{F}[f_1(x, y)] = F_1(p, q)$$

and

$$\mathscr{F}[f_2(x, y)] = F_2(p, q),$$

then

$$\mathscr{F}[C_1 f_1(x, y) + C_2 f_2(x, y)] = C_1 F_1(p, q) + C_2 F_2(p, q). \qquad (1.28)$$

Proof: The proof of the property can be directly derived from the definition of the Fourier transform, eq. (1.8).

2 *Translation Theorem*

If $f(x, y)$ is Fourier transformable, so that $\mathscr{F}[f(x, y)] = F(p, q)$, then

$$\mathscr{F}[f(x - x_0, y - y_0)] = F(p, q)\, exp\, [-i(px_0 + qy_0)], \qquad (1.29)$$

where x_0 and y_0 are arbitrary real constants.

Proof: The proof of this theorem follows from the definition

$$\mathscr{F}[f(x - x_0, y - y_0)] = \iint_{-\infty}^{\infty} f(x - x_0, y - y_0) \exp[-i(px + qy)]\, dx\, dy.$$

If we let the variables $x' = x - x_0$ and $y' = y - y_0$, then from the above equation it can readily be shown that

$$\mathscr{F}[f(x - x_0, y - y_0)] = \exp[-i(px_0 + qy_0)] \iint_{-\infty}^{\infty} f(x', y') \times \exp[-i(px' + qy')]\, dx'\, dy' = \exp[-i(px_0 + qy_0)]\, F(p, q).$$

Essentially, this theorem tells us that the translation of a function in the spatial domain causes a linear phase shift in the spatial frequency domain.

3 *Reciprocal Translation Theorem*

If $f(x, y)$ is Fourier transformable, so that $\mathscr{F}[f(x, y)] = F(p, q)$, then

$$\mathscr{F}\{f(x, y) \exp[-i(xp_0 + yq_0)]\} = F(p + p_0, q + q_0), \qquad (1.30)$$

where p_0 and q_0 are arbitrary real constants.

Proof: The proof of this theorem can be obtained immediately by application of the Fourier transformation

$$\mathscr{F}\{f(x, y) \exp[-i(xp_0 + yq_0)]\} = \iint_{-\infty}^{\infty} f(x, y) \exp\{-i[(p + p_0)\, x + (q + q_0)\, y]\}\, dx\, dy = F(p + p_0, q + q_0).$$

4 *Scale Change in a Fourier Transform*

If $f(x, y)$ is Fourier transformable, so that $\mathscr{F}[f(x, y)] = F(p, q)$, then

$$\mathscr{F}[f(ax, by)] = \frac{1}{ab} F\left(\frac{p}{a}, \frac{q}{b}\right), \tag{1.31}$$

where a and b are arbitrary complex constants.

Proof:

$$\begin{aligned}\mathscr{F}[f(ax, by)] &= \iint_{-\infty}^{\infty} f(ax, by) \exp[-i(px + qy)]\, dx\, dy \\ &= \frac{1}{ab} \iint_{-\infty}^{\infty} f(ax, by) \exp\left[-i\left(\frac{p}{a} ax + \frac{q}{b} by\right)\right] d(ax)\, d(by) \\ &= \frac{1}{ab} F\left(\frac{p}{a}, \frac{q}{b}\right).\end{aligned}$$

This theorem shows that a magnification of the spatial domain results in demagnification in the spatial frequency domain, and an overall reduction in the amplitude spectrum.

5 *Parseval's Theorem*

If $f(x, y)$ is Fourier transformable, so that $\mathscr{F}[f(x, y)] = F(p, q)$, then

$$\iint_{-\infty}^{\infty} |f(x, y)|^2\, dx\, dy = \frac{1}{4\pi^2} \iint_{-\infty}^{\infty} |F(p, q)|^2\, dp\, dq. \tag{1.32}$$

Proof:

$$\iint_{-\infty}^{\infty} |f(x, y)|^2\, dx\, dy = \iint_{-\infty}^{\infty} f(x, y) f^*(x, y)\, dx\, dy,$$

where $*$ denotes the complex conjugate. This equation can be written as

$$\begin{aligned}&\iint_{-\infty}^{\infty} |f(x, y)|^2\, dx\, dy \\ &\qquad = \iint_{-\infty}^{\infty} dx\, dy \left\{\frac{1}{4\pi^2} \iint_{-\infty}^{\infty} F(p', q') \exp[i(xp' + yq')]\, dp'\, dq'\right\} \\ &\qquad\quad + \left\{\frac{1}{4\pi^2} \iint_{-\infty}^{\infty} F^*(p'', q'') \exp[-i(xp'' + yq'')]\, dp''\, dq''\right\},\end{aligned}$$

which can be reduced to

$$\iint_{-\infty}^{\infty} |f(x, y)|^2 \, dx \, dy$$

$$= \frac{1}{4\pi^2} \iint_{-\infty}^{\infty} F(p', q') \, dp' \, dq' \iint_{-\infty}^{\infty} F^* (p'', q'') \, dp'' \, dq''$$

$$\times \frac{1}{4\pi^2} \iint_{-\infty}^{\infty} \exp \{i[x(p' - p'') + y(q' - q'')]\} \, dx \, dy$$

$$= \frac{1}{4\pi^2} \iint_{-\infty}^{\infty} F(p', q') \, dp' \, dq' \iint_{-\infty}^{\infty} F^*(p'', q'') \, dp'' \, dq'' \; \delta(p' - p'', q' - q'')$$

$$= \frac{1}{4\pi^2} \iint_{-\infty}^{\infty} |F(p, q)|^2 \, dp \, dq.$$

It should be clear to the reader that Parseval's theorem implies the conservation of energy.

6 Convolution Theorem

If $f_1(x, y)$ and $f_2(x, y)$ are Fourier transformable, so that

$$\mathscr{F}[f_1(x, y)] = F_1(p, q)$$

and

$$\mathscr{F}[f_2(x, y)] = F_2(p, q).$$

then

$$\mathscr{F}\left[\iint_{-\infty}^{\infty} f_1(x, y) \, f_2(\alpha - x, \beta - y) \, dx \, dy\right] = F_1(p, q) \, F_2(p, q). \tag{1.33}$$

We may write this as

$$\mathscr{F}[f_1(x, y) * f_2(x, y)] = F_1(p, q) \, F_2(p, q), \tag{1.34}$$

where $*$ denotes the convolution operation.

Proof:

$$\mathscr{F}\left[\iint_{-\infty}^{\infty} f_1(x, y) \, f_2(\alpha - x, \beta - y) \, dx \, dy\right]$$

$$= \iint_{-\infty}^{\infty} f_1(x, y) \; \mathscr{F}[f_2(\alpha - x, \beta - y)] \, dx \, dy$$

$$= \left\{\iint_{-\infty}^{\infty} f_1((x, y) \exp [-i(px + qy)] \, dx \, dy\right\} F_2(p, q) = F_1(p, q) \, F_2(p, q).$$

This convolution theorem will be found very useful in our study of linear spatially invariant systems.

7 Autocorrelation Theorem (Wiener–Khinchin Theorem)

If $f(x, y)$ is Fourier transformable, so that $\mathcal{F}[f(x, y)] = F(p, q)$, then

$$\mathcal{F}\left\{\iint_{-\infty}^{\infty} f(x, y)\, f^*(x - \alpha, y - \beta)\, dx\, dy\right\} = |F(p, q)|^2. \tag{1.35}$$

We may write this as

$$\mathcal{F}[R(\alpha, \beta)] = \mathcal{F}[f(x, y) \circledast f^*(x, y)] = |F(p, q)|^2, \tag{1.36}$$

where

$$R(\alpha, \beta) = \iint_{-\infty}^{\infty} f(x, y)\, f^*(x - \alpha, y - \beta)\, dx\, dy$$

is known as the *autocorrelation function,* and $\circledast$ denotes the correlation operation.

Conversely, we have

$$\mathcal{F}^{-1}[|F(p, q)|^2] = R(x, y). \tag{1.37}$$

Equations (1.36) and (1.37) constitute a Fourier transform pair. In other words, the theorem states that the autocorrelation function and power spectral density are Fourier transforms of each other.

Proof:

$$\mathcal{F}\left[\iint_{-\infty}^{\infty} f(x, y)\, f^*(x - \alpha, y - \beta)\, dx\, dy\right]$$

$$= \mathcal{F}\left[\iint_{-\infty}^{\infty} f(\xi + \alpha, \eta + \beta)\, f^*(\xi, \eta)\, d\xi\, d\eta\right]$$

$$= \iint_{-\infty}^{\infty} d\xi\, d\eta\, f^*(\xi, \eta)\, \mathcal{F}[f(\xi + \alpha, \eta + \beta)]$$

$$= \iint_{-\infty}^{\infty} d\xi\, d\eta\, f^*(\xi, \eta) \exp[i(p\xi + q\eta)]\, F(p, q) = F^*(p, q)\, F(p, q).$$

8 Crosscorrelation Theorem

If $f_1(x, y)$ and $f_2(x, y)$ are Fourier transformable, so that $\mathcal{F}[f_1(x, y)] = F_1(p, q)$ and $\mathcal{F}[f_2(x, y)] = F_2(p, q)$, then

$$\mathscr{F}[R_{12}(\alpha, \beta)] = \mathscr{F}[f_1(x, y) \circledast f_2^*(x, y)] = F_1(p, q)\, F_2^*(p, q) \qquad (1.38)$$

and

$$\mathscr{F}[R_{21}(\alpha, \beta)] = \mathscr{F}[f_1^*(x, y) \circledast f_2(x, y)] = F_1^*(p, q)\, F_2(p, q), \qquad (1.39)$$

where

$$R_{12}(\alpha, \beta) = \iint_{-\infty}^{\infty} f_1(x + \alpha, y + \beta)\, f_2^*(x, y)\, dx\, dy,$$

and

$$R_{21}(\alpha, \beta) = \iint_{-\infty}^{\infty} f_1^*(x - \alpha, y - \beta)\, f_2(x, y)\, dx\, dy$$

are the *crosscorrelation functions*.

Proof: The proof of this theorem is similar to that of the autocorrelation theorem.

9 Some Symmetric Properties

(a) If $f(x, y)$ is real, i.e., if $f^*(x, y) = f(x, y)$, then

$$F^*(-p, -q) = F(p, q), \qquad (1.40)$$

and

$$F^*(p, q) = F(-p, -q). \qquad (1.41)$$

(b) If $f(x, y)$ is real and even, i.e., if $f(x, y) = f^*(x, y) = f^*(-x, -y)$, then $F(p, q)$ is real and even.

$$F(p, q) = F^*(-p, -q) = F^*(p, q). \qquad (1.42)$$

(c) If $f(x, y)$ is real and odd, i.e., if $f(x, y) = f^*(x, y) = -f(-x, -y)$, then

$$F(p, q) = -F(-p, -q) = F^*(-p, -q). \qquad (1.43)$$

Proof: The proof of these properties can be obtained by direct application of the definition of the Fourier transformation, eq. (1.8).

The Fourier transform properties that we have just mentioned may not be adequate for all the applications in modern optical information processing. However, we use these properties frequently in the course of this book, and thereby avoid much tedious calculation.

1.4 RESPONSE OF A LINEAR SPATIALLY INVARIANT SYSTEM

In sec. 1.1 we discussed the basic properties of a linear system from a physical point of view. In this Section we develop the mathematical viewpoint more fully. It is well known that a linear system may be represented by a linear operator L, as shown in the block diagram of fig. 1.1. Since the operator is linear, we have the following property:

$$L[C_1 f_1(x, y) + C_2 f_2(x, y)] = C_1 L[f_1(x, y)] + C_2 L[f_2(x, y)], \quad (1.44)$$

where C_1 and C_2 are arbitrary complex constants.

If the system has spatial invariance, then the linear operator must also have spatial invariance.

$$L[f(x - x_0, y - y_0)] = g(x - x_0, y - y_0), \quad (1.45)$$

where $g(x, y)$ is the output response.

It may be interesting to note that, if the input excitation to a linear spatially invariant system is the Dirac delta function $\delta(x, y)$, then the *output impulse response* is

$$h(x, y) = L[\delta(x, y)]. \quad (1.46)$$

Now if the input excitation is an arbitrary function of $f(x, y)$, which may be represented by the *convolution integral*

$$f(x, y) = \iint_{-\infty}^{\infty} f(x', y')\, \delta(x - x', y - y')\, dx'\, dy', \quad (1.47)$$

then the output response may be written

$$g(x, y) = L\left[\iint_{-\infty}^{\infty} f(x', y')\, \delta(x - x', y - y')\, dx'\, dy'\right]$$

$$= \iint_{-\infty}^{\infty} f(x', y')\, L[\delta(x - x', y - y')]\, dx'\, dy'. \quad (1.48)$$

It follows that

$$g(x, y) = \iint_{-\infty}^{\infty} f(x', y')\, h(x - x', y - y')\, dx'\, dy'. \quad (1.49)$$

Equation (1.49) shows the remarkable property of the output response of a linear spatially invariant system: the response is the convolution of the impulse response with the input excitation.

Furthermore, from the *Fourier convolution theorem* in the previous section, eq. (1.49) can be written in the Fourier transform form,

$$G(p, q) = F(p, q)\, H(p, q), \tag{1.50}$$

where $G(p, q) = \mathscr{F}[g(x, y)]$, $F(p, q) = \mathscr{F}[f(x, y)]$, and $H(p, q) = \mathscr{F}[h(x, y)]$. That is, the Fourier transform of the output response of a linear spatially invariant system is the product of the Fourier transform of the input excitation and the Fourier transform of the impulse response. We may emphasize the fact that, if the impulse response of a linear system is known, then the corresponding Fourier transform is the *system transfer function*.

1.5 DETECTION OF SIGNAL BY MATCHED FILTERING

A problem of considerable importance in optical information processing is the detection of a signal corrupted by random noise. It is our aim in this section to discuss a special type of optimum linear filter, known as a *matched filter*, which is useful in complex spatial filtering and in optical pattern or character recognition. We first derive an expression of the spatial filter transfer function on a somewhat general basis, namely, for a stationary additive noise. Then we immediately move to a *white* (i.e., of uniform spectral density over the spatial frequency domain) stationary additive noise.

It is well known in communicative theory that the signal-to-noise ratio at the output end of a correlator can be improved to a large degree. Let us consider the input excitation to a linear filtering system to be an additive mixture of signal $s(x, y)$ and a stationary random noise $n(x, y)$, i.e.,

$$f(x, y) = s(x, y) + n(x, y). \tag{1.51}$$

Let the output response of the linear filter due to the signal $s(x, y)$ alone be $s_0(x, y)$, and that due to the random noise $n(x, y)$ alone be $n_0(x, y)$. The figure of merit, on which the filter design is based, is the output signal-to-noise ratio at $x = y = 0$,

$$\frac{S}{N} \triangleq \frac{|s_0(0, 0)|^2}{\sigma^2}, \tag{1.52}$$

where σ^2 is the mean-square value of the output noise.

In terms of the filter transfer function $H(p, q)$ and the Fourier transform $S(p, q)$ of the input signal $s(x, y)$, these quantities may be written

$$s_0(0, 0) = \frac{1}{4\pi^2} \iint_{-\infty}^{\infty} H(p, q)\, S(p, q)\, dp\, dq \tag{1.53}$$

and

$$\sigma^2 = \frac{1}{4\pi^2} \iint_{-\infty}^{\infty} |H(p, q)|^2\, N(p, q)\, dp\, dq, \tag{1.54}$$

where $|H(p, q)|^2 N(p, q)$ is the power spectral density of the noise at the output end of the filter, and $N(p, q)$ is the power spectral density of the noise at the input end. Thus the output signal-to-noise ratio may be expressed explicitly in terms of the filter function $H(p, q)$,

$$\frac{S}{N} = \frac{1}{4\pi^2} \frac{\left| \iint_{-\infty}^{\infty} H(p, q)\, S(p, q)\, dp\, dq \right|^2}{\iint_{-\infty}^{\infty} |H(p, q)|^2\, N(p, q)\, dp\, dq}. \tag{1.55}$$

The objective of the filter designer is to specify a filter function such that the output signal-to-noise ratio is a maximum. To obtain such a filter transfer function, the designer may apply the Schwarz inequality, which states that

$$\frac{\left| \iint_{-\infty}^{\infty} u(x, y)\, v^*(x, y)\, dx\, dy \right|^2}{\iint_{-\infty}^{\infty} |u(x, y)|^2\, dx\, dy} \leq \iint_{-\infty}^{\infty} |v^*(x, y)|^2\, dx\, dy, \tag{1.56}$$

where $u(x, y)$ and $v(x, y)$ are arbitrary functions and * denotes the complex conjugate. The equality in eq. (1.56) holds if and only if $u(x, y)$ is proportional to $v(x, y)$.

To make the Schwartz inequality applicable to eq. (1.55), it may be expedient to express the output noise spectral density as the product of two conjugate factors:

$$N(p, q) = N_1(p, q)\, N_1^*(p, q). \tag{1.57}$$

Then eq. (1.55) may be written as

$$\frac{S}{N} = \frac{1}{4\pi^2} \frac{\left| \iint_{-\infty}^{\infty} [H(p, q)\, N_1(p, q)] \left[\frac{S^*(p, q)}{N_1^*(p, q)} \right]^* dp\, dq \right|^2}{\iint_{-\infty}^{\infty} |H(p, q)\, N_1(p, q)|^2\, dp\, dq} \tag{1.58}$$

If we identify the bracketed quantities of eq. (1.58) as $u(x, y)$ and $v(x, y)$, then, in view of the Schwarz inequality, we have

$$\frac{S}{N} \leq \frac{1}{4\pi^2} \iint_{-\infty}^{\infty} \frac{|S(p, q)|^2}{N(p, q)}\, dp\, dq\,. \tag{1.59}$$

The equality in eq. (1.59) holds if and only if the filter function is

$$H(p, q) = K\frac{S^*(p, q)}{N(p, q)}, \tag{1.60}$$

where K is a proportionality constant. The corresponding value of the output signal-to-noise ratio is therefore

$$\frac{S}{N} = \frac{1}{4\pi^2} \iint_{-\infty}^{\infty} \frac{|S(p, q)|^2}{N(p, q)}\, dp\, dq. \tag{1.61}$$

It may be interesting to note that, if the stationary additive input noise is white, then the optimum filter function is

$$H(p, q) = KS^*(p, q), \tag{1.62}$$

which is proportional to the conjugate of the signal spectrum. This opimum filter is then said to be matched to the input signal $s(x, y)$. The output spectrum of the matched filter is therefore proportional to the power spectral density of the input signal, i.e.,

$$G(p, q) = K|S(p, q)|^2. \tag{1.63}$$

Consequently, we see that the phase variation at the output end of the matched filter vanishes. In other words, the matched filter has the capability of eliminating all the phase variations of $S(p, q)$ across the spatial frequency domain.

In concluding this chapter, we would like to emphasize again that the Fourier treatment presented in this chapter makes no attempt at mathematical rigor and completeness, but it is rather introduced for the operational purpose of this book. For the reader interested in a more complete treatment, we would suggest the excellent texts by Papoulis (ref. 1.1), and by Bracewell (ref. 1.2). Between these two books, a complete treatment of Fourier transform theory and some of the applications of Fourier analysis can be found.

PROBLEMS

1.1 Determine the Fourier transforms of the following functions:

(a) $f(x) = \exp(ipx)$.

(b) $f(x) = \cos px$.

(c) $f(x) = \Sigma_{n=-\infty}^{\infty} \delta(x - na)$.

(d) $f(x) = \begin{cases} 1, & |x| \leq a/2, \\ 0, & |x| > a/2. \end{cases}$

(e) $f(x) = \exp(-ax^2)$.

Here p is an angular spatial frequency, δ is the Dirac delta function, and α is an arbitrary positive constant.

1.2 Suppose the Fourier transforms of $f(x, y)$ and $g(x, y)$ exist, i.e., $\mathscr{F}[f(x, y) = F(p, q)$ and $\mathscr{F}[g(x, y)] = G(p, q)$, respectively. Show that

$$\iint_{-\infty}^{\infty} f(x, y)\, g^*(x, y)\, dx\, dy = \frac{1}{4\pi^2} \iint_{-\infty}^{\infty} F(p, q)\, G^*(p, q)\, dp\, dq.$$

1.3 If $f(x, y)$ is Fourier transformable, i.e., if $\mathscr{F}[f(x, y)] = F(p, q)$ exists, show that

$$\frac{\displaystyle\iint_{-\infty}^{\infty} f(x, y)\, dx\, dy}{f(0, 0)} = \frac{4\pi^2 F(0, 0)}{\displaystyle\iint_{-\infty}^{\infty} F(p, q)\, dp\, dq}.$$

1.4 With reference to the Hankel transform pair of eqs. (1.21) and (1.22), if $f(r)$ and $g(r)$ are Hankel transformable, i.e., if $\mathscr{H}[f(r)] = F(\rho)$ and $\mathscr{H}[g(r)] = G(\rho)$ exist, derive the following properties:

(a) $\mathscr{H}[f(ar)] = \dfrac{1}{a^2} F\left(\dfrac{\rho}{a}\right)$,

where a is an arbitrary constant.

(b) $\mathscr{H}[f(r) + g(r)] = F(\rho) + G(\rho)$.

(c) $\mathscr{H}\left[\int_0^{\infty}\int_0^{2\pi} f(r')\, g(\alpha)\, r'\, dr'\, d\theta\right] = F(\rho)\, G(\rho)$.

where $\alpha^2 = r^2 + r'^2 - 2rr' \cos\theta$.

(d) $2\pi \int_0^{\infty} f(r)\, r\, dr = F(0)$.

(e) $\dfrac{1}{2\pi} \int_0^{\infty} r^2 f(r)\, r\, dr = -\dfrac{1}{2\pi^2} \dfrac{d^2F(\rho)}{d\rho^2}\bigg|_{\rho=0}$.

1.5 Consider a one-dimensional band-limited function $f(x)$ sampled at Nyquist's rate (i.e., minimum sampling rate at p_0/π). Show that the function $f(x)$ may be written as

$$f(x) = \sum_{n=-\infty}^{\infty} f(nx_0) \frac{\sin p_0(x - nx_0)}{p_0(x - nx_0)},$$

where p_0 is the highest angular spatial frequency of the function $f(x)$ and

$x_0 = \pi/p_0$, the Nyquist sampling interval. (This equation is also known as Shannon's sampling theorem.)

1.6 We extend the sampling theorem in the previous problem to a two-dimensional band-limited function $f(x, y)$. Show that the function $f(x, y)$ may be written as

$$f(x, y) = \sum_{n=-\infty}^{\infty} \sum_{m=-\infty}^{\infty} f(nx_0, my_0) \frac{\sin p_0(x - nx_0) \sin q_0(y - my_0)}{p_0(x - nx_0)\, q_0(y - my_0)},$$

where p_0 and q_0 are the highest angular spatial frequency with respect to the x and y coordinates of $f(x, y)$, and $x_0 = \pi/p_0$ and $y_0 = \pi/q_0$ are the respective Nyquist sampling intervals.

1.7 In a manner similar to that of the previous problem, the sampling theorem can also be applied to a circularly band-limited function, i.e., one where the spatial frequency is band limited within a circle of radius r_0 in the spatial frequency domain. Show that

$$f(x, y) = \frac{\pi}{2} \sum_{n=-\infty}^{\infty} \sum_{m=-\infty}^{\infty} f\left(\frac{n\pi}{r_0}, \frac{m\pi}{r_0}\right) \frac{J_1[\sqrt{(r_0x - n\pi)^2 + (r_0y - m\pi)^2}]}{\sqrt{(r_0x - n\pi)^2 + (r_0y - m\pi)^2}},$$

where J_1 is the first-order Bessel function of the first kind.

1.8 Given a spatial random **noise** $n(x, y)$, which is assumed to have a zero mean over the space Σ of (x, y) coordinates, i.e.,

$$\lim_{\Sigma\to\infty} \frac{1}{\Sigma} \iint_{\Sigma} n(x, y)\, dx\, dy = 0.$$

Let us denote $s(x, y)$ as a useful spatial signal. If $s(x,y)$ and $n(x, y)$ are assumed uncorrelated, show that

$$\lim_{\Sigma\to\infty} \frac{1}{\Sigma} \iint_{\Sigma} s(x', y')\, n(x' - x, y' - y)\, dx'\, dy' = 0$$

for every (x, y).

1.9 Show that the autocorrelation function of the sum of a given signal $s(x, y)$ and an additive random noise $n(x, y)$ is the sum of the individual autocorrelations and their crosscorrelations.

1.10 Suppose a useful signal is described by

$$s(x, y) = \exp\,[(-a + ib)\,(x^2 + y^2)],$$

where a and b are arbitrary positive constants. This signal is assumed to be embedded in random additive noise $n(x, y)$ of zero mean:

$$f(x, y) = s(x, y) + n(x, y).$$

The received signal $f(x, y)$ is fed into an input-output linear spatially invariant system that has a spatial impulse response of

$$h(x, y) = s^*(-x, -y).$$

Calculate the output response. Make a rough sketch of the functions $|s(x, y)|$ and

$$g(x, y) = \iint_{-\infty}^{\infty} s(x', y')\, s^*(x' - x, y' - y)\, dx'\, dy'.$$

Compare the envelopes of these two plots and give a brief discussion of their significance in regard to signal detection.

1.11 Let us consider a short symmetric pulse $f(x)$ with a spatial duration of Δx, which has its maximum value at $x = 0$, i.e., $f(0) \geq f(x)$. Since this symmetric pulse $f(x)$ is Fourier transformable, show that the following uncertainty relationship holds:

$$\Delta x\, \Delta p \geq 2\pi,$$

where Δx and Δp are the nominal spatial duration and nominal spatial bandwidth of the $f(x)$.

1.12 Let us assume an optical signal $s(x - x_0, y - y_0)$ is corrupted by an additive Gaussian noise $n(x, y)$ with zero mean and variance σ^2, where x_0 and y_0 are arbitrary constants. If the signal is fed into an optical processor for matched filtering, as shown in the linear system diagram of fig. 1.2, where K is an arbitrary constant, $S(p, q)$ is the Fourier spectrum of the signal $s(x, y)$, and * denotes the complex conjugate, then:

(a) Evaluate the output complex light distribution.

(b) Sketch an output signal coordinate system where the correlation peak can be identified.

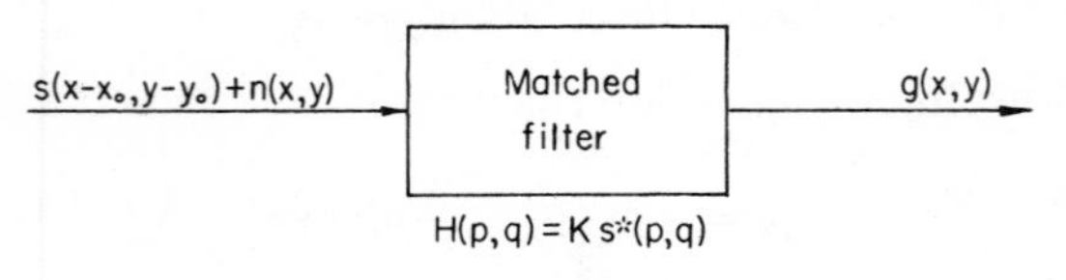

Fig. 1.2

REFERENCES

1.1 A. Papoulis, *The Fourier Integral and Its Applications,* McGraw-Hill, New York, 1962.

1.2 R. N. Bracewell, *The Fourier Transform and its Applications,* McGraw-Hill, New York, 1965.

1.3 W. M. Brown, *Analysis of Linear Time-Invariant Systems.* McGraw-Hill, New York, 1963.

1.4 D. K. Cheng, *Analysis of Linear Systems,* Addison-Wesley, Reading, Mass., 1959.

1.5 W. B. Davenport, Jr. and W. L. Root, *An Introduction to the Theory of Random Signals and Noise,* McGraw-Hill, New York, 1958.

1.6 M. J. Lighthill, *Introduction to Fourier Analysis and Generalized Functions,* Cambridge University Press, New York, 1960.

2

Introduction to Diffraction

2.1 GENERAL ASPECTS

In fig. 2.1 a source of light is shown illuminating a screen, except for the shadow of an opaque object placed between source and screen. Let us examine the sharpness of the edge of the shadow. It is clear that, if the source of light subtends an appreciable angle, then the shadow's edge will fade gradually from full illumination to total darkness, since there are points on the screen from which a part but not all of the source can be seen. To avoid this effect, let us make the source so small that we can consider it a single point of light. Under this condition, and with the usual assumption of geometrical optics that rays of light in a homogeneous medium travel in straight lines, we would expect the edge of the shadow to be quite sharp, so that points on the screen would either be fully illuminated or lie in total darkness.

If we set up such a system, the edge of the shadow at first glance appears to be sharp. If we examine it more closely, however, we find that the illumination on the screen fades gradually over a short distance. In addition, if the source of light is monochromatic, there will be seen narrow light and dark bands parallel to the edge of the geometrical shadow. These bands are called *fringes*. It is apparent that the light, in passing the edge of the obstacle, deviates from a straight-line propagation. This effect is called *diffraction*.

Diffraction can be explained from the theory of wave motion. Historically, it was the observation of diffraction that led to general acceptance of the wave theory of light, as opposed to the corpuscular theory. The wave theory shows that the magnitude of the diffraction effect, i.e., the angle of deviation from straight-line propagation, is directly proportional to the wavelength. Thus it is that with a sound wave we are usually not conscious of a shadow at all, because of the large diffraction angle in the case of the long sound waves. Even with sound, however, shadows can be demonstrated if supersonic frequencies are used. Since the wavelengths of visible light are extremely small, the diffraction

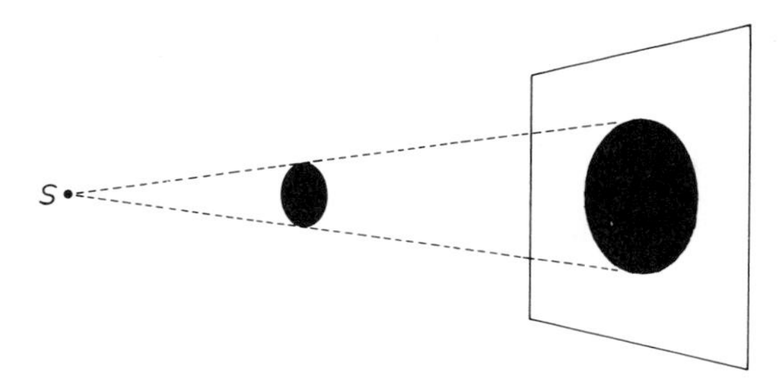

Fig. 2.1 Shadow formed by an opaque object and a point source S.

is also small. The straight-line propagation assumed in geometrical optics is only the limit approached as the wavelength is allowed to approach zero.

The theory of diffraction is treated in the next chapter on the basis of *Huygens' principle,* in the form given by Kirchhoff. This predicts results very close to the effects actually observed and gives at least a qualitative understanding of the phenomenon. It is also a much less difficult approach than that which uses the wave equation in a field bounded by the shadowing object. The latter approach is generally satisfactory only in some special cases in which simplifying assumptions can be made. Nevertheless, some interesting conclusions about diffraction can be deduced from the complete wave theory, as are elaborated in the remaining sections of this chapter.

2.2 FRAUNHOFER AND FRESNEL DIFFRACTION

In the preceding section we have spoken of the diffraction of light as it passes the edge of an obstacle. Diffraction may be treated more simply if we consider the light as passing through one or more small apertures in a diffracting obstacle. The former case is then the limit as the dimensions of the aperture become infinite.

It is customary to divide diffraction into two cases, depending on the distances of light source and viewing screen from the diffracting screen, and these cases have been given the names of two early investigators of diffraction. If the source and viewing point are so far from the diffracting screen that lines drawn from the source or viewing point to all points of the apertures do not differ in length by more than a small fraction of a wavelength, the phenomenon is called *Fraunhofer diffraction*. If these conditions do not hold, it is called *Fresnel diffraction*. The boundary between these two cases is somewhat arbitrary, and depends upon the accuracy desired in the results. In most cases it is sufficient to use the Fraunhofer methods if the difference in distances does not exceed one-twentieth of a wavelength. Of course, Fraunhofer diffraction may be achieved without a great physical distance of the source, if a collimating lens is used to make the rays of light from the source nearly parallel.

Figure 2.2 is drawn to illustrate the above considerations. A source of mono-

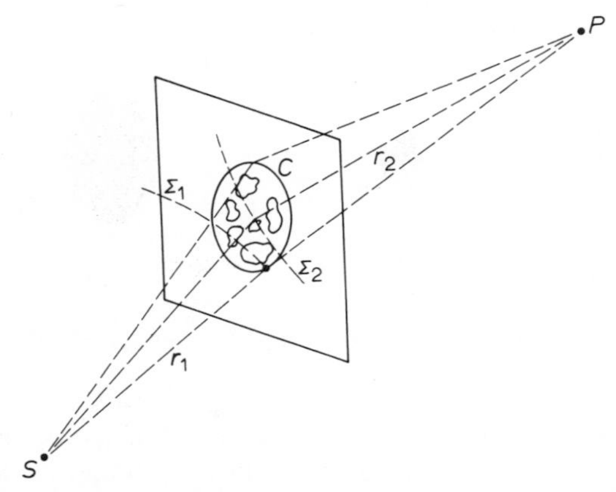

Fig. 2.2 Geometry for defining Fraunhofer and Fresnel diffraction.

chromatic light is at S, and a viewing point at P, and between them is an opaque screen having a finite number of apertures. Let a circle C be drawn on the plane of the screen, and let it be as small as possible while still enclosing all the apertures. Let C be the base of cones with S and P as vertices. Also draw spherical surfaces Σ_1 and Σ_2, having S and P as centers, and as radii r_1 and r_2, the shortest distances from S and P to the circle C. If the largest distances from C to Σ_1 and to Σ_2 are not more than one-twentieth of the wavelength of the light used, then the diffraction is Fraunhofer, and the light falling on the observing screen at P forms a *Fraunhofer diffraction pattern*.

On the other hand, if by reason of the large size of C, or the shortness of the distances to S or P, the distances between C and Σ_1 or Σ_2 are greater than one-twentieth of the wavelength, then we must speak of Fresnel diffraction and a *Fresnel diffraction pattern* at P.

The radius of the circle C of fig. 2.3 is denoted by ρ, the shortest distance of S from the screen is l, and the greatest separation of sphere and screen is Δl. From the definition of Fraunhofer diffracton, Δl must be a small fraction of a wavelength. However, ρ may be many wavelengths long (as it can also be in the Fresnel case). In the right triangle of fig. 2.3 we have

$$(l + \Delta l)^2 = l^2 + \rho^2, \tag{2.1}$$

and, because of the small size of $(\Delta l)^2$ in comparison with the other quantities, we may make the approximation

$$l \simeq \frac{\rho^2}{2\Delta l}. \tag{2.2}$$

As an example, suppose that ρ is 1 cm, and that the source has the longest wavelength in the visible red, i.e., about 8×10^{-5} cm. Let Δl be one-twentieth of this, or 0.4×10^{-5} cm. The distance l would then be approximately 1.25 km.

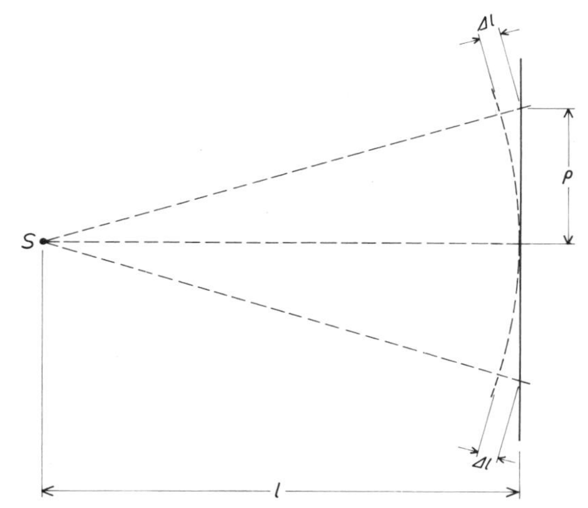

Fig. 2.3 Geometry for determining the value of ℓ for Fraunhofer diffraction.

If violet light of half the wavelength were used, l would be 2.5 km. The requirements for Fraunhofer diffraction can thus be rather stringent.

2.3 FRAUNHOFER DIFFRACTION FROM MULTIPLE APERTURES

Let the diffracting screen be in the form of a plane, and let it have n apertures. Furthermore, let the apertures be identical in size, shape, and orientation. Choose an origin O and axes of coordinates in the plane, by which the location of the apertures may be specified, as shown in fig. 2.4. Each aperture is identified by a point Q_i. The position of this point with respect to the ith aperture is the same for all i.

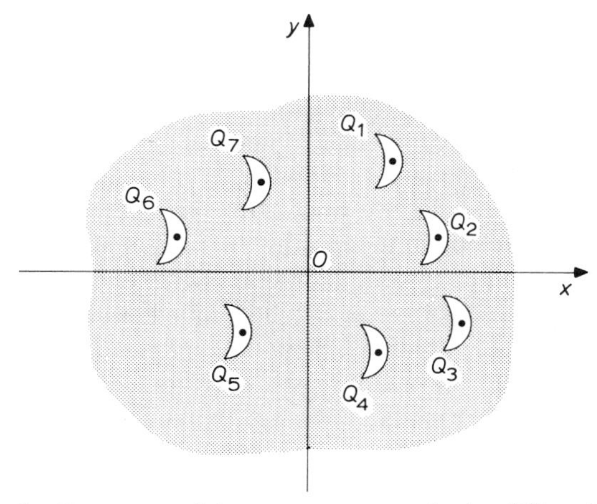

Fig. 2.4 Geometry of the open apertures in the diffracting screen.

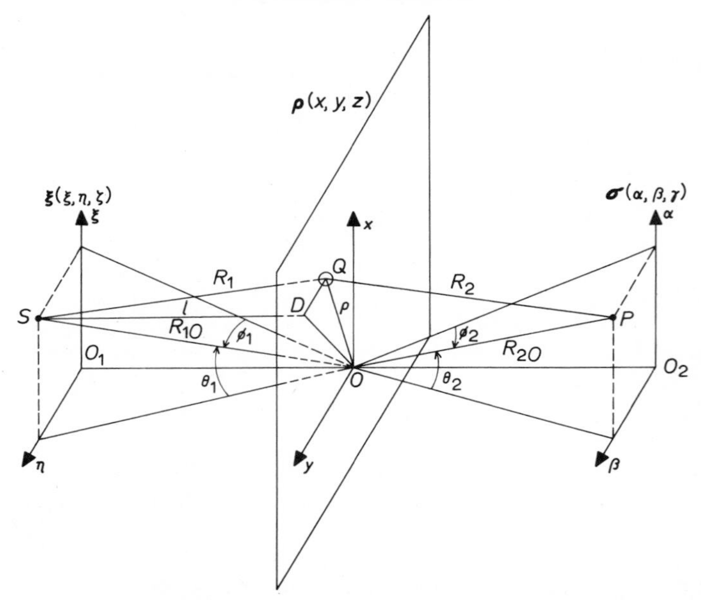

Fig. 2.5 The defined coordinate systems. The points (ξ,η) define the source plane; (x,y), the diffracting screen; and (α,β), the image or observing plane.

This diffracting screen is located between the source S and the viewing point P, as shown in fig. 2.5. (Only one of the many apertures is indicated in this figure.) Draw planes through S and P parallel to the plane of the diffracting screen, and draw a line through O perpendicular to this screen. Let the intersections of this line with the other planes, at O_1 and O_2, be the origins of coordinates in these planes, and let the ξ and η axes of the source plane and the α and β axes of the viewing plane be parallel to the x and y axes of the diffracting plane. The line from O to S makes angles ϕ_1 with the O_1Ox plane and θ_1 with the O_1Oy plane. Similarly, the line OP makes angles ϕ_2 and θ_2 with these planes. Arrows in the figure indicate the positive directions of these angles.

Consider for the moment the illumination of the viewing screen due to a single aperture in the diffracting screen. The geometrical spot of light is larger than the size of the aperture—on the order of twice as large if S and P are at equal distances from the aperture. In addition, however, there is a spreading due to diffraction at the aperture. Even though the diffraction angle is small, the large distance to P required in Fraunhofer diffraction permits the spread of light to be great, so that the diffraction pattern at P will be much larger than the geometrical spot. The angle of spread, in radians, is given approximately by the ratio of the wavelength of the light to the width of the aperture.

According to Huygen's principle (which is explained more fully at the end of this chapter), the different points in the plane of the aperture may be treated as if they were secondary sources of light, each source radiating in all directions into the space beyond the aperture. The secondary sources all have the same amplitude, since their distances from the primary source S are all very nearly the same under the Fraunhofer conditions. Their slight differences in distance, however, amount to appreciable fractions of a wavelength of the light; hence these secondary sources differ in phase.

All points in the receiving plane are equally illuminated by the various secondary sources in the aperture, and, again because of the large distance, the amplitudes at the different points in the plane of P are substantially equal. If the aperture were reduced in size to a single point, the irradiance* over the receiving plane would be uniform. However, because of the different phases in the secondary sources in an aperture of appreciable size, and the further phase changes brought about by the slight differences in distance from aperture to receiving plane, there are considerable phase differences at different points on this plane. At some points there will be increased irradiance because of favorable phase combination in the light from the secondary sources, while at other points there will be destructive interference. This is the reason for the light and dark fringes in the diffraction pattern.

Now let us suppose that a second aperture is opened up on the diffraction screen. This will be in a shifted position with respect to the first, but will still lie within the circle C of fig. 2.2. Then all that we have said above about phase differences but lack of appreciable amplitude differences applies equally well to the pattern from the new aperture. The pattern from the new aperture alone would be like that from the first, but of course in a shifted position, as determined by drawing lines from S through the two apertures and continuing to the receiving plane. This shift will be small compared with the size of the diffraction pattern from either aperture. The combined pattern from both apertures will depend on the phase combinations from the two separate patterns.

The strength of the irradiance at P due to an aperture Q may be expressed in terms of that which would result if the aperture were at the origin O on the diffracting screen (see fig. 2.5). Thus we may suppose that the electric field at P, when the aperture is at O, is given by the complex number $Ae^{i\phi}$, where A is a constant and ϕ is a phase angle that varies with the position of the point P. The phase angle ϕ also changes if the aperture is moved to a point other than O. Now if the aperture at Q is considered instead of the one at O, the difference of paths from S to P may be reduced to a phase shift by dividing it by the wavelength λ, and multiplying by 2π. Thus in fig. 2.5, let $d_1 = R_1 - R_{1O}$ and $d_2 = R_2 - R_{2O}$. Then the complex electric field at P due to the aperture at Q is given by

$$Ae^{i\phi} \exp\left[ik(d_1 + d_2)\right] \tag{2.3}$$

where $k = 2\pi/\lambda$. The quantity $(d_1 + d_2)$ may be called the geometrical path difference between the apertures at Q and at O, and $k(d_1 + d_2)$ is the phase difference at P for the two positions.

*The definition of *irradiance* follows eq. 2.13.

The geometrical path difference may also be expressed in terms of the coordinates of S, Q, and P of fig. 2.5. Suppose a line is drawn from S perpendicular to the diffraction plane. This line is parallel to O_1O, and has the same length. Call this length l. Call the point of intersection with the diffraction plane D. It has the same coordinates as S, i.e., ξ and η. Draw lines from D to $Q(x, y)$ and to $O(0, 0)$. Two right triangles are formed, having R_1 and R_{1O} as hypotenuses and having l as a common leg. Also notice that

$$(DO)^2 = \xi^2 + \eta^2 \qquad \text{and} \qquad (DQ)^2 = (\xi - x)^2 + (\eta - y)^2.$$

Then

$$l^2 = R_1^2 - (\xi - x)^2 - (\eta - y)^2 = R_{1O}^2 - \xi^2 - \eta^2, \tag{2.4}$$

from which

$$R_1^2 - R_{1O}^2 = x^2 + y^2 - 2(x\xi + y\eta). \tag{2.5}$$

Now $R_1 - R_{1O}$ has been defined as d_1, and $x^2 + y^2$, in fig. 2.5, is ρ^2. Also, R_1 and R_{1O} are so nearly equal that $R_1 + R_{1O}$ may be taken as $2R_{1O}$. Equation (2.5), then, becomes

$$d_1 \simeq \frac{\rho^2}{2R_{1O}} - \frac{x\xi + y\eta}{R_{1O}}. \tag{2.6}$$

The first term on the right side of eq. (2.6) represents the path difference when ξ and η are both zero, i.e., when S coincides with O_1. According to our requirements for Fraunhofer diffraction, as already stated, this difference might be as much as one-twentieth of the wavelength of the light used. In terms of phase shift, this would be an angle of $\pi/10$. We propose to neglect this first term of eq. (2.6), and if a phase shift of $\pi/10$ seems too large an angle to neglect, then we will make our conditions for Fraunhofer diffraction stricter. In other words, we can always arrange matters so that the wave front arriving at the diffraction screen is essentially plane.

As for the remaining term in eq. (2.6), it is clear that d_1 will not represent an appreciable phase shift unless either ξ or η, or both, are considerably larger in absolute value than the corresponding x and y when Q lies close to O, as is usually the case. Note also from fig. 2.5 that $\xi/R_{1O} = \sin\theta_1$, and $\eta/R_{1O} = \sin\phi_1$. Equation (2.6) then reduces to

$$d_1 \simeq -x \sin\theta_1 - y \sin\phi_1. \tag{2.7}$$

In an exactly similar manner, and with the same strict Fraunhofer conditions, we can treat the geometrical path difference d_2,

$$d_2 \simeq -x \sin \theta_2 - y \sin \phi_2. \tag{2.8}$$

The complete path difference for the apertures at Q and at O is the sum of eqs. (2.7) and (2.8). Note that it can be either positive or negative, depending on the signs of the angles. We would expect this ambiguity of sign from our definitions of $d_1 = R_1 - R_{1O}$ and $d_2 = R_2 - R_{2O}$. The angles ϕ_1, θ_1, ϕ_2, θ_2 are not restricted, up to values of $\pm\pi/2$.

Substituting eqs. (2.7) and (2.8) into eq. (2.3) gives the electric field at P due to the single aperture Q in terms of the coordinates of Q, S, and P. Let us go a step further, and assume now that there are N identical apertures (except for their coordinates) and add the respective fields. Note that each aperture is treated here as so small in itself that there is no appreciable phase shift involved when different points within the same aperture are considered. Then

$$\begin{aligned} E_p = {} & Ae^{i\phi} \exp(-i\omega t) \sum_{n=1}^{N} \exp\{-ik[x_n(\sin \theta_1 + \sin \theta_2) \\ & + y_n(\sin \phi_1 + \sin \phi_2)]\}, \end{aligned} \tag{2.9}$$

where (x_n, y_n) are the coordinates of the defining point Q_n of the nth aperture. It is consistent with our previous assumption of long distances for S and P that we use the same angles for all the apertures. The factor $\exp(-i\omega t)$ is introduced to provide the time variation of the light signal; ω has the usual value, $\omega = 2\pi c/\lambda$, where c is the velocity of light.

The E_p of eq. (2.9) is the time-varying magnitude of the electric vector of the light wave at P. We may also write down the magnetic vector:

$$\mathbf{H}_p = \frac{\mathbf{k}}{k} \times \mathbf{E}_p. \tag{2.10}$$

The $\mathbf{k}$ in eq. (2.10) is the wave vector, and $\times$ denotes the vector product.

We can now obtain the time average of the Poynting vector of the wave, given by

$$\langle \mathbf{S} \rangle = \mathrm{Re}[\mathbf{S}] = \tfrac{1}{2}\,\mathrm{Re}(\mathbf{E} \times \mathbf{H}^*). \tag{2.11}$$

In this equation $\mathbf{S}$ is the Poynting vector, and $\langle\ \rangle$ represents the time average; Re is the real part. Combining eq. (2.11) with eqs. (2.10) and (2.9),

$$\langle \mathbf{S}_p \rangle = \langle \mathbf{S}_p \rangle_1 \left| \sum_{n=1}^{N} \exp\{-ik[x_n(\sin \theta_1 + \sin \theta_2) + y_n(\sin \phi_1 + \sin \phi_2)]\} \right|^2, \tag{2.12}$$

where $\langle \mathbf{S}_p \rangle_l = (\mathbf{k}/2k)\, A^2$ is the time average of the Poynting vector at P when there is a single aperture at the origin. For convenience, let us abbreviate eq. (2.12) as

$$I = I_1|G|^2, \tag{2.13}$$

where I and I_1 represent the time averages of the Poynting vectors, and

$$G = \sum_{n=1}^{N} \exp \{-ik[x_n(\sin \theta_1 + \sin \theta_2) + y_n(\sin \phi_1 + \sin \phi_2)]\}. \tag{2.14}$$

The I in equation (2.13) represents the *irradiance* at P, that is, the average time rate of the energy falling on a unit area at P, perpendicular to the direction of wave propagation. This I combines light from all apertures of the diffraction screen. The irradiance I_1 is that due to light that has passed through the central aperture only.

It may be noted that the x and y axes of fig. 2.5 were freely chosen, the only restriction being that the origin O should be close to the system of apertures. A different set of axes would change the x's and y's of eq. (2.12), and along with them the θ's and ϕ's, but the value of I would be unchanged. It is therefore expedient to choose the axes in such a way as to make the calculation of $|G|^2$ simple.

2.4 SPECIAL CASE OF TWO APERTURES

In 1801, Thomas Young performed an experiment that is credited with bringing about general acceptance of the wave theory of light. This experiment was a demonstration of the diffraction of light by two close apertures. He provided a very small source by using the sunlight passing through a pinhole in an otherwise opaque screen, then letting the light from this source fall on a screen containing two pinholes. Of course his light was not monochromatic, and the fringes produced by the different wavelengths fell in different places on the viewing screen.

Let us assume a single wavelength λ from a point source, and let the light fall on two similar apertures separated by a distance d. Let one of these be at the origin O, and the other on the x axis, as shown in fig. 2.6. We can now apply the theory of the preceding section. The summation in eq. (2.14) will have only two terms, with x and y coordinates of (0, 0) and (d, 0). It therefore reduces to

$$G = 1 + \exp [-ikd (\sin \theta_1 + \sin \theta_2)]. \tag{2.15}$$

The angles θ_1 and θ_2 are unknown, since in our choice of coordinate axes we have made no assumption with regard to the positions of S and P.

In eq. (2.15), $k = 2\pi/\lambda$, so that $kd = 2\pi d/\lambda$, and $d(\sin \theta_1 + \sin \theta_2)$ is the geometrical path difference $SQ_1P - SQ_2P$, found by the sum of eqs. (2.7) and

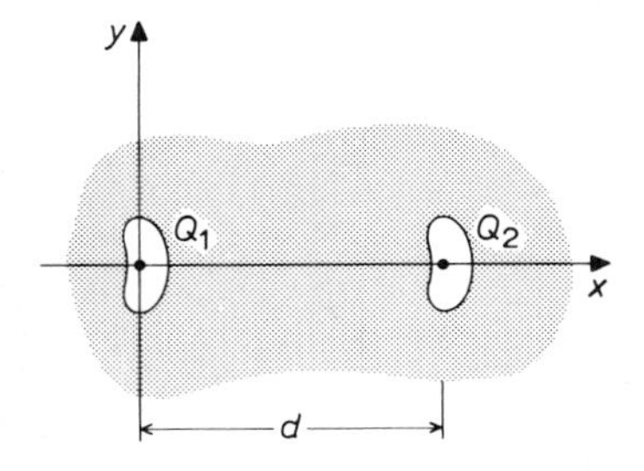

Fig. 2.6 Two identical apertures.

(2.8) (with the signs reversed). Let us define the *optical path difference* δ as the geometrical path difference (without regard to sign) divided by the wavelength. In other words, δ is the path difference from S to P, for light passing through Q_1 or through Q_2, as measured in wavelengths. In this case it is given by

$$\delta = \frac{d}{\lambda}(\sin\theta_1 + \sin\theta_2), \tag{2.16}$$

and eq. (2.15) reduces to

$$G = 1 + e^{-i2\pi\delta} = 2e^{-i\pi\delta}\cos(\pi\delta). \tag{2.17}$$

Then,

$$|G|^2 = 4\cos^2(\pi\delta). \tag{2.18}$$

A vector representation of eq. (2.17) is given in fig. 2.7. Equation (2.18) may be substituted into eq. (2.13), thus

$$I = 4I_1\cos^2(\pi\delta). \tag{2.19}$$

As the point P is moved about in the observing plane, the angle θ_2 changes if the motion of P is parallel with the x axis, that is, parallel to the line between the two apertures. By eq. (2.16) this changes the value of δ. We see then from eq. (2.19) that the irradiance I goes through a series of maxima and minima, having a maximum of $4I_1$ when δ is zero or any integer (positive or negative), and a minimum of zero when δ is an odd multiple of one-half. In other words, when the path difference is an exact number of wavelengths, there is construc-

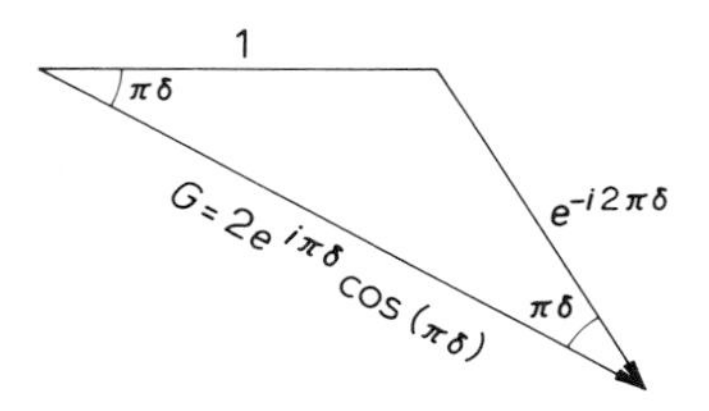

Fig. 2.7 Vector representation of eq. (2.17).

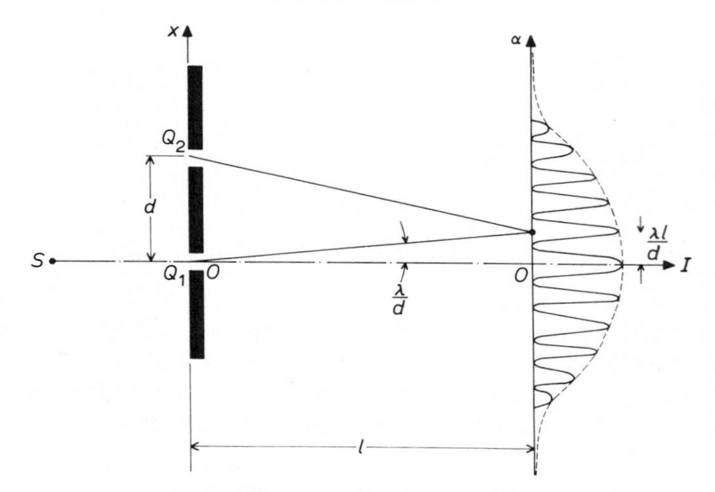

Fig. 2.8 Diffraction due to two identical apertures.

tive interference between the waves from the two apertures; when the difference is an odd number of half-wavelengths, the interference is destructive.

The variation of irradiance for two apertures is illustrated in fig. 2.8. The solid curve along the α axis is a plot of eq. (2.19), while the dashed envelope represents the variation of $4I_1$.

For a photograph of the actual pattern produced on an observing screen by a single circular aperture of appreciable size, refer to fig. 3.11. The dark ring surrounding the central bright region has an angular diameter (with the aperture as center) of approximately λ/d radians, where d is the diameter of the aperture. The center of the circular spot is at the point on the observing plane that is directly opposite the aperture in the diffraction screen. If l is the distance between these planes, the linear diameter of the dark ring is approximately $\lambda l/d$.

Finally, with the two circular apertures discussed in this section, the diffraction pattern looks like that of fig. 2.9. The variation of irradiance vertically across this figure corresponds to the curve drawn on the α axis in fig. 2.8. In the other direction the $|G|^2$ of eq. (2.18) does not apply.

2.5 THE RECIPROCITY THEOREM

The reciprocity theorem states that, if the source is placed at an image point P, the same irradiance will be observed at the source point S as appeared at P when the source was at S. Thus there is a symmetry between a motion of P and a motion of S. The theorem is equivalent to saying that the diffraction system is both *linear* and *bilateral*. The proof of the theorem lies in showing that there is no change in the I in eq. (2.13) if ϕ_1 and ϕ_2 are interchanged, along with θ_1 and

Fig. 2.9 Diffraction pattern of two circular apertures.

θ_2. That this is true for the factor $|G|^2$ of eq. (2.13) is at once apparent from an examination of eq. (2.14). The determination of I_1 has not yet been explained in detail. It is shown in the next chapter that I_1 is also unchanged by the exchange of source point and image point.

The reciprocity theorem suggests a method of designating the irradiance at P as the source is moved about over a surface Σ near S. Let a source of unit intensity be placed at P, and let the resulting irradiance be observed at each point of the pattern on Σ. Call this the *sensitivity,* a function of the surface, $\Lambda(\Sigma)$. Then if the source is placed at any point of Σ, the resulting irradiance at P can be calculated by multiplying the Λ of the point where the source is placed by the intensity of the source.

2.6 HUYGENS' PRINCIPLE

By means of Huygens' principle it is possible to obtain by graphical methods the shape of a wave front at any instant if the wave front at an earlier instant is known. The principle may be stated as follows: Every point of a wave front may be considered as the source of a small secondary wavelet, which spreads in all directions from the point at the wave propagation velocity. A new wave front is found by constructing a surface tangent to all the secondary wavelets. If the

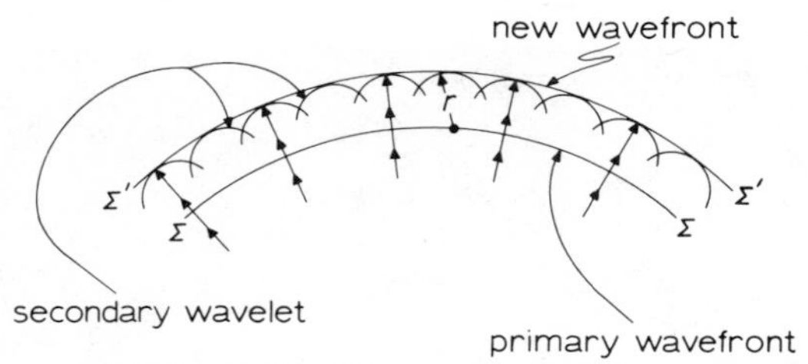

Fig. 2.10 Huygens' principle.

velocity of propagation is not constant for all parts of the wave front, then each wavelet must be given its appropriate velocity.

An illustration of the use of Huygens' principle is given in fig. 2.10. The known wave front is shown by the arc $\Sigma\Sigma$, and the directions of propagation are indicated by small arrows. To determine the wave front after an interval Δt, with a wave velocity v, simply construct a series of spheres of radius $r = v\Delta t$ from each point of the original front $\Sigma\Sigma$. These spheres represent the secondary wavelets. The envelope enclosing the surfaces of the spheres represents the new wave front. This is the surface marked $\Sigma'\Sigma'$ in the figure. In this example the magnitude of the wave velocity is considered to be the same at all points.

It may be noted that Huygens' principle predicts the existence of a backward wave, which, however, is never observed. The explanation for this discrepancy can be found by examining the interference of the secondary wavelets throughout the space surrounding Σ. The main use of Huygens' principle is that it enables us to predict diffraction patterns. This is elaborated in later chapters. The principle when first stated was a useful method for finding the shape of a new wave front, and little physical reality was attached to the secondary wavelets at that time. Later, as the wave nature of light came to be more fully understood, Huygens' principle took on a deeper significance than was at first supposed.

2.7 A SCALAR TREATMENT OF LIGHT

Up to this point we have treated the diffraction of light from a representation of Maxwellian electromagnetic theory. In studying diffraction in this way we have been able to show how the interaction between the light from different apertures affects the complete diffraction pattern. This, however, is not a complete solution of the diffraction problem, since it does not explain in detail the role played by the sizes and shapes of the apertures. Thus the factor $|G|^2$ of eq. (2.13) is written out explicitly in eq. (2.14). The factor I_1, however, has only been introduced as being the irradiance due to a single aperture, without explaining how this factor is affected by the characteristics of the aperture. In most cases it is extremely difficult to reach such a complete solution by the Maxwell technique. One problem that has been solved in this way is that of monochromatic light passing the straight edge of a thin conducting sheet. The solution was obtained by Sommerfeld (ref. 2.5, p. 247). However, most cases remain unsolved by the vector method.

Fortunately, however, there are scalar methods that can be applied, and that in most cases give diffraction results of high accuracy. It is not to be expected that these scalar methods can be effectively employed where the *polarization* of the light is an important factor. The early theories of Huygens and Young on the propagation of light waves lead readily to scalar methods of treatment. However, it was not until after the introduction of the Maxwell theory, in 1856, that Kirchhoff brought out, in 1882, a generally satisfactory scalar method. This is the method we propose to follow here. Readers interested in further details of the historical development are referred to *Principles of Optics* by Born and Wolf (ref. 2.1). In Maxwell's theory a *wave equation* is derived in the form

$$\nabla^2 u - \frac{1}{c^2}\frac{\partial^2 u}{\partial t^2} = 0, \tag{2.20}$$

where c is the velocity of the wave and ∇^2 is the Laplacian differential operator

$$\nabla^2 = \frac{\partial^2}{\partial x^2} + \frac{\partial^2}{\partial y^2} + \frac{\partial^2}{\partial z^2}.$$

The u in eq. (2.20) may represent either the electric vector **E** or the magnetic vector **H** of the field in free space. In the scalar treatment, we assume that the same eq. (2.20) applies, but with u in this case representing a purely scalar quantity, the *amplitude* of the wave.

In the following, we attempt to find solutions of eq. (2.20) that may be applied to diffraction phenomena. The results are not exact, but in many cases are very close to the facts disclosed by experiment. Note also that the treatment is not confined to light alone, but may also be applied to other forms of wave propagation (such as sound) in which the wavelength is small compared with aperture dimensions and in which polarization is unimportant.

Given a point source that radiates uniformly in all directions (i.e., an *isotropic* source), the amplitude in the free field at a distance r from the source may be expressed as

$$u(r, t) = \frac{1}{r} f(r - ct). \tag{2.21}$$

It may be verified that the solution (2.21) satisfies eq. (2.20), except at $r = 0$. This singularity is unimportant, since in fact a physical source cannot have a zero radius. The f in eq. (2.21) represents some function of the quantity in parentheses. The form of f depends upon the nature of the source. If we are dealing with what we call a monochromatic wave, then f is a sine or cosine function, or in general

$$f(r - ct) = a \cos [k(r - ct) + \theta], \tag{2.22}$$

where a is a positive constant, k is the wave number $2\pi/\lambda$, and θ is a phase angle. If now eq. (2.22) is written in complex form and then substituted into eq. (2.21), but with $ae^{i\theta}$ replaced by the complex constant A,

$$u(r, t) = \frac{A}{r} \exp [i(kr - \omega t)]. \tag{2.23}$$

The ω in eq. (2.23) is the angular frequency of the wave; $\omega = kc$. To recapitulate, eq. (2.23) gives the scalar amplitude of the wave from a point source of light that is monochromatic and isotropic, at a distance r from the source, in a free (i.e., unbounded) field.

It may also be said for the scalar eq. (2.20) that, if two different forms of u are solutions, then their sum is also a solution. This is called the *principle of superposition,* which holds for the scalar treatment of light, as it is in the vector theory.

2.8 KIRCHHOFF'S INTEGRAL

Suppose that we wish to know the scalar amplitude at a point P that is in the combined field of a number of monochromatic sources, all of the same wavelength. Suppose also that none of the sources coincides with P. Then if we have complete knowledge of all the sources—their positions, amplitudes, and phases—the amplitude at P can be found from the principle of superposition. Kirchhoff, however, proposed that such a knowledge of the sources is unnecessary, provided that at each point of an arbitrary surface Σ, surrounding P but not enclosing any source, the amplitude and its normal spatial derivatives are known.

The above idea was developed by Kirchhoff into what is known as Kirchhoff's integral, which is very useful in the scalar theory of diffraction. The development makes use of Green's second identity, which may be described as follows: Let U and V be any two complex scalar fields, functions of position only, and let a closed surface Σ enclose a region σ in the space where U and V exist. Now if U and V are continuous and have continuous second derivatives both on and within the surface Σ, then

$$\iint_{\Sigma} \left(V\frac{dU}{dn} - U\frac{dV}{dn} \right) d\Sigma = \iiint_{\sigma} (V\nabla^2 U - U\nabla^2 V)\, d\sigma, \tag{2.24}$$

where d/dn is a derivative normal to Σ in the outward direction, $d\Sigma$ is an element of surface, and $d\sigma$ an element of volume.

The space σ has the outer boundary Σ. Let us also give it an inner boundary Σ', which we set arbitrarily as a small sphere with its center at P, the point where

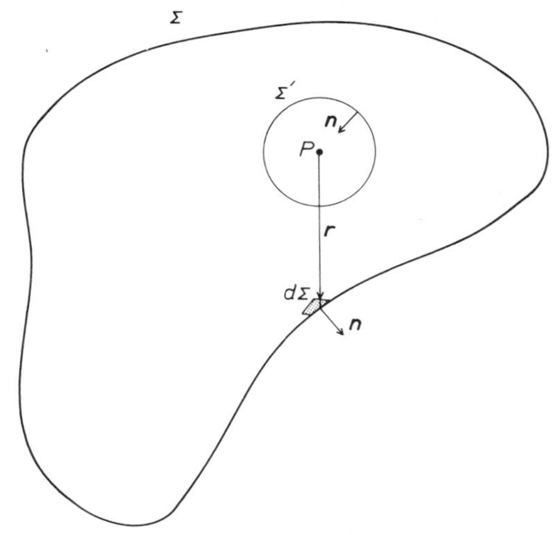

Fig. 2.11 The enclosed surfaces used in determining Kirchhoff's integral from Green's second identity.

the field amplitude is to be found. Figure 2.11 illustrates this situation. Here σ is the space between Σ and Σ' only, and the positive normals on the bounding surfaces are oriented away from σ, as indicated by the small arrow.

We have assumed that the entire scalar light wave is monochromatic. We can therefore express its amplitude u in complex form

$$u = u_0 e^{-i\omega t}, \tag{2.25}$$

where u_0 is a function of position only. It follows from eq. (2.20) that u_0 satisfies the equation

$$\nabla^2 u_0 = -k^2 u_0, \tag{2.26}$$

which is known as the Helmholtz equation. Here $k = \omega/c$, just as in eq. (2.23).

It was said in connection with eq. (2.24) that V and U could be any two scalar fields. Let us take them as

$$V = u_0, \tag{2.27}$$

$$U = \frac{1}{r} e^{ikr}, \tag{2.28}$$

where r is the radial distance from P. Note that the V and U given in eqs. (2.27) and (2.28) satisfy the continuity conditions on solutions of eq. (2.24).

Just as with u_0 in eq. (2.26), U satisfies the wave equation

$$\nabla^2 U = -k^2 \frac{1}{r} e^{ikr}, \tag{2.29}$$

except at $r = 0$. But since the point has been excluded from the space σ, eq. (2.29) is true for all of this space.

We can now substitute eqs. (2.27), (2.28), (2.29), and (2.26) into eq. (2.24), remembering that the surface integral must now be taken over both Σ and Σ'. We also note that U has been defined in such a way that the volume integral in eq. (2.24) becomes zero. Then

$$\iint\limits_{\Sigma} \left[u_0 \frac{d}{dn} \left(\frac{1}{r} e^{ikr} \right) - \frac{1}{r} e^{ikr} \frac{du_0}{dn} \right] d\Sigma + \iint\limits_{\Sigma'} \left[-u_0 \frac{d}{dr} \left(\frac{1}{r} e^{ikr} \right) + \frac{1}{r} e^{ikr} \frac{\partial u_0}{\partial r} \right] d\Sigma' = 0. \tag{2.30}$$

In the last term of the Σ' integral in eq. (2.30), d/dn has been replaced by $-\partial/\partial r$, since Σ' is a spherical surface. If the derivative of the first term of the Σ' integral in eq. (2.30) is carried out, this integral becomes

$$\iint\limits_{\Sigma'} \left(\frac{u_0}{r^2} - \frac{iku_0}{r} + \frac{1}{r} \frac{\partial u_0}{\partial r} \right) e^{ikr} \, d\Sigma'.$$

Let us now make the secondary spherical surface Σ' very small, but still with the point P at its center. The value of u_0 at every point of Σ' will then be practically the same as at P itself, so that u_0 may be called $u_0(P)$ everywhere in the Σ' integral. We also note that with r very small, e^{ikr} is approximately 1. Now the second and third terms of the Σ' integral, as expressed above, approach zero as r does, and the entire integral depends on the first term. The integral then reduces to $4\pi u_0(P)$. Putting this value into eq. (2.30) gives

$$u_0(P) = -\frac{1}{4\pi} \iint\limits_{\Sigma} \left[u_0 \frac{d}{dn} \left(\frac{1}{r} e^{ikr} \right) - \frac{1}{r} e^{ikr} \frac{du_0}{dn} \right] d\Sigma. \tag{2.31}$$

This is Kirchhoff's integral, and it says that the scalar field at a point P can be found, provided the field and its outward derivative are known at every point on a surface that surrounds P but does not contain any sources. In this equation, u_0 is the known field at each elementary area $d\Sigma$, and r is the distance of $d\Sigma$ from P.

Let us apply the Kirchhoff integral to the case of a single monochromatic point source, but for the moment without a diffraction screen. The situation is pictured in fig. 2.12a, where r_1 and r_2 represent the distances of any point on the surface Σ from the source S and from the point P, respectively. Figure 2.12b shows how the angle between $\mathbf{r}_1$ and $\mathbf{n}$, the normal to Σ, and the angle between $\mathbf{r}_2$ and $\mathbf{n}$ are designated.

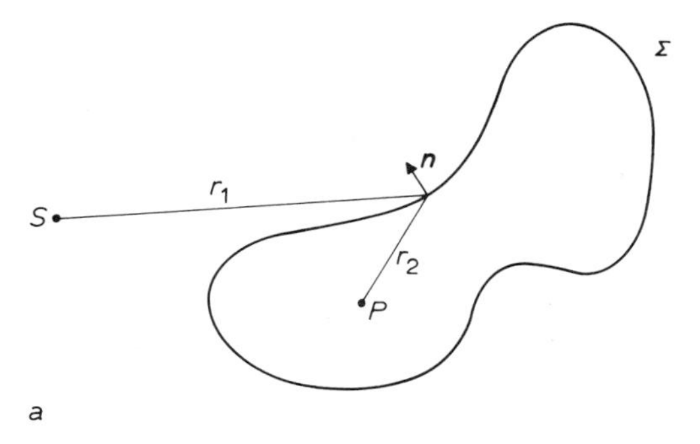

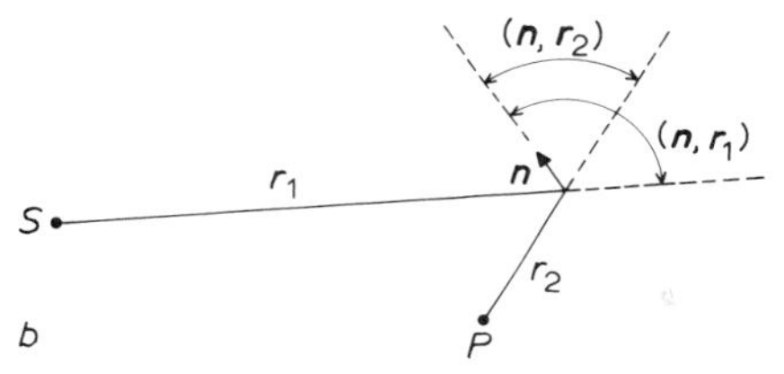

Fig. 2.12 Application of Kirchhoff's theorem to a single point source, without diffraction screen.

The complex amplitude u_0 at the element $d\Sigma$ of the surface may be expressed as

$$u_0 = \frac{A}{r_1} \exp\,(ikr_1), \tag{2.32}$$

where A is a complex constant, and $k = \omega/c$. We can also express the normal derivative at $d\Sigma$ in terms of the angles designated in fig. 2.12*b*:

$$\frac{du_0}{dn} = \frac{du_0}{dr_1} \cos\,(\mathbf{n},\,\mathbf{r}_1) = A \cos\,(\mathbf{n}\,,\,\mathbf{r}_1) \left(-\frac{1}{r_1^2} + \frac{ik}{r_1}\right) \exp\,(ikr_1). \tag{2.33}$$

The r of eq. (2.31) is now r_2, and in a manner similar to the above,

$$\frac{d}{dn}\left[\frac{1}{r_2} \exp\,(ikr_2)\right] = \cos\,(\mathbf{n}\,,\,\mathbf{r}_2) \left(-\frac{1}{r_2^2} + \frac{ik}{r_2}\right) \exp\,(ikr_2). \tag{2.34}$$

In eqs. (2.32), (2.33), and (2.34) we have evaluated the parts of eq. (2.31). Before the substitution, however, let us note that in most situations r_1 and r_2 are many wavelengths long. Suppose $r_1 = N\lambda$, where N is a large number. Then

$1/r_1^2 = 1/N^2\lambda^2$. The constant k can be expressed as $2\pi/\lambda$, so that $k/r_1 = 2\pi/N\lambda^2$. Since N is large, $1/r_1^2$ may be neglected in comparison with k/r_1 and the same is true of $1/r_2^2$. The Kirchhoff integral (2.31) in this case may then be written

$$u_0(P) = -\frac{ikA}{4\pi}\iint\limits_{\Sigma} \frac{1}{r_1 r_2}[\cos(\mathbf{n}, \mathbf{r}_2) - \cos(\mathbf{n}, \mathbf{r}_1)] \exp[ik(r_1 + r_2)]\, d\Sigma. \tag{2.35}$$

The quantity $[\cos(\mathbf{n}, \mathbf{r}_2) - \cos(\mathbf{n}, \mathbf{r}_1)]$ is known as the *obliquity factor*.

The actual shape of the enclosure around P is at our disposal, and in order to lead up to the use of a plane diffraction screen let us give Σ the form shown in fig. 2.13. The figure shows a cross section of a spherical surface of radius R, intersected by a plane that we later will make the plane of the diffraction screen. P is at the center of the spherical part of Σ. Let us call this spherical section Σ_0, and that part of the integral in eq. (2.35) contributed by Σ_0 is called W. As shown in Fig. 2.13, the normal to Σ_0 at any point is called $\mathbf{n}'$, and the distance from S is called r_1'. Note also that the angle $(\mathbf{n}', \mathbf{R})$ is zero. Then the Σ_0 part of the integral in eq. (2.35) is

$$W = \iint\limits_{\Sigma_0} \frac{1}{r_1' R}[1 - \cos(\mathbf{n}', \mathbf{r}_1')] \exp[ik(r_1' + R)]\, d\Sigma_0. \tag{2.36}$$

The obliquity factor in this equation can be expressed in terms of the distances involved, including the distance between S and P, which we call l. For, by the law of cosines,

$$\begin{aligned} l^2 &= r_1'^2 + R^2 - 2r_1' R \cos(\mathbf{n}', \mathbf{r}_1') \\ &= (r_1' - R)^2 + 2r_1' R[1 - \cos(\mathbf{n}', \mathbf{r}_1')]. \end{aligned}$$

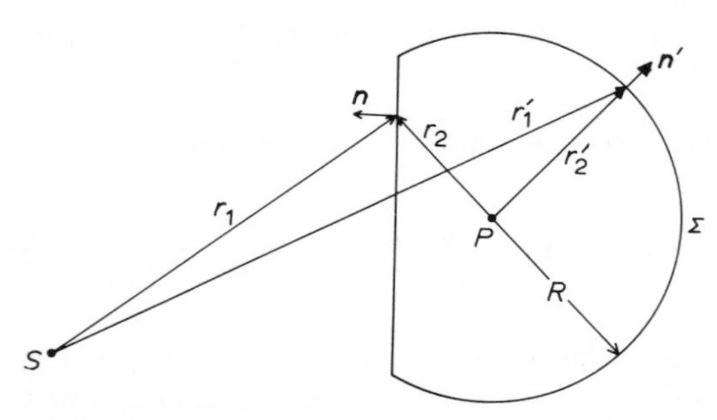

Fig. 2.13 The surface Σ in the form of a sphere intersected by a plane.

Then,

$$1 - \cos(\mathbf{n}', \mathbf{r}_1') = \frac{l^2 - (r_1' - R)^2}{2r_1'R}. \tag{2.37}$$

Substituting this in eq. (2.36), and considering only the absolute value (to reduce the exponential to unity), we obtain

$$|W| = \iint_{\Sigma_0} \frac{l^2 - (r_1' - R)^2}{2r_1'^2R^2}\, d\Sigma_0.$$

It can be seen from eq. (2.37) that $l^2 - (r_1' - R)^2$ is always a positive number, hence l^2 must be the greater term, and $l^2 - (r_1' - R)^2 < l^2$. We can therefore put

$$|W| < \iint_{\Sigma_0} \frac{l^2}{2r_1'^2R^2}\, d\Sigma_0.$$

We now let R become infinitely large. This can be done without including the source S inside the surface Σ, even though l is a fixed finite distance, provided the plane part of Σ lies between S and P. Note also that as R increases, r_1' also increases and in the limit can be taken as equal to R. Then

$$|W| < \frac{l^2}{2R^4} \iint_{\Sigma_0} d\Sigma_0 < \frac{2\pi l^2}{R^2}. \tag{2.38}$$

The last inequality comes from the fact that the integral over Σ_0 is less than that over a whole sphere, and the latter is given by $4\pi R^2$. Now as R becomes infinite, $|W|$, and therefore the whole contribution of Σ_0 to the integral of eq. (2.35), becomes zero. The scalar amplitude at P can then be found from the amplitudes and their normal derivatives at all points of a plane of infinite extent, lying between S and P.

It may now be seen how the scalar amplitude due to a plane diffracting screen may be evaluated. Let the plane of the screen coincide with the plane part of the surface Σ of fig. 2.13, and let the screen be opaque everywhere except for a finite region in which there are apertures. So far as the point P is concerned, u_0 is zero on all the opaque parts of the screen, and the integration of eq. (2.35) need be taken only over the areas where the apertures are located.

Figure 2.14 illustrates the situation for a single source and a diffraction screen with a single aperture. The surface Σ is shown as lying behind the screen, to emphasize the fact that the amplitude of the light reaching P from the opaque part of the screen is zero. In the case of Fraunhofer diffraction, both r_1 and r_2, and also the shortest distances of S and P from the screen, are to be taken as large compared with the wavelength of light. Over the unshadowed parts of Σ we assume that the complex wave amplitude is given by eq. (2.32). Then, by the

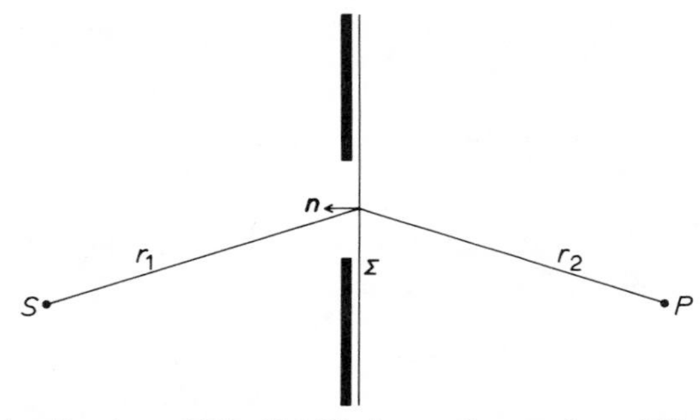

Fig. 2.14 Application of Kirchhoff's integral to a plane diffracting screen.

Kirchhoff theorem, the amplitude over an element of surface $d\Sigma$ is given by

$$Kd\Sigma = -\frac{ikA}{4\pi r_1}[\cos(\mathbf{n}, \mathbf{r}_2) - \cos(\mathbf{n}, \mathbf{r}_1)] \exp(ikr_1)\, d\Sigma, \quad (2.39)$$

and its contribution to the amplitude at P is

$$du_0(P) = \frac{Kd\Sigma}{r_2} \exp(ikr_2). \quad (2.40)$$

This is to be integrated only over the unshadowed parts of Σ. Application of this method to diffraction is made in the next chapter.

It should be mentioned that the choice of a plane surface Σ is arbitrary, and that other surfaces may be chosen. For example, in fig. 2.15, with a circular aperture and S on the axis of the aperture, it might be convenient to take Σ as a segment of a sphere with S at its center. In this case r_1 is constant, and the angle $(\mathbf{n}, \mathbf{r}_1)$ is always π. Also, the angle θ of the figure is the same as angle $(\mathbf{n}, \mathbf{r}_2)$. Hence eq. (2.35) simplifies to

$$u_0(P) = -\frac{ikA}{4\pi r_1} \exp(ikr_1) \iint_{\Sigma} \frac{1}{r_2}[1 + \cos\theta] \exp(ikr_2)\, d\Sigma. \quad (2.41)$$

However, in most cases to be treated, the plane form of Σ will be more convenient.

It will be recognized that the Kirchhoff scalar method, using the wave amplitude at each element of area on an intermediate surface, agrees with Huygens' principle, which regards each point of a wave front as a new source. However, if we ask whether the Kirchhoff method agrees with the electromagnetic theory, the answer must be "not quite." We have assumed that the amplitude at all points of an aperture is the same as it would be without the presence of the screen. This is true for all parts of the aperture that are more than a few wavelengths of light distant from an edge of the screen. For points very close to an edge, however,

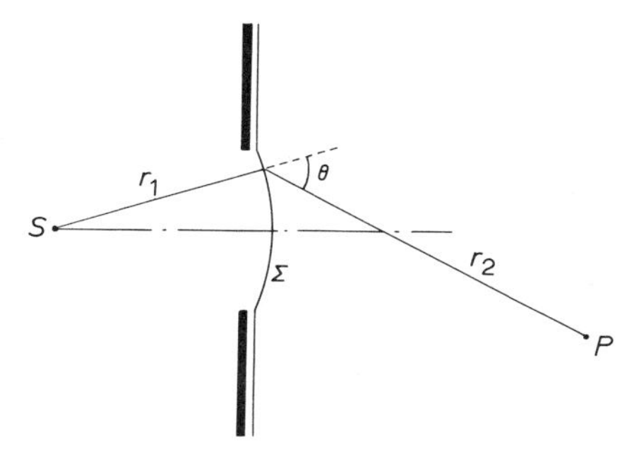

Fig. 2.15 Integration surface in the form of a spherical segment.

the field is affected, and it is also different from zero close to the edge on the shadowed side. Now an aperture does not have to be very large for most of its area to be far from the edges, and in these cases the result of using the scalar method is accurate, and agrees with experiment. If the aperture is a very small hole or very narrow slit, however, the scalar calculation does not give the real conditions. The scalar treatment of diffraction remains an approximate method.

It may be seen that for the scalar method embodied in eq. (2.35), the reciprocity theorem mentioned in sec. 2.5 still holds. An interchange of S and P, the source and the point of observation, involves the interchange of r_1 and r_2 in eq. (2.35) and a reversal in the direction of the normal **n**. These changes leave the complete equation exactly as it was, and the reciprocity is confirmed.

2.9 BABINET'S PRINCIPLE

Let us assume two diffraction screens such that when one is superimposed on the other, the open region of one of the screens is the opaque region of the other. Such two diffraction screens are said to be *complementary* to each other. Let one of the diffraction screens be illuminated by a monochromatic point source of light; then the complex amplitude $U_{01}(P)$ of the diffraction pattern may be found. Now if the diffraction screen is replaced by its complementary screen, then the complex light field $U_{02}(P)$ may be obtained without direct calculation by the following subtraction:

$$U_{02}(P) = U_{00}(P) - U_{01}(P), \tag{2.42}$$

where $U_{02}(P)$ and $U_{00}(P)$ are the respective complex light amplitudes at P due to the complementary screen and due to no screen. The equation is known as *Babinet's principle*. If any two of the complex amplitudes in eq. (2.42) were obtained, say, from the evaluation of Kirchhoff's integral, then the third could be found by simple addition or subtraction of the complex quantities. It may be emphasized that the generalization of Babinet's principle will be extremely useful in a number of cases, particularly in the Fraunhofer diffractions.

PROBLEMS

2.1 Given a diffraction screen that consists of a linear array of five identical apertures, which are located on the diffraction screen at $x = 0$, $x = d$, $x = 2d$, $x = 4d$, and $x = 8d$, respectively.

(a) Draw a vector diagram to find the function G.

(b) Make a rough sketch of $|G|^2$ as a function of the path difference δ, as described in sec. 2.4.

2.2 Consider a rectangular array of $m \times n$ identical apertures, as shown in fig. 2.16.

(a) Determine the function G for the Fraunhofer diffraction pattern.

(b) Write out all the terms of the function $|G|^2$ for the case $m = n = 2$.

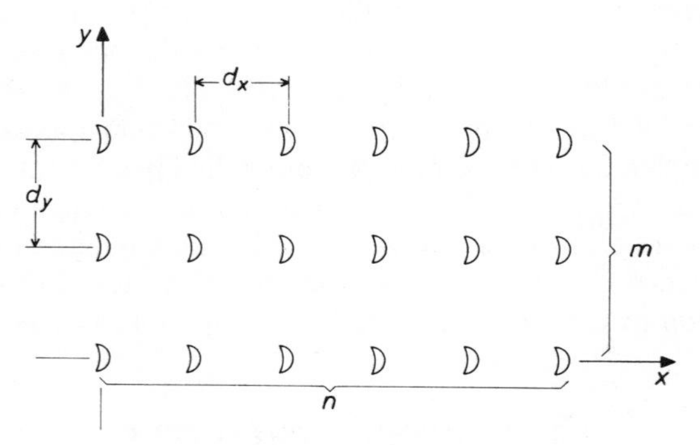

Fig. 2.16

2.3 A monochromatic plane wave is normally incident on an array of identical apertures. The diffraction pattern from any one of the apertures is confined to small values of θ_2 and ϕ_2 (see fig. 2.5). The diffraction screen is then stretched in the x direction such that an aperture originally situated at (x, y) is shifted to (mx, y). Show that the function $|G|^2$ changes after such a deformation of the diffraction screen. Also, discuss briefly the changes in the fringe pattern.

2.4 Show that, when a rectangular diffracting array is deformed to a square array, or vice versa, the change in shape of the principal maximum of the diffraction pattern is in agreement with the result of the preceding problem.

2.5 Given a large number of identical apertures placed at random in a diffraction screen. Assume that no two apertures are so far apart that the condition of Fraunhofer diffraction is not satisfied or so close that they do not contribute independently to the diffraction. Assume that a monochromatic plane wave is incident normally on the diffraction screen. Determine the probabilistic distribution of $|G|$.

2.6 Consider an array of N identical circular apertures, of separation d, normally illuminated by a monochromatic plane wave.

(a) Evaluate the far field irradiance distribution.

(b) Determine the Fourier transform of part (a).

2.7 Refer to problem 2.6; assume we let the number of apertures N become exceedingly large and the separation between the apertures d become vanishingly small. Determine the corresponding far field irradiance distribution.

2.8 The analysis of prob. 2.6 can also be applied to an array of narrow slits.

(a) Evaluate the far field irradiance distribution for N slits of slitwidth W and separation d.

(b) Sketch the corresponding $|G|^2$ as N becomes exceedingly large.

2.9 An array can be interesting not only in itself but also as it applies to the problem in radio astronomy. We consider a crossed array of small circular apertures, as shown in fig. 2.17. Assume that the two arms contain an odd number N of apertures.

(a) Evaluate the function $|G|^2$ for $N = 9$ and $N = 13$.

(b) Calculate the Fourier transforms of part (a).

(c) Show that the crosscorrelations among the apertures can be determined from part (b).

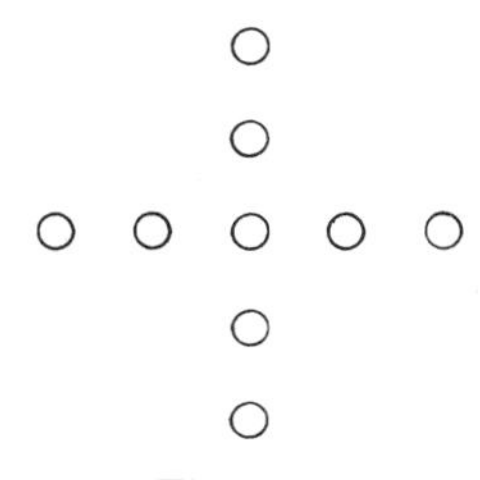

Fig. 2.17

2.10 Show that, by the application of the Kirchhoff integral of eq. (2.35) to a plane diffraction screen, the integral can be reduced to the form

$$U(p) = C \iint_{\Sigma} e^{ikr_2} \, d\Sigma$$

where C is a complex constant, and the integration is over the whole diffraction screen Σ.

2.11 With reference to the result of prob. 2.10, we assume that the observation point is over a two-dimensional coordinate plane parallel to the diffraction screen.

(a) Develop the coordinate relationship between the diffraction screen and the observation plane.

(b) If the diffraction screen Σ is spatially limited to some degree, show that a new form of Kirchhoff integral can be developed.

(c) From this result of part (b), discuss some significant aspects that are related to the linear system theory of sec. 1.1.

2.12 Let us consider that the complex amplitude transmittance function of a finite diffraction screen is $T(x, y)$.

(a) Develop a Kirchhoff integral to determine the complex light distribution over the observation plane.

(b) Draw an input-output system analog diagram to represent this integration.

2.13 Given a normalized amplitude diffraction distribution of a rectangular aperture as

$$Q(p) = \frac{\sin \pi\mu}{\pi\mu} \frac{\sin \pi\nu}{\pi\nu},$$

where $\mu = ka/2\pi\,(\sin\theta_1 + \sin\theta_2)$, $\nu = kb/2\pi\,(\sin\phi_1 + \sin\phi_2)$, a and b are the dimensions of the aperture. Apply Babinet's principle, and evaluate the diffraction pattern of a rectangular opaque screen that is complementary to the open one.

REFERENCES

2.1 M. Born and E. Wolf, *Principles of Optics,* 2nd rev. ed., Pergamon Press, New York, 1964.

2.2 B. Rossi, *Optics*, Addison-Wesley, Reading, Mass., 1957.

2.3 J. M. Stone, *Radiation and Optics,* McGraw-Hill, New York, 1963.

2.4 F. W. Sears, *Optics,* Addison-Wesley, Reading, Mass., 1949.

2.5 A. Sommerfeld, *Optics* (Lectures on Theoretical Physics, vol. 4), Academic Press, New York, 1954.

2.6 M. Kline and I. W. Kay, *Electromagnetic Theory of Geometrical Optics,* Interscience, New York, 1965.

2.7 M. Françon, *Diffraction Coherence in Optics,* Pergamon Press, New York, 1966.

2.8 W. K. H. Panofsky and M. Phillips, *Classical Electricity and Magnetism,* 2nd ed., Addison-Wesley, Reading, Mass., 1962.

2.9 J. A. Stratton, *Electromagnetic Theory,* McGraw-Hill, New York, 1941.

2.10 H. Margenau and G. M. Murphy, *The Mathematics of Physics and Chemistry,* vol. 1, 2nd ed., Van Nostrand, New York, 1956.

3

Fraunhofer and Fresnel Diffraction

Our understanding of both Fraunhofer and Fresnel diffraction (defined in chap. 2) may be improved by considering the application of Kirchhoff's integral. This application is made in the present chapter. In both cases it is assumed that aperture dimensions are large compared with the wavelengths of the incident light, and the results apply only where this assumption is valid. Rectangular and circular apertures are treated in both cases. In the Fraunhofer case it is found that a simplified form of the Kirchhoff integral may be used. Finally, the resolving power for a general imaging system is discussed.

3.1 FRAUNHOFER DIFFRACTION

Assume a situation like that of fig. 3.1, where S is a point source of monochromatic light, P is the point of observation, and between them is a plane

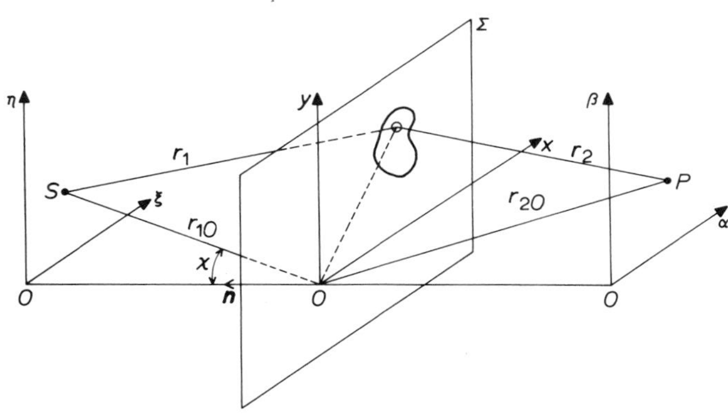

Fig. 3.1 Relation of source, screen, and observation point for the application of Kirchhoff's integral to Fraunhofer diffraction.

diffraction screen having an open aperture. A strict Fraunhofer case is assumed; i.e., the distances of both S and P from the screen are large compared with the aperture dimensions, and no lenses or other devices are used to increase the effective distances. The surface for Kirchhoff integration, Σ, is assumed to lie very closely behind the diffracting screen. That is to say, the plane of Σ is the same as the plane of the screen, but Kirchhoff integration takes place only where there are open apertures. The coordinate axes are indicated in the figure.

Under the Fraunhofer conditions, some simplifying approximations can be made. Thus **n**, the normal to the screen at O, makes very nearly the same angle with r_1 as it does with r_{1O}, and the same is true on the P side of the screen. That is,

$$\cos(\mathbf{n}, \mathbf{r}_1) \simeq \cos(\mathbf{n}, \mathbf{r}_{1O})$$

and

$$\cos(\mathbf{n}, \mathbf{r}_2) \simeq \cos(\mathbf{n}, \mathbf{r}_{2O}).$$

These approximations may be substituted in the Kirchhoff integral, eq. (2.35). Also, in the same equation we can put

$$\frac{1}{r_1 r_2} \simeq \frac{1}{r_{1O} r_{2O}}. \tag{3.1}$$

The $r_1 + r_2$ in the exponent of eq. (2.35) may be written

$$r_1 + r_2 = r_{1O} + r_{2O} + (r_1 - r_{1O}) + (r_2 - r_{2O}).$$

But by eqs. (2.7) and (2.8), the difference of path $r_1 - r_{1O}$ may be replaced by $-(x \sin\theta_1 + y \sin\phi_1)$, and $r_2 - r_{2O}$ by $-(x \sin\theta_2 + y \sin\phi_2)$, where

$$\sin\theta_1 = \frac{\xi}{r_{1O}}, \qquad \sin\phi_1 = \frac{\eta}{r_{1O}}, \qquad \sin\theta_2 = \frac{\alpha}{r_{2O}}, \qquad \text{and} \qquad \sin\phi_2 = \frac{\beta}{r_{2O}}.$$

Then

$$r_1 + r_2 = r_{1O} + r_{2O} - (x \sin\theta_1 + y \sin\phi_1) - (x \sin\theta_2 + y \sin\phi_2). \tag{3.2}$$

We can also take advantage of the fact that the deviation of the diffracted light is small. Thus if we take point (α_0, β_0) in the observing plane for which $\theta_2 = -\theta_1$ and $\phi_2 = -\phi_1$, only points close to (α_0, β_0) in angular distance as seen from the aperture will receive appreciable irradiance. Then all points that lie within the appreciable diffraction pattern will be *almost* in line with S and the aperture. Thus if we compare the angle χ, as shown in fig. 3.1, with angles of fig. 2.12 under the condition of r_2 opposite to r_1, we see that $\chi \simeq (\mathbf{n}, \mathbf{r}_{2O})$ and that $(\mathbf{n}, \mathbf{r}_{1O}) \simeq \pi - \chi$. We then have

$$\cos(\mathbf{n}, \mathbf{r}_{2O}) - \cos(\mathbf{n}, \mathbf{r}_{1O}) \simeq 2 \cos \chi. \tag{3.3}$$

The angle χ may be called the *angle of incidence*.

Now by substituting the approximations (3.1), (3.2), and (3.3) in Kirchhoff's integral, eq. (2.35), we have

$$U_0(P) = -\frac{ikA}{4\pi r_{1O} r_{2O}} 2 \cos \chi \exp[ik(r_{1O} + r_{2O})]$$
$$\times \iint\limits_{\substack{\text{over the} \\ \text{apertures}}} \exp\{-ik[x(\sin \theta_1 + \sin \theta_2) + y(\sin \phi_1 + \sin \phi_2)]\}\, dx\, dy. \tag{3.4}$$

It will be convenient to make the integral in eq. (3.4) apply over the entire Kirchhoff surface Σ by introducing a complex transmission function

$$T(x, y) = |T(x, y)| e^{i\phi(x, y)}, \tag{3.5}$$

which would be defined in the present case by

$$T(x, y) = \begin{cases} 1, & \text{over the apertures} \\ 0, & \text{otherwise} \end{cases}.$$

Equation (3.4) is then written

$$U_0(P) = K \iint\limits_{\Sigma} T(x, y) \exp\{-ik[x(\sin \theta_1 + \sin \theta_2)$$
$$+ y(\sin \phi_1 + \sin \phi_2)]\}\, dx\, dy, \tag{3.6}$$

where

$$K = -\frac{ikA}{4\pi r_{1O} r_{2O}} 2 \cos \chi \exp[ik(r_{1O} + r_{2O})]$$

is a complex constant.

Now at the point (α_0, β_0) of the observing plane, where $\theta_2 = -\theta_1$ and $\phi_2 = -\phi_1$, the exponential in the integral of eq. (3.6) would be unity, and

$$U_0(\alpha_0, \beta_0) = K \iint\limits_{\Sigma} T(x, y)\, dx\, dy.$$

If we now define another function $Q(P)$, such that

$$Q(P) = \frac{\iint_{\Sigma} T(x, y) \exp\{-ik[x(\sin\theta_1 + \sin\theta_2) + y(\sin\phi_1 + \sin\phi_2)]\}\, dx\, dy}{\iint_{\Sigma} T(x, y)\, dx\, dy} \tag{3.7}$$

then eq. (3.6) may be rewritten as

$$U_0(P) = U_0(\alpha_0, \beta_0)\, Q(P). \tag{3.8}$$

Note that $Q(\alpha_0, \beta_0) = 1$. Since eq. (3.8) deals in amplitudes, the average rate of energy per unit area (irradiance) may be obtained by squaring absolutes,

$$I(P) = I_0 |Q(P)|^2, \tag{3.9}$$

where $I_0 = |U_0(\alpha_0, \beta_0)|^2$, the irradiance in the undeviated direction of light from the source. When this is known, eq. (3.9) tells us that irradiance elsewhere in the diffraction pattern can be found by multiplying I_0 by the factor $|Q(P)|^2$, which, from eq. (3.7), depends on the overall disposition of the apertures. Equation (3.9) is very similar to eq. (2.13). In fact, for identical apertures the two equations are identical.

3.2 THE FOURIER TRANSFORM IN FRAUNHOFER DIFFRACTION

Let us return to the simplified form of Kirchhoff's integral for Fraunhofer diffraction, eq. (3.6).

If we recall the following relations in the defined coordinate systems:

$$\sin\theta_1 = \frac{\xi}{r_{10}}, \qquad \sin\phi_1 = \frac{\eta}{r_{10}}$$

and

$$\sin\theta_2 = \frac{\alpha}{r_{20}}, \qquad \sin\phi_2 = \frac{\beta}{r_{20}},$$

it is clear that eq. (3.6) can be thought of as a two-dimensional Fourier integral. For example, if we let the source be a monochromatic point source, located at (ξ_0, η_0) of the source plane, then eq. (3.6) may be written as

$$U_0(P) = K \iint_{\Sigma} T(x, y) \exp[-i(\mu x + \nu y)]\, dx\, dy, \tag{3.10}$$

where

$$\mu = k\left(\frac{\xi_0}{r_{10}} + \frac{\alpha}{r_{20}}\right), \qquad \nu = k\left(\frac{\eta_0}{r_{10}} + \frac{\beta}{r_{20}}\right).$$

Equation (3.10) is the Fourier transform of $T(x, y)$, which occurs in the observation plane. This occurrence of the Fourier transformation in Fraunhofer diffraction can be seen from the examples of rectangular and circular apertures, which are given in the next section. More specifically, we can define two new variables

$$\mu' = k\left(\frac{\xi}{r_{10}} + \frac{\alpha}{r_{20}}\right), \qquad \nu' = k\left(\frac{\eta}{r_{10}} + \frac{\beta}{r_{20}}\right).$$

Then eq. (4.6) can be written

$$U_0(P) = K \iint_{\Sigma} T(x, y) \exp[-i(\mu' x + \nu' y)]\, dx\, dy. \tag{3.11}$$

3.3 EXAMPLES OF FRAUNHOFER DIFFRACTION

Rectangular Aperture

In fig. 3.2 we show a rectangular aperture with dimensions a and b. In this case the transmission function for the diffraction screen will be

$$T(x, y) = \begin{cases} 1 & \left(-\frac{a}{2} \le x \le \frac{a}{2}, \quad -\frac{b}{2} \le y \le \frac{b}{2}\right). \\ 0 & \text{otherwise.} \end{cases} \tag{3.12}$$

For convenience in notation, let us define two new variables.

$$\mu = \frac{ka}{2\pi}(\sin\theta_1 + \sin\theta_2), \tag{3.13}$$

$$\nu = \frac{kb}{2\pi}(\sin\phi_1 + \sin\phi_2). \tag{3.14}$$

By substituting eqs. (3.12), (3.13), and (3.14) in eq. (3.7), we have for the normalized complex amplitude

$$Q(P) = \frac{1}{ab}\int_{-a/2}^{a/2} \exp\left(-i\frac{2\pi\mu x}{a}\right) dx \int_{-b/2}^{b/2} \exp\left(-i\frac{2\pi\nu y}{b}\right) dy$$

$$= \frac{\sin \pi\mu}{\pi\mu}\cdot\frac{\sin \pi\nu}{\pi\nu}. \tag{3.15}$$

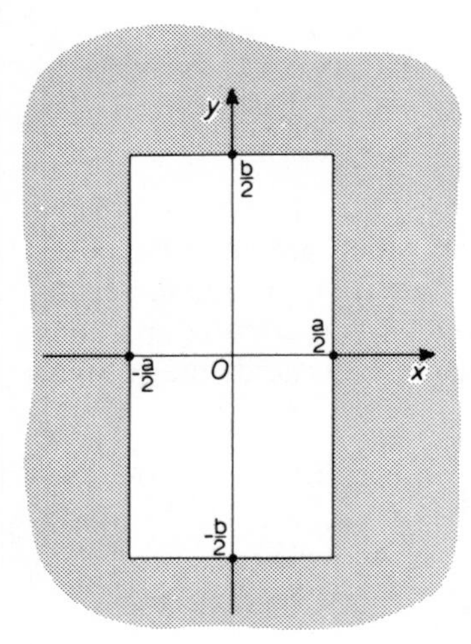

Fig. 3.2 Geometry of a rectangular aperture.

This normalized amplitude is plotted in fig. 3.3 as a function of μ, for a fixed value of ν. It has a maximum when $\mu = 0$. Although $Q(P)$ is in general complex, by proper choice of the coordinate system in this special problem $Q(P)$ can be made real for all values of μ and ν. The manner in which $Q(P)$ alternates between positive and negative values is shown in fig. 3.4. On the lines separating the regions in this figure, $Q(P)$ is zero.

From the amplitude eq. (3.15), the irradiance can be written as

$$I(P) = I_0 \frac{\sin^2 \pi\mu}{(\pi\mu)^2} \frac{\sin^2 \pi\nu}{(\pi\nu)^2}. \tag{3.16}$$

where I_0 is the irradiance when both μ and ν are zero. $I(P)$ is plotted in fig. 3.5, again for a fixed value of ν. The actual Fraunhofer diffraction pattern for a rectangular aperture is shown in fig. 3.6.

The variables μ and ν, defined in eqs. (3.13) and (3.14) and used in (3.15) and (3.16), are dimensionless. Actual positions of the pattern on the observing plane, with reference to source and aperture, can be found from the θ's and ϕ's of eqs. (3.13) and (3.14). The coordinates α and β on the observing screen, as used in fig. 3.1, may be found from

$$\alpha = \frac{\lambda l}{a} \mu, \qquad \beta = \frac{\lambda l}{b} \nu, \tag{3.17}$$

where l is the separation between the diffraction aperture and the observing screen.

The variation of irradiance along the α axis of the diffraction pattern is shown in fig. 3.7. It shows a strong central maximum, having an angular half-width of λ/a radians, and a series of much smaller secondary maxima on each side. The photograph of fig. 3.6 shows the same effect. The same fig. 3.7 also holds for the β axis, if α and a in the figure are replaced by β and b.

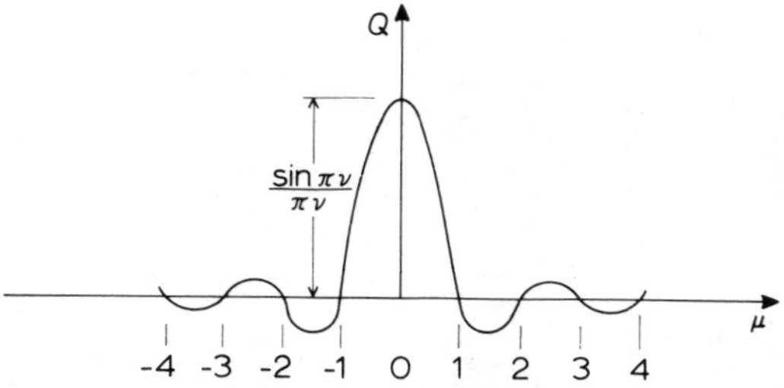

Fig. 3.3 The diffraction amplitude $Q(P)$ as a function of μ for a given value of ν.

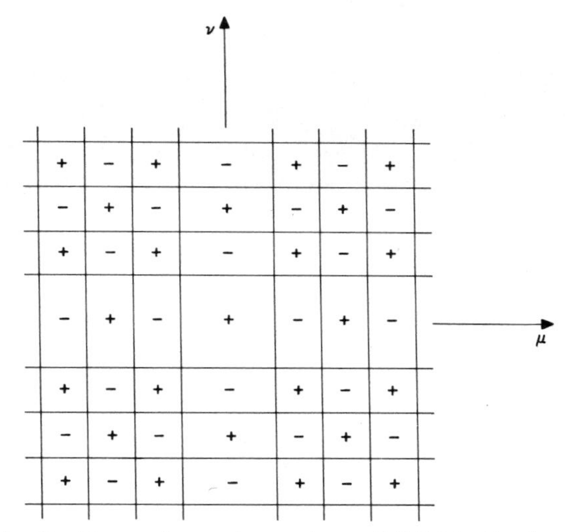

Fig. 3.4 Positive and negative regions of $Q(P)$ in the (μ, ν) plane.

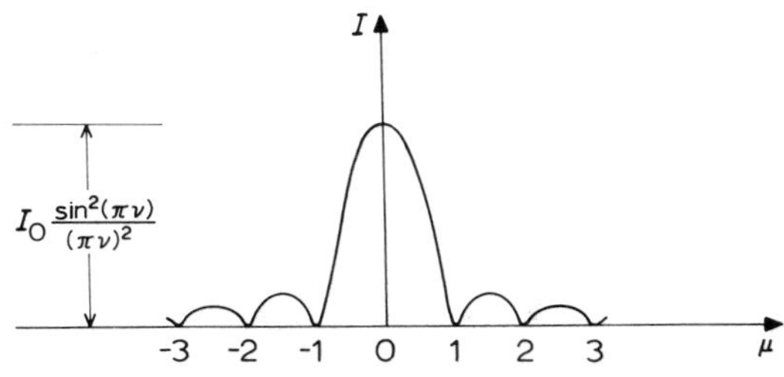

Fig. 3.5 Irradiance $I(P)$ as a function of μ for a given value of ν.

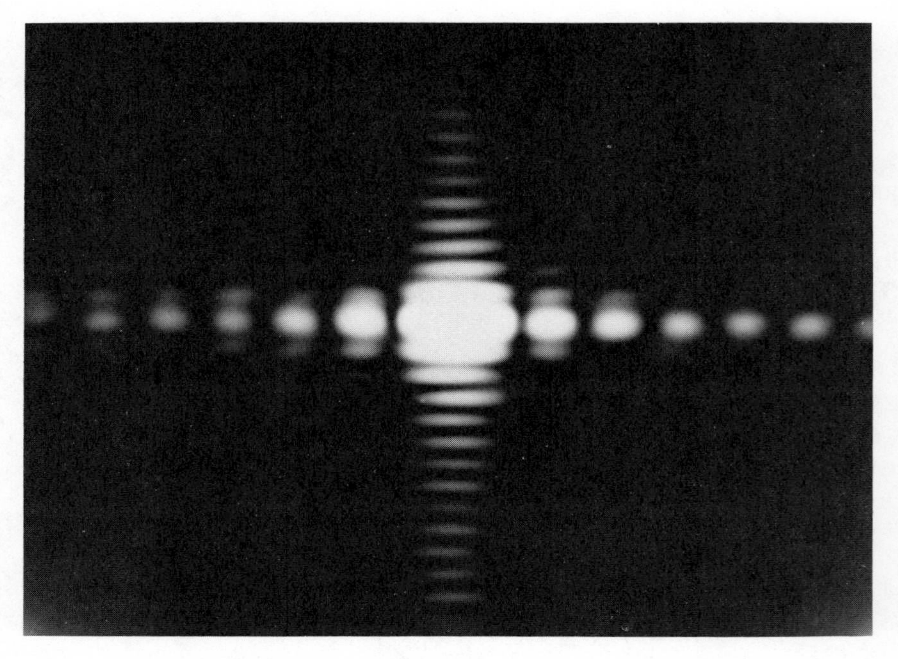

Fig. 3.6 Fraunhofer diffraction pattern of a rectangular aperture.

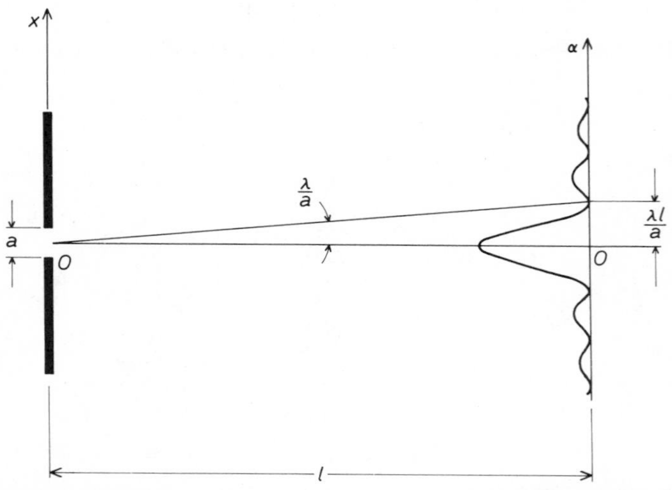

Fig. 3.7 Irradiance along one axis for a rectangular aperture, Fraunhofer diffraction.

Circular Aperture

In examining the Fraunhofer diffraction pattern from a circular aperture, it will be assumed that the incident light is normal to the plane of the aperture. This is not the most general case, but it simplifies the calculation. Let a circular aperture of diameter D be centered at the origin of the diffraction screen [the (x, y) plane] of fig. 3.1. To satisfy the normal incidence assumption, the source S in fig. 3.1 would lie at the origin of the (ξ, η) coordinates. This results in making both θ_1 and ϕ_1, as defined in eq. (3.2), equal to zero, and these values can be substituted in eq. (3.7).

It is apparent that, with the circular aperture and the normal incidence of light, there must be complete symmetry of diffraction about the normal line. That is, if we find the variation of pattern along one radius from O in the observing plane, it will be the same along every other radius from O. Let us then find this pattern along the α axis, for which we can put $\phi_2 = 0$, and we need only consider positive values of θ_2.

Let us remind ourselves that the double integrals over Σ in eq. (3.7), together with the factor $T(x, y)$ in the integrand, mean that integration occurs only over the area of the aperture. In this case both x and y vary from $-D/2$ to $+D/2$. The denominator of eq. (3.7) means simply the area of the aperture, in this case $\pi D^2/4$. In order to obtain the variation along the α axis, we write eq. (3.7) as

$$Q(P) = \frac{4}{\pi D^2} \iint_{\Sigma} \exp[-ikx \sin \theta_2] \, dx \, dy.$$

Considering for the moment a constant x, the corresponding limits of y are

$$\pm \sqrt{\frac{D^2}{4} - x^2},$$

and the entire integration with respect to y gives the factor

$$2\sqrt{\frac{D^2}{4} - x^2}.$$

If we make the further substitution

$$\mu = \frac{kD}{2\pi} \sin \theta_2,$$

then

$$Q(P) = \frac{4}{\pi D^2} \int_{-D/2}^{D/2} 2\left(\frac{D^2}{4} - x^2\right)^{1/2} \exp\left[-i \frac{2\pi}{D} \mu x\right] dx.$$

Now let

$\tau = 2x/D$, so that $x = D\tau/2$, $dx = (D/2)\,d\tau$, and $\tau = 1$ when $x = D/2$. Then

$$Q(P) = \frac{2}{\pi}\int_{-1}^{1} (1 - \tau^2)^{1/2} \exp[-i\pi\mu\tau]\,d\tau. \tag{3.19}$$

The variable of integration here is τ, which is directly related to x, which is one coordinate in the circular aperture. The quantity μ is related by eq. (3.18) to θ_2, which gives angular position in the diffracted beam, and is not related explicitly to x.

If we divide the integral (3.19) into two, from -1 to 0, and from 0 to 1, we notice that $(1 - \tau^2)$ remains the same for each $\pm$ value of τ, and that the sign is changed in the exponential. Now the sum of $e^{i\pi\mu\tau}$ and $e^{-i\pi\mu\tau}$ is $2\cos(\pi\mu\tau)$, which is a real quantity. Equation (3.19) therefore becomes

$$Q(P) = \frac{4}{\pi}\int_{0}^{1} (1 - \tau^2)^{1/2} \cos(\pi\mu\tau)\,d\tau. \tag{3.20}$$

Integrating, we obtain the first-order Bessel function

$$Q(P) = \frac{2J_1(\pi\mu)}{\pi\mu}, \qquad \mu \geq 0. \tag{3.21}$$

The amplitude Q in the observing plane is plotted as a function of μ in fig. 3.8, with the variation shown along a complete radius of the pattern. The pattern will be the same along every other radius, so that the complete pattern is a series of concentric rings about O in the observing plane. The amplitude is a maximum

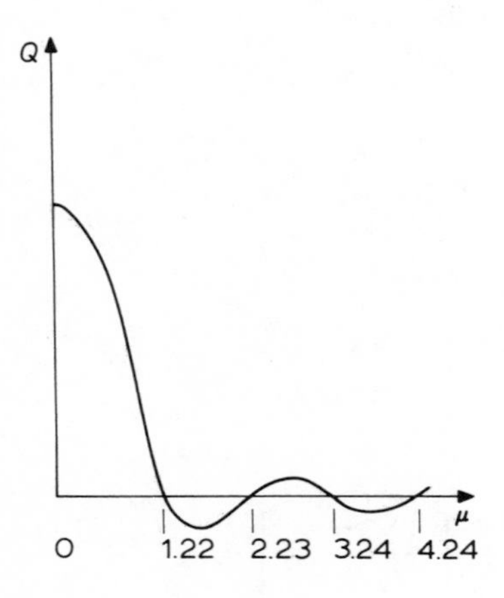

Fig. 3.8. Diffraction amplitude $Q(P)$ as a function of μ for a circular aperture.

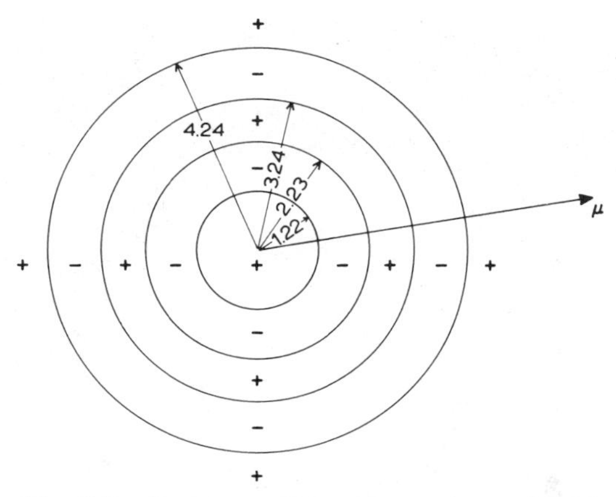

Fig. 3.9 Positive and negative regions of $Q(P)$.

where μ (and therefore θ_2) is zero. Figure 3.9 indicates how the amplitudes in the pattern vary between positive and negative values. Note that μ is directly proportional to the radius of the pattern. For, if we call such a radius ρ, then $\rho = l \tan \theta_2$, where l is the separation between the aperture plane and the observing plane. But under Fraunhofer conditions, θ_2 is so small that the tangent may be put equal to the sine, and by eq. (3.18) we have

$$\mu = \frac{kD}{2\pi l}\rho = \frac{D}{\lambda l}\rho.$$

Corresponding to the amplitude eq. (3.21), the irradiance in the diffraction pattern from a circular aperture is given by

$$I(P) = I_0\left[\frac{2J_1(\pi\mu)}{\pi\mu}\right]^2, \qquad \mu \geq 0. \tag{3.22}$$

The variation of $I(P)$ across a radius of the pattern is plotted in fig. 3.10.

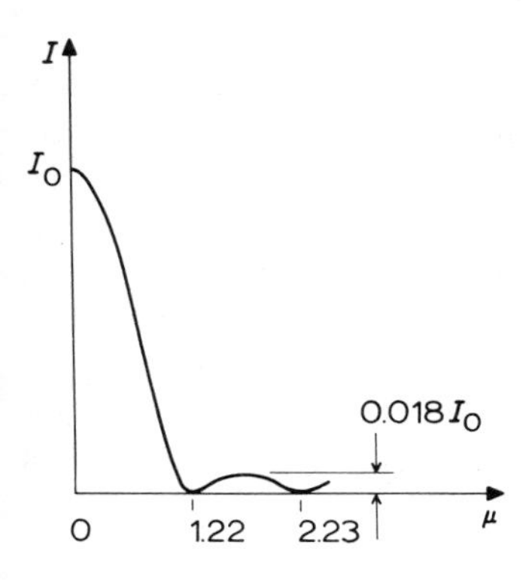

Fig. 3.10 Irradiance $I(P)$ as a function of μ for a circular aperture.

Fig. 3.11 Airy pattern (Fraunhofer diffraction for circular aperture).

Equation (3.22) was first derived by G. B. Airy, and the pattern formed by diffraction from a circular aperture is known as the *Airy pattern.* A photograph of the Airy pattern is given in fig. 3.11. The *Airy disk*, as the central circle of irradiance in the pattern is called, extends outward from the center to $\mu = 1.22$, or to a radius $\rho = 1.22\lambda l/D$. Eighty-four percent of the total power in the pattern is within this central disk. The first null in the pattern comes at a somewhat larger radius than with the rectangular aperture, where (by fig. 3.7) it is given as $\lambda l/a$, the a being comparable with D in the circular case.

3.4 FRESNEL DIFFRACTION

As explained in chap. 2, we apply the term "Fresnel" when the distances from source to diffraction screen, or from the latter to the observing plane, or both, are not great compared with dimensions of the diffracting apertures. We are forced to abandon the assumption that the rays from a point source to the different points of an aperture are essentially parallel, or that the geometrical rays from these points to the point of observation are parallel. Thus the simplified equations that we have found for Fraunhofer diffraction can no longer be used, and the problem of finding the diffraction pattern becomes much more difficult. However, the patterns from apertures of certain shapes may be treated, at least in part. In particular, we take up the case of the rectangular aperture, and then the more difficult case of the circular aperture.

Rectangular Aperture

In fig. 3.12 a rectangular aperture with dimensions a and b is shown. A monochromatic point source is at S, and SO is the perpendicular from S to the plane of the aperture. Let this line SO be projected to the observing plane at P, and let us observe the irradiance at P. To explore a pattern, we should of course examine other points about P. Instead of this, let us find the changes in irradiance at P as the aperture is moved to different positions in the diffraction screen, with S, O, and P remaining fixed. The pattern found in this way will not differ greatly from that with a fixed aperture and P moved, provided the displacements given the aperture are small compared with the distances SO and OP (r_{1O} and r_{2O}). There are three cases that can be treated, for different assumptions with regard to the sizes of a and b.

For the first case, let a and b both be small compared with r_{1O} and r_{2O}. This would seem to be a return to the conditions assumed for the Fraunhofer diffraction. However, the treatment here is somewhat different. For one thing, we assume that the incidence of light on the plane of the aperture is nearly normal. If we compared fig. 3.12 with fig. 3.1, we see that the angle χ in the latter has been made zero in the present case. Thus we now have a further simplification of eq. (3.3),

$$\cos(\mathbf{n}, \mathbf{r}_{2O}) - \cos(\mathbf{n}, \mathbf{r}_{1O}) \simeq 2.$$

We can, however, make the same assumption as we did in eq. (3.1), that

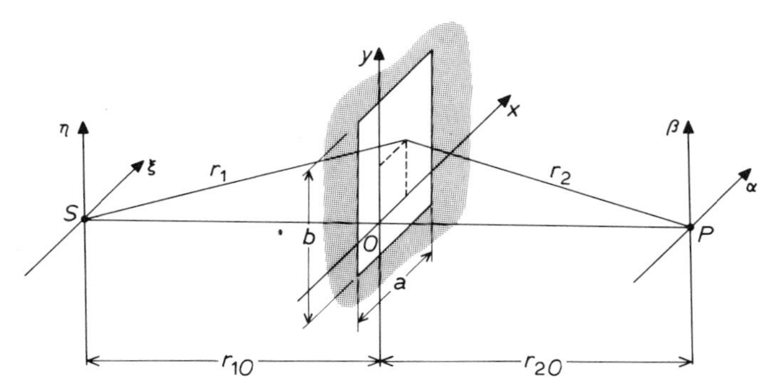

Fig. 3.12 Geometry for the application of Kirchhoff's integral to Fresnel diffraction of a rectangular aperture.

$$\frac{1}{r_1 r_2} \simeq \frac{1}{r_{1O} r_{2O}}.$$

If these conditions are substituted in the Kirchhoff equation (2.35), we can then write down the complex amplitude at the observation point:

$$U_0(P) = -\frac{iA}{\lambda r_{1O} r_{2O}} \int_{y_1}^{y_2} \int_{x_1}^{x_2} \exp[ik(r_1 + r_2)]\, dx\, dy, \tag{3.23}$$

where λ in the constant part has been substituted for its equivalent $2\pi/k$, and the surface integral over Σ has been split into the two integrals over x and y.

Our estimate of $r_1 + r_2$ must be a little different than in the Fraunhofer case, eq. (3.2). If we compare fig. 3.12 with fig. 2.5, which defines the θ's and ϕ's, we see that now all these angles are zero, and eq. (3.2) could be written $r_1 + r_2 = r_{1O} + r_{2O}$. This comes about by the neglect of certain quantities in the formulation of eq. (3.2), and is a more extreme simplification than we wish to make in the present case. Instead, we point out that $r_1^2 = r_{1O}^2 + x^2 + y^2$ (see fig. 3.12), and it is a sufficiently accurate approximation to set

$$r_1 \simeq r_{1O} + \frac{x^2 + y^2}{2r_{1O}}. \tag{3.24}$$

Similarly, on the other side of the aperture,

$$r_2 \simeq r_{2O} + \frac{x^2 + y^2}{2r_{2O}}. \tag{3.25}$$

Adding these,

$$r_1 + r_2 \simeq r_{1O} + r_{2O} + (x^2 + y^2)\frac{r_{1O} + r_{2O}}{2r_{1O} r_{2O}}. \tag{3.26}$$

Before substituting eq. (3.26) in eq. (3.23), it is convenient to introduce two new dimensionless variables,

$$\mu = x\left[\frac{2(r_{1O} + r_{2O})}{\lambda r_{1O} r_{2O}}\right]^{1/2}, \qquad \nu = y\left[\frac{2(r_{1O} + r_{2O})}{\lambda r_{1O} r_{2O}}\right]^{1/2}. \tag{3.27}$$

Equation (3.23) now becomes

$$U_0(P) = -\frac{iA}{2(r_{1O} + r_{2O})} \exp[ik(r_{1O} + r_{2O})] \int_{\mu_1}^{\mu_2} \exp\left(\frac{i\pi\mu^2}{2}\right) d\mu$$
$$\times \int_{\nu_1}^{\nu_2} \exp\left(\frac{i\pi\nu^2}{2}\right) d\nu. \tag{3.28}$$

If we designate by $U_{00}(P)$ the amplitude of the irradiance at P when no screen is interposed between S and P, we have

$$U_{00}(P) = \frac{A}{r_{1O} + r_{2O}} \exp[ik(r_{1O} + r_{2O})].$$

and the constant factor in eq. (3.28) is thus $-(i/2)\, U_{00}(P)$.

The exponentials under the integral signs in eq. (3.28) may be separated into real and imaginary parts by means of Euler's equation:

$$\exp\left(\frac{i\pi t^2}{2}\right) = \cos\left(\frac{\pi t^2}{2}\right) + i \sin\left(\frac{\pi t^2}{2}\right),$$

where t is used for either μ or ν. Let us also use the variable ν to represent any of the limits of integration, and define the quantities

$$C(v) = \int_0^v \cos\left(\frac{\pi t^2}{2}\right) dt, \qquad S(v) = \int_0^v \sin\left(\frac{\pi t^2}{2}\right) dt. \tag{3.29}$$

These are called *Fresnel integrals*. Tables of the Fresnel integrals may be found in *Tables of Higher Functions* by Jahnke, Emde, and Lösch (ref. 3.6).

Using the notation of (3.29), eq. (3.28) can be written

$$U_0(P) = -\frac{i}{2} U_{00}(P)\, \{C(\mu_2) - C(\mu_1) + i[S(\mu_2) - S(\mu_1)]\}$$
$$\times \{C(\nu_2) - C(\nu_1) + i[S(\nu_2) - S(\nu_1)]\}. \tag{3.30}$$

Equation (3.30) can be evaluated by a graphical method. Let v be given different values, both positive and negative, and let the corresponding values of $C(v)$ and $S(v)$, from eq. (3.29), be plotted against each other in the complex plane, with $S(v)$ on the imaginary axis. The resulting curve is called *Cornu's spiral* (fig. 3.13). The values of v used in calculating the C and S coordinates are plotted along the curve. When v is zero both $C(v)$ and $S(v)$ are zero, and as v is increased either positively or negatively the curve spirals inwardly to the points F and E, which are given by $C(\infty) = S(\infty) = \frac{1}{2}$, and $C(-\infty) = S(-\infty) = -\frac{1}{2}$. The curve is symmetrical about the origin. It may be pointed out that equal increments in v, anywhere along the curve, add equal lengths to the curve. The turns of the curve become more and more tightly packed together as points E and F are approached.

In making use of Cornu's spiral for finding the irradiance at P, we must first find the limits of μ and ν. Let the center of the rectangular aperture have the coordinates x_0 and y_0. The limits of x are then $x_0 - (a/2)$ and $x_0 + (a/2)$, and those of y are $y_0 - (b/2)$ and $y_0 + (b/2)$. These four limits are then substituted

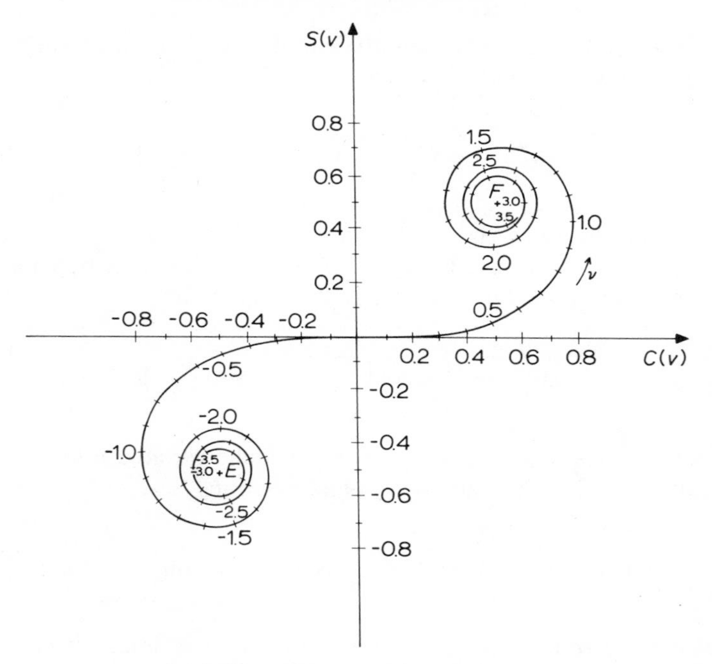

Fig. 3.13 Cornu's spiral.

in the appropriate parts of eqs. (3.27) to get μ_1, μ_2, ν_1, and ν_2. Each of these, used as the value of v on the curve, gives a point from which the C and S values are obtained. Putting these into eq. (3.30) gives $U_0(P)$ in terms of the unrestricted amplitude $U_{00}(P)$. Changing the position of the aperture just means the choice of a different x_0 and y_0.

The foregoing procedure may be simplified as follows. Let the two points on the curve corresponding to $v = \mu_1$ and $v = \mu_2$ be joined by a line, as shown in fig. 3.14. This line represents a vector of magnitude A (length of the line) and phase angle ϕ (angle the line makes with the real axis). A and ϕ are found graphically, by ruler and protractor, and they can be put into the equation

$$C(\mu_2) - C(\mu_1) + i[S(\mu_2) - S(\mu_1)] = Ae^{i\phi}. \tag{3.31}$$

The same process can be used for ν_1 and ν_2, finding the amplitude B and the phase ψ. Then

$$C(\nu_2) - C(\nu_1) + i[S(\nu_2) - S(\nu_1)] = Be^{i\psi}. \tag{3.32}$$

Putting eqs. (3.31) and (.32) into eq. (3.30) gives a new expression for the amplitude and phase of the irradiance at P. Going from this to the absolute value of the energy (eliminating the phase) we have, for the irradiance at P.

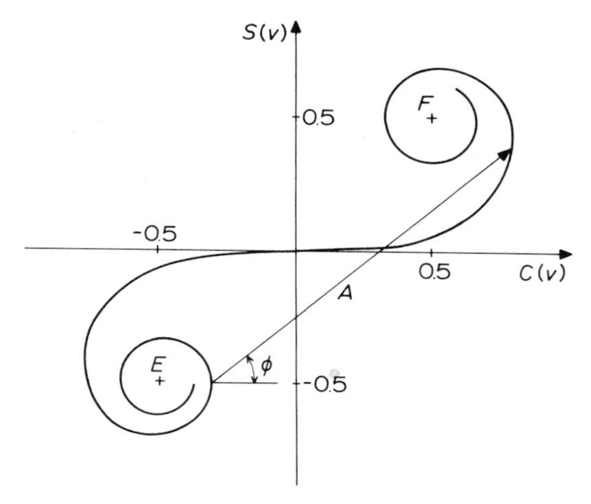

Fig. 3.14 Using Cornu's spiral to find $Ae^{i\phi}$.

$$I(P) = \tfrac{1}{4} A^2 B^2 I_0(P), \tag{3.33}$$

where $I_0(P)$ is the unobstructed irradiance at P.

We can now examine the variation that takes place in the irradiance at P as the aperture is moved to different positions in the diffraction screen relative to the point O (fig. 3.12). Let this motion first be only in the x direction, affecting the limits of μ but not those of ν, and therefore causing changes in A but not in B [eqs. (3.31) and (3.32)]. It will be convenient to define two new quantities,

$$\Delta v = \mu_2 - \mu_1 = a\left[\frac{2(r_{1O} + r_{2O})}{\lambda r_{1O} r_{2O}}\right]^{1/2} \tag{3.34}$$

and

$$\mu_0 = \tfrac{1}{2}(\mu_2 + \mu_1) = x_0\left[\frac{2(r_{1O} + r_{2O})}{\lambda r_{1O} r_{2O}}\right]^{1/2}. \tag{3.35}$$

Equation (3.34) follows at once from the method of finding μ_1 and μ_2 of eq. (3.27). For a given width of aperture a, Δv is a constant independent of how the aperture is moved about. In fact, Δv is the arc length along Cornu's spiral between the points on the curve corresponding to the limits $v = \mu_1$ and $v = \mu_2$.

In eq. (3.35), x_0 is the x coordinate of the center of the aperture, and μ_0 is the value of μ corresponding to this value of x. Actually, μ_0 is the midpoint of the section of Cornu's spiral lying between the points for μ_1 and μ_2. When x_0 is zero it means that the origin O (which lies on the direct line between S and P) is midway between two sides of the aperture. The quantity μ_0 is zero at this value of x_0. When $x_0 = -a/2$, the origin is at one edge of the aperture, and when

$x_0 = a/2$, it is at the opposite edge. The corresponding values of μ_0 are $\pm\Delta v/2$. Between these limits there is a clear path between S and P, but μ_0 must be extended beyond the limits to find the extent of bending of light by diffraction into the geometrical shadow of the screen.

From the distances r_{1O} and r_{2O}, the width a of the aperture, and the wavelength λ of the light, the value of Δv is found by eq. (3.34). Keeping this constant, let x_0 be varied, the corresponding μ_0 being given by eq. (3.35). Also let the points on Cornu's spiral corresponding to μ_1 and μ_2 be found, and the straight-line distance A between them be measured as shown in fig. 3.14 [this measurement must be in the same units used for plotting $C(v)$ and $S(v)$]. If a given motion of x_0 causes the point for μ_1 to move a certain distance along Cornu's spiral, the points for μ_0 and for μ_2 move the same distance along the curve. The length A, however, will in general be different. Now let the square of the A found be plotted against μ_0 for the different positions of the center line (x_0) of the aperture. The result is a curve like one of those in fig. 3.15. The curves of fig. 3.15 have been drawn for six different values of Δv. The heavy lines along the μ_0 axis show the extension of the clear path from S to P, within the region $-\Delta v/2 \leq \mu_0 \leq \Delta v/2$. Beyond this is geometrical shadow.

If the motion of the aperture is in the y direction instead of x, we have only to replace a, μ, and x_0 in eqs. (3.34) and (3.35) by b, ν, and y_0. Curves exactly like those of fig. 3.15 would be found, with the coordinates now ν_0 and B^2.

The cross section of the diffraction pattern in the x direction, for an aperture of width a, is given by finding the Δv corresponding to a and then using the correct curve like those of fig. 3.15. Similarly, the cross section in the y direction is found from the height b of the aperture and the curve for the corresponding Δv. If patterns not on the axes are desired, the product of A^2 and B^2 must be used.

Most of the curves of fig. 3.15 show a considerable diffraction effect in the region of the geometrical shadow (beyond the heavy lines). When Δv is 10 or more, however, the light diffracted into the shadow is very small, and even the fringes outside of the geometrical shadow are small in their fluctuation except near the edge of the shadow. No matter how large Δv is made, the fringes near the shadow edge will always appear. When Δv is small, say less than unity, the extension of the pattern into the shadow is very marked. As Δv is decreased, the shape of the Fresnel pattern approaches that of the Fraunhofer pattern.

To obtain some notion of the relation of dimensions required for a small Δv, let us take a particular case. Suppose r_{1O} and r_{2O} are each 100 cm, and that λ is 0.8×10^{-4} cm, which is the longest wavelength of the visible spectrum. By eq. (3.34) we would have $\Delta v = 22.4a$, with a in centimeters. For $\Delta v = 1$, the distance a would be less than half a millimeter. To obtain an a of half a centimeter, we would have to have a Δv of approximately 11.

Rectangular Aperture, *a* or *b* Large

In the last section, the dimensions of the aperture were assumed small enough to permit some approximations to be used in evaluating the Kirchhoff integral.

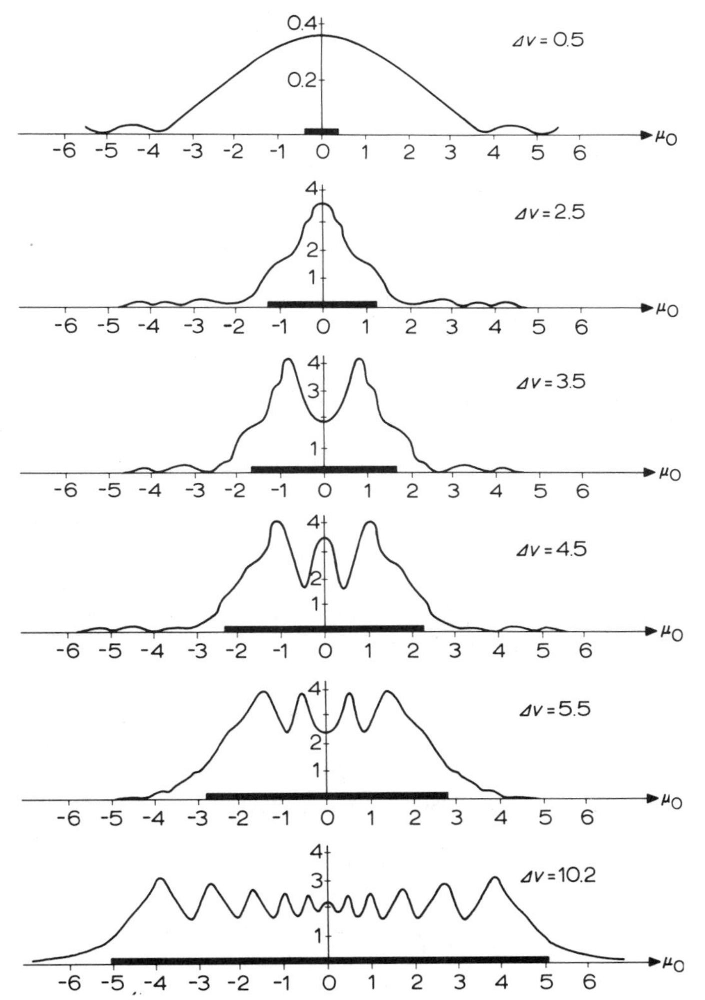

Fig. 3.15 The variation of A^2 along one axis of a rectangular aperture for six different values of Δv. Heavy lines extend over those values of μ_0 for which there is a clear geometric path through the aperture from S to P.

Let us now suppose that a and b is too large for these approximations to be used. At the upper limit (a and b infinite), there would be no screen at all between light source and observation point, so that the irradiance $I(P)$ would simply be the unobstructed value $I_0(P)$. Rather surprisingly, eq. (3.33) takes this form if we substitute infinite values of the μ and ν limits in eqs. (3.31) and (3.32). For

$$Ae^{i\phi} = C(\infty) - C(-\infty) + i[S(\infty) - S(-\infty)] = 1 + i = \sqrt{2}e^{i\pi/4}. \tag{3.36}$$

It will be noted that $\sqrt{2}$ and $\pi/4$ are, respectively, the length of the straight line from E to F in fig. 3.13, and the angle that this line makes with the real axis. Similarly, with the ν limits infinite,

$$Be^{i\psi} = \sqrt{2}e^{i\pi/4}. \tag{3.37}$$

Using the absolute values of A and B from eqs. (3.36) and (3.37) and substituting in eq. (3.33), $I(P) = i_0(P)$, which is the exact value already stated. We can, in fact, use eqs. (3.36) and (3.37) in eq. (3.30), and obtain

$$U_0(P) = U_{00}(P). \tag{3.38}$$

We have thus shown that eq. (3.30), derived for the case of small aperture dimensions, also holds true when these dimensions are infinite.

From this fact alone it cannot be concluded that eq. (3.30) is also exact for intermediate values of a and b. However, if we limit λ to wavelengths of visible light, while r_{10} and r_{20} have convenient practical values, we find that the limits of μ and ν [see eqs. (3.27)] and the values of Δv [eq. (3.34)] can become quite large while x, y, a, and b are still small enough for the approximations used for eq. (3.30) to remain valid. Thus we can say that, as a and b increase, they are still small enough for eq. (3.30) to be accurate when the limits of μ and ν become near enough infinity for eq. (3.30) to be correct for that reason. Then for small wavelengths, eq. (3.30) can be expected to hold, approximately, for any size apertures.

Let us now consider an aperture in the shape of a long, narrow slit. In this case we take b large, but a small. For any motion of the aperture in the y direction, the B in eq. (3.33) would be constant, following eq. (3.37). The variation in $I(P)$ therefore depends only on A, which varies as the aperture is moved in the x direction, as shown by the curves of fig. 3.15. The amplitude equation is

$$U_0(P) = \frac{-i}{\sqrt{2}}\,U_{00}(P)\,A\,\exp\left\{i\left(\phi + \frac{\pi}{4}\right)\right\},$$

and the corresponding irradiance is

$$I(P) = \frac{A^2}{2}\,I_0(P).$$

The factor A^2 can be obtained from fig. 3.14, using the curve for the correct Δv. The pattern consists of long fringes parallel to the slit, and unvarying in this direction except where the ends of the split are approached.

A third case arising from the rectangular aperture may also be treated: that of the diffraction caused by a straight edge. Let the edge run parallel to the y axis, so that in the y direction the limits are $-\infty$ and $+\infty$. In the x direction the space is semi-infinite, with the limits $x = -\infty$ and $x = x_2$, where x_2 is the position of

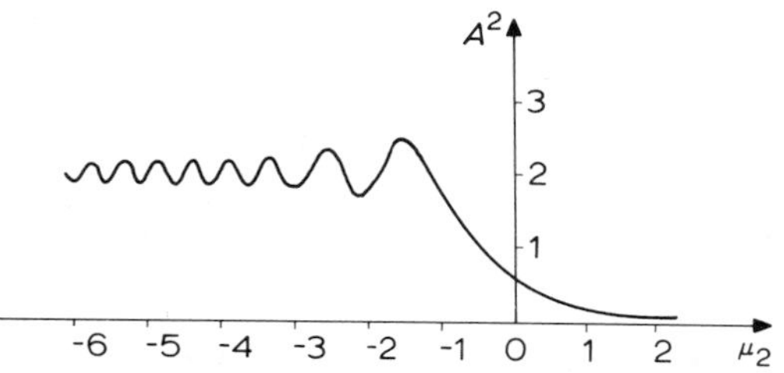

Fig. 3.16 The quantity A^2 as a function of μ_2 for diffraction of a straight edge. The value $\mu_2 = 0$ corresponds to the edge of the geometric shadow.

the screen edge, and $x > x_2$ corresponds to opaque parts of the screen. Let the point O (fig. 3.12) be so close to the screen edge that $|x_2|$ is always small compared with r_{1O} and r_{2O}. The point P will then always be close to the geometrical shadow. By eqs. (3.27), there will be a finite μ_2 corresponding to x_2, but $\mu_1 = -\infty$, $\nu_1 = -\infty$, and $\nu_2 = +\infty$. In eq. (3.32), the coefficient B will have the constant value given by eq. (3.37), while from eq. (3.31) we have

$$Ae^{i\phi} = C(\mu_2) - C(-\infty) + i[S(\mu_2) - S(-\infty)].$$

One end of the line A in fig. 3.14 will always be at the point E (where $C = S = -\frac{1}{2}$), while the other end varies according to the position chosen for x_2, which determines μ_2. It is now possible to draw a curve for A^2 versus μ_2, from graphical measurements on fig. 3.14; this has been done in fig. 3.16. This curve is similar to those of fig. 3.15, but does not have the symmetrical quality of the latter. The pattern is one of fringes parallel to the straight edge, mostly on the clear side of the geometrical shadow but decreasing in amplitude as distance from the edge increases, and with some irradiance extending into the shadow. An example of the Fresnel diffraction pattern from a straight edge is photographed in fig. 3.17.

Circular Aperture

In discussing Fresnel diffraction by a circular aperture, we deal only with the situation where the line from the light source to the point of observation is perpendicular to the plane of the aperture, and passes through the center of the circle. Figure 3.18 illustrates this situation. Let c be the radius of the aperture, and let ρ be the radial distance to any point in the aperture.

Take first the case where c is much smaller than r_{1O} or r_{2O}. We can then make the same assumptions that we used in arriving at eq. (3.23). Also, with the ciruclar form of the surface Σ, we can take $d\Sigma$ to be the ring-shaped element $2\pi\rho\, d\rho$, and carry the integration from $\rho = 0$ to $\rho = c$. Thus the Kirchhoff integral is

$$U_0(P) = -\frac{iA}{\lambda r_{1O} r_{2O}} \int_0^c \exp[ik(r_1 + r_2)]\, 2\pi\rho\, d\rho. \tag{3.39}$$

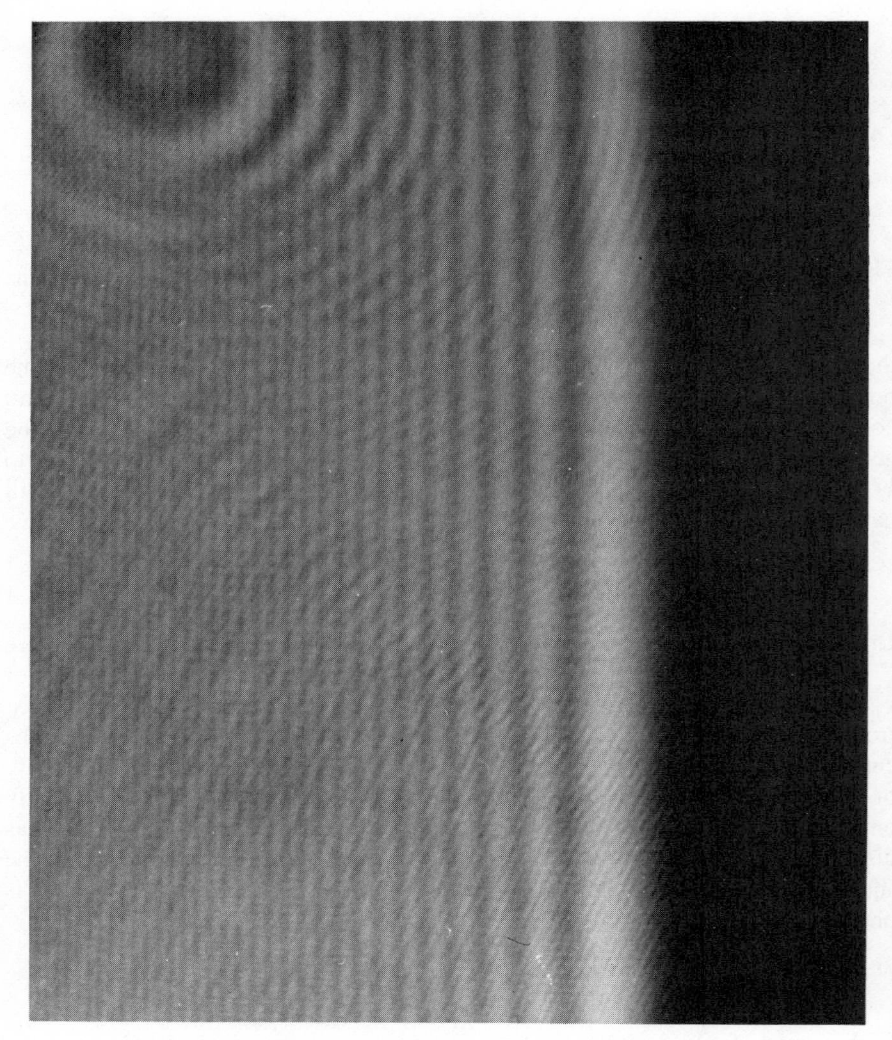

Fig. 3.17 Fresnel diffraction from a straight edge.

From fig. 3.18 it can be seen that $r_1^2 = r_{1O}^2 + \rho^2$ and $r_2^2 = r_{2O}^2 + \rho^2$. Since r_{1O} and r_{2O} are constants, the differentiation of these equations and division by 2 gives

$$\rho \, d\rho = r_1 \, dr_1 = r_2 \, dr_2.$$

Thus $dr_1 = (1/r_1) \, \rho \, d\rho$ and $dr_2 = (1/r_2) \, \rho \, d\rho$, and, adding, we obtain

$$d(r_1 + r_2) = \left(\frac{1}{r_1} + \frac{1}{r_2}\right) \rho \, d\rho.$$

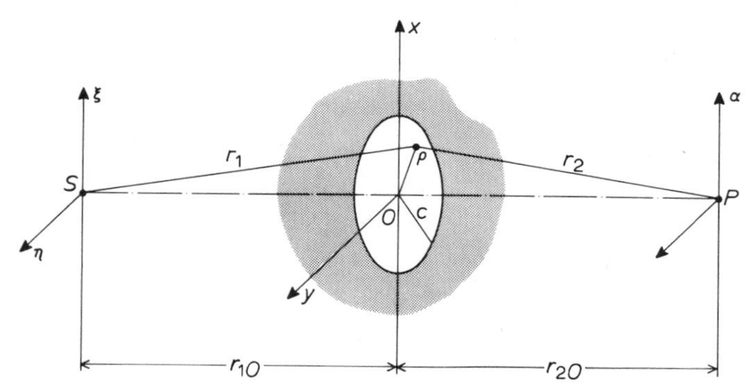

Fig. 3.18 Geometry for the application of Kirchhoff's integral to Fresnel diffraction of a circular aperture.

Then

$$\rho\, d\rho = \frac{r_1 r_2}{r_1 + r_2}\, d(r_1 + r_2).$$

Since c has been assumed small, it will not be far wrong to make the approximation

$$\rho\, d\rho \simeq \frac{r_{1O} r_{2O}}{r_{1O} + r_{2O}}\, d(r_1 + r_2).$$

Making the substitution $l = r_1 + r_2$, we can now write eq. (3.39) in the form

$$U_0(P) = -i\frac{2\pi A}{\lambda(r_{1O} + r_{2O})} \int_{l(0)}^{l(c)} e^{ikl} dl, \tag{3.40}$$

where $l(0) = r_{1O} + r_{2O}$.

The path l, from S to P, is always slightly longer than the direct distance $r_{1O} + r_{2O}$. Let us express the difference in the form first introduced by Fresnel.

$$l = r_{1O} + r_{2O} + \Delta\frac{\lambda}{2}. \tag{3.41}$$

Since λ here is the wavelength of light, the path difference is expressed in half-wavelengths. The variable Δ is dimensionless, and $\Delta\lambda/2$ expresses in half-wavelengths the increase of path from S to P as ρ is increased from zero. This notation will be found useful later, when we discuss the *Fresnel zone plate*.

In applying eq. (3.41) to (3.40) let us note that $dl = (\lambda/2)\, d\Delta$, and that $\exp(ikl) = \exp[ik(r_{1O} + r_{2O})] \cdot \exp(i\pi\Delta)$, since $k\lambda = 2\pi$. The first of these exponentials is constant and may be removed from the integral, and it combines with other constants to form $U_{00}(P)$, the unobstructed amplitude of the irradiance at P[see the equation after eq. (3.28)]. Removing $\lambda/2$ from the integral, and noting that $\Delta = 0$ at $l(0)$, eq. (3.40) becomes

$$U_0(P) = -i\pi U_{00}(P) \int_0^{\Delta(c)} \exp(i\pi\Delta)\, d\Delta, \tag{3.42}$$

where $\Delta(c)$ is the value of Δ where $\rho = c$.

By integrating eq. (3.42), we have

$$U_0(P) = U_{00}(P)\, \{1 - \exp[i\pi\Delta(c)]\}, \tag{3.43}$$

and the corresponding irradiance is

$$I(P) = |U_0(P)|^2 = 2I_0(P)\, [1 - \cos\, \pi\Delta(c)] = 4I_0(P) \sin^2 \tfrac{1}{2}\pi\Delta(c). \tag{3.44}$$

By eq. (3.41), the quantity $\Delta(c)$ expresses in half-wavelengths the excess of the path from S to P by way of the aperture perimeter, rather than straight through the center. As c varies from zero to moderate values, $\sin^2 \frac{1}{2}\pi\Delta(c)$ and therefore $I(P)$ passes through a series of values that fluctuate between zero and a maximum. For $I(P)$, the maximum is four times the irradiance when no diffraction screen is present. Thus depending on the radius of the aperture, the center of the diffraction pattern may be either dark or bright. Figure 3.19 shows the patterns for different values of $\Delta(c)$.

The value of $\Delta(c)$, in terms of c and the other dimensions shown in fig. 3.18, plus the wavelength λ, may be deduced from eq. (3.41). When the radius is c, the l in this equation becomes

$$l = (r_{1O}^2 + c^2)^{1/2} + (r_{2O}^2 + c^2)^{1/2}.$$

Or, approximately, since c is small,

$$l \simeq r_{1O} + \frac{c^2}{2r_{1O}} + r_{2O} + \frac{c^2}{2r_{2O}}.$$

Using this value for l in eq. (4.41),

$$\Delta(c) = \frac{c^2}{\lambda}\left(\frac{1}{r_{1O}} + \frac{1}{r_{2O}}\right) = \frac{c^2(r_{1O} + r_{2O})}{\lambda r_{1O} r_{2O}} \tag{3.45}$$

Our derivation of the irradiance from a circular aperture, expressed in eq. (3.44), has dealt only with this irradiance at the center of the pattern, with light from the source striking perpendicularly at the center of the aperture. Nothing is disclosed, in this equation alone, regarding the diffraction pattern away from

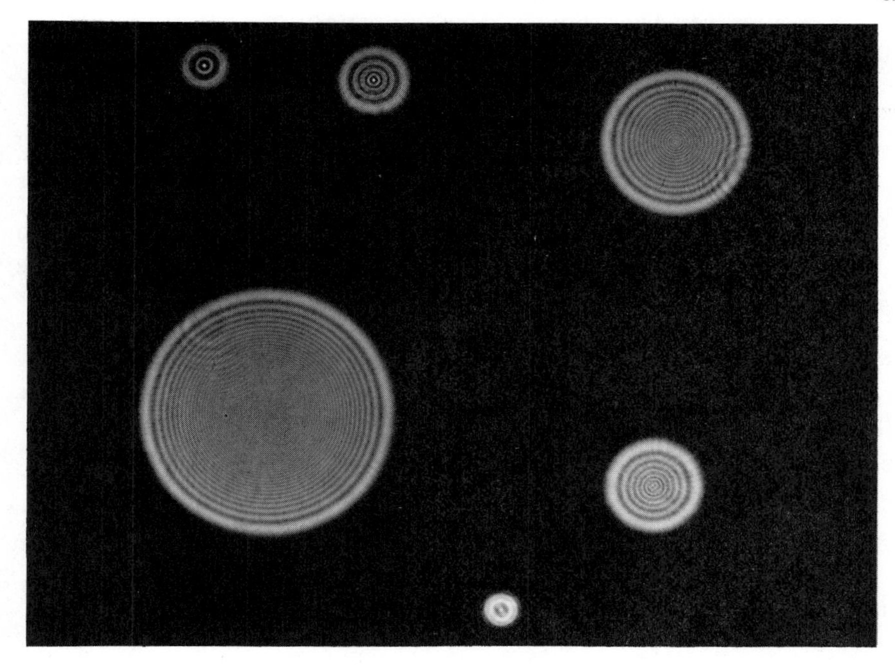

Fig. 3.19 Fresnel diffraction patterns of circular apertures for various values of $\Delta(c)$.

the central point P. Experimentally, the pictures of fig. 3.19 show what the pattern is like. We would certainly expect the fringes to be ring-shaped, about P as center. We would also expect that as $\Delta(c)$ is decreased toward unity, the fringes would become less extensive in total area, but more widely spaced, as they do in fig. 3.15 for the rectangular aperture as Δv decreases. If we compare $\Delta(c)$ in eq. (3.45) with Δv in eq. (3.34), we find that $\Delta(c)$ is comparable to $(\Delta v)^2$, provided that c^2 is comparable to $2a^2$. For Δv and $\Delta(c)$ both to be one, we would have $c = \sqrt{2}\, a$.

It is to be noticed from eq. (3.44) that $\Delta(c) = 1$ is the lowest value of $\Delta(c)$ that will give the maximum irradiance at the center of the observed pattern. As $\Delta(c)$ is decreased below one, we would expect the diffraction to spread further into the geometrical shadow (as in fig. 3.15), but with decreased irradiance. The pattern would approach that of the Fraunhofer case [eq. (3.22)]. Lord Rayleigh concluded that $\Delta(c) = 0.9$ gave the most distinct image from a pinhole camera: i.e., the least spreading of the image by diffraction.

Large Circular Aperture

It is interesting to compare the irradiance at the center of the pattern from a circular aperture, as the radius of the aperture increases, with that from a square aperture of comparable dimensions. We can equate the width of the square with the diameter of the circle: $a = 2c$. Put the values $x_1 = y_1 = -a/2 = -c$ and $x_2 = y_2 = a/2 = c$ into eq. (3.27); thus

$$\mu_2 = c\left[\frac{2(r_{1O} + r_{2O})}{\lambda r_{1O} r_{2O}}\right]^{1/2} = \nu_2 \tag{3.46}$$

and $\mu_1 = \nu_1 = -\mu_2$. We also note that for the Fresnel integrals, eq. (3.29), $C(-\mu_2) = -C(\mu_2)$ and $S(-\mu_2) = -S(\mu_2)$. The two main factors of eq. (3.30) are alike, and each becomes $2[C(\mu_2) + iS(\mu_2)]$. Thus eq. (3.30) is now

$$U_0(P) = -2iU_{00}(P)\,[C(\mu_2) + iS(\mu_2)]^2.$$

The irradiance is the square of the absolute value of this equation, i.e.,

$$I_s(P) = 4I_0(p)\,[C(\mu_2) + iS(\mu_2)]^2, \tag{3.47}$$

where the subscript s is used to denote that the equation applies to the square.

If we compare eqs. (3.45) and (3.46), we see that $\Delta(c) = \frac{1}{2}\mu_2^2$. Substituting this into eq. (3.44), and using a subscript c to refer to the circle,

$$I_c(P) = 4I_0(P)\sin^2 \tfrac{1}{4}\pi\mu_2^2. \tag{3.48}$$

Equations (3.47) and (3.48) are both stated in terms of μ_2, which by eq. (3.46) is proportional to the radius c of the circular aperture. The irradiances $I_s(P)$ and $I_c(P)$ are plotted against μ_2 in fig. 3.20, where it is seen that, while the irradiance in the center of the pattern from the square converges toward the unobstructed value I_0 as size increases, that from the circle continues to oscillate between zero and its maximum value. This comparison is valid for moderate values of μ_2, and can be observed. Of course, we realize that as μ_2 approaches

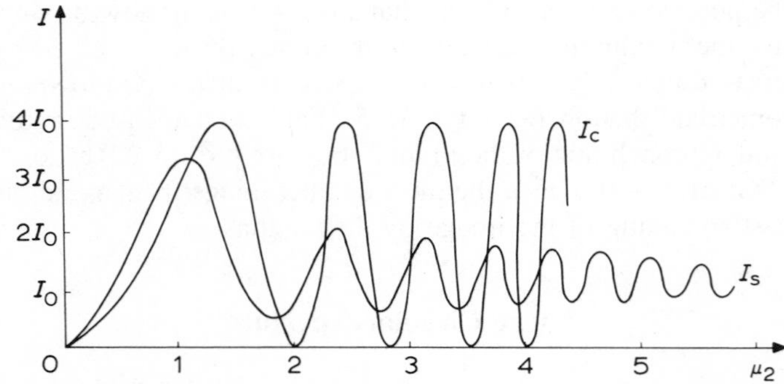

Fig. 3.20 Irradiance at the center of the diffraction pattern as a function of μ_2: I_c, irradiance for a circular aperture of radius c;I_s, irradiance for a circular aperture of edge $2c$.

infinity the two cases must be the same. This only means that for extreme size of the apertures, the approximations that have been used in both cases are no longer valid, and a more exact way to treat the Kirchhoff integral in this region must be found.

3.5 FRESNEL ZONE PLATE

In eq. (3.42), instead of taking the integral over the whole open circle from $\Delta = 0$ to $\Delta = \Delta(c)$, let us consider the contribution to $U_0(P)$ of a ring-shaped zone from $\Delta = \Delta_1$ to $\Delta = \Delta_2$:

$$U_0(P) = -i\pi U_{00}(P) \int_{\Delta_1}^{\Delta_2} \exp(i\pi\Delta)\, d\Delta = U_{00}(P)\,[\exp(i\pi\Delta_1) - \exp(i\pi\Delta_2)]$$

$$= U_{00}(P)\,[\cos \pi\Delta_1 - \cos \pi\Delta_2 + i(\sin \pi\Delta_1 - \sin \pi\Delta_2)].$$

Now let us suppose that Δ_1 is an even integer and that $\Delta_2 = \Delta_1 + 1$, and is therefore odd. We see that in this case $U_0(P) = 2U_{00}(P)$. Alternatively, when Δ_1 is odd and Δ_2 the next even number, $U_0(P) = -2U_{00}(P)$.

Suppose the complete circular aperture is divided into such zones, the boundaries of which have Δ equal to successive integers. The corresponding radii from the center of the aperture are found by solving eq. (3.45) for c, with $\Delta(c)$ being replaced by n:

$$c_n = \left[\frac{n\lambda r_{1O} r_{2O}}{r_{1O} + r_{2O}}\right]^{1/2}, \qquad n = 1, 2, 3, \ldots. \tag{3.49}$$

Remembering the significance of Δ, we see that each integral increase in n in eq. (3.49) means that the light path from S to P has been increased by one half-wavelength. Such zones are called *Fresnel half-period zones*, or simply *Fresnel zones*. Each successive zone, in the open circular aperture, cancels the preceding zone in its effect on the amplitude at P. The net $U_0(P)$ can be considered as coming from the part of the circle remaining after the greatest even value of Δ.

With the circular aperture divided into Fresnel zones, as above, let alternate zones be covered by opaque material. We have then what is called the *Fresnel zone plate*. This is illustrated in fig. 3.21, where the central zone is shown open. The effect would be the same if we started with a closed zone in the center. Suppose there are N of these zones, with the screen entirely opaque for all parts outside the last open zone. The total number of open zones is then $N/2$ if the central zone is blocked, or $(N + 1)/2$ if the central zone is open. Each open zone contributes 2 $U_{00}(P)$ to $U_0(P)$. (We need pay no attention to the negative sign if the central zone is opaque.) The total effect is then, with the central zone open,

$$U_0(P) = (N + 1)\, U_{00}(P), \tag{3.50}$$

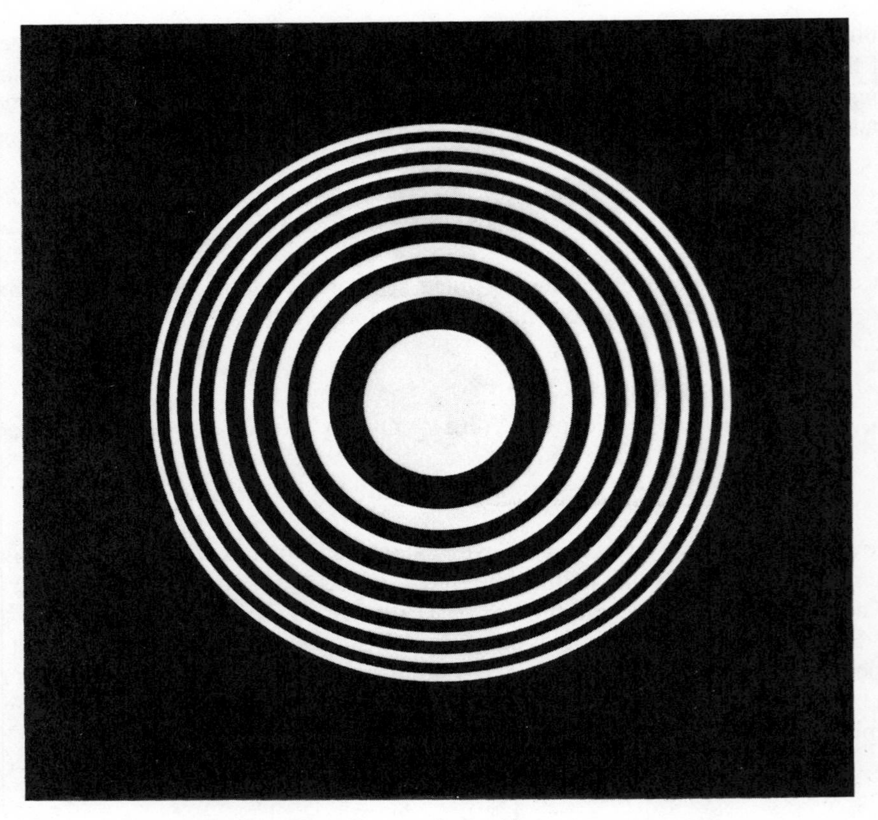

Fig. 3.21 Fresnel zone plate with center zone open.

and the irradiance is

$$I(P) = (N + 1)^2 \, I_0(P). \tag{3.51}$$

With a large number of zones, the brightness of the spot at p is quite high. There is, in effect, a focusing of the source S at the point P. A more exact treatment for the complex light amplitude distribution can be obtained by means of the Fresnel–Kirchhoff integral, as is shown in the chapter on wave front reconstructions. It is also shown later that the focusing effect of the Fresnel zone plate is the foundation of modern interference photography, i.e., holography.

We can compare the focusing effect of the Fresnel zone plate, which is due to diffraction, with that produced by refraction in the ordinary lens. If we make r_{1O} the object distance and r_{2O} the image distance, the focal length f of the lens is given by

$$\frac{1}{f} = \frac{1}{r_{1O}} + \frac{1}{r_{2O}} \qquad \text{or} \qquad f = \frac{r_{1O} r_{2O}}{r_{1O} + r_{2O}}.$$

The physical meaning of the focal length of the lens, defined by this equation, is that f is the value taken by the image distance r_{2O} when the object distance r_{1O} is made infinite. It can be given exactly the same meaning for the zone plate, and the same equation substituted into eq. (3.49), which on squaring becomes

$$c_n^2 = n\lambda f \qquad \text{or} \qquad f = \frac{c_n^2}{n\lambda}.$$

Since c_n^2 and n are proportional, let us make $n = 1$ and $c_n = c_1$. The focal length of the zone plane is then

$$f = \frac{c_1^2}{\lambda}. \tag{3.52}$$

The focal length is thus dependent on λ, and if white light is used there will be a great deal of chromatic aberration. For this reason, monochromatic light is much to be preferred for use with the Fresnel zone plate.

No restriction has been placed on the value of λ, and thus zone plates can be designed to focus ultraviolet light or even X rays. This is not possible with lenses because of the lack of a material that has the refractive properties needed for a lens at these wavelengths.

3.6 RAYLEIGH'S CRITERION AND ABBE'S SINE CONDITION

Thus far we have been considering diffraction simply as a phenomenon in the propagation of light. It must also be considered as a source of imperfection in the performance of optical systems.

For example, in fig. 3.22 a simple lens is shown, projecting the images P_1 and P_2 of two sources, S_1 and S_2. The optical system could just as well be more complicated, with multiple lenses and mirrors, but forming, as here, two images of two sources. We assume, also, that it is a "stigmatic" system, by which is meant that any optical ray coming from a source S and striking the system at any

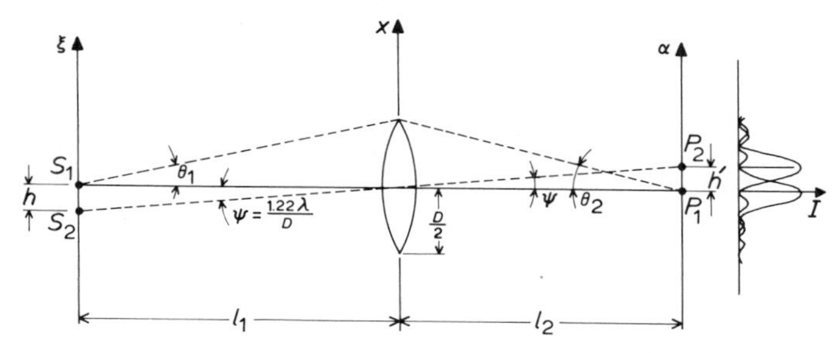

Fig. 3.22 Rayleigh criterion for two point sources.

point must eventually also pass through the same image point P, and that the optical lengths (lengths measured in wavelengths of light) of all such paths from S to P are the same. Obviously, the geometrical length of the ray passing through the center of the lens in fig. 3.22 is less than the lengths of other rays from S to P. However, the lens is thicker in the center, and since the velocity of light is less inside the lens, the number of wavelengths in this section of the lens is increased.

Now in addition to geometrical-ray properties of the system, diffraction also takes place. Thus the space occupied by the lens in fig. 3.22 may be regarded as a circular aperture in a diffracting screen, and because of the stigmatic character of the system, the diffraction may be taken as Fraunhofer. The result is that the images at P_1 and P_2 are not sharp, but each has a diffraction pattern around it. This tends to confuse the images of the points when they are close together, and the question arises as to how close they can be and still be *resolved*.

Rayleigh's criterion is that the images are resolved if the central maximum in the pattern of one point coincides with the first dark fringe in the pattern of the second point. Irradiance curves for P_1 and P_2 are drawn at the right of fig. 3.22, with the spacing just corresponding to Rayleigh's criterion. These curves are given by eq. (3.22), for the Fraunhofer diffraction from a circular aperture. As was explained in connection with this equation, the first dark fringe comes at $\mu = 1.22$, and this can be put into eq. (3.18). Let θ_2 of that equation be called here the angle ψ, and with it quite small we have

$$\psi_{\min} \simeq \frac{1.22\lambda}{D}. \tag{3.53}$$

This $\psi_{\min}$ is the angle at the center of the lens subtended by the distance between images just resolvable under Rayleigh's criterion. The diameter of the lens is D, and λ is the wavelength of light. For an angle less then $\psi_{\min}$, the images would not be resolved.

It can be seen from fig. 3.22 that the angles between the central rays are equal on the two sides of the lens. If the lens is the objective lens of an astronomical telescope, the distance l will be indefinitely great, and the least angular separation given by eq. (3.53) will be satisfactory to rate the resolving power. For much closer sources, as with a compound microscope, it may be more significant to replace ψ by h/l, and express the least h as

$$h_{\min} = \frac{1.22\lambda l}{D}. \tag{3.54}$$

In applying eq. (3.54) to a particular optical instrument, it is customary to use another relation, known as *Abbe's sine condition*. This is developed as follows. In fig. 3.23, again let a focusing optical system be represented by a single convex lens, although the system may be more complicated than this. Let a source S_1

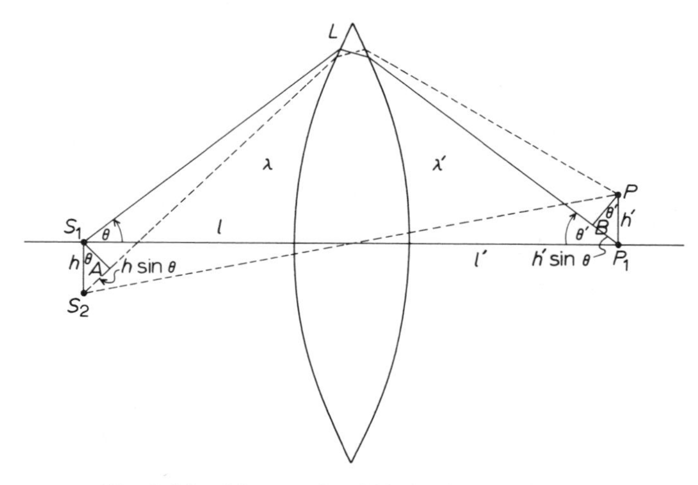

Fig. 3.23 Diagram for Abbe's sine condition.

be located on the axis of the lens, with its image at P_1. Let a second source S_2 be located at a very small distance h off the axis, and let its image at P_2 be at a small distance h from P_1. The system is assumed to be stigmatic, but let it be assumed that there is a different medium on the image side of the lens than that on the source side. Although the frequency of the light remains constant, it will have a different velocity and therefore a different wavelength on the two sides. Call the wavelengths λ and λ' on the source and image sides, respectively.

Now draw a set of near-axial rays from S_1 to P_1, and from S_2 to P_2. Let the optical lengths of these rays be b_{1O} and b_{2O}. Draw other rays of the optical length b_1 and b_2 from S_1 and S_2 at very nearly the same angle as the first set, but intersecting those rays inside the lens. The emission angles for the two sets of rays, being very nearly the same, may be assigned the same value θ. Both sets of rays may then be said to arrive at P_1 and P_2 at the same angle θ'. Since the system is stigmatic, we have at once that

$$b_1 = b_{1O} \qquad \text{and} \qquad b_2 = b_{2O}.$$

But since h and h' are very small, we also have $b_{1O} \simeq b_{2O}$, and therefore,

$$b_1 \simeq b_2. \tag{3.55}$$

These are the total optical distances from S_1 and S_2 to P_1 and P_2.

Let us drop a perpendicular line from S_1 to A on the line S_2L. The geometrical distances S_1L And AL are very nearly the same, so that the line S_2L exceeds S_1L in length by the quantity $h \sin \theta$. In the same way on the image side, LP_1 exceeds LP_2 by $h' \sin \theta'$. The optical excess lengths are found from these by dividing by their respective wavelengths, and in view of eq. (3.55) we must have

$$\frac{h \sin \theta}{\lambda} \simeq \frac{h' \sin \theta'}{\lambda'}. \tag{3.56}$$

Equation (3.56) is known as *Abbe's sine condition*. It is very nearly true for all values of θ if h and h' are very small and the optical system is stigmatic.

Let us use this sine condition to find the resolving power of a microscopic objective. In the first case, let there be air on both sides of the lens, so that λ and λ' are equal. Notice also, from fig. 3.23, that $h/h' = l/l'$. Putting these conditions into eq. (3.56),

$$l \sin \theta = l' \sin \theta'. \tag{3.57}$$

In the microscope the angle θ' is quite small, so that we can put

$$\sin \theta' \simeq \theta' = \frac{D}{2l'}. \tag{3.58}$$

Putting this into eq. (3.57) we have

$$\frac{D}{l} = 2 \sin \theta,$$

and this can be applied to eq. (3.54), giving

$$h_{\min} = \frac{1.22\lambda}{2 \sin \theta} = \frac{0.61\lambda}{\text{N.A.}}, \tag{3.59}$$

where the initials N.A. stand for "numerical aperture," which is equal to sin θ. It is the value of the N.A. that is usually given by manufacturers for the rating of the microscopic objective. It represents the angular radius of the pencil of rays from an object in focus, to the objective lens. The least separation of points to be resolved is found from eq. (3.59).

There is an instrument, called the oil immersion microscope, for which λ and λ' are not equal. Although the medium on the image side of the objective lens is air, the object and the space between it and the lens is immersed in oil with an index of refraction η. The index of refraction of air may be taken as very close to one. Now the velocity of light in a medium, and therefore also the wavelength of light of a given frequency, is inversely proportional to the index of refraction of the medium. In the above case, then, we have

$$\frac{\lambda'}{\lambda} = \frac{\eta}{1} = \eta.$$

The sine condition eq. (3.56) then becomes

$$\eta h \sin \theta \simeq h' \sin \theta'. \tag{3.60}$$

We must modify the conditions for resolution, based on fig. 3.20. The angles ψ and ψ' are now not equal, but with an index η on the object side we must put $\eta\psi = \psi'$. Equation (3.53) still holds, if we put

$$\psi'_{\min} \simeq \frac{1.22\lambda'}{D},$$

but in terms of ψ it is

$$\psi_{\min} = \frac{1.22\lambda'}{\eta D}. \tag{3.61}$$

Since we may replace ψ by h/l, the minimum separation of objects is

$$h_{\min} \simeq \frac{1.22\lambda' l}{\eta D}. \tag{3.62}$$

We continue to use λ', the wavelength in air, to designate the character of the light. Since $n\psi = \psi'$, we also have

$$\frac{\eta h}{l} = \frac{h'}{l'},$$

and, if we put this into eq. (3.60), we get, for the sine condition,

$$l \sin \theta \simeq l' \sin \theta', \tag{3.63}$$

which is the same as eq. (3.57). We can also use eq. (3.58) with this, and when the result is substituted in eq. (3.62),

$$h_{\min} \simeq \frac{0.61\lambda'}{\eta \sin \theta}. \tag{3.64}$$

This may be expressed as in eq. (3.59), provided we set, in this case,

$$\text{N.A.} = \eta \sin \theta. \tag{3.65}$$

From eq. (3.65), it may be seen that, for the oil immersion microscope, it is possible to have a numerical aperture greater than unity.

All of the equations that have been derived for resolving power, such as eqs. (3.53), (3.54), and (3.59), are significant and useful relations. There are conditions, however, such as an inhomogeneous or turbulent medium, under which they would not serve accurately. It must also be said that our assumption has been the separation of two close but independent sources. Although these may emit light of equal wavelength, they are assumed to be incoherent, in the sense that there is no fixed phase link between them. Our conclusions are not strictly applicable to separate objects that are illuminated by a common source of light, since in that case there is some degree of coherence in the light from the objects. Equation (3.59) would then require some modification.

3.7 FRESNEL–KIRCHHOFF THEORY

In sec. 2.8, we derived the Kirchhoff integral with the application of *Green's theorem* to the *scalar wave theory*. We saw that the approach was rather involved mathematically. However, in this section we approach the Fresnel–Kirchhoff theory by a *linear system concept*.

According to Huygens' principle, the complex amplitude observed at a point p' of a coordinate system $\sigma(\alpha, \beta, \gamma)$, due to a light source located in another coordinate system $p(x, y, z)$, as shown in fig. 3.24, may be calculated by assuming that each point of light source is an infinitesimal, spherical radiator. Thus, the complex light amplitude $E_l^{\dagger}$ $(\rho; k)$ contributed by a point p in the ρ coordinate system can be considered to be that from an unpolarized monochromatic point source, such that

$$E_l^{\dagger} = -\frac{1}{\lambda r} \exp[i(kr - \omega t)], \tag{3.66}$$

where λ, k, ω, are the wavelengths, wave number, and angular frequency, respectively, of the point source, and r is the distance between the point source and the point of observation:

$$r = [l + \gamma - z)^2 + (\alpha - x)^2 + (\beta - y)^2]^{1/2} \tag{3.67}$$

If the separation l, of the two coordinate systems is assumed to be large com-

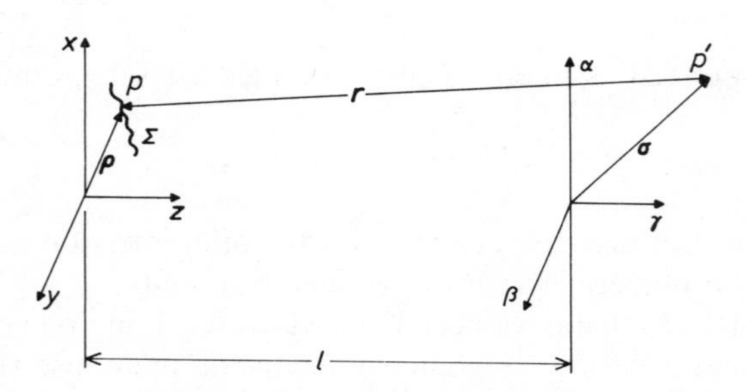

Fig. 3.24 The coordinate systems.

pared to the magnitude of ρ and σ, then r may be approximated by l in the denominator of eq. (3.66) and by

$$r = l + \gamma - z + \frac{(\alpha - x)^2}{2l} + \frac{(\beta - y)^2}{2l} \tag{3.68}$$

in the exponent. Therefore, eq. (3.66) may be written as

$$E_l^+(\boldsymbol{\sigma} - \boldsymbol{\rho}; k) \simeq -\frac{i}{\lambda l} \exp\left\{ ik\left[l + \gamma - z + \frac{(\alpha - x)^2}{2l} + \frac{(\beta - y)^2}{2l} \right] \right\}, \tag{3.69}$$

where the time-dependent exponential has been dropped for convenience. Since eq. (3.69) represents the freespace radiation from a monochromatic point source, it is known as freespace or spatial impulse response. In other words, the complex amplitude produced at the $\boldsymbol{\sigma}$ coordinate system by a monochromatic radiating surface located in the $\boldsymbol{\rho}$ coordinate system can be written as

$$U(\boldsymbol{\sigma}) = \iint_{\Sigma} T(\boldsymbol{\rho})\, E_l^+ (\boldsymbol{\sigma} - \boldsymbol{\rho}; k)\, d\Sigma, \tag{3.70}$$

where $T(\boldsymbol{\rho})$ is the complex light field of the monochromatic radiating surface, Σ denotes the surface integral, and $d\Sigma$ is the incremental surface element. Needless to say, eq. (3.70) is essential to the Kirchhoff's integral in eq. (2.35).

As an illustration, given a complex monochromatic radiating field $T(x, y)$ at (x, y), the complex light disturbances at (α, β) can be obtained by the following convolution integral:

$$U(\alpha, \beta) = T(x, y) * E_l(x, y), \tag{3.71}$$

where * denotes the convolution operation,

$$E_l(x, y) = C \exp[i \frac{k}{2l} \rho^2], \tag{3.72}$$

is the spatial impulse response between the spatial coordinate systems (x, y) and (α, β), $C = i/\lambda l \exp(ikl)$ a complex constant, and $\rho^2 = x^2 + y^2$. We note that eq. (3.71) can be represented by a block box system diagram as shown in fig. 3.25.

On the other hand, if the complex light disturbances at (α, β) are provided,

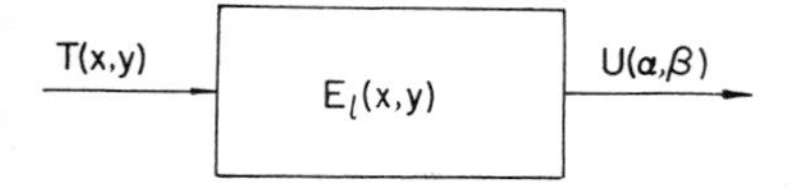

Fig. 3.25 A linear system representation.

the monochromatic radiating field of $T(x, y)$ can be determined,

$$T(x, y) = U(\alpha, \beta) * E_l^*(\alpha, \beta), \tag{3.73}$$

where the superscript * denotes the complex conjugate,

$$E_l^*(\alpha, \beta) = C^* \exp[-i \frac{k}{2l} \rho^2], \tag{3.74}$$

and

$$C^* = \frac{i}{\lambda l} \exp(-ikl).$$

PROBLEMS

3.1 Given a diffraction screen containing a pair of rectangular apertures, as shown in fig. 3.26. Let a monochromatic plane wave be normally incident on this diffraction screen. Determine the irradiance of the corresponding Fraunhofer diffraction pattern. Make a rough plot of the corresponding irradiance along the vertical axis of the observation screen.

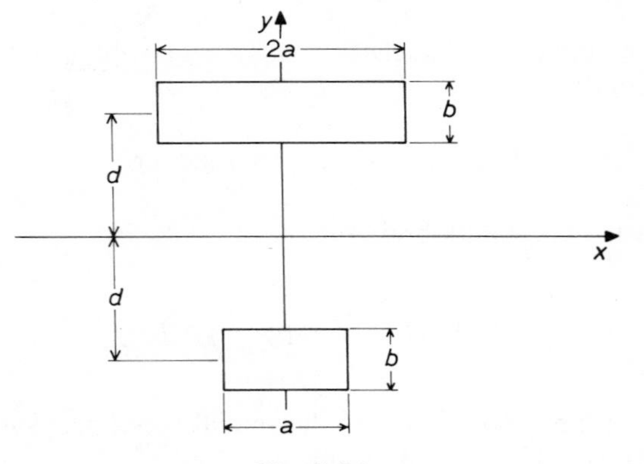

Fig. 3.26

3.2 The expression of the normalized amplitude diffraction pattern $Q(\rho)$ of a circular aperture in the Fraunhofer case is given as

$$Q(\rho) = \frac{2J_1(\pi\delta)}{\pi\delta}, \qquad \delta \geq 0,$$

where J_1 is the Bessel function of order one, $\delta = (kD/2\pi) \sin \theta_2 =$

α/r_{20}, and D is the radius of the circular aperture. Given a screen having an annular aperture as shown in fig. 3.27, determine the normalized irradiance of the Fraunhofer diffraction pattern.

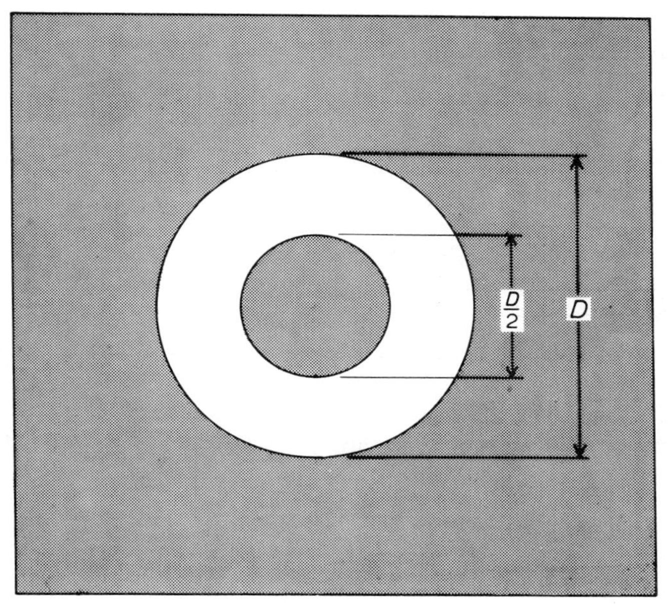

Fig. 3.27

3.3 A diffraction screen contains an $n \times m$ rectangular array of identical circular apertures of radius c, arranged as shown in fig. 3.28. Determine the irradiance of the corresponding Fraunhofer diffraction pattern.

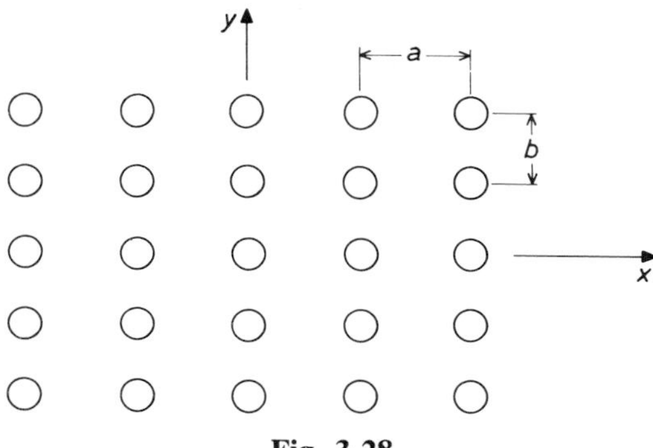

Fig. 3.28

3.4 If a diffraction screen contains a linear array of five rectangular apertures, as shown in fig. 3.29, determine the complex light field of the Fraunhofer diffraction pattern. Make a rough sketch of the corresponding irradiance.

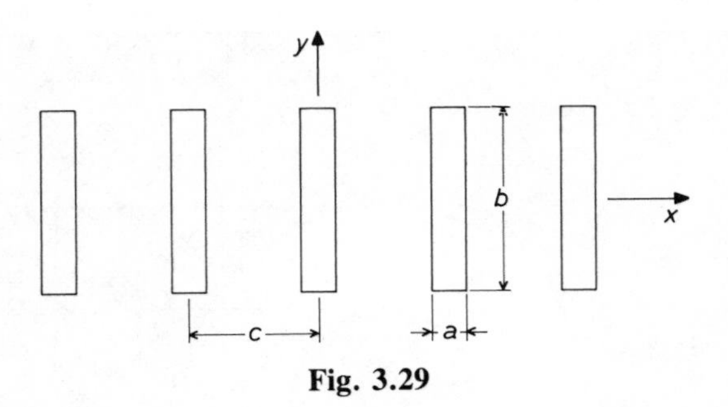

Fig. 3.29

3.5 Let a plane wave of wavelength $\lambda = 5000$ Å and irradiance I be normally incident on a diffraction screen that has the open aperture shown in fig. 3.30. Determine the complex amplitude and the irradiance at a point P on the axis of the circles about 2 meters behind the diffraction screen.

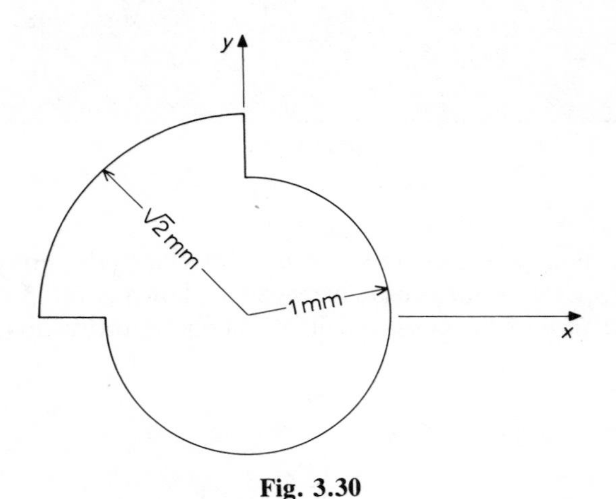

Fig. 3.30

3.6 By interchange of the opaque and transparent regions of the diffraction screen in the previous problem, show that the sum of the disturbances observed with the two complementary screens satisfies Babinet's principle.

3.7 A plane wave of wavelength $\lambda = 5000$ Å is normally incident upon a semi-infinite straight-edge diffraction screen. By the application of Cornu's spiral, determine the locations of the maxima and minima of the diffraction pattern at a distance of 2 meters behind the screen.

3.8 It is assumed that a plane wave of wavelength λ is normally incident on a diffraction screen that has a transmission function

$$T(x, y) = \tfrac{1}{2} + m \sin p_0 x,$$

where $T(x, y) \leq 1$, and p_0 is the angular spatial frequency.

(a) Determine the Fresnel pattern behind the diffraction screen.

(b) If $m \ll \frac{1}{2}$, determine the distances behind the diffraction screen at which the amplitude and phase modulations take place.

3.9 In fig. 3.18 let r_{1O} be infinitely large. Show that the diffraction pattern goes from Fresnel to Fraunhofer as the radius of the aperture becomes very small.

3.10 **(a)** Calculate the minimum angular separation between two distant stars that may be barely resolved by a 75-cm diameter telescope, when a filter that selects the violet light of wavelength $\lambda = 4000$ Å is used.

(b) If the focal length of the objective is assumed to be 8 meters long, compute the minimum resolvable separation between the two images of the distant stars.

3.11 A diffraction screen contains five narrow slits of slightly different widths, as shown in fig. 3.31. If the screen is normally incident by a monochromatic plane wave of $\lambda = 5000$ Å and the optical disturbances produced by the individual slits, at a point P 2 meters behind the screen, are assumed to be in phase, then:

(a) Determine the separations of the narrow slits.

(b) Compute the corresponding irradiance at P in terms of irradiance without the screen.

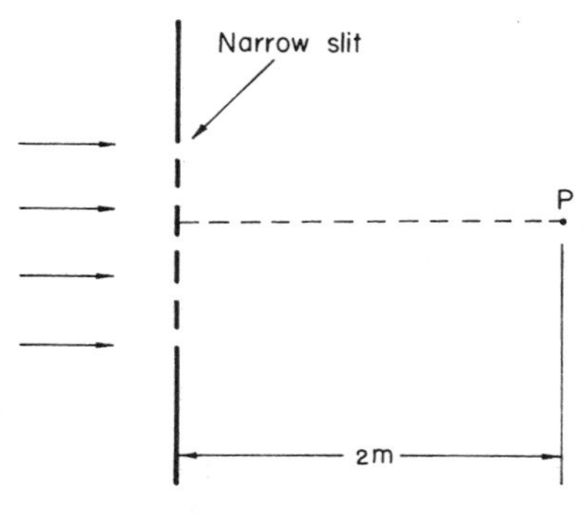

Fig. 3.31

3.12 Given two identical circular apertures of diameter D with a separation $2d$, where $d \geq D$, as shown in fig. 3.32. If these two apertures are normally illuminated by a monochromatic plane wave, then determine the corresponding normalized Fraunhofer diffraction pattern.

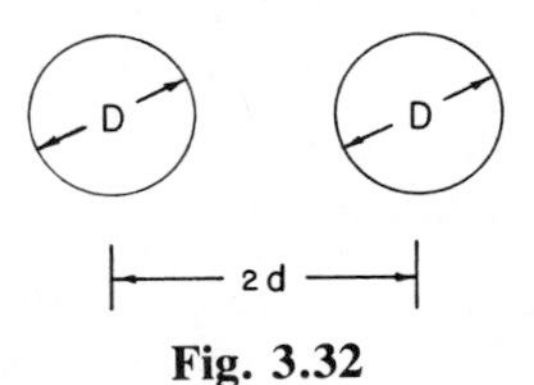

Fig. 3.32

3.13 It is assumed that two equally bright incoherent point objects are located at the minimum distance of 250 mm of distinct vision of an unaided human eye. The diameter of the pupil is assumed to be 2 mm.

(a) Determine the numerical aperture (N.A.) of the eye.

(b) Determine the Rayleigh's minimum resolution distance of the two object points, as a function of λ.

(c) If the refractive index of the liquid within the eyeball is $\eta' = 1.33$ and diameter of eyeball is 25 mm, then compute the minimum resolution distance of the two object points on the retina of the eye.

3.14 Given a monochromatic point source S of $\lambda = 6000$ Å that is located at the optical axis of an opaque circular disk, as shown in fig. 3.33. If it is assumed that the disk covers the first two Fresnel zones (i.e., half-wave zones), then compute the diameter of the disk.

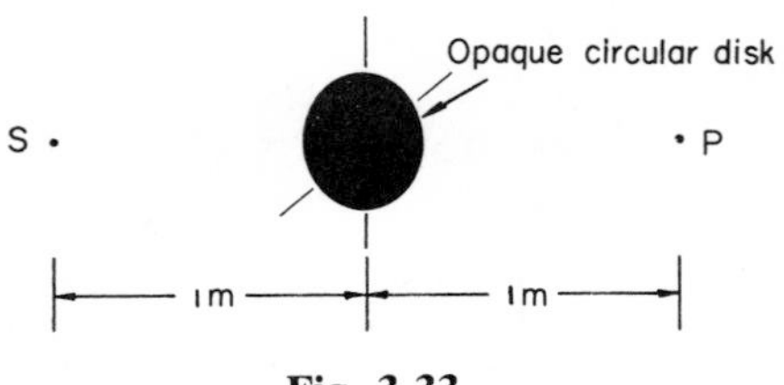

Fig. 3.33

3.15 A narrow slit of W-mm width is normally illuminated by a plane wave front, which contains two wavelengths λ_1 and λ_2. The corresponding diffraction pattern can be observed at the focal plane of a lens, where the lens is placed immediately behind the slit. It is found that the fourth minimum corresponding to λ_1 and fifth minimum corresponding to λ_2 occur at a location that is d mm away from the central maximum. Compute λ_1 and λ_2 as a function of d and the focal length f of the lens.

3.16 Given an opaque screen with a circular aperture 0.4 cm in diameter. Assume that the screen is illuminated by a plane monochromatic wave of $\lambda = 5000$ Å.

(a) Determine the location along the optical axis where the maximum and minimim irradiances occur.

(b) How far from the diffraction screen, along the optical axis, does the last minimum occur?

3.17 With reference to the geometry of fig. 3.18, we assume that $r_{1O} = 100$ cm, $r_{2O} = 200$ cm, and the wavelength of the light source is $\lambda = 6328$ Å. Sketch the Fresnel diffraction pattern for a single slit with the following slitwidths:

(a) $W = 0.69$ mm.

(b) $W = 1.15$ mm.

(c) $W = 5.52$ mm.

3.18 Given a 40× microscopic objective. Its numerical aperture at object space is equal to 0.45; then determine the resolution limits at the object and the image space, respectively.

REFERENCES

3.1 J. M. Stone, *Radiation and Optics,* McGraw-Hill, New York, 1963.

3.2 M. Born and E. Wolf, *Principles of Optics,* 2nd rev. ed., Pergamon Press, New York, 1964.

3.3 A. Sommerfeld, *Optics* (Lectures on Theoretical Physics, vol. IV), Academic Press, New York, 1954.

3.4 F. W. Sears, *Optics,* Addison-Wesley, Reading, Mass., 1949.

3.5 B. Rossi, *Optics,* Addison-Wesley, Reading, Mass., 1957.

3.6 E. Jahnke, F. Emde, and F. Lösch, *Tables of Higher Functions,* 6th ed. McGraw-Hill, New York, 1960.

4

Introduction to Partial Coherence Theory

4.1 GENERAL ASPECTS OF MUTUAL COHERENCE

The development and widespread application of the laser has made a discussion of the principles of coherence in radiation more important.

If the radiations from two point sources maintain a fixed phase relation between them, they are said to be mutually coherent. An extended source is coherent if all points of the source have fixed phase differences between them. In this chapter we discuss some elementary concepts of coherence theory, as it applies to optical information processing and wave front reconstruction. In the process, we find it necessary to modify and extend the foregoing definitions.

In a classical discussion of electromagnetic radiation, as in the development of Maxwell's equations, it is usually assumed that the electric and magnetic fields at any position are at all times measurable. In this case no account need be taken of coherence or incoherence. There are problems, however, in which this assumption of known fields cannot be made; in these cases it is often helpful to apply coherence theory. For example, if a diffraction pattern as a result of radiation from several sources is to be worked out, an exact result cannot be obtained unless the degree of coherence of the separate sources is taken into account. It may be desirable, in such a case, to obtain an average that would represent the statistically most likely result from any such combination of sources. It may be more useful to provide a statistical description than to follow the dynamical behavior of a system in detail.

Our treatment of coherence will be on such an averaging basis. In particular, following Born and Wolf (ref. 4.1, pp. 499–503) we choose the second-order moment as the quantity to be averaged. Thus what we call the *mutual coherence function* is set down as

$$\Gamma_{12}(\tau) = \langle u_1(t + \tau)\, u_2^*(t) \rangle, \tag{4.1}$$

where $u_1(t)$ and $u_2(t)$ are the respective complex fields at points P_1 and P_2, and $\Gamma_{12}(\tau)$ is the mutual coherence function between these points for a time delay τ; the symbols $\langle\ \rangle$ indicate a time average. From eq. (4.1) we can define a *normalized mutual coherence function*.

$$\gamma_{12}(\tau) = \frac{\Gamma_{12}(\tau)}{[\Gamma_{11}(0)\, \Gamma_{22}(0)]^{1/2}}; \tag{4.2}$$

$\gamma_{12}(\tau)$ may also be called the complex degree of coherence or the degree of correlation.

A clearer idea of $\Gamma_{12}(\tau)$ and a demonstration of how it can be measured may be obtained from a consideration of Young's experiment on inteference (see sec. 2.4). In fig. 4.1, Σ is an extended source of light, which is assumed to be incoherent, but nearly monochromatic; that is, its spectrum is of finite width, but narrow. The light from this source falls upon a screen at a distance r_{1O} from the source, and upon two small apertures (pinholes) in this screen, Q_1 and Q_2, separated by a distance d. On an observing screen at a distance r_{2O} from the diffracting screen, an interference pattern is formed by the light passing through Q_1 and Q_2. Now let us suppose that the changing characteristics of the interference fringes are observed as the parameters of fig. 4.1 are changed. As a measurable quantity, let us adopt Michelson's (ref. 4.1, p. 267) *visibility* $\mathcal{V}$ of the fringes, which he defines as

$$\mathcal{V} = \frac{I_{\max} - I_{\min}}{I_{\max} + I_{\min}}, \tag{4.3}$$

where $I_{\max}$ and $I_{\min}$ are the maximum and minimum irradiances of the fringes.

For the visibility to be measurable, the conditions of the experiment, such as narrowness of spectrum and closeness of optical path lengths, must be such as

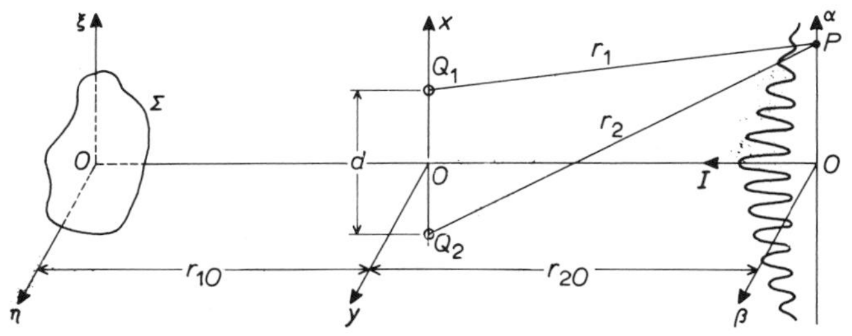

Fig. 4.1 Young's experiment. Here Σ is an extended but nearly monochromatic source.

to permit I_{max} and I_{min} to be clearly defined. Let us assume that these ideal conditions exist.

As we begin our parameter changes, we find, first, that the average visibility of the fringes increases as the size of the source Σ is made smaller. Next, as the distance apart of the pinholes Q_1 and Q_2 is changed with Σ constant (and circular in form), the visibility changes in the manner shown in fig. 4.2. When Q_1 and Q_2 are very close together the irradiance between the fringes falls to zero, and the visibility is unity. As d is increased, the visibility falls rapidly and reaches zero as I_{max} and I_{min} become equal. Further increase in d causes a reappearance of fringes, although they are shifted on the screen by half a fringe; that is, the previously light areas are now dark, and vice versa. Still further increase in d causes the repeated fluctuations in visibility shown in the figure. A curve similar to that of fig. 4.2 is obtained if the hole spacing is kept constant while the size of Σ is changed. These effects can be predicted from the theorem of Van Cittert–Zernike (ref. 4.1, p. 507). The visibility versus pinhole separation curve is sometimes used as a measure of spatial coherence, as discussed in the next few sections. The screen separations r_{1O} and r_{2O} are both assumed to be large compared with the aperture spacing d and with the dimensions of the source. Beyond this limitation, changes in r_{1O} or r_{2O} change the scale of effects such as are shown in fig. 4.2, without changing their general character.

As the point of observation P (see fig. 4.1) is moved away from the center of the observing screen, the visibility decreases as the path difference $\Delta r = r_2 - r_1$ increases, eventually becoming zero. The effect also depends upon how nearly monochromatic the source is. It is found that the visibility of the fringes is appreciable only for a path difference

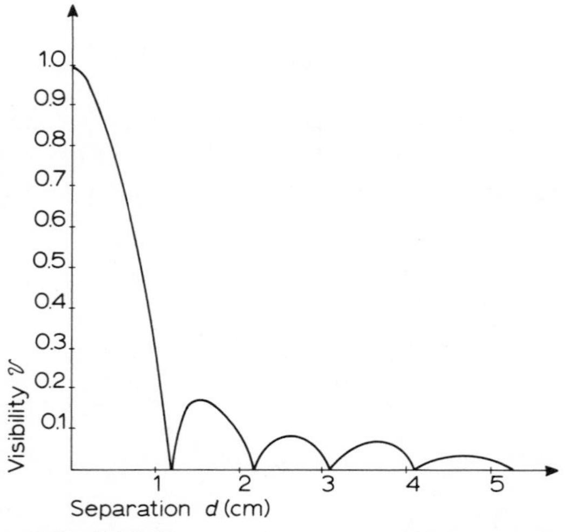

Fig. 4.2 Visibility as a function of pinhole separation.

$$\Delta r \ll \frac{2\pi c}{\Delta\omega}, \tag{4.4}$$

where c is the velocity of light and $\Delta\omega$ is the width of the spectrum of the source of light. The inequality (4.4) is often used to define the *coherence length* of the source.

The foregoing example shows that it is not necessary to have completely coherent light to produce an interference pattern, but that under the right conditions such a pattern may be obtained from an incoherent source. This effect may properly be called one of partial coherence, and we need a method for defining and measuring it.

A further development of the preceding equations will be helpful. Thus $u_1(t)$ and $u_2(t)$ of eq. (4.1), the complex fields at the points Q_1 and Q_2, are subject to the scalar wave equation in free space

$$\nabla^2 u = \frac{1}{c^2}\frac{\partial^2 u}{\partial t^2}. \tag{4.5}$$

This is a linear equation, and the field at the point P on the observing screen is a sum of those from Q_1 and Q_2:

$$u_p(t) = c_1 u_1\left(t - \frac{r_1}{c}\right) + c_2 u_2\left(t - \frac{r_2}{c}\right), \tag{4.6}$$

where c_1 and c_2 are appropriate complex constants. The corresponding irradiance at P may be written

$$I_p = \langle u_p(t)\, u_p^*(t)\rangle = I_1 + I_2 + 2\,\mathrm{Re}\left\langle c_1 u_1\left(t - \frac{r_1}{c}\right) c_2^*\, u_2^*\left(t - \frac{r_2}{c}\right)\right\rangle, \tag{4.7}$$

where I_1 and I_2 are proportional to the squares of the magnitudes of $u_1(t)$ and $u_2(t)$. In eq. (4.7) let us put

$$t_1 = \frac{r_1}{c} \quad \text{and} \quad t_2 = \frac{r_2}{c}, \tag{4.8}$$

and rewrite the irradiance at P,

$$I_p = I_1 + I_2 + 2c_1c_2^*\,\mathrm{Re}\langle u_1(t - t_1)\, u_2^*(t - t_2)\rangle, \tag{4.9}$$

since c_1 and c_2 are not time variable. The quantity averaged in eq. (4.9) is the crosscorrelation of the two complex fields.

If we put $t_2 - t_1 = \tau$, eq. (4.9) can be written

$$I_p = I_1 + I_2 + 2c_1c_2^* \operatorname{Re}\langle u_1(t + \tau)\, u_2^*(t)\rangle,$$

and combining this with eq. (4.1), we obtain

$$I_p = I_1 + I_2 + 2c_1c_2^* \operatorname{Re}[\Gamma_{12}(\tau)]. \tag{4.10}$$

The autocorrelations (i.e., the self-coherence functions) of the radiations from the two pinholes are

$$\Gamma_{11}(0) = \langle u_1(t)\, u_1^*(t)\rangle \qquad \text{and} \qquad \Gamma_{22}(0) = \langle u_2(t)\, u_2^*(t)\rangle. \tag{4.11}$$

We can also put

$$|c_1|^2\, \Gamma_{11}(0) = I_1 \qquad \text{and} \qquad |c_2|^2\, \Gamma_{22}(0) = I_2.$$

The irradiance at P, eq. (4.10), can then be put in terms of the degree of complex coherence, eq. (4.2),

$$I_p = I_1 + I_2 + 2(I_1I_2)^{1/2} \operatorname{Re}[\gamma_{12}(\tau)]. \tag{4.12}$$

Let us put $\gamma_{12}(\tau)$ in the form

$$\gamma_{12}(\tau) = |\gamma_{12}(\tau)| \exp[i\phi_{12}(\tau)], \tag{4.13}$$

and assume also that $I_1 = I_2 = I$, which may be called the best condition. Then eq. (4.12) becomes

$$I_p = 2I[1 + |\gamma_{12}(\tau)| \cos \phi_{12}(\tau)]. \tag{4.14}$$

The maximum of I_p is $2I[1 + |\gamma_{12}(\tau)|]$, while the minimum is $2I[1 - |\gamma_{12}(\tau)|]$. Putting these into the visibility equation, eq. (4.3), we find that

$$\mathcal{V} = |\gamma_{12}(\tau)|. \tag{4.15}$$

That is, under the best conditions the visibility of the fringes is a measure of the absolute value of degree of coherence.

4.2 MUTUAL COHERENCE FUNCTION

The complex crosscorrelation is usually defined as the second-moment time average. In these terms,

$$\Phi_{12}(\tau) \stackrel{\triangle}{=} \langle f_1^*(t)\, f_2(t + \tau)\rangle = \lim_{T\to\infty} \frac{1}{2T} \int_{-T}^{T} f_1^*(t)\, f_2(t + \tau)\, dt. \tag{4.16}$$

Accordingly, we define the mutual coherence function as

$$\Gamma_{12}(t) \stackrel{\Delta}{=} \langle u_1(t + \tau)\, u_2^*(t)\rangle = \lim_{T\to\infty} \frac{1}{2T} \int_{-T}^{T} u_1(T, t + \tau)\, u_2^*(T, t)\, dt. \tag{4.17}$$

It is now possible, however, to give a more general definition, fundamental to the theory of coherence. This can be done by working from two different aspects of the phenomenon: first, from an ensemble viewpoint; and later, from time-averaging.

To obtain an ensemble average, we define the mutual coherence function as

$$\Gamma_{12}(\boldsymbol{\rho}_1, t_1; \boldsymbol{\rho}_2, t_2) = \lim_{N\to\infty} \frac{1}{N} \sum_{n=1}^{N} u_{1n}(\boldsymbol{\rho}_1, t_1)\, u_{2n}^*(\boldsymbol{\rho}_2, t_2). \tag{4.18}$$

The summation is over an ensemble of N systems, and $\boldsymbol{\rho}_1$ and $\boldsymbol{\rho}_2$ are the respective position vectors.

If the statistics of the ensemble of systems are stationary, then

$$\Gamma_{12}(\boldsymbol{\rho}_1, t_1; \boldsymbol{\rho}_2, t_2) = \Gamma_{12}(\boldsymbol{\rho}_1, \boldsymbol{\rho}_2, \tau), \tag{4.19}$$

where $\tau = t_1 - t_2$. Equation (4.18) can then be rewritten,

$$\Gamma_{12}(\boldsymbol{\rho}_1, \boldsymbol{\rho}_2, \tau) = \lim_{N\to\infty} \frac{1}{N} \sum_{n=1}^{N} u_{1n}(\boldsymbol{\rho}_1, t_2 + \tau)\, u_{2n}^*(\boldsymbol{\rho}_2, t_2). \tag{4.20}$$

It is well known from the theory of random processes (ref. 4.6, p. 15) that when the statistics are stationary the time and ensemble averages may be equated. Thus we have

$$\Gamma_{12}(\boldsymbol{\rho}_1, \boldsymbol{\rho}_2, \tau) = \Gamma_{12}(\tau) = \langle u_1(\boldsymbol{\rho}_1, t + \tau)\, u_2^*(\boldsymbol{\rho}_2, t)\rangle. \tag{4.21}$$

This definition of the mutual coherence function as a time average has been made dependent on having stationary statistics. It therefore is valid where the source of radiation is periodic. In general, however, the time-average definition may be written as

$$\Gamma_{12}(\tau) = \langle u(\boldsymbol{\rho}_1, t + \tau)\, u^*(\boldsymbol{\rho}_2, t)\rangle. \tag{4.22}$$

The *complex degree of coherence* will be defined as in eq. (4.2),

$$\gamma_{12}(\tau) = \frac{\Gamma_{12}(\tau)}{[\Gamma_{11}(0)\, \Gamma_{22}(0)]^{1/2}}, \tag{4.23}$$

with the limits $0 \le |\gamma_{12}(\tau)| \le 1$. The lower limit represents complete incoherence, the upper limit complete coherence between the radiations at $\boldsymbol{\rho}_1$ and $\boldsymbol{\rho}_2$. Note that $|\gamma_{12}(\tau)|$ is a function of τ. It is therefore possible that the radiations at two points may be coherent at one value of τ, but incoherent at another value.

The *self-coherence,* or autocorrelation, function is given by

$$\Gamma_{11}(\tau) = \langle u_1(\boldsymbol{\rho}_1, t + \tau)\, u_1^*(\boldsymbol{\rho}_1, t)\rangle. \tag{4.24}$$

The self-coherence function is very important in the analysis of the operation of Michelson's interferometer. Its zero value, $\Gamma_{11}(0)$, is the highest irradiance at the point $\boldsymbol{\rho}_1$, i.e.,

$$\Gamma_{11}(0) \geq \Gamma_{11}(\tau). \tag{4.25}$$

The zero value, $\Gamma_{12}(0)$, of the mutual coherence function is called the *mutual irradiance function*. It is essential in an examination of the "stellar" form of the interferometer.

It will also be found useful to take the Fourier transforms of both the mutual and self-coherence functions. That of the mutual function is called the *mutual power spectrum,* and is given by

$$\hat{\Gamma}_{12}(\omega) = \begin{cases} \int_{-\infty}^{\infty} \Gamma_{12}(\tau)\, e^{-i\omega\tau}\, d\tau, & \omega > 0 \\ 0, & \omega < 0 \end{cases}. \tag{4.26}$$

That the second of these conditions holds can be seen from the fact that $\hat{\Gamma}_{12}(\omega)$ is analytic. The Fourier transform of a self-coherence function is the power spectrum of that particular radiation,

$$\hat{\Gamma}_{11}(\omega) = \begin{cases} \int_{-\infty}^{\infty} \Gamma_{11}(\tau)\, e^{-i\omega\tau}\, d\tau, & \omega > 0 \\ 0, & \omega < 0 \end{cases}, \tag{4.27}$$

and of course a similar equation for $\hat{\Gamma}_{22}(\omega)$ in terms of $\Gamma_{22}(\tau)$.

A further remark must be made concerning coherence. The phrase *spatial coherence* is applied to those effects that are due to the size, in space, of the source of radiation. If we consider a point source, and look at two points at equal light-path distances from the source, the radiations reaching these points will be exactly the same. The mutual coherence will be equal to the self-coherence at either point. That is, if the points are Q_1 and Q_2,

$$\Gamma_{12}(Q_1, Q_2, \tau) = \langle u(Q_1, t + \tau)\, u^*(Q_2, t)\rangle = \Gamma_{11}(\tau). \tag{4.28}$$

As the source is made larger we can no longer claim an equality of mutual coherence and self-coherence. The lack of complete coherence is a *spatial* effect. *Temporal coherence* is an effect due to the finite spectral width of the source. The coherence is complete for strictly monochromatic radiation, but becomes only partial as other wavelengths are added, giving a finite spectral width to the source. It is never possible to completely separate the two effects, but it is well to name them and point out their significance.

4.3 PROPAGATION OF THE MUTUAL COHERENCE FUNCTION

Equations that describe the manner in which the mutual coherence function is propagated can be found by first assuming that the field can be represented by a complex scalar function $u(t)$ that satisfies the scalar wave equation

$$\nabla^2 u(t) = \frac{1}{c^2}\frac{\partial^2 u(t)}{\partial t^2}. \tag{4.29}$$

The mutual coherence function, as we have previously defined it, is

$$\Gamma_{12}(\tau) = \langle u_1(t + \tau)\, u_2^*(t)\rangle. \tag{4.30}$$

The Laplacian, taken at the point Q_1 of eq. (4.30), may be written

$$\nabla_1^2\, \Gamma_{12}(\tau) = \langle \nabla_1^2\, u_1(t + \tau)\, u_2^*(t)\rangle. \tag{4.31}$$

From eq. (4.29) we can write

$$\nabla_1^2 u_1(t + \tau) = \frac{1}{c^2}\frac{\partial^2 u_1(t + \tau)}{\partial (t + \tau)^2}. \tag{4.32}$$

But

$$\frac{\partial^2 u_1(t + \tau)}{\partial (t + \tau)^2} = \frac{\partial^2 u_1(t + \tau)}{\partial \tau^2},$$

and therefore eq. (4.32) becomes

$$\nabla_1^2 \Gamma_{12}(\tau) = \left\langle \frac{\partial^2 u_1(t + \tau)}{c^2 \partial \tau^2}\, u_2^*(t) \right\rangle. \tag{4.33}$$

The field $u_2(t)$ is independent of τ, and we can therefore make the second partial derivative apply to the whole time-average function,

$$\nabla_1^2 \Gamma_{12}(\tau) = \frac{1}{c^2}\frac{\partial^2}{\partial \tau^2}\langle u_1(t + \tau)\, u_2^*(t)\rangle = \frac{1}{c^2}\frac{\partial^2 \Gamma_{12}(\tau)}{\partial \tau^2}. \tag{4.34}$$

This may be regarded as the fundamental equation describing the propagation of the mutual coherence function.

In the same way, we can take the Laplacian of eq. (4.30) with respect to the coordinates of the point Q_2, and it can be shown that it also reduces to

$$\nabla_2^2 \Gamma_{12}(\tau) = \frac{1}{c^2} \frac{\partial^2 \Gamma_{12}(\tau)}{\partial \tau^2}. \tag{4.35}$$

Each of the eqs. (4.34) and (4.35) contains only four independent variables, while $\Gamma_{12}(\tau)$ contains seven—the three spatial coordinates of each of the two points, plus the delay τ. A complete propagation equation may be obtained by combining (4.34) and (4.35):

$$\nabla_1^2 \nabla_2^2 \Gamma_{12}(\tau) = \frac{1}{c^4} \frac{\partial^4 \Gamma_{12}(\tau)}{\partial \tau^4}. \tag{4.36}$$

Suppose that the source of radiation is a surface of finite but bounded extent. For each pair of points on the surface a mutual coherence function, $\Gamma_{12}(\tau)$, can be specified. We then apply Sommerfeld's radiation condition at infinity, which says that $u_1(t)$ and $u_2^*(t)$ behave asymptotically as point radiators

$$\frac{A_1}{r_1} \exp(ikr_1) \qquad \text{and} \qquad \frac{A_2}{r_2} \exp(ikr_2) \tag{4.37}$$

as r_1 and r_2 (the distances from a pair of points on the source surface) approach infinity. These conditions may be applied to eqs. (4.34) and (4.35), and these solved as simultaneous wave equations. Since each equation contains four variables, their solutions can be written as four-dimensional Green's functions. The number of variables in each can, however, be reduced to three by first taking the Fourier transform of $\Gamma_{12}(\tau)$, which we designate by $\hat{\Gamma}_{12}(\omega)$. Now $\Gamma_{12}(\tau)$, as an analytic function, contains only positive frequencies, and we can write

$$\Gamma_{12}(\tau) = \int_0^\infty \hat{\Gamma}_{12}(\omega) \exp(i\omega\tau)\, d\omega \tag{4.38}$$

and

$$\hat{\Gamma}_{12}(\omega) = \begin{cases} \int_{-\infty}^{\infty} \Gamma_{12}(\tau) \exp(i\omega\tau)\, d\tau, & \omega > 0 \\ 0, \quad \omega < 0 \end{cases} \tag{4.39}$$

By substituting eq. (4.38) into eqs. (4.34) and (4.35), and reversing the order of integrations and differentiations, we have the equations

$$\int_0^\infty [\nabla_1^2 + k^2(\omega)]\, \hat{\Gamma}_{12}(\omega) \exp(i\omega\tau)\, d\omega = 0 \tag{4.40}$$

and

$$\int_0^\infty [\nabla_2^2 + k^2(\omega)]\, \hat{\Gamma}_{12}(\omega) \exp(i\omega\tau)\, d\omega = 0. \tag{4.41}$$

These two equations must hold for every value of τ, and therefore

$$[\nabla_1^2 + k^2(\omega)]\, \hat{\Gamma}_{12}(\omega) = 0 \tag{4.42}$$

and

$$[\nabla_2^2 + k^2(\omega)]\, \hat{\Gamma}_{12}(\omega) = 0. \tag{4.43}$$

From these equations it can be seen that the Fourier transform of the mutual coherence function satisfies the scalar Helmholtz equations.

Now we set up a Green's function $G_1(P_1, P_1', \omega)$ such that

$$[\nabla_1^2 + k^2(\omega)]\, G_1(P_1, P_1', \omega) = -\delta(P_1 - P_1'), \tag{4.44}$$

with the boundary

$$G_1(P_1, P_1', \omega)\big|_{P'_1 = S_1} = 0, \tag{4.45}$$

where P and S are the position coordinates, and δ is the Dirac delta function.

The solution of eq. (4.42) may be put in terms of this Green's function,

$$\hat{\Gamma}(P_1, S_2, \omega) = -\int_{S_1} \hat{\Gamma}(S_1, S_2, \omega)\, \frac{\partial G_1(P_1, P_1', \omega)}{\partial n_{S_1}}\bigg|_{P'_1 = S_1} dS_1. \tag{4.46}$$

With another Green's function, $G_2(P_2, P_2', \omega)$, defined as in eqs. (4.44) and (4.45), and noting the fact that eq. (4.46) provides a boundary condition, we see that the solution of (4.43) becomes

$$\hat{\Gamma}(P_1, P_2, \omega) = -\int_{S_2} \hat{\Gamma}(P_1, S_2, \omega)\, \frac{\partial G_2(P_2, P_2', \omega)}{\partial n_{S_2}}\bigg|_{P'_2 = S_2} dS_2. \tag{4.47}$$

By substituting eq. (4.46) into (4.47), we obtain the combined solution

$$\hat{\Gamma}(P_1, P_2, \omega) = \int_{S_2}\int_{S_1} \hat{\Gamma}(S_1, S_2, \omega)\, \frac{\partial G_1(P_1, P_1', \omega)}{\partial n_{S_1}}\bigg|_{P'_1 = S_1} \times \frac{\partial G_2(P_2, P_2', \omega)}{\partial n_{S_2}}\bigg|_{P'_2 = S_2} dS_1\, dS_2. \tag{4.48}$$

Let us note that $\hat{\Gamma}(P_1, S_2, \omega)$ in eq. (4.46) is the measure of coherence between a point on the source surface and an arbitrary point in space. The phrase *longitudinal direction* is used for space points that lie on perpendiculars to the surface points. The so-called longitudinal coherence is quite different from that between points on the surface. An elaboration of this is beyond the scope of this book, but an interested reader can refer to the excellent texts by Born and Wolf (ref. 4.1, pp. 491–555) and by Beran and Parrent (ref. 4.2, chap. 3).

The Fourier transform $\hat{\Gamma}(P_1, P_2, \omega)$ is a power spectrum, and if P_1 and P_2 are the same, it is the power spectrum of a single point in space. Equation (4.48) then shows that this power spectrum is not determined by the power spectrum of the whole source surface, but rather by $\hat{\Gamma}(S_1, S_2, \omega)$, which is the cross-correlation between two points of the surface.

According to the Sommerfeld radiation condition, eq. (4.37), the asymptotic behavior of $\hat{u}_1(\omega)$ and $\hat{u}_2^*(\omega)$ as r_1 and r_2 approach infinity should follow

$$\frac{a_1(\theta_1, \phi_1, \omega)}{r_1}\exp(ikr_1) \quad \text{and} \quad \frac{a_2(\theta_2, \phi_2, \omega)}{r_2}\exp(ikr_2). \tag{4.49}$$

If we assume that the statistics between $u_1(t)$ and $u_2(t)$ are those of a stationary random process, then $\hat{\Gamma}(P_1, P_2, \omega)$ may be expressed as a time average,

$$\hat{\Gamma}(P_1, P_2, \omega) = \lim_{T\to\infty} \frac{1}{2T}\, \hat{u}_1(T, \omega)\, \hat{u}_2^*(T, \omega), \tag{4.50}$$

and by (4.49) and (4.50),

$$\hat{\Gamma}(P_1, P_2, \omega) \to a_{12}(\theta_1, \theta_2, \phi_1, \phi_2, \omega)\, \frac{\exp[ik(r_1 - r_2)]}{r_1 r_2} \tag{4.51}$$

as $r_1, r_2 \to \infty$.

With $\hat{\Gamma}(P_1, P_2, \omega)$ given by eq. (4.48), the solution for $\Gamma(P_1, P_2, \tau)$ is

$$\Gamma(P_1, P_2, \tau) = \int_0^\infty \hat{\Gamma}(P_1, P_2, \omega)\, e^{i\omega\tau}\, d\omega. \tag{4.52}$$

4.4 SOME PHYSICAL CONSTRAINTS ON MUTUAL COHERENCE

The coherence functions have certain limits, which are discussed in this section. We have already, in discussing eq. (4.23), called attention to a set of limits [0, 1] for the function $|\gamma_{12}(\tau)|$, with 0 representing complete incoherence and 1 representing complete coherence. We have pointed out that the degree of coherence depends upon the value of τ; but it is also true that it depends upon the particular pair of points chosen for comparison. Thus we may expect $|\gamma_{12}(\tau)|$ to be zero for *some* points and *some* delays; but we would not expect it to vanish in general. The question remains, however, whether it is *possible* for an extended field (which might be the source itself) to have the property that $|\gamma_{12}(\tau)| = 0$ or $|\gamma_{12}(\tau)| = 1$ for every pair of points in the field and for any time delay τ. If so, it would seem to be proper to call the entire field incoherent or coherent, respectively.

In this connection we quote three well-known theorems, proofs of which the reader can find in the text by Beran and Parrent (ref. 4.2, pp. 47–52).

Theorem 1

An electromagnetic field has a unity degree of coherence (i.e., $|\gamma_{12}(\tau)| = 1$) for every pair of points in the field and every time delay τ, if and only if the field is monochromatic.

Theorem 2

A non-null electromagnetic field, for which $|\gamma_{12}(\tau)| = 0$ for every pair of points in the field and for every time delay τ, cannot exist in free space. Conversely, if $|\gamma_{12}(\tau)| = 0$ for every pair of points on a continuous closed surface, then the surface does not radiate.

Theorem 3

If spectral filtering is used on the radiation from two source points whose degree of coherence is zero, the degree of coherence will remain zero.

With regard to Theorem 1, it must be said that strictly monochromatic fields do not exist in practice, since all fields have some finite spectral bandwidth. However, it is possible for a field to have a spectral bandwidth $\Delta\omega$ that is small compared with the center frequency ω_0 of the radiation. Such a field is called *quasi-monochromatic*. If path differences in the radiation are small, a theory for quasi-monochromatic fields may be developed. Of course the concept of quasi-monochromatic fields is intended to give a practical approach to real monochromaticity, but there are some respects in which they differ widely. For example, although $|\gamma_{12}(\tau)| = 1$ is a requirement for all pairs of points in a strictly monochromatic field, this need not be the case for quasi-monochromaticity. In fact, there may be pairs of points for which $|\gamma_{12}(\tau)| = 0$. In order to help the reader obtain a more realistic feeling for the quasi-monochromatic field, we discuss it a little further.

We take as the condition for quasi-monochromatic radiation

$$\frac{\Delta\omega}{\omega_0} \ll 1. \tag{4.53}$$

It then follows that $\hat{\Gamma}_{12}(\omega)$ is essentially zero for all frequencies outside the band. That is, for an appreciable $\hat{\Gamma}_{12}(\omega)$ we must have

$$|\omega - \omega_0| < \Delta\omega. \tag{4.54}$$

From eq. (4.38), the mutual coherence function can then be written

$$\Gamma_{12}(\tau) = \exp(-i\omega_0\tau) \int_0^\infty \hat{\Gamma}_{12}(\omega) \exp[-i(\omega - \omega_0)\tau] \, d\omega. \tag{4.55}$$

By considering only small values of τ, such that $\Delta\omega|\tau| \ll 1$, eq. (4.55) reduces to

$$\Gamma_{12}(\tau) = \Gamma_{12}(0) \exp(-i\omega_0\tau). \tag{4.56}$$

Instead of zero for the value of τ [giving $\Gamma_{12}(0)$], the standard point can be taken as any other value, which we call τ_0, and make $\tau = \tau_0 + \tau'$, $\Delta\omega|\tau'| \ll 2\pi$. Then instead of eq. (4.56), we write

$$\Gamma_{12}(\tau_0 + \tau') = \Gamma_{12}(\tau_0) \exp(-i\omega_0\tau'). \tag{4.57}$$

The condition $\Delta\omega|\tau'| \ll 2\pi$ is essential in the theory of quasi-monochromatic fields. We can then define coherence in a limited way, by saying that a field is coherent if a τ_0 can be found for every pair of points, such that $|\gamma_{12}(\tau)| = 1$ when $\Delta\omega|\tau'| \ll 2\pi$. For very narrow bandwidths ($\Delta\omega/\omega_0$ very small), the field may be considered monochromatic for all values of τ that are practical in a given problem.

In a strictly monochromatic field, the mutual coherence function may be stated as

$$\Gamma_{12}(\tau) = [\Gamma_{11}(0) \, \Gamma_{22}(0)]^{1/2} \exp[i\phi_{12}(\tau)], \tag{4.58}$$

and for $\Delta\omega|\tau'| \ll 1$ this can be reduced to

$$\Gamma_{12}(\tau) = u(P_1) \, u^*(P_2) \exp(-i\omega_0\tau'), \tag{4.59}$$

where u is the field at a particular point (P_1 or P_2).

We do not want to give the impression that all fields with a narrow bandwidth are necessarily coherent, even when τ' is kept small. That is to say, that with $\Delta\omega/\omega_0 \ll 1$ and $\Delta\omega|\tau'| \ll 2\pi$, it still is not necessary that $|\gamma_{12}(\tau)| = 1$. In fact, this quantity can have any value between 0 and 1 under these conditions. However, if the quasi-monochromatic field at every point can be expressed as

$$u(t) = A(t) \{-i[\omega_0 t + \alpha(t)]\}, \tag{4.60}$$

where the functions $A(t)$ and $\alpha(t)$ are slow in their variation with time, compared with $2\pi/\omega_0$, then the field can be said to behave in a coherent manner.

Finally, we note that the treatment of coherent theory in this chapter is by no means complete. For a more intensive treatment of this topic, the reader is referred to the excellent book *Theory of Partial Coherence*, by Beran and Parrent (ref. 4.2).

PROBLEMS

4.1 With reference to eq. (4.12) of sec. 4.1, show that the condition of obtaining a maximum visibility of fig. 4.1 is $I_1 = I_2$.

4.2 Let us consider the Young's interference experiment as shown in fig. 4.3 where $d \ll l_1, l_2$. Show that the degree of coherence $|\gamma_{12}| = 1$ when the source converges to a point source, and $|\gamma_{12}| = 0$ as the monochromatic surface source Σ becomes infinitely large.

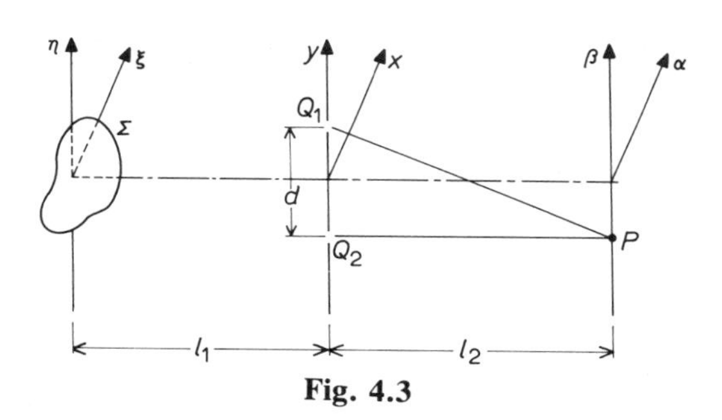

Fig. 4.3

4.3 Instead of using an extended monochromatic light source as in fig. 4.3, assume that the light source is a narrow slit monochromatic source perpendicular to the η axis. Assume also that the wavelength of the light source is 5000 Å, the separation between the source plane and the diffraction screen is $l_1 = 1$ meter, the separation between Q_1 and Q_2 is $d = 2.5$ mm, and the minimum-to-maximum irradiance ratio at the observation plane (α, β) is $I_{\min}/I_{\max} = 0.22$. Calculate the slit width of the light source.

4.4 With reference to Young's experiment of fig. 4.3, we assume that the light source is a circular source of radius ρ. Show that the visibility at the observation plane is dependent upon the product of radius ρ and the separation d between Q_1 and Q_2, for a fixed value of l_1.

4.5 We consider the illumination of two monochromatic plane waves on an observations plate. It produces an interference fringe pattern that may be described by the following equation:

$$I(x, y) = K_1 + K_2 \cos(p_0 x),$$

for all y, where K_1 and K_2 are arbitrary real positive constraints, p_0 is the spatial frequency in rad/mm, and (x, y) is the coordinate system of the observation plate. Calculate the degrees of coherence between these two light waves.

4.6 Consider two sources both emitting two discrete frequencies, ω_1 and ω_2. One source emits radiation of the form

$$u_1 = a_{11} \exp[-i(\omega_1 t + \phi_{11})] + a_{12} \exp[-i(\omega_2 t + \phi_{12})]$$

and the other emits radiation of the form

$$u_2 = a_{21} \exp[-i(\omega_1 t + \phi_{21})] + a_{22} \exp[-i(\omega_2 t + \phi_{22})].$$

(a) Determine the degree of coherence $|\gamma_{12}(\tau)|$.

(b) From the result obtained in (a), show that $|\gamma_{12}(\tau)| \leq 1$ for all τ. If $\omega_1 = \omega_2$, so that we may set a_{12} and a_{22} equal to zero, show that the degree of coherence is unity.

4.7 Let us consider an infinitely large plane source of fig. 4.3, which is superimposed with a sinusoidal grating of spatial frequency p_0 rad/mm, i.e., $T(x, y) = \frac{1}{2}[1 + \cos(p_0 \eta)]$. There is a condition between this spatial frequncy p_0 and the separation d between Q_1 and Q_2 such that high visibility fringe patterns can be observed. Determine this relationship.

4.8 Given an experiment setup as shown in fig. 4.4. The light source is assumed to be a uniform circular source of radius ρ and it is a quasi-monochromatic source. If the two pinholes Q_1 and Q_2 have the same radius r, where $r \ll d$, and $\rho < d$, then evaluate the corresponding irradiance at the observation plane P.

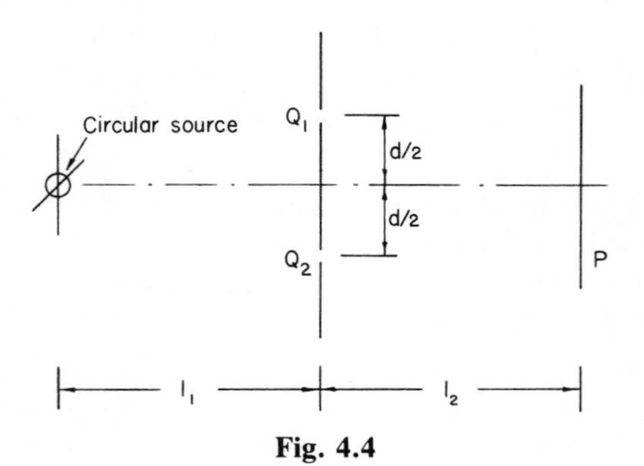

Fig. 4.4

4.9 In order to maintain a high spatial coherence between Q_1 and Q_2 for a fixed separation d, it is desirable to sample the extended monochromatic source with narrow slits of fig. 4.3.

(a) Calculate the locations of these narrow slits to fulfill this requirement.

(b) Evaluate the corresponding spatial coherence function.

(c) Show that the mutual coherence function is invariant for any two vertical points with separation d over the diffraction screen.

4.10 Given Young's experiment of fig. 4.3. We assume that the source is a monochromatic slit source 0.1 mm wide, the separation between this source and the two narrow slits is 1 meter, separation between the sets is 3 mm, and $\lambda = 6000$ Å.

(a) Determine the complex degree of coherence between the slits.

(b) Calculate the visibility of this fringe pattern.

4.11 Let us consider two monochromatic point radiators separated by distance $\Delta\eta$. With reference to Young's experiment of fig. 4.3:

(a) Calculate the mutual coherence Γ_{12} for the case where the radiators are derived from a common monochromatic source.

(b) Repeat part (a), if these point radiators are derived from two independent monochromatic sources.

REFERENCES

4.1 M. Born and E. Wolf, *Principles of Optics,* 2nd rev. ed., Pergamon Press, New York, 1964.

4.2 M. J. Beran and G. B. Parrent, Jr., *Theory of Partial Coherence,* Prentice-Hall, Englewood Cliffs, N. J., 1964.

4.3 M. Françon, *Diffraction Coherence in Optics,* Pergamon Press, New York, 1966.

4.4 E. L. O'Neill, *Introduction to Statistical Optics,* Addison-Wesley, Reading, Mass., 1963.

4.5 J. B. DeVelis, and G. O. Reynolds, *Theory and Applications of Holography,* Addison-Wesley, Reading, Mass., 1967.

4.6 A. M. Yaglom, *An Introduction to the Theory of Stationary Random Functions,* Prentice-Hall, Englewood Cliffs, N. J., 1962.

5

Basic Properties of Recording Materials

Photographic film (or plate) occupies an important place as an optical element in modern optical processing systems. Although it serves primarily as a recording medium, it can also be used in the synthesis of complex spatial filters and signal transparencies. There are several other optical elements (photochromic film, thermoplastic tape, etc.), some of whose optical properties are similar to those of photographic film. However, these materials at the present time are still in the research and development stage. It is doubtful whether these new materials can ever totally replace photographic film and plate. For this reason we devote this chapter to a discussion of the basic properties of photographic film. However, the presentation is by no means complete. More thorough treatments of the photographic process will be found in the books by Mees and James (refs. 5.1, 5.2).

5.1 PHOTOGRAPHIC FILM AS A RECORDING MEDIUM

We first discuss photographic film considered as a recording medium. A photographic detector is generally composed of a base, which may be made of a transparent glass plate or acetate film, and a layer of photographic emulsion (fig. 5.1). The photographic emulsion consists of a large number of tiny photosensitive silver halide particles, which are suspended more or less uniformly in a gelatin for support. When the photographic emulsion is exposed to light, some of the silver halide grains absorb optical energy and undergo a complex physical change. Some of the grains that absorb sufficient light energy are immediately reduced, forming tiny metallic silver particles. (These are the so-called development centers.) The reduction to silver is completed by the chemical process of *development*. Those grains that were not exposed or that have not absorbed sufficient optical energy will remain unchanged. If this developed film is then

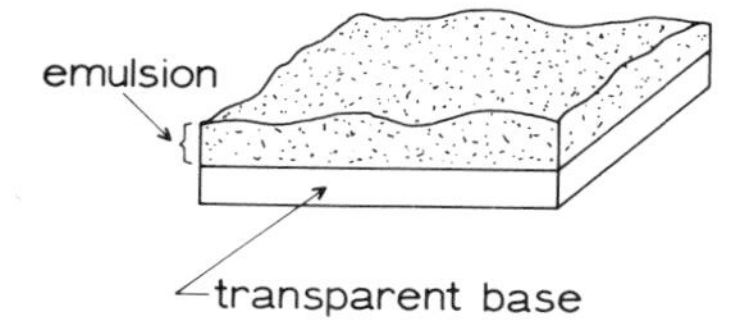

Fig. 5.1 A section of photographic film. The emulsion is composed of silver halide particles embedded in gelatin.

subjected to some chemical *fixing* process, then the unexposed silver halide grains will be removed, leaving only the metallic silver particles in the gelatin. These metallic silver grains remaining in the emulsion are largely opaque at optical frequencies. The transmittance of developed film therefore depends on the density of the metallic silver grains (i.e., the exposed and developed grains) in the gelatin. The relation of the intensity transmittance and the density of the developed grains was first demonstrated in a classic article written in 1890 by F. Hurter and V. C. Driffield (ref. 5.3). They showed that the area density of the metallic silver particles of a developed film should be proportional to $-\log T_i$; thus they defined the *photographic density* by

$$D = -\log T_i, \tag{5.1}$$

where T_i is the *intensity transmittance*. The intensity transmittance is in turn defined as

$$T_i(x, y) = \left\langle \frac{I_o(x, y)}{I_i(x, y)} \right\rangle, \tag{5.2}$$

where $\langle\ \rangle$ denotes the localized ensemble average and $I_i(x, y)$ and $I_o(x, y)$ denote the input and output irradiances, respectively, at the point (x, y).

One of the most commonly used descriptions of the photosensitivity of a given photographic film is that given by the Hurter–Driffield curve (or, for short, "H and D" curve), as shown in fig. 5.2. It is the plot of the density of the developed grains versus the logarithm of the exposure E. The plot shows that if the exposure is below a certain level, then the photographic density is quite independent of the exposure; this minimum density is usually referred as *gross fog*. As the exposure increases beyond the *toe* of the curve, the density begins to increase in direct proportion to $\log E$. The slope of the straight-line portion of the H and D curve is usually referred to as the *film gamma*, γ. If the exposure is further increased beyond the straight-line portion of the H and D curve, then the density saturates, after an intermediate region called the *shoulder*. In the saturated region, there is no further increase of developed-grain density with an increase of exposure.

It is the straight-line portion of the H and D curve that is used in conventional photography. However, we show in chap. 12 that, for the linear optimization in

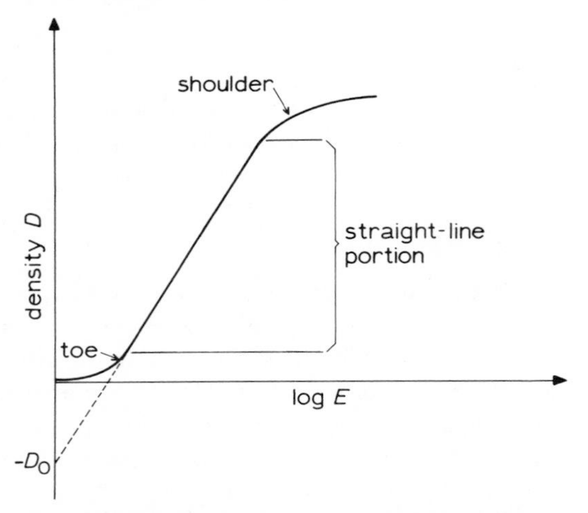

Fig. 5.2 The Hurter–Driffield ("H and D") curve.

wave front recording and in complex spatial filter synthesis, the optimum recording is not restricted to the linear region of the H and D curve.

A photographic film with a large value of γ is generally referred to as high-contrast film. Conversely, a film that has a low value of γ is referred to as low-contrast film. The value of γ is generally affected not only by the type of photographic emulsion used, but also by the chemical of the developer and the time of the developing process. Therefore, in practice, it is possible to achieve a prescribed value of γ, with a fair degree of accuracy, if suitable types of film, developer, and developing time are used.

In some cases, particularly in incoherent processing, the intensity transmittance of a developed film is a more appropriate parameter than the film density. We first consider the properties of the photographic film under incoherent illumination, and then move toward the coherent case.

If a given film is recorded in the straight-line region of the H and D curve, then the photographic density may be written as

$$D = \gamma_n \log E - D_0, \tag{5.3}$$

where the subscript n means that a negative film is being used, and $-D_0$ is the intercept of the projection of the straight-line portion of the H and D curve with the density ordinate. By substitution of eq. (5.1) in eq. (5.3) we have

$$\log T_{in} = -\gamma_n \log(It) + D_0, \tag{5.4}$$

where I is the incident irradiance, and t is the exposure time. (Note that the energy delivered to the film is given by $E = It$.) Equivalently, eq. (5.4) can be written,

$$T_{\text{in}} = K_{\text{n}} I^{-\gamma_{\text{n}}}, \tag{5.5}$$

where $K_{\text{n}} = 10^{D_0} t^{-\gamma_{\text{n}}}$, a positive constant.

From eq. (5.5) it is clear that the intensity transmittance is nonlinear with respect to the incident irradiance. However, it may be possible to achieve a positive power-law relation between the intensity transmittance and the incident irradiance. To do so may require the two-step process called *contact printing*.

In the first step, a negative film is recorded. Then in the second step another negative film is laid under the developed film and an incoherent light is transmitted through the first film to expose the second. By developing the second film so as to obtain the prescribed value of γ, a positive transparency with a linear relationship between the intensity transmittance and the incident irradiance may be obtained. In order to illustrate this two-step process, let the resultant intensity transmittance of the second developed film (i.e., the positive transparency) be

$$T_{\text{ip}} = K_{\text{n2}} I_2^{-\gamma_{\text{n2}}}, \tag{5.6}$$

where K_{n2} is a positive constant. The subscript n2 denotes the second-step negative. The irradiance I_2 is that incident on the second film; this may be written as

$$I_2 = I_1 T_{\text{in}}, \tag{5.7}$$

where

$$T_{\text{in}} = K_{\text{n1}} I^{-\gamma_{\text{n1}}},$$

with the subscript n1 meaning the first-step negative. The illuminating irradiance of the first film during contact printing is I_1, and I is the irradiance originally incident on the first film. By substitution of eq. (5.7) into eq. (5.6), we have

$$T_{\text{ip}} = K I^{\gamma_{\text{n1}} \gamma_{\text{n2}}}, \tag{5.8}$$

where

$$K = K_{n2} K_{\text{n1}}^{-\gamma_{\text{n2}}} I_1^{-\gamma_{\text{n2}}}$$

is a positive constant. A linear relationship between the intensity transmittance of the positive transparency and the incident recording irradiance may be achieved if the overall gamma is made to be unity, i.e., if we make $\gamma_{\text{n1}} \gamma_{\text{n2}} = 1$.

If the film is used as an optical element in a coherent system, then it is more appropriate to use the complex amplitude transmittance rather than the intensity transmittance. The complex amplitude transmittance is defined by

$$T(x, y) = [T_{\text{i}}(x, y)]^{1/2} \exp[i\phi(x, y)], \tag{5.9}$$

where T_i is the intensity transmittance, and $\phi(x, y)$ represents random phase retardations. Such phase retardations are primarily due to emulsion thickness variations. These thickness variations are of two sorts. One is the coarse "outer scale" variation, which is a departure from optical flatness of the emlusion and base. The fine "inner scale" variation is a result of random fluctuations in the density of developed silver grains, which has been found to produce nonuniform swelling of the surrounding gelatin. This fine scale variation of emulsion thickness obviously depends upon the exposure of the film.

In most practical applications, phase retardations due to emulsion thickness variations can be removed by means of an *index-matching liquid gate* (fig. 5.3). Such a gate consists of two parallel optically flat glass plates with the space between them filled with refractive index-matching liquid, i.e., a liquid whose refractive index is very close to that of the film emulsion. If a developed film is submerged in the liquid gate, then the overall complex amplitude transmittance may be written as

$$T(x, y) = [T_i(x, y)]^{1/2}, \tag{5.10}$$

which is a real function. The random phase retardation is therefore removed.

Recall the negative and positive transparencies of the two-step process, i.e., eqs. (5.5) and (5.8). The amplitude transmittances of these two transparencies can be written as

$$T_n = [T_{in}]^{1/2} = K_n^{1/2} I^{-\gamma_n/2} = K_n^{1/2}(uu^*)^{-\gamma_n/2}, \tag{5.11}$$

and

$$T_p = [T_{ip}]^{1/2} = K^{1/2} I^{(\gamma_{n1}\gamma_{n2})/2} = K^{1/2}(uu^*)^{(\gamma_{n1}\gamma_{n2})/2}, \tag{5.12}$$

where u is the complex amplitude of the incident recording field. Thus the amplitude transmittance of the two-step contact process can be written as

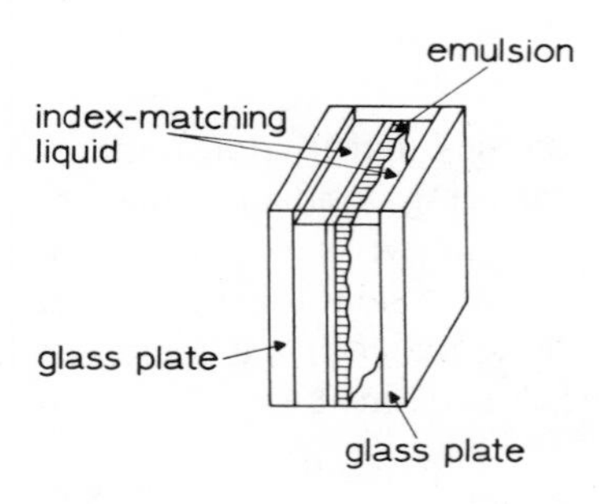

Fig. 5.3 Refractive index-matching liquid gate. The outward-facing surfaces of the glass are optically flat.

$$T = K_1 \, |u|^{\gamma}, \tag{5.13}$$

where K_1 is a positive constant and $\gamma = \gamma_{n1}\gamma_{n2}$. Therefore, a linear relationship between the amplitude transmittance and the amplitude of the recording light field may be achieved by making the overall gamma equal to unity. This is identical to the case for incoherent illumination. It may be pointed out that, for convenience, the first gamma is frequently chosen to be less than unity (e.g., $\gamma_{n1} = \frac{1}{2}$), and the second gamma is then chosen to be larger than *two* (e.g., $\gamma_{n2} = 4$). The overall gamma is therefore equal to two. This provides a square-law relation rather than linearity for the intensity transmittance versus incident irradiance. By eq. (5.10), however, it becomes a linear relation for amplitude transmittance.

In most coherent information processing it is more convenient to use the transfer characteristic directly, rather than the H and D curve. This direct transfer characteristic is frequently referred to as the *T-E* curve (fig. 5.4). It can be seen from fig. 5.4 that if the film is properly biased at an operating point that lies well within the linear region of the transfer characteristic of the *T-E* curve, then within a certain dynamic range the film will offer the best linear transfer amplitude transmittance. If E_Q and T_Q ("quiescent") denote the corresponding bias exposure and amplitude transmittance, then within the linear region of the transfer characteristic the amplitude transmittance may be written as

$$T \simeq T_Q + \alpha(E - E_Q) = T_Q + \alpha' |\Delta u|^2, \tag{5.14}$$

where α is the slope measured at the quiescent point of the *T-E* curve, $|\Delta u|$ is the incremental amplitude variation, and $\alpha' = \alpha t$, where t is the exposure time. For further details of the use of the *T-E* curve and the direct transfer description, see ref. 5.4, p. 48, and ref. 5.5.

To conclude this section, it may be emphasized that the constraint of the signal recorded within the linear region of the *T-E* curve may not be necessary

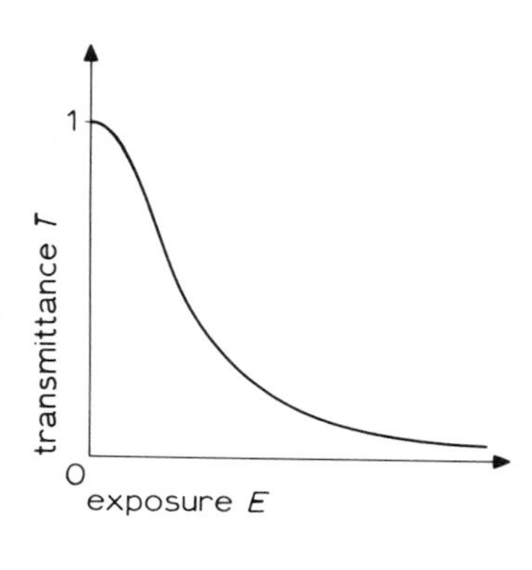

Fig. 5.4 Amplitude transmittance versus exposure of a photographic negative (*T-E* curve).

in some coherent optical systems, particularly in wave front reconstructions. The removal of this linear constraint in wave front reconstruction is treated in chap. 10. It is shown that for optimal linear image reconstruction, the wave front recording is actually required to exceed the linear region of the *T-E* curve!

5.2 MARKOV PHOTOGRAPHIC NOISE

Studies of the granularity of photographic emulsions have appeared in a number of papers. In particular, we refer to those by Jones (ref. 5.6), Zweig (refs. 5.7–5.9), Picinbono (ref. 5.10), and Savelli (ref. 5.11). The work of Jones and Zweig was based on empirical data relating rms density to average density of a photographic film that had been uniformly exposed and developed. Picinbono and Savelli assumed that the probability of reduction of photographic grains follows a stationary Poisson distribution. In this section we study photographic noise as a very special type of stochastic process. We show that the noise behavior of photographic emulsions can be interpreted as a continuous-parameter Markov chain (ref. 5.12; ref. 5.13, p. 288). This model predicts a new relationship between rms density and average density. Furthermore, the distribution of the number of developed grains is shown to be not a Poisson distribution, but one that corresponds to a spatial nonstationary probability.

In order to determine the probability distribution of the undeveloped grains (as well as developed grains), the following assumptions are made:

1. Grain size is uniform and the grains are uniformly distributed throughout the film.
2. A given particle is either perfectly opaque, if the grain has been developed, or perfectly transparent, if the grain has remained undeveloped.
3. The minimum resolvable image [Altman and Zweig (ref. 5.14) call this a cell] contains a large number of grains.

In the following, a conditional-probability distribution of the undeveloped grains is calculated. The corresponding mean and variance are determined, and the application to a realistic photographic film is illustrated.

From the assumptions stated, it is possible to derive a conditional-probability distribution of the undeveloped grains defined in the volume covered by the minimum resolvable point image. Let us define the conditional probability as

$$P_m(t) = P[x(t) = m \mid x(0) = M], \qquad x(t) = 0, 1, 2, \ldots, M, \tag{5.15}$$

where t is the exposure time, $x(t)$ is the number of undeveloped grains, and $P_m(t)$ is the conditional probability that m undeveloped grains exist at time t. The quantity M is the total number of grains enclosed in the minimum resolvable point image. If the probability of development of more than one grain is an interval Δt is assumed to be the order of $(\Delta t)^2$, then a set of nonhomogeneous Kolmogorov differential equations can be obtained (ref. 5.13):

$$\frac{dP_M(t)}{dt} = -Mq(t)\, P_M(t) \qquad \text{for} \qquad m = M, \tag{5.16}$$

$$\frac{dP_m(t)}{dt} = -mq(t)\, P_m(t) + (m+1)\, q(t)\, P_{m+1}(t) \qquad \text{for} \qquad m < M, \tag{5.17}$$

with the initial conditions

$$P_M(0) = 1, \tag{5.18}$$

$$P_m(0) = 0 \qquad \text{for} \qquad m < M, \tag{5.19}$$

where $mq(t)\, dt$ is the probability that one of the m grains will be developed at time t, and $mq(t)$ is the rate of the development, which is also called the transition intensity. The solution of eqs. (5.16) and (5.17) can be shown to be

$$P_m(t) = \frac{M!}{(M-m)!m!}\,[1 - e^{-\rho(t)}]^{M-m}\, e^{-m\rho(t)} \qquad \text{for} \qquad m \le M, \tag{5.20}$$

where

$$\rho(t) = \int_0^t q(t')\, dt'. \tag{5.21}$$

The corresponding mean and variance of $P_m(t)$ are

$$\bar{m} = Me^{-\rho(t)}, \tag{5.22}$$

$$\sigma^2 = Me^{-\rho(t)}\,[1 - e^{-\rho(t)}]. \tag{5.23}$$

The expression relating rms density fluctuation to the average density is therefore

$$\sigma = \left[\bar{m}\left(1 - \frac{\bar{m}}{M}\right)\right]^{1/2} = \left[\bar{n}\left(1 - \frac{\bar{n}}{M}\right)\right]^{1/2}, \tag{5.24}$$

where $\bar{n} = M - \bar{m}$ is the average density of the developed grains.

The conditional-probability distribution of the developed grains can be obtained by simply substituting the random variable $m = M - n$ in eq. (5.20), where n is the number of developed grains included in the minimum resolvable point image.

In order to determine the conditional probability $P_m(t)$ of a typical photographic film, we must know $q(t)$. To determine $q(t)$, we can start from the H and

D curve of a given photographic film. The density of developed grains increases with exposure (that is, the irradiance times exposure time). For a given irradiance, we can plot the density of the developed grains against the exposure time t. In fig. 5.5, a set of D versus t curves for different values of irradiance I are plotted. The corresponding slopes of the D versus t curves are also plotted in dashed lines in the same figure. Assuming that the dashed lines in this figure follow the Rayleigh probability distribution law, then the corresponding slopes can be written as

$$\frac{dD}{dt} = \begin{cases} M \dfrac{t}{\alpha^2} \exp\left[-\dfrac{1}{2}\left(\dfrac{t}{\alpha}\right)^2\right], & t \geq 0 \\ 0, & \text{otherwise} \end{cases}, \tag{5.25}$$

where the parameter α corresponds to the exposure time when dD/dt is maximum.

We may consider that the photographic density is equal to the number of developed grains,

$$D = M - \bar{m}. \tag{5.26}$$

By substituting eq. (5.22) in the above equation, and then differentiating with respect to t, we have

$$\frac{dD}{dt} = Mq(t)\, e^{-\rho(t)}. \tag{5.27}$$

By comparing eqs. (5.27) and (5.25), we conclude that

$$q(t) = \frac{t}{\alpha^2} \tag{5.28}$$

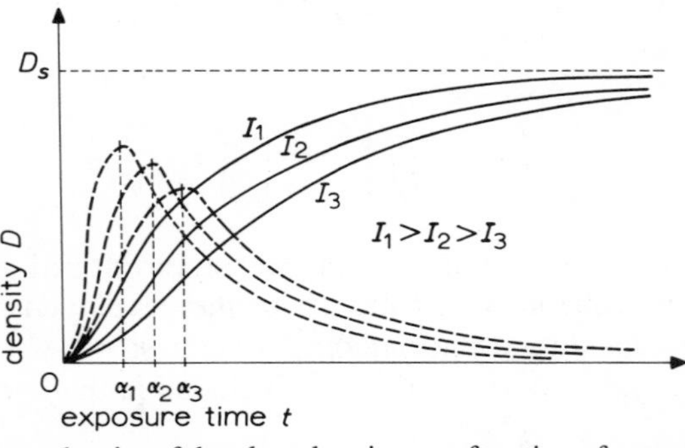

Fig. 5.5 Solid curves: density of developed grains as a function of exposure time for various values of irradiance I. Dashed curves: slope of the solid curves.

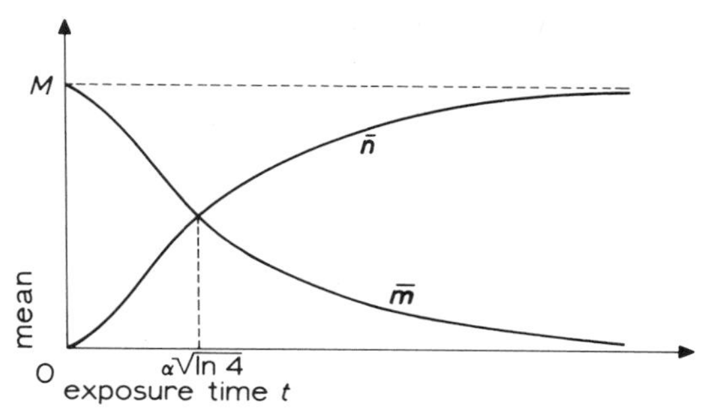

Fig. 5.6 Mean of the developed and undeveloped grains as a function of exposure time.

and

$$\rho(t) = \frac{1}{2}\left(\frac{t}{\alpha}\right)^2. \tag{5.29}$$

From eq. (5.24) we can see that the random behavior of the photographic noise is dependent on the average density while the probability of grain development is nonstationary (see ref. 5.15, p. 300).

Furthermore, from eq. (5.20) the conditional-probability distribution of the undeveloped (as well as the developed) grains is more similar to a continuous-parameter binomial distribution than to a Poisson distribution.

Accordingly, from eqs. (5.22), (5.23), and (5.29), we can conclude that the mean of undeveloped (developed) grains is a monotonically decreasing (increasing) function of t, as shown in fig. 5.6. However, the variance is not a monotonic function of t (fig. 5.7). The maximum value of noise, $\sigma^2 = M/4$, occurs at $t = \sqrt{\alpha(\ln 4)}$ and the minimum values, $\sigma^2 = 0$, occur at $t = 0$ and $t = \infty$. This figure shows that a variation of the standard deviation is quite compatible with the experimental results in the range obtained in fig. 6 of ref. 5.16. Equation (5.24) also shows that the standard deviation of density increases with respect to the value of M. That is, for an emulsion of smaller grain size, the rms fluctuation in density is higher. Moreover, it will be shown in the following that the informational sensitivity is also an increasing function of M.

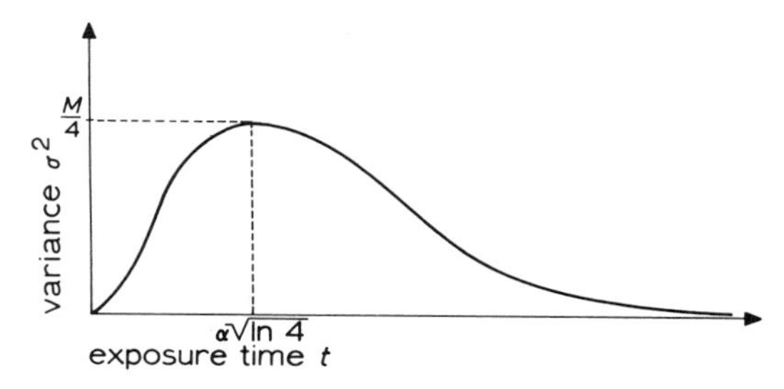

Fig. 5.7 Mean-square fluctuation of developed grains as a function of exposure time.

Informational Sensitivity

The signal of interest in conventional photography may be specified by the linear gradient (ref. 5.2, p. 540)

$$g = \frac{dD}{dE}, \tag{5.30}$$

where E is the exposure.

Then from eqs. (5.22) and (5.26), the signal is

$$g = \frac{M}{I} \frac{t}{\alpha^2} \exp\left[-\frac{1}{2}\left(\frac{t}{\alpha}\right)^2 \right], \tag{5.31}$$

where I is the irradiance of the minimum resolvable image. Therefore, the informational sensitivity for the effectiveness of information transmission is defined (see ref. 5.2, p. 540) by

$$\text{informational sensitivity} \triangleq \frac{g}{\sigma} = \frac{\sqrt{M}}{I} \frac{t}{\alpha^2} \left[\frac{1}{e^{\rho(t)} - 1} \right]^{1/2}. \tag{5.32}$$

The informational sensitivity as a function of exposure time given by eq. (5.32) is sketched in fig. 5.8. The informational sensitivity is not a monotonic function of t, but has a maximum at some finite exposure time. The variation of the informational sensitivity is compatible with the shape of the graphs in fig. 8 of ref. 5.16. Equation (5.32) also shows that the informational sensitivity increases as the square root of M. That is, for any particular resolvable cell, a smaller grain emulsion would be expected to have a higher informational sensitivity than one of larger grain size.

In addition, from eq. (5.31) and the corresponding dashed curves of fig. 5.5,

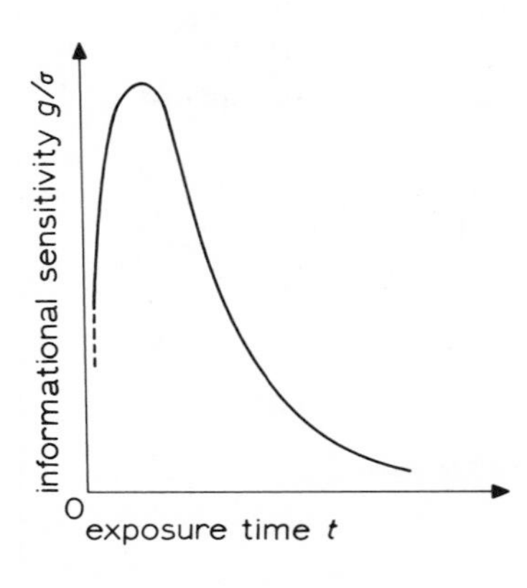

Fig. 5.8 Information sensitivity as a function of exposure time.

we can readily see that the variation of signal as a function of average density is quite similar to the experimental results obtained in fig. 7 of ref. 5.16.

This stochastic model predicts a new relation between rms noise and average developed-grain density. It shows the probability of grain development to be nonstationary and the distribution of developed grains to be non-Poisson. If the effects due to random variations in grain size and the stochastic behavior of its sensitivity are considered, then the rms noise will be higher than we have obtained.

5.3 CLASSIFICATIONS OF THE *T-E* CURVES

The Hurter–Driffield curve, or H and D curve, is the most commonly used method of describing the characteristics of photographic films, and the slope γ is the most widely used index to describe the relative contrast of the emulsions. Other indexes, such as gradient and contrast index, are also sometimes used, but they too are based on the H and D curve. It has been pointed out by quite a few authors that in many applications of coherent optics such as holography (chap. 10), the use of the H and D curve is not appropriate since the linear region of the *T-E* curve (amplitude transmittance versus linear exposure) is used instead (refs. 5.4, 5.5). Nevertheless, the index γ is still being used in such applications simply because it is the most available parameter. The usefulness of the γ value is primarily in the field of conventional photography; it provides very little insight or meaning to the *T-E* characteristic of the film. There is a definite need to explore better ways of classifying film characteristics for applications where the linear region of the *T-E* curve is utilized. In this section we discuss some possible classifying methods and their applications in spectrum analysis (sec. 6.5) and holography (chap. 10). We look at two separate cases. One is for applications such as spectrum analysis where it is imperative that the output transmittance be linearly proportional to the input exposure (refs. 5.17, 5.18). That is, the exposure range is entirely within the linear region of the *T-E* curve. The other is for holographic applications where an exposure range beyond that of the linear region is utilized to achieve maximum diffraction efficiency (refs. 5.19–5.21). Similar to the γ value, the classifying indexes that we discuss are independent of the absolute value of exposures and are obtainable by simple graphical procedures (refs. 5.22, 5.23). We start with the simplest case, where a linear recording is imperative.

In fig. 5.9, we have a typical *T-E* curve, characteristic of silver halide emulsions with a linear region between E_1 and E_2. In the middle is the bias point $E_b = (E_1 + E_2)/2$. In order to produce the best linear recording with this film characteristic, the exposure time should be adjusted such that the bias exposure is equal to E_b. We can then define an index to describe the contrast of this *T-E* characteristic. We first note that the slope of the *T-E* curve, measured in transmittance per absolute unit of energy, cannot be used directly since it is more of a measure of film speed than of contrast. Contrast, as applied to applications like

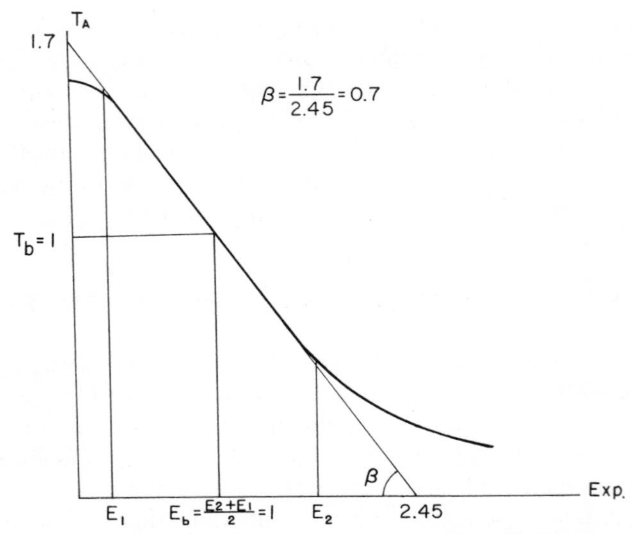

Fig. 5.9 Classification of the linear region of the T-E curve.

holography and spectrum analysis, should be a measure of the amount of relative transmittance variation obtained for a given variation of exposure. In other words, a film with a higher contrast should produce a transparency with a larger percentage of modulation when used to record a given illumination pattern. We thus propose an index β, defined as the slope of the T-E curve measured with the bias levels of transmittance T_b and exposure E_b both normalized to the value of 1. For example, the T-E curve in fig. 5.9 would have a β value of 1.7/2.44 or 0.7.

To demonstrate the meaning of such an index, we can look at a simple example. In fig. 5.10, we have characteristics of two films, one of which has a β value twice as large as the other. They represent two films having a different relative contrast but the same film speed. If these two films are used to record a given input illumination pattern, we shall find that the percentage of modulation of the output transmittance is twice as high for the film with the larger value of β. Thus we see that the index β, as defined, is equal to the ratio of the percentages of modulation between the output transmittance and input exposure. That is,

$$\beta \triangleq \frac{\%\text{ modulation of output amplitude transmittance}}{\%\text{ modulation of input exposure}}.$$

It should be emphasized that this relationship holds true only under the condition of best linear recording. That is, the bias exposure is adjusted to the middle of the linear region and the exposure range confined to within this region.

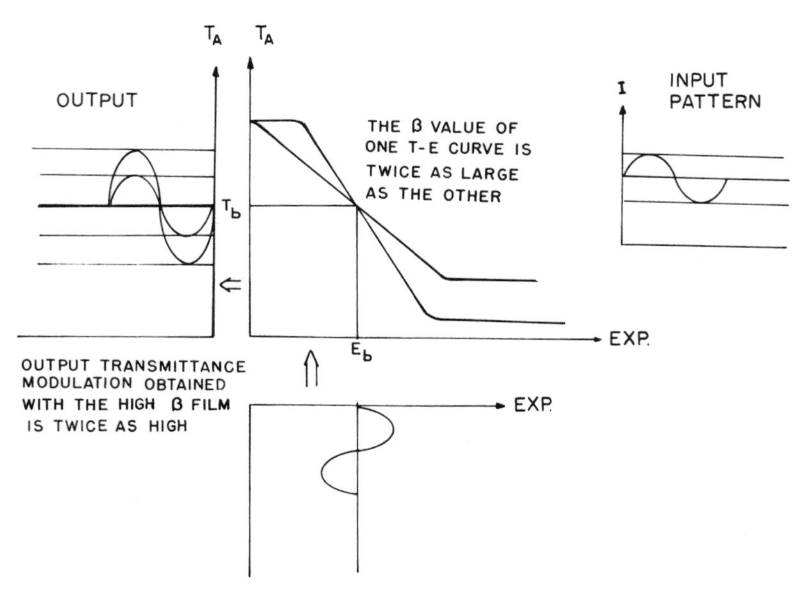

Fig. 5.10 Linear recording with two films having different β values.

In the previous example, the two films whose characteristics are compared have the same bias transmittance. In fig. 5.11 we show the charcteristics of two films that have the same β value but whose bias transmittances are different. If these two films are used to record a given illumination pattern, the resulting transparencies would have the same percentage of modulation, but the average transmittance of one would be lower. In spectrum analysis applications, the side lobes of the dc-bias term are the major source of noise. The main concern in such an application would be the percentage of modulation of the output transmittance, since a higher percentage of modulation would result in a higher SNR. The level of bias may not be as important a consideration. In other applications such as holography, however, with the presence of the carrier, the first- and zero-order diffractions are well separated. The main concern in this case would be the diffraction efficiency* and average transmittance of the output transparency must also be taken into account. For simplicity, let us consider a hologram made with two plane waves. The illumination pattern in one-dimension can be written as

$$I(x) = C(1 + M \cos px), \tag{5.33}$$

where M is the modulation index of the illumination pattern, $p = \sin \theta/\lambda$ is the

*The diffraction efficiency is defined as

$$\text{D.E.} \triangleq \frac{\text{output (hologram image irradiance)}}{\text{input (incident irradiance)}}.$$

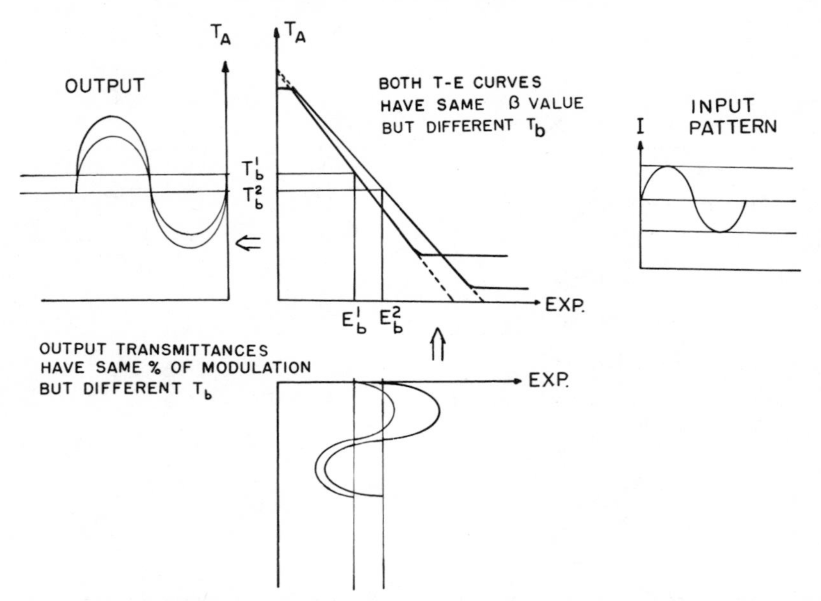

Fig. 5.11 Linear recording with two films having the same β value but different bias transmittances.

angular spatial frequency of the pattern, and C is the appropriate constant. The value M can be calculated from the reference-to-object beam intensity ratio K by the equation

$$M = \frac{2\sqrt{K}}{1 + K}. \tag{5.34}$$

If this pattern is recorded linearly with contrast index β, and bias transmittance T_b (without normalization), the resulting amplitude transmittance can be described as

$$T_A = T_b\,(1 + \beta M \cos px). \tag{5.35}$$

The diffraction efficiency of the hologram would be

$$\text{D.E.} = \left(\frac{T_b \beta M}{2}\right)^2 \times 100. \tag{5.36}$$

The frequency response of the film generally decreases at high spatial frequency; the diffraction efficiency as a function of spatial frequency would be

$$\text{D.E.}(p) = \left[\frac{T_b \beta H(p)\, M}{2}\right]^2 \times 100\%, \tag{5.37}$$

where $H(p)$ is the relative response of the film. If we are to compare two films with parameters β_1, T_{b1} and β_2, T_{b2} respectively, the ratio of the diffraction efficiencies achievable with the two films can be expressed as

$$\frac{\text{D.E.}(2)}{\text{D.E.}(1)} = \frac{[H_2(p)\beta_2 M T_{b2}/2]^2}{[H_1(p)\beta_1 M T_{b1}/2]^2}. \tag{5.38}$$

And if $H(p)$ of the two films are assumed the same, the ratio would simply be

$$\frac{\text{D.E.}(2)}{\text{D.E.}(1)} = \frac{(\beta_2 T_{b2})^2}{(\beta_1 T_{b1})^2}, \qquad \text{for all } M. \tag{5.39}$$

Thus we see that $(T_b\beta)^2$ is another useful index for the *T-E* curve, providing a comparative value for the maximum diffraction efficiency expected from the film characteristics under the linear recording condition.

The definition of β, as we described earlier, is not appropriate for many holographic applications. In off-axis holography, the spurious higher order images are spatially separated from the desired first-order reconstruction (sec. 11.3). A much higher degree of nonlinearity can therefore be tolerated. The main concern in this case is the maximum diffraction efficiency achievable with the film characteristics. It is therefore acceptable and desirable to utilize an exposure range beyond that of the linear region to achieve maximum diffraction efficiency. For such applications, one approach is to first obtain a linear approximation of the *T-E* curve over the exposure range utilized (sec. 11.5). The values of β and the corresponding T_b can then be obtained from this linear approximation. However, the values of β and T_b are dependent on the linear approximation, which in turn is a function of the exposure range utilized. One possible index that can be used is the maximum value of $(\beta T_b{}^2)$ obtainable with the film characteristic. It would provide a comparative index for the maximum diffraction efficiencies to be expected with the films. To obtain this value, however, would require cumbersome mathematic computations using, for example, the least-square method to obtain the linear approximation (sec. 11.5). Such an approach would not satisfy our requirement of being obtainable by simple graphical procedures. We therefore take another approach to obtain a similar value. We take the case where the reference-to-object beam intensity ratio K is equal to 1. The input modulation M_{in} is therefore also equal to 1. Since $\beta = M_{\text{out}}/M_{\text{in}}$, for $M_{\text{in}} = 1$ we have

$$\beta = M_{\text{out}}, \qquad \text{for } K = 1.$$

And since

$$M_{\text{out}} = \frac{T_{\text{max}} - T_{\text{min}}}{T_{\text{max}} + T_{\text{min}}},$$

we can then rewrite the quantity $(\beta T_b)^2$ as

$$(\beta T_b)^2 = \left(\frac{T_{\max} - T_{\min}}{T_{\max} + T_{\min}} \times T_b\right)^2. \tag{5.40}$$

This quantity can be obtained quite easily by graphical procedures as illustrated in fig. 5.12. The value of β here is no longer defined for the slope of the linear region but for a linear approximation of the *T-E* curve.

The quantity $[(T_{\max} - T_{\min})/(T_{\max} + T_{\min}) \times T_b]^2$ is not a unique value but a function of exposure range $E_{\max}$. In fig. 5.13 we show the values of this quantity for different exposure ranges. It behaves very much like diffraction efficiencies as a function of exposures. We define an index α as the maximum value of $[(T_{\max} - T_{\min})/(T_{\max} + T_{\min}) \times T_b]^2$. This index α would then be a measure of the maximum diffraction efficiency that can be obtained with the film characteristic. An estimate of the maximum diffraction efficiency can also be obtained with the index by the equation

$$\text{D.E.} = \frac{\alpha M^2}{4} \times 100, \qquad \text{where } M = \frac{2\sqrt{K}}{1 + K}. \tag{5.41}$$

Thus a film with a higher value of α can be expected to provide a higher diffraction efficiency for any reference-to-object beam ratio.

To test the accuracy of the index α as a comparative value for diffraction efficiency, we first measured the *T-E* curves of the Kodak 649F plate developed for 2 and 6 minutes in D-19. The values for the index α are then obtained for the *T-E* curves corresponding to the two development times. Holograms were

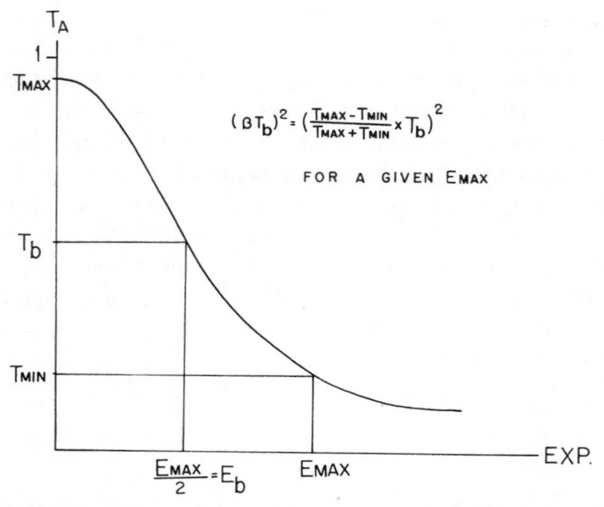

Fig. 5.12 Classification of the *T-E* curve for cases where the exposure range exceeds the linear region.

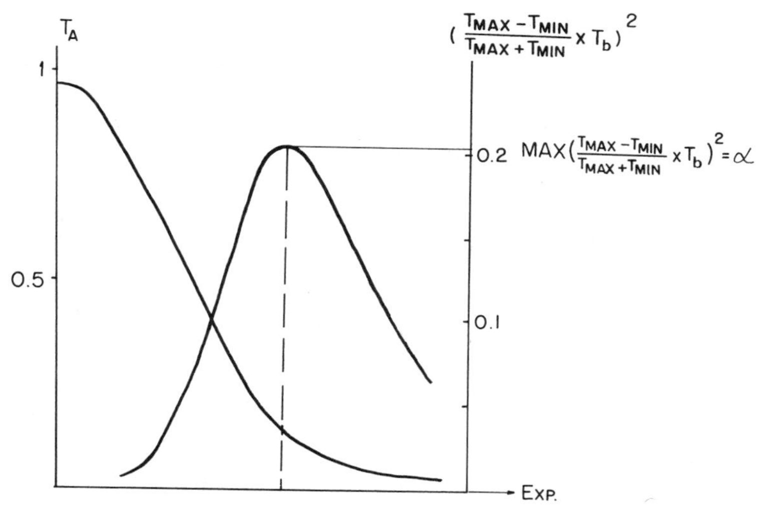

Fig. 5.13 Plot of $[(T_{\max} - T_{\min})/(T_{\max} + T_{\min}) \times T_b]^2$ as a function of exposure range $E_{\max}$.

then constructed with two plane waves of different K ratio using a He-Ne laser. The offset angle was kept small to assume that the frequency response would be close to the maximum and the requirement for a plane absorption hologram is satisfied. For each K ratio, a set of exposures were made and developed separately for 2 and 6 minutes in D-19. The maximum diffraction efficiencies were then measured for the different K ratios. In fig. 5.14 we compare the calculated

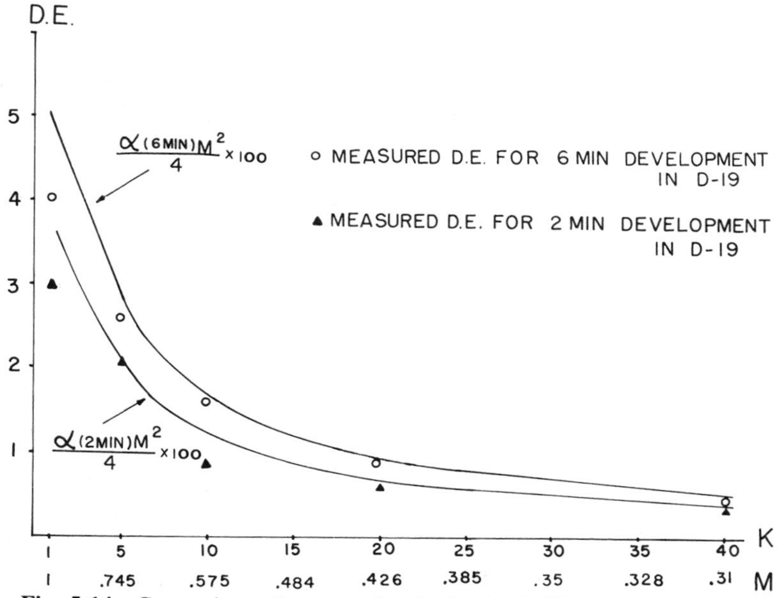

Fig. 5.14 Comparison of measured and calculated diffraction efficiencies.

diffraction efficiencies [with $H(p)$ assumed equal to 1] and the measured maximum diffraction efficiencies for various K ratios. They correspond quite well. The purpose of the index, however, is not to estimate the actual achievable diffraction efficiency but to serve as a comparative value. In fig. 5.15 we compare the ratio of the α index and the ratio of diffraction efficiencies obtained with the two different development times for various K ratios. The correspondence between the two ratios is also very compatible. One reason for the close match is because the same type of film was used in the measurements, varying only in development time. If two different types of films are compared, their difference in frequency response must be taken into account.

In concluding this section, we would like to point out that it is always desirable to have a numerical index to classify the characteristic of a photographic film. The traditional index γ is clearly inappropriate in many applications where the linear region of the T-E characteristic is utilized. While there is an obvious need for a classifying index for the T-E curve, it has not been carefully looked into. It is unlikely that any single classifying index can provide a meaningful comparative value for all applications with their differing requirements. What we have presented in this section are some possible classifying methods that might be used for applications where the linear region of the T-E curve is used. The index β is actually a measure of contrast much like the traditional γ index except that it applies to the T-E curve instead of the H and D curve. A film with a higher β value can be expected to produce a transparency with a larger percentage of output modulation for a given input pattern. The index $(\beta T_b)^2$ is a measure of diffraction efficiency obtainable with the film characteristic under the linear recording condition. The index α provides a comparative value for the upper limit of diffraction efficiency achievable with

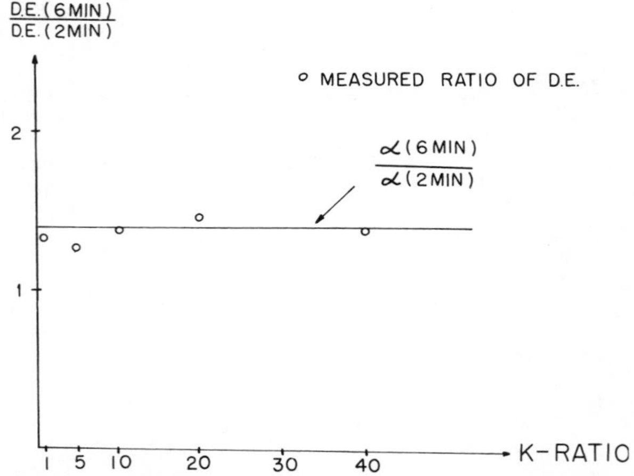

Fig. 5.15 Comparison of the ratio of the α index and the ratio of diffraction efficiencies.

the film characteristic without the linear recording restriction. A film with a higher α index can be expected to produce a higher diffraction efficiency. As with the traditional γ value, these proposed indexes can all be obtained directly from the film characteristics by simple graphical procedures.

5.4 FILM-GRAIN NOISE MODELS AND MEASUREMENT TECHNIQUES

In the application of photographic film to optical information processing (chap. 6) and holography (chap. 10), the T-E characteristic of the photographic film is generally treated as a deterministic quantity. As we have seen in a previous section, photographic films are made up of individual grains of finite sizes and they are generally distributed randomly over the film emulsion. If we measure the transmittance with a very small scanning aperture over a uniformly exposed film, we find that the transmittance fluctuates randomly around a mean value that corresponds to the transmittance measured by a larger scanning aperture. Thus we can describe the transmittance of a uniformly exposed emulsion in one dimension as

$$T(x) = B + f(x), \tag{5.42}$$

where B is the mean transmittance, $f(x)$ is the random fluctuation in transmittance, and x is the corresponding spatial coordinate. When the photographic transparency is illuminated by a coherent light, this variation in transmittance results in a random scattering of the incident light field called the *grain noise*.

Different mathematical models have been proposed to describe the stochastic properties of photographic emulsions. With one of the simplest models, the film grains are assumed to be evenly distributed in a checkerboard pattern, as shown in fig. 5.16. That is, for a given area A with M grains, the probability of m undeveloped grains, after the film is exposed to a level of E exposure, is given by the binomial distribution (ref. 5.24),

Fig. 5.16 Checkerboard model of film-grain noise.

$$P_M(m) = \frac{M!}{M!(M - m)!} P^m(1 - p)^{M-m}, \tag{5.43}$$

where p is the probability of each individual grain remaining undeveloped; p is assumed to be equal to the macroscopic transmittance corresponding to the particular exposure level E obtained from the T-E curve. Therefore, with $p = \bar{T}$, the mean and variance of P_M would be

$$\bar{m} = pM = \bar{T}M, \tag{5.44}$$

and

$$\sigma^2 = p(1 - p)\, M = \bar{T}(1 - \bar{T})M. \tag{5.45}$$

A second model, proposed by Yu (sec. 5.2), also assumes that the photographic grains are evenly distributed and the film-grain noise is treated as a Markov process. The probability of each individual grain remaining undeveloped would then be a function of the rate of development [eq. (5.21)],

$$p = \exp\left[-\int_0^E q(E')\, dE'\right], \tag{5.46}$$

where $mq(E)$ is the rate of development at exposure level E. The mean and variance of $P_M(m)$ would be [eqs. (5.22) and (5.23)]

$$\bar{m} = pM = \exp\left[-\int_0^E q(E')\, dE'\right]M, \tag{5.47}$$

and

$$\sigma^2 = p(1 - p)\, M - \exp\left[-\int_0^E q(E')\, dE'\right]\left\{1 - \exp\left[-\int_0^E q(E')\, dE'\right]\right\}M. \tag{5.48}$$

A third model assumes that the film grains are distributed uniformly and continously across the aperture. Thus wherever a photon strikes the film, an opaque circular spot would be developed. We note that, with this model, there is no restraint placed on the location of the film grains as with the checkerboard model. An overlapping of developed grains is therefore allowed, as illustrated in fig. 5.17. This model is also referred to as the overlapping circles model (refs. 5.10, 5.11). Since photon shower follows a Poisson statistic, it is therefore assumed that the developed grains are Poisson distributed. The probability of having n developed grains in an area A would be

$$P_A(n) = \frac{(\bar{n})^n e^{-n}}{n!}, \tag{5.49}$$

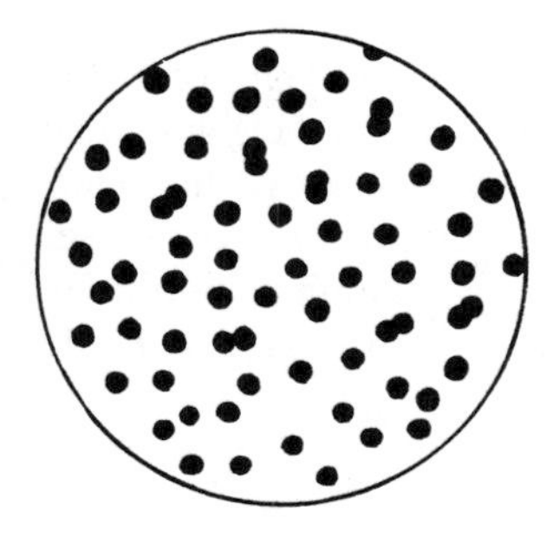

Fig. 5.17 Overlapping circles model of film-grain noise.

where $n = (M - m)$ and $\bar{n} = (M - \bar{m})$. With this assumption, the variance is found to be

$$\sigma^2 = 8 \int_0^1 [T^{2-F(\xi)} - T^2]\xi \, d\xi, \tag{5.50}$$

where $\xi = l/2R$, R is the radius of the circular grain, l is the distance variable between two circular grains, and

$$F(\xi) = \frac{2}{\pi} [\cos^{-1} \xi - \xi(1 - \xi^2)^{1/2}]. \tag{5.51}$$

With reference to eqs. (5.45), (5.48), and (5.50), the film-grain noise as a function of amplitude transmittance predicted by these three models is plotted in fig. 5.18. Over a larger aperture, the film grains observed from a large scale can

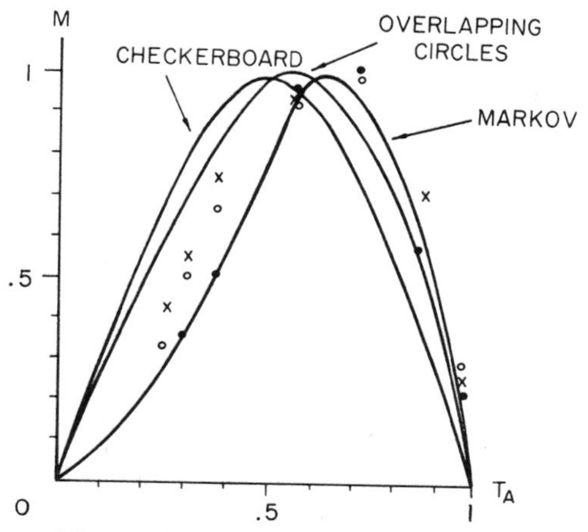

Fig. 5.18 Comparison of film-grain noise as a function of amplitude transmittance obtained with the three mathematical models and with different measurement techniques. ×, lens measurement technique; ○; direct scattering techinque; ●, pinhole scattering technique.

be approximated to be uniformly and continuously distributed. Therefore, the assumption of the overlapping circles grain model (or Poisson model) would be closer to reality for a large input aperture. Thus we may expect that the overlapping circles grain model would be an appropriate model for *outer-scale* noise. On the other hand, with the Markov model, the grains within a minimum resolvable image point (or cell) are considered. The Markov model would therefore be a better model for *inner-scale* noise.

Grain Noise Measurements and Signal-to-Noise Ratio

Since film-grain noise is aperture size dependent, the grain noise measurements must be made for specific aperture sizes (ref. 5.25). Different techniques may be employed for the grain noise measurement, depending on the aperture size of interest. As we have pointed out, the amplitude transmittance of a uniformly exposed film can be described as $T(x) = B + f(x)$, and the magnitudes of the power spectra $|\mathcal{F}[f(x)]|^2$ and $|\mathcal{F}[B]|^2$ would correspond to the power of the noise and the power of the signal, respectively, where $\mathcal{F}$ denotes the Fourier transformation.

For a large input aperture, the noise power can be measured using the Fourier transform property of a positive lens. The optical arrangement of the grain noise measurement is illustrated in fig. 5.19*a*. The spectrum of the signal (dc term) would be focused to a small center spot. The random transmittance fluctuation $f(x)$ would cause the light field to diffract off the optical axis in the higher spatial frequency space. The total noise power can therefore be measured by simply blocking out the zero-order term with a dc stop. The signal-to-noise ratio (SNR)

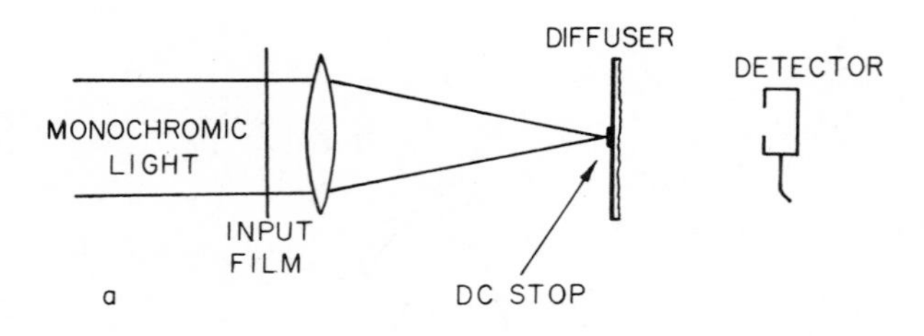

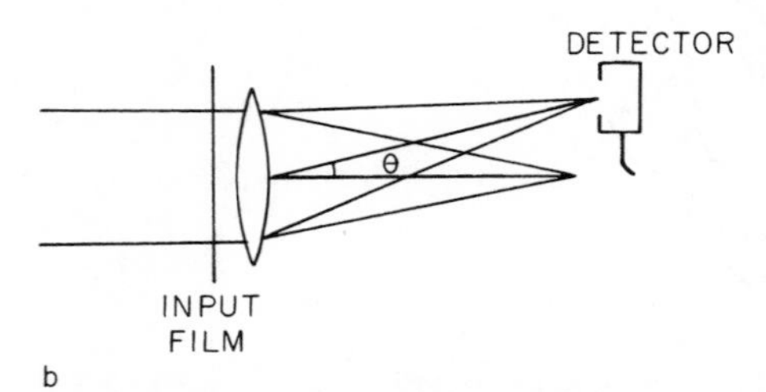

Fig. 5.19 Measurement of film-grain noise with a transform lens.

would then correspond to the ratio of the light power at the center spot to light power at the high spatial frequency plane. Since the signal power is in general much larger than the noise power, the SNR can be approximated by the ratio of the power of the transmitted light to the power of the scattered light; that is, the ratio of the measured light power with and without the dc stop. This technique can also be used to measure relative noise intensity as a function of spatial frequency by scanning the output noise spectrum with a small detector as shown in fig. 5.19*b*. The detector aperture should be small enough to provide adequate frequency resolution but large enough to include sufficient numbers of speckles to give consistent average values.

In the preceding discussions, the scattered light is assumed to be caused only by the film grains and there is no other source of noise. However, other factors could also produce such random scattering. The most notable is the random thickness variation of the film emulsion. Nevertheless, this effect can be minimized by the use of a liquid gate. Other sources of noise are the scratches and dust on the optical system. To obtain a meaningful measurement of film-grain noise, the light scattered by these other effects must be subtracted from the measurement of total noise power.

If we assume that the diffraction by the film grain is the only source of noise, then the output should be noise-free for zero exposure, since there would not be any film grain remaining in the emulsion. Thus whatever amount of scattered noise is measured for zero exposure would be due to other physical effects, for example, emulsion thickness variation. It is reasonable to assume that these effects are independent of the number of developed grains. The scattered noise contributed by these other effects for different transmittance levels would therefore be equal to $n_0\bar{T}_1$, where n_0 is the noise power for zero exposure and $\bar{T}_1$ is the normalized intensity transmittance. The corrected grain noise power would be

$$N = n - n_0\bar{T}_1, \tag{5.52}$$

where n is the total scattered noise measured for the particular transmittance level.

For small input apertures, a much simpler technique utilizing direct scattering can be used for the measurement of film-grain noise. An unexpanded laser beam is used to illuminate the exposed photographic film and the scattered light field is measured at a distance d from the film, as shown in fig. 5.20*a*. If the beam diameter and the distance d satisfy the far field Fraunhofer diffraction condition, the light field at the detection plane would correspond to the Fourier transform of the input transmittance of the illuminated area. The grain noise can therefore be measured as the power of the scattered light and the SNR would be the ratio of the power of the transmitted beam to the power of the scattered light field. This technique can also be used to measure film-grain noise as a function of spatial frequency by scanning the intensities of the scattered light circularly, as shown in fig. 5.20*b*. The greatest disadvantage of this technique is that the size

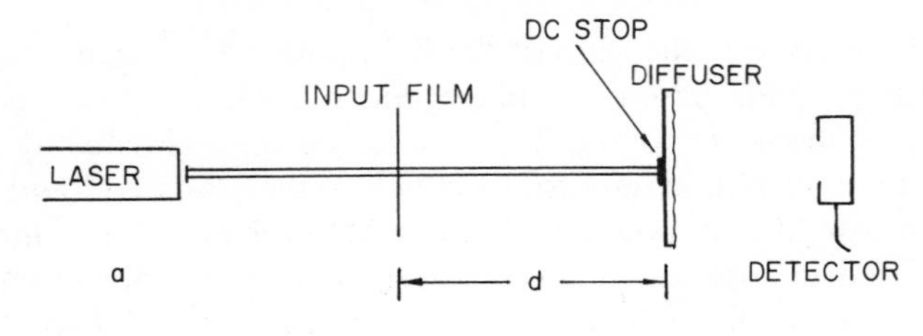

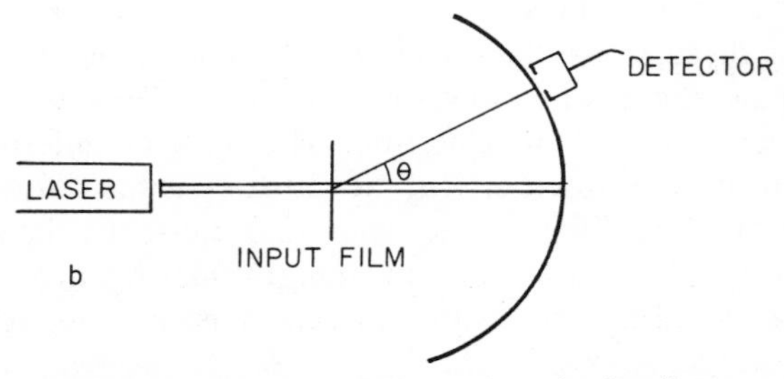

Fig. 5.20 Measurement of film-grain noise by direct scattering.

of the laser beam is difficult to define or control. Since film-grain noise is aperture size dependent, it is important to define the size of the input aperture precisely. Therefore, for applications where an absolute value of grain noise is required, the direct scattering technique cannot be conveniently applied. An alternative technique can be used for such applications. An expanded and collimated laser beam is used to illuminate the photographic film through a pinhole aperture of precisely known size as shown in fig. 5.21. We note that the beam is expanded to assure uniform illumination across the pinhole aperture. At far field diffraction, the light distribution would be in the form of an Airy disc. If there is no transmittance variation in the input aperture, the intensity of the output pattern would be attenuated according to the transmittance of the input aperture but the diffraction pattern would remain unchanged. However, if there is a transmittance variation at the input aperture, more of the input light field would be diffracted away from the center of the Airy disc. With the scattering technique, there would not be any extra noise components contributed by the optics in the measurement setup. However, the side lobes of the diffraction pattern spread over a large frequency range. Grain noise measurement is made by measuring the total diffracted light power with the central disc of the diffraction pattern blocked. Grain noise is once again defined as $N = n - n_0\bar{T}_1$. We note that this is similar to the constant noise factor contributed by the optics in the transform technique using a lens. The component of light contributed by the side lobes is subtracted out.

The measurements obtained by this technique are defined for a particular input aperture. The maximum size of the pinhole aperture that can be used with the technique is limited by the Fraunhofer diffraction condition and the minimum

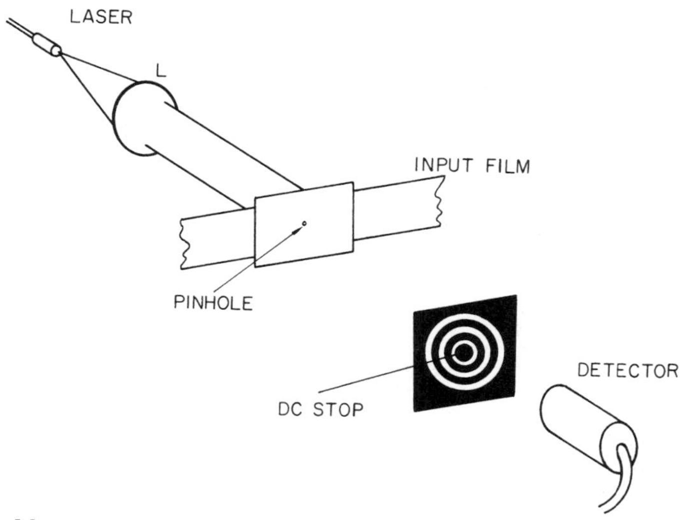

Fig. 5.21 Measurement of inner-scale film-grain noise by scattering through a pinhole.

size is determined by the sizes of the photographic grains. The pinhole aperture should include a sufficient number of film grains to produce consistent results. We note that the technique utilizing a transform lens corresponds to an *outer-scale grain noise* measurement, while the technique utilizing a pinhole corresponds to an *inner-scale grain noise* measurement.

To compare the various grain noise measurement techniques for the various mathematical models, we exposed a strip of Kodak Ortho Type III film and processed it for 30 seconds in Ethol 90 developer and 2 minutes in Kodak Rapid Fixer. The input aperture for the transform lens technique is about 1 cm^2, the beam diameter for the direct scattering technique is about 2 mm, and the aperture of the pinhole scattering technique is 0.5 mm in diameter. The normalized measured noise powers as a function of amplitude transmittance obtained with the different measurement techniques are also plotted in fig. 5.18. From this figure, we see that the experimental results are very compatible with the theoretical models. We see that the predicted relative noise level is lower for the inner-scale (Markov) noise at low transmittance (or high exposure) levels. Thus we would expect that with a small diffracting aperture (inner-scale noise), the optimum SNR would occur at a higher exposure level than with a large diffracting aperture (outer-scale noise). To see if such is the case, we measure the SNR of a set of holographic gratings recorded with different exposure levels using a large (10 mm diameter) and a small (0.12 mm diameter) input aperture using the transform and pinhole scattering techniques. The measured SNR with the two apertures are plotted in fig. 5.22. The SNR is defined as

$$\mathrm{SNR} = \frac{(\mathrm{D.E.})I_{\mathrm{in}}}{N}, \tag{5.53}$$

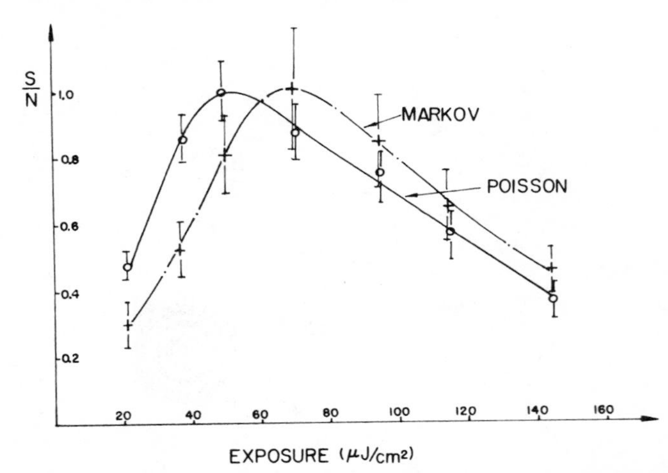

Fig. 5.22 Normalized curves for signal-to-grain-noise ratio as a function of exposure for 10-mm diameter and 0.12-mm diameter input apertures.

where I_{in} is the input light power, N is the measured noise power, and D.E. denotes diffraction efficiency. The noise power was measured with both the zero- and first-order diffracted spots blocked. The experimental results appear to verify the predictions of the models; the optimum SNR occurs at a higher exposure level if a small diffracting aperture is used.

In ending this section, we stress again that the Poisson model (i.e., overlapping circles grain model) is suitable for the *outer-scale* noise classification, while the Markov model is appropriate for the *inner-scale* noise classification.

5.5 DICHROMATED GELATIN FILMS

Dichromated gelatin exhibits excellent properties for thick emulsion volume phase holograms; it has low scattering, low absorption, high resolution, and high refractive index modulation, which are very suitable for the fabrication of holographic optical elements and phase diffraction grating. The earlier work on dichromated gelatin for volume phase holograms was reported by Shankoff (ref. 5.26) in 1968. The detailed fabrication of dichromated phase holograms can also be found in the excellent text by Collier et al. (ref. 5.27), in a doctoral dissertation by Zech (ref. 5.28), and in a recent article by Chang and Leonard (ref. 5.29).

In this section we discuss the basic holographic characteristics of the dichromated gelatin. However, we do not attempt to discuss here the detailed physical and chemical properties of dichromated gelatin. We refer the interested reader to the cited references. Before we discuss the dichromatic gelatin formation, we would like to highlight some of the properties that have been reported:

1 The refractive index modulation capacity of dichromated gelatin is the largest among all the recording materials available today. An index modulation of 0.08 has been reported.

2 Dichromated gelatin has a high resolution capacity. It has been reported that the spatial frequency is relatively uniform from 100 to 5000 lines/mm.

3 Dichromated gelatin holograms can be reprocessed so that the desired refractive index modulation can be obtained.

4 The thickness of dichromated gelatin can be controlled by pre-processing and post-processing techniques.

5 Dichromated gelatin has high SNR and low absorption over the visible to near infrared wavelengths.

6 Dichromated gelatin has good environmental stability with a cover plate or in a dry atmospheric environment.

7 Although dichromated gelatin is sensitive to ultraviolet to blue-green wavelengths, it can be made to respond to red wavelengths by a dye sensitization technique.

We note that the exact mechanism of dichromated gelatin formation in holography is not fully understood, and there is considerable disagreement among the investigators. However, we describe a model postulated by Chang and Leonard (ref. 5.29), which seems to agree with the experimental results they obtained.

In dichromated gelatin hologram formation, the gelatin film is sensitized with ammonium dichromate solution and dried. Then it is exposed to actinic radiation; the hexavalent chromium ion C_r^{6+} is either directly or indirectly photo-induced to trivalent C_r^{3+}, which forms several intermediate chromium compounds, as shown in the block diagram of fig. 5.23. The chromic ion C_r^{3+} is

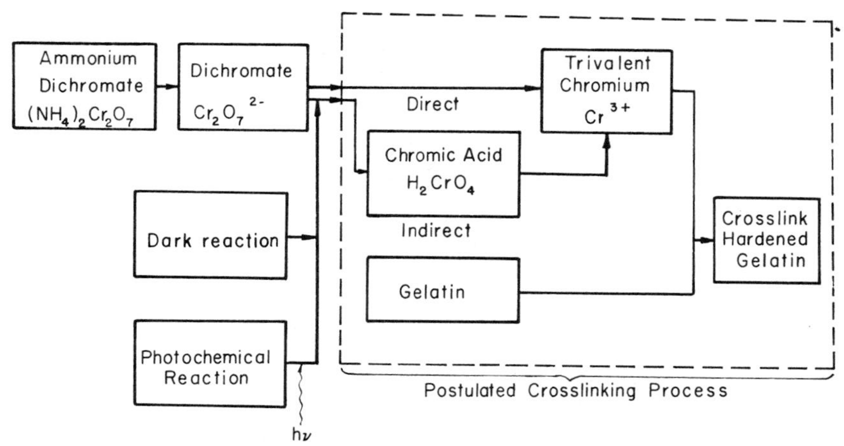

Fig. 5.23 Cross-linking and hardness change of gelatin.

thought to form a cross-link bond between the gelatin molecules. This cross-link hardens the gelatin, creating a hardness differential between highly exposed regions and relatively unexposed regions, and the hardness differential forms the latent image of the hologram. During the development process, the residual chemical compounds are removed, and the gelatin film swells to a significant amount. This is due to absorption of more water in the unexposed region. For example, the dichromated gelatin hologram can be formed more efficiently inside a water tank before the final alcohol development, as shown in fig. 5.24. The developed dichromated gelatin holograms do not lose their diffraction efficiency inside a xylene liquid gate. Thus we see that the refractive index modulation is caused by the rearrangement of gelatin molecule chains during the development steps due to the hardness differential. As shown in fig. 5.23, the C_r^{6+} ion can be reduced to C_r^{3+} in the dark, changing its color from yellowish to reddish-brown. Since the reaction takes place in the dark wihout exposure to actinic radiation, it is called the *dark reaction*. This dark reaction quickly hardens the dichromated gelatin layer, reducing the refractive index modulation capacity when the environmental condition is met. In practice, the dark reaction reduces the shelf life of dichromated gelatin layers and the repeatability of experimental results. We note that the dark reaction is very sensitive to humidity as well as to temperature. In other words, if the relative humidity is low, the shelf life of dichromated gelatin can be extended for a long period of time, for example, several months under refrigeration.

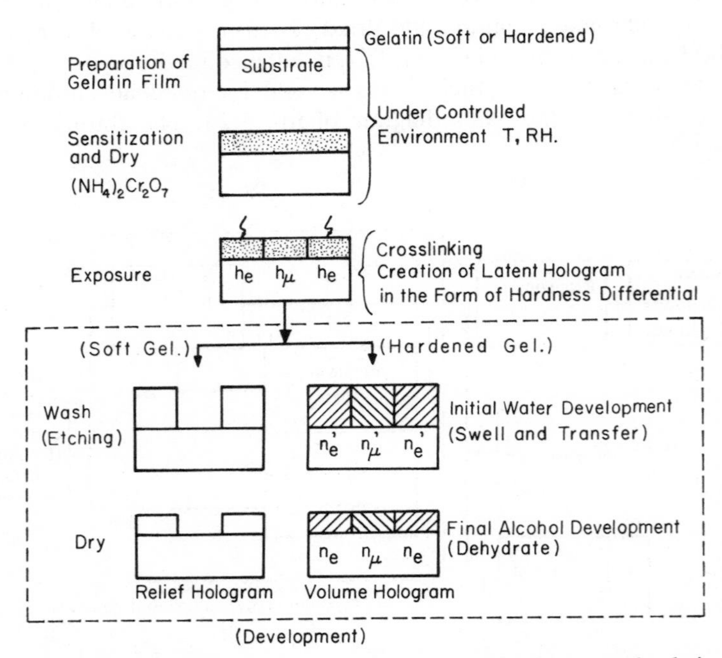

Fig. 5.24 Recording procedures of holograms in dichromated gelatin.

We now proceed to discuss the fabrication of a dichromated gelatin hologram, as shown in fig. 5.24. From this figure, we see that the process is divided into four steps: preparation of gelatin film, sensitization, hologram exposure, and development. We note that the hologram recorded in dichromated gelatin film can either be a relief surface or a volume-type hologram. The type of dichromated gelatin hologram depends primarily upon the initial bias hardness of the dichromated gelatin. Thus to produce a volume-type hologram, the bias hardness should be higher than a threshold hardness (ref. 5.29). There is, however, a limit to the bias hardness; if the bias hardness is too high, the refractive index modulation will be decreased. In practice, it is desirable to have the initial bias hardness slightly higher than a threshold level because soft gelatin film often produces a noisy effect.

It is interesting to note that the development procedures can be divided into two parts, that is, the initial water development and the final alcohol development. The water development creates a refractive index modulation by swelling the film so that some gelatin from regions of low relative exposure may be transferred to highly exposed regions. Thus a higher refractive index modulation volume hologram can be formed, as shown in fig. 5.25*a*.

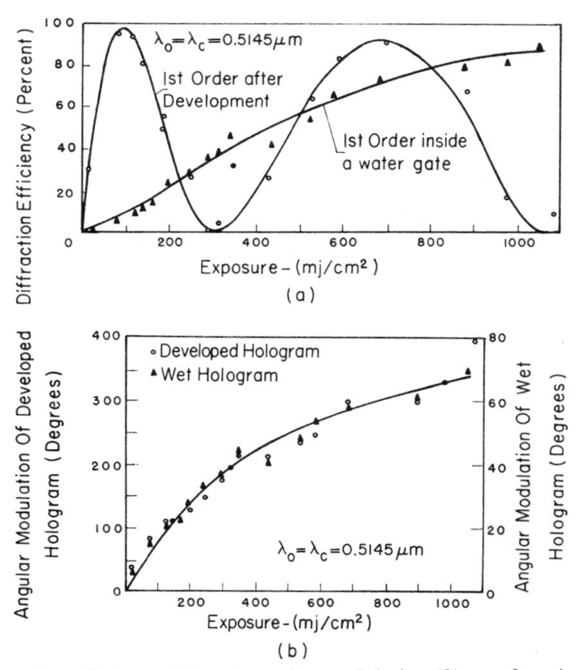

Fig. 5.25 Diffraction efficiency (*a*) and angular modulation (*b*) as a function of exposure for dichomated gelatin holograms in air and in water. The angular modulation is defined by eq. (5.54), and the angular modulations of 90° and 270° correspond to the first and second peak diffraction efficiencies, respectively. The diffraction efficiency is defined by the ratio of the diffracted intensity to the incident intensity.

Figure 5.25*b* shows the angular modulation m as a function of exposure, which is the argument of a sine function of Kogelnik's diffraction efficiency formula (ref. 5.30), i.e.,

$$m = \left(\frac{\pi}{\lambda \cos \theta_0}\right) \Delta\eta d, \tag{5.54}$$

where λ is the reconstruction wavelength in air, θ_0 is the internal angle between the illuminating beam and the fringe planes, d is the thickness of the gelatin film, and $\Delta\eta$ is the refractive index modulation. We note that the initial refractive index modulation in water, $\Delta\eta_w$, is dependent upon the hardness differential Δh, that is, $\Delta\eta_w = G_1\Delta h$, where G_1 is a gain factor depending upon the swelling of the gelatin. This index modulation is further amplified with final alcohol development, i.e.,

$$\Delta\eta = G_2\Delta\eta_w = G_1G_2\Delta h, \tag{5.55}$$

where G_2 is a gain factor. As shown in fig. 5.25*a*, the final effective index modulation $\Delta\eta_d$ is about five times higher than the initial effective index modulation in water $\Delta\eta_w d_w$, where d_w is the swollen thickness of the hologram in water.

Dichromated gelatin film can be prepared by several methods. The simplest procedure begins with the removal of the silver halides and dyes from a photographic emulsion. However, as commercial films are not prepared for recording of dichromated gelatin holograms, they generally require additional porcedures to generate a high quality dichromated gelatin emulsion. For a detailed discussion of dichromatic gelatin film, we refer the reader to the work by Zech (ref. 5.28), and by Chang and Leonard (ref. 5.29).

Although the dichromated gelatin volume phase hologram exhibits excellent diffraciton efficiency and little scattering and absorption, it has two basic drawbacks. The first is the limited spectral response; dichromated gelatin is primarily sensitive to blue-green to ultraviolet radiation. The second is the relatively high exposure requirement; a typical exposure at 515 nm is about 50–100 mJ/cm^2. Although dichromated gelatin can be made to respond to red wavelengths with proper dye sensitization (ref. 5.31), the exposure requirement is considerably higher.

These drawbacks may be alleviated with silver halide gelatin holograms as reported by Chang and Winick (ref. 5.32). Since silver halide gelatin uses silver halide (or other silver salts including dye-sensitized silver salts) as sensitizer, its spectral and exposure sensitivities are equivalent to those of conventional holographic plates. The basic difference between the dichromated gelatin process and the silver halide process is the creation of a hologram latent image, which gives the form of hardness differential between exposed regions and unexposed regions. The latent image of a dichromated gelatin hologram is created by the

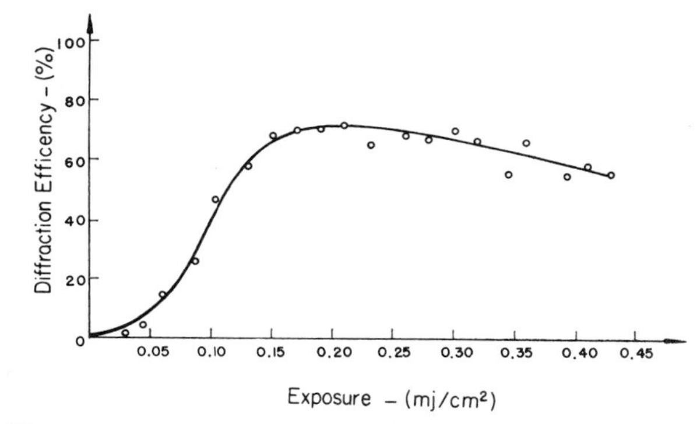

Fig. 5.26 Diffraction efficiency as a function of exposure for silver halide holograms recorded in Kodak 649F plates.

photochemical hardening reaction mechanism between dichromates and gelatin during the exposure, while the latent image of a silver halide gelatin hologram is formed by the reaction products of either tanning development or tanning bleach, which harden their gelatin by forming cross-link bonds between the chain of gelatin molecules. In concluding this section, we show, in fig. 5.26, the result obtained by Chang and Winick (ref. 5.32); from this figure we see that a diffraction efficiency can reach as high as 70 percent for silver halide gelatin holograms recorded in Kodak 649F plates.

5.6 SOME REAL-TIME RECORDING MATERIALS

Silver halide photographic film is still the most commonly used and best developed recording material. Even with the advent of many new recording materials, photographic film is still unsurpassed in many parameters such as resolution, consistency, and unit cost. It is likely that conventional photographic film will remain the prime recording material for some time. However, photographic film possesses one major weakness; its wet development process is cumbersome, and more importantly, nonreal-time. For applications where the amount of data to be processed is very large, optical systems still hold an edge in performance and cost effectiveness. However, the access time delay in the development of the photographic film represents a major bottleneck in many optical processing systems. In order to compete with the rapidly growing and increasingly sophisticated digital systems, the photographic film must be replaced by a real-time reusable recording material. New materials that can fulfill this requirement are being developed by various laboratories. However, except for a few special applications, these materials still require further developments before they can be competitively adapted into optical processing systems. In this section we examine three very different types of recording materials that are potential candidates for use in a real-time optical spectrum analyzer.

Pockel's Read-Out Optical Modulator

Pockel's Read-Out Optical Modulaor (PROM) is the most promising of all the real-time recording materials currently under development (ref. 5.33). It can be fabricated from various electro-optical crystals such as ZnS, ZnSe, and $Bi_{12}SiO_{20}$.

The basic construction of the PROM device is shown in fig. 5.27. The electro-optical crystal wafer is sandwiched between two transparent electrodes and spaced by an insulator. The crystal wafer is oriented in such a way that the field applied between the electrodes produces a longitudinal electro-optic effect. The operation of the PROM device is also illustrated in fig. 5.27. An applied dc voltage with an erase light pulse is used to create mobile carriers that cause the voltage V_0 in the active crystal to decay to zero. The polarity of the applied voltage is brought to zero and then reversed. A total voltage of $2V_0$ will appear across the crystal. The device is then exposed to the illumination pattern of blue light. The voltage in the area exposed to the bright part of the input pattern would decay, while the voltage in the dark area would remain unchanged. The intensity pattern is thereby converted into a voltage pattern. The relationship between input exposure and voltage across the crystal is given by

$$V_c = V_0 \exp(-KE), \tag{5.56}$$

where V_0 is the applied voltage and K is a positive constant. The read-out is performed with a red linear polarized light (e.g., He-Ne laser). For $Bi_{12}SiO_{20}$, the crystal is 200 times more sensitive in the blue region (400 nm) than in the

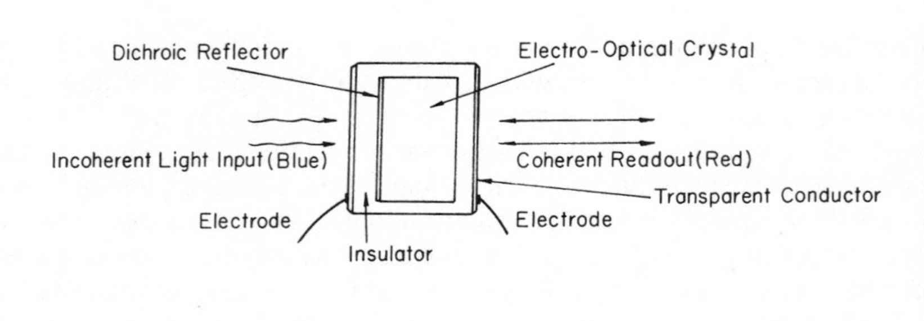

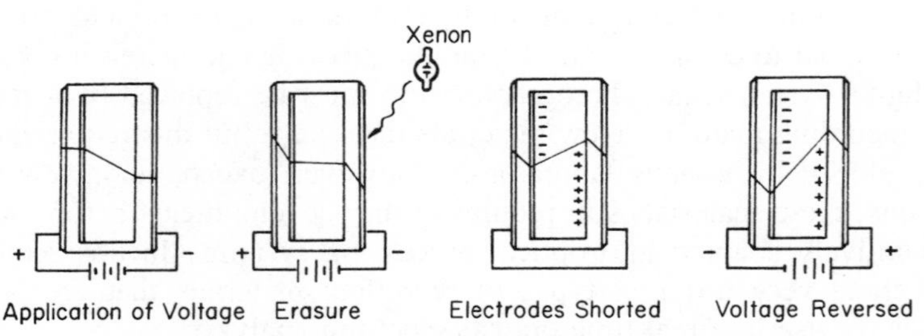

Fig. 5.27 Composition of the PROM and its operation.

red region (633 nm). Therefore, reading with a He-Ne light source in real-time would not produce significant voltage decay over a period of time.

The read-out is performed by reflection. In this read-out mode, the area where the voltage across the crystal that has not been decayed by the input light intensity would act like a half-wave retardation plate. The angle of polarization of the linearly polarized laser light input reflected by such an area would be rotated to 90°. Thus the light reflected by the area corresponding to the bright region has a polarization perpendicular to that of the dark region. The reflected light is then passed through a polarizer; the amplitude of the transmitted light would be attenuated according to the polarization of the input. The amplitude of the light transmitted through the polarizer (i.e., read-out amplitude) as a function of the voltage across the crystal can be written as

$$A = A_0 \sin\left[\pi\left(\frac{V_C}{V_{1/2}}\right)\right], \tag{5.57}$$

where A_0 is the input light amplitude and $V_{1/2}$ is the half-wave voltage. In the reflection read-out mode, $V_{1/2} \simeq 2V_0$. The output coherent light amplitude, as a function of input exposure, would be

$$A = A_0 \sin\left[\frac{\pi V_0}{V_{1/2}} \exp(-KE)\right]. \tag{5.58}$$

The theoretical curve for A versus E is plotted in fig. 5.28. We can see that it

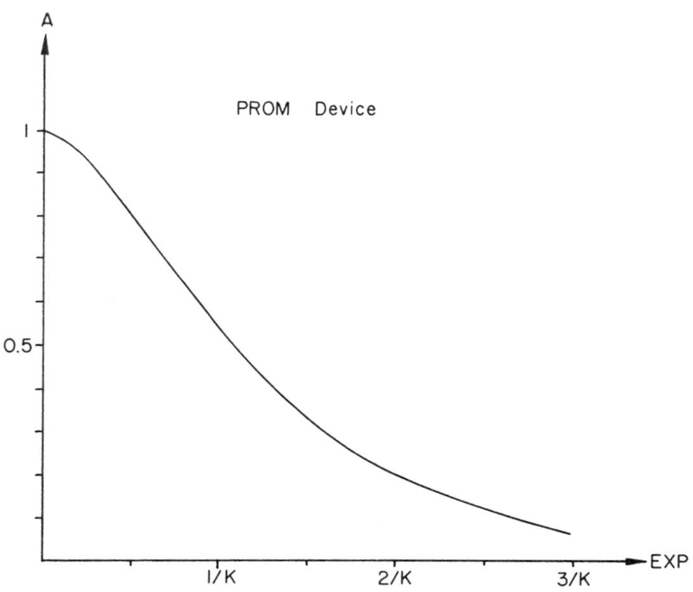

Fig. 5.28 Amplitude reflectance of the PROM device as a function of input exposure.

has a characteristic quite similar to that of photographic film. There is a linear region with a range between $E = 2/K$ and $E = 0$ and a bias point at $E = 1/K$. In the following table, we list some of the specifications of the PROM device produced by the Itek Corporation:

Optical aperture	2.5×2.5 cm
Read-in sensitivity (voltage decay to $1/e$)	5 μJ/cm^2 at 404 nm
Read-out energy density (voltage decay to $1/e$)	1000 μJ/cm^2 at 633 nm
Resolution (incoherent read-in)	60 line-pairs/mm
Recycling time	30 Hz
Lifetime	No known limit

The read-in sensitivity and resolution are both compatible with those of photographic films, and the recycling time of 30 Hz provides the system with true real-time capability. In fig. 5.29 we show a possible system design utilizing the PROM device as the recording medium (see chap. 6). Since the device is relatively insensitive to the red light of the He-Ne laser, the read-out light can be illuminating the device as the cathode ray tube (CRT) scan is being read into the device. The output spectrum would, therefore, appear the instant each scan line is completed and the entire spectrogram would be displayed when the electron beam of the CRT completes the raster scan across the screen.

There are some drawbacks, however, with the PROM device. The fabrication techniques of the device still require further refinement to improve the uniformity and consistency of the crystal. A more narrow bandpass characteristic is required in its spectral sensitivity. With the Itek PROM device, for example, an exposure to the read-out light energy of 1000 μJ/cm^2 would decay the crystal voltage to $1/e$ of the peak value. Using a 2-mW He-Ne laser as the processing

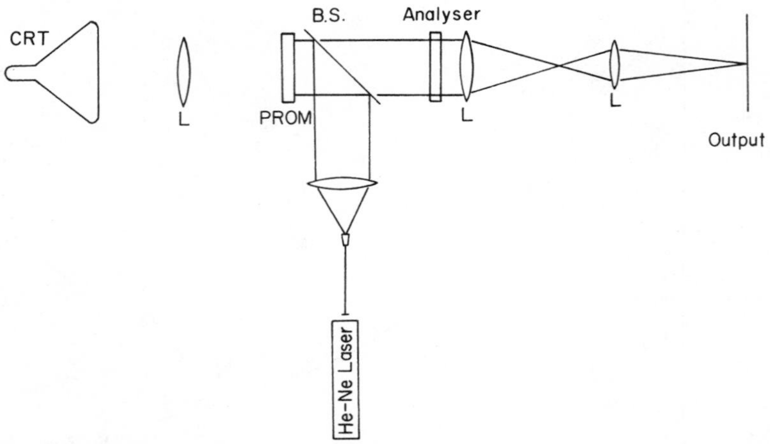

Fig. 5.29 Possible optical system utilizing the PROM device for signal conversion.

light source, if we assume that half the laser power is illuminating the PROM aperture and the illumination is approximately uniform within the aperture, the intensity of illumination would then be 160 $\mu W/cm^2$. Thus in 6.25 seconds of illumination by the read-out beam, the crystal voltage would drop to $1/e$ of the peak value. The effect would be to shift the output amplitude distribution down the A-E curve of fig. 5.28. Therefore, the device cannot be read over an extended period of time under strong illumination. Another drawback of the PROM device is its high cost at the present time. Nevertheless, the PROM device is expected to be improved and the cost lowered with further developments.

Photoplastic Device

Photoplastics are classified as phase recording materials, with the recording of the input intensity in the form of surface deformation (refs. 5.34, 5.35). The basic components of a photoplastic recording device are shown in fig. 5.30. It is composed of a clear substrate (usually glass), coated with a transparent conductive layer (tin oxide or indium oxide) over which is a layer of photoconductive material and then a layer of thermoplastic. For the photoconductor, poly-n-vinyl carbazole (PVK) sensitized with trinitrifluorenone (TNF) can be used with an ester resin thermoplastic (Hurculus Floral 105). Before the exposure, the device is charged either by corona discharge or with a charging plate made of another transparent conductive material, as shown in fig. 5.31. The charging plate is spaced with a 100-micron Mylar tape. After the charging process, the device is then exposed to light, causing a variation of charge pattern proportional to the input light intensity. The illuminated region will exhibit a displacement of charge from the transparent conductive layer to the photoplastic

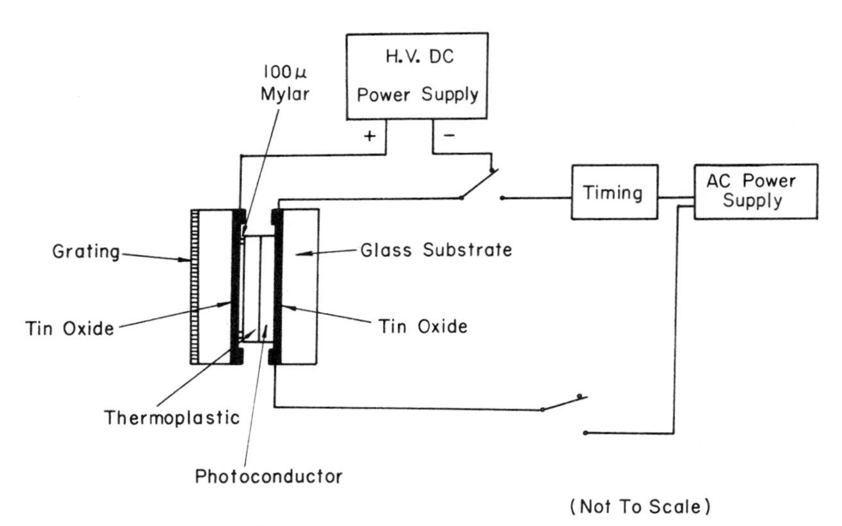

Fig. 5.30 Composition of a photoplastic recording device.

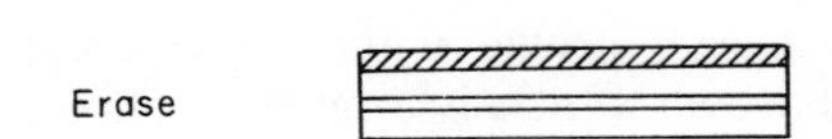

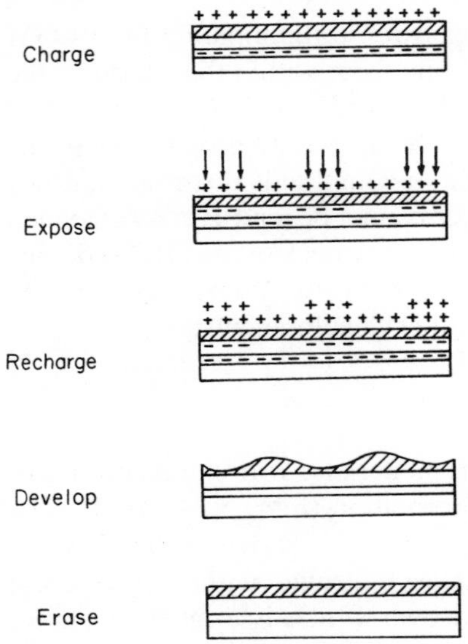

Fig. 5.31 Operation of the photoplastic device for recording of light intensity pattern.

interface, which reduces the surface potential of the outer surface of the thermoplastic as shown in fig. 5.31. The device is then recharged to the original surface potential and developed by raising the temperature of the thermoplastic to the softening point and then lowering the temperature rapidly to room temperature. This can be done by passing an ac voltage pulse through the conductive layer. Surface deformation caused by the electrostatic force produces a phase recording of the input light intensity. The recording can be erased by raising the temperature of the thermoplastic above the plastic point; the surface tension of thermoplastic will smooth out the surface deformation and erase the recording. The frequency response of the photoplastic device is very poor at low frequencies. In order to use the device for the recording of signals with low spatial frequencies, the signal must be modulated with a sinusoidal signal of a spatial frequency that corresponds to the peak of the frequency response of the device. This can be done by putting a sinusoidal grating directly in front of the thermoplastic device. By using a thick thermoplastic layer, the frequency response of the device can be made to peak at 50 lines/mm or below. Grating of that frequency is readily available. We note that the sensitivity of the photoplastic device is fairly low, and it would not be able to record a CRT scan.

The greatest advantage of the photoplastic device is its relatively low cost, making it a very practical recording device for use in holography. However, its performance is less than ideal in many respects. The cycling time is relatively slow, its lifetime is limited (about 500 cycles, depending on laboratory conditions), and the SNR is relatively low due to the random thickness variation of developed thermoplastic.

Liquid Crystal Light Valve

One of the most useful real-time electro-optical light modulators is the liquid crystal light valve (LCLV) (ref. 5.36). The LCLV is capable of converting an incoherent optical image into a coherent image suitable as input for nearly any type of coherent optical processing system (ref. 5.37). Most of us are familiar with liquid crystal materials from their use in digital wristwatch and calculator displays. In fact, liquid crystal displays have found their way into portable laboratory equipment, kitchen appliances, automobile dashboards, and many other applications requiring good visibility, low power consumption, and/or special display geometries. Most of these applications make use of the liquid crystal in a twisted nematic cell, in which a segment of the cell is either opaque or transparent, depending on the value of the applied electrical voltage. However, in applications related to optical signal processing, the hybrid field effect LCLV is more desirable. The LCLV combines the properties of the twisted nematic cell in the "off" state with the property of electrically tuneable birefringence to control the transmittance of the LCLV over a wide continuous range in the "on" state. The range of real-time optical signal processing applications of the LCLV has extended from real-time tracking and recognition of military vehicles (airplanes, tanks, etc.) (ref. 5.38) to robot controlled automated assembly-line manufacturing to providing real-time optical feedback in optical processing systems (ref. 5.39).

The optical behavior of the LCLV can be described by two effects: twisted nematic effect and pure optical birefringence. A simplified sketch of a transmission-type twisted nematic cell is shown in fig. 5.32*a*. A thin layer (1–20 μm) of nematic liquid crystal is sandwiched between two transparent electrode coated glass plates. The electrode surfaces are treated to provide a preferential direction of alignment of the liquid crystal molecules. The plates are arranged so that, with no electrical voltage applied across the layer, the layer of liquid

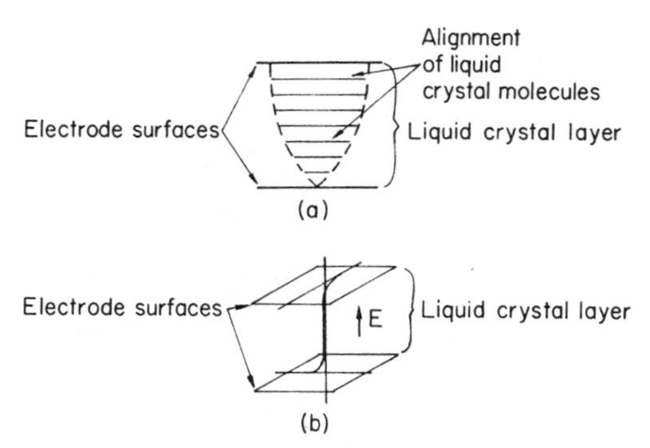

Fig. 5.32 (*a*) Twisted nematic liquid crystal layer (no electric field applied). (*b*) Molecular alignment in direction of applied electric field **E** (greater than threshold value).

crystal molecules is twisted continuously by 90° in going from one electrode surface to the other. A sheet of polarizing material is placed in front of (polarizer) and behind (analyzer) the sandwiched cell. The direction of polarization of light passing through the polarizer must be the same as the direction of molecular alignment at the front electrode surface. As this light passes through the twisted liquid crystal layer, its direction of polarization will also be twisted by 90°. If we make the direction of polarization of the analyzer parallel to the direction of molecular alignment at the back electrode surface, the polarized light will pass through the cell unaffected. If, on the other hand, we choose to make the direction of polarization of the analyzer perpendicular to that of the back surface molecular alignment, the polarized light will not exit the cell and the device will be in its dark state. With slight modification, this latter mode of operation of the twisted nematic cell is utilized in the LCLV.

An interesting phenomenon occurs when a low frequency voltage is applied across the twisted nematic liquid crystal layer. The molecules tend to align themselves along the direction of the applied electric field—that is, perpendicular to the electrode surfaces. The resulting splay and bend of the molecules is shown in fig. 5.32*b*. Thus if the polarizer and the analyzer have parallel directions of the polarization, the polarized light will pass through the liquid crystal layer and the analyzer unaffected. This "on" state behavior is not exhibited in the LCLV, however, because of the reflection-type configuration.

The LCLV takes advantage of the pure birefringence of the liquid crystal material in order to modulate the output laser beam while in the "on" state. Operation of the LCLV is clarified in the side view of fig. 5.33. An incoherent image is focused onto the photoconductor layer to gate the applied alternating

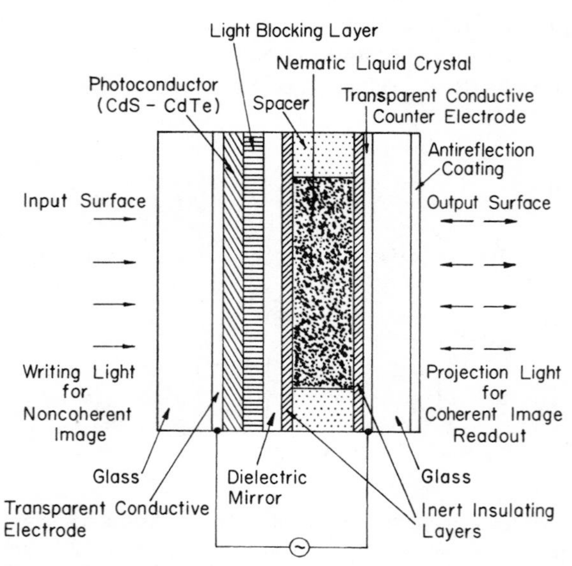

Fig. 5.33 Side view of hybrid field effect liquid crystal light value (LCLV).

voltage to the liquid crystal layer in response to the input intensity of every point in the input space. Laser light illuminating the back of the LCLV is reflected back but modulated by the birefringence of the liquid crystal layer at every point. In order to achieve this effect, the molecular alignment has a 45° twist between the two surfaces of the liquid crystal layer. The dielectric mirror plays an important role by providing optical isolation between the input incoherent light and the coherent read-out beam. Important optical and electrical properties of the hybrid field effect LCLV are summarized in the following (ref. 5.40):

Aperture size	1 in.^{2}
Sensitivity (full contrast)	160 μW/cm^2 at 525 nm
Resolution	60 lines/mm at 50 percent Modulation Transfer Function (MTF)
Contrast	>100 : 1
Grayscales	9
Speed	
Excitation (0–90 percent)	10 msec
Extinction (100–10 percent)	15 msec
Projection light throughput	>100 mW/cm^2
Reflectivity	>90 percent
Optical quality	<λ/4 wavelength curvature at 632.8 nm
Voltage	6 V_{rms} at 10 kHz

It is worthwhile noting that the maximum resolution of the LCLV (60 lines/mm) is insufficient for holographic displays by approximately one order of magnitude.

The optical setup depicted in fig. 5.34 illustrates the use of a LCLV in a typical real-time optical signal processing application (refs. 5.38–5.41). A changing scene or object illuminated with incoherent light is imaged (written) onto the photoconductor side of the LCLV. The intensity distribution is converted from an incoherent to a coherent image on the liquid crystal side of the

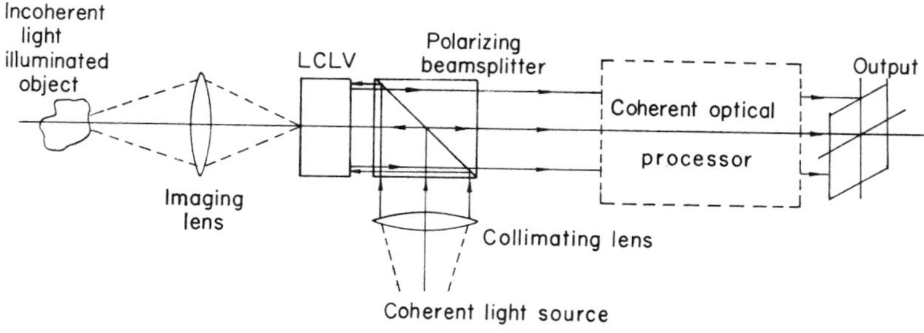

Fig. 5.34 Typical application of the LCLV in optical signal processing.

valve. This image is read out on the coherent (laser) beam and used as an input image for the coherent optical processing system.

Commercial availability of the LCLV has provided a great impetus for real-time optical signal processing applications. The resolution, sensitivity, contrast, and general versatility of the device have induced researchers to continue improving and developing new uses for the LCLV. Recent advances have included the following: a reduction of the response time to below 1 msec, the mating of a channel intensifier tube to the input side of a LCLV for low-intensity-level input images, and the incorporation of charge-coupled-devices with a LCLV for TV-type input operation.

PROBLEMS

5.1 Convert the H and D curve of fig. 5.35 to one giving amplitude transmittance as a function of exposure (i.e., construct a *T-E* curve).

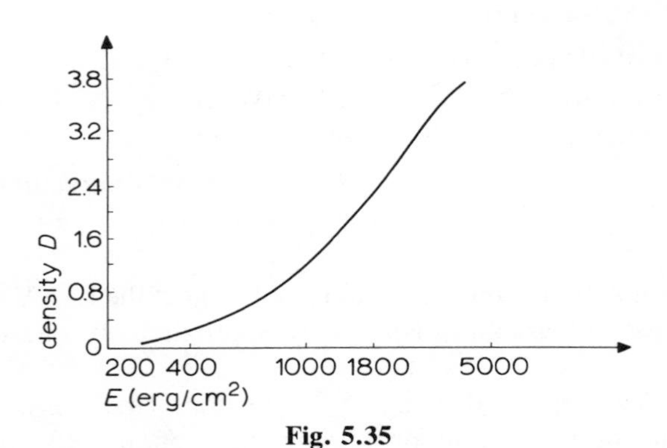

Fig. 5.35

5.2 If the contrast of the amplitude transmittance is defined as

$$\frac{T_{\max} - T_{\min}}{T_{\max} + T_{\min}},$$

then from the *T-E* transfer characteristic obtained in prob. 5.1, determine an approximate linear exposure in which the contrast is optimized.

5.3 Consider a certain photographic plate that is recorded by a low-contrast exposure over the plate. The exposure is written as

$$E(x, y) = E_Q + E_1(x, y), \qquad \text{with} \qquad |E_1(x, y)| \ll E_Q,$$

where E_Q is the bias exposure and $E_1(x, y)$ is the variable incremental exposure. If the recorded photographic plate is biased in the linear region of the H and D curve, show that the contrast of the amplitude transmittance is linearly proportional to $E_1(x, y)$, provided $|E_1(x, y)| \ll E_Q$.

5.4 Suppose that the T-E transfer characteristic of a certain photographic emulsion is as shown in fig. 5.36, and that the input exposure is given by a sinusoidal grating.

$$E(x, y) = 1000 + 800 \cos(100\,\pi x)\ \text{erg/cm}^2, \qquad \text{for every } y.$$

Use a graphical method to plot the corresponding output transmittance $T(x, y)$.

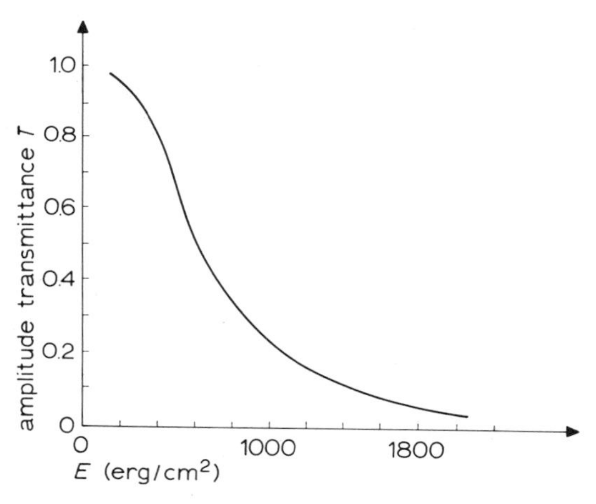

Fig. 5.36

5.5 Using fig. 5.35, explain why a hard-clipped grating gives the optimum transmission irradiance-to-noise ratio.

5.6 Plot the diffraction efficiency of a holographic recording as a function of exposure range (E_{max}) using the data in fig. 5.9.

(a) Assume a linear recording.

(b) Assume the recording extends into the nonlinear range.

The maximum non-normalized value of T_A in fig. 5.9 is 0.95, $K = 4$, and $H(p) = 1$.

5.7 A color transparency illuminated with a white-light source serves as the input for a two-step contact printing utilizing two black-and-white (B&W) negative films. If the resulting transparency is to be used in an optical processing system, describe in sufficient detail at least five relevant characteristics of the B&W films to produce a successful linear recording.

5.8 What type of a B&W photographic film is most suitable for recording a halftone screen image? Why?

5.9 The sawtooth grating shown in fig. 5.37 is to be contact printed onto a B&W film to generate a sinusoidal grating in one step.

(a) Plot the required T-E curve for this film if the resulting transmittance is given by

$$T_{A\,\sin} = 0.5 - 0.4\,\sin\left(\frac{\pi}{8}\,x\right),$$

and $E_{\sin}$ (normalized) $= (2T_{A\,\text{saw}})^2 = 1.0$ when $T_{A\,\text{saw}} = 0.5$. Does this curve have the shape of a typical T-E curve?

(b) Calculate the diffraction efficiency of the resulting transparency using eq. (5.41) with $K = 1$.

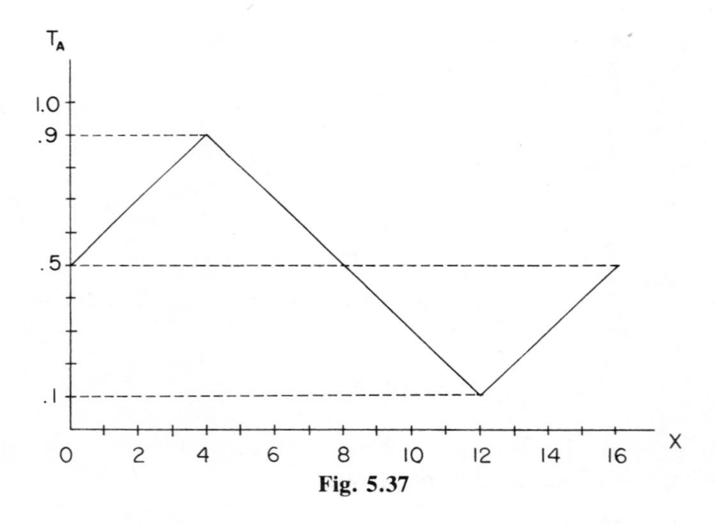

Fig. 5.37

5.10 A total of N images are to be contact printed sequentially onto the same piece of black and white photographic film. All N input images have identical density ranges extending over 2 log units. Calculate the required input exposure range of this film, i.e., $E_{\max}/E_{\min}$.

REFERENCES

5.1 C. E. K. Mees, *The Theory of the Photographic Process,* rev. ed., Macmillan, New York, 1954.

5.2 C. E. K. Mees and T. H. James, *The Theory of the Photographic Process,* 3rd ed., Macmillan, New York, 1966.

5.3 F. Hurter and V. C. Driffield, "Photochemical Investigations and a New Method of Determination of the Sensitiveness of Photographic Plates," *J. Soc. Chem. Ind.,* **9,** 445 (1890).

5.4 E. L. O'Neill, (Ed.), *Communication and Information Theory Aspects of Modern Optics,* General Electric Company, Electronics Laboratory, Syracuse, N.Y., 1962.

5.5 A. Kozma, "Photographic Recording of Spatially Modulated Coherent Light," *J. Opt. Soc. Am.,* **56,** 428 (1966).

5.6 R. C. Jones, "New Method of Describing and Measuring the Granularity of Photographic Materials," *J. Opt. Soc. Am.,* **45,** 799 (1955).

5.7 H. J. Zweig, "Autocorrelation and Granularity, Part I. Theory," *J. Opt. Soc. Am.,* **46,** 805 (1956).

5.8 H. J. Zweig, "Autocorrelation and Granularity, Part II. Results or Flashed Black-and-White Emulsions," *J. Opt. Soc. Am.,* **46,** 812 (1956).

5.9 H. J. Zweig, "Autocorrelation and Granularity, Part III. Spatial Frequency Response of the Scanning System and Granularity Correlation Effect Beyond the Aperture," *J. Opt. Soc. Am.,* **49,** 238 (1959).

5.10 M. B. Picinbono, "Modèle Statistique Suggéré par la Distribution de Grains D'Argent Dans les Films Photographiques," *Compt. Rend.,* **240,** 2206 (1955).

5.11 M. M. Savelli, "Résultats Pratiques de l'Étude d'un Modèle à Trois Paramètres pour la Représentation des Propriétés Statistiques de la Granularité des Films Photographiques Notamment des Propriétés Spectrales," *Compt. Rend.,* **246,** 3605 (1958).

5.12 F. T. S. Yu, "Markov Photographic Noise," *J. Opt. Soc. Am.,* **5,** 342 (1969).

5.13 E. Parzen, *Stochastic Processes,* Holden-Day Publishing Company, San Francisco, 1962.

5.14 J. H. Altman and H. J. Zweig, "Effect of Spread Function on the Storage of Information on Photographic Emulsions," *Phot. Sci. Eng.,* **7,** 173 (1963).

5.15 A Papoulis, *Probability, Random Variables and Stochastic Processes,* McGraw-Hill, New York, 1965.

5.16 H. J. Zweig, G. C. Higgins, and D. L. MacAdam, "On the Information-Detecting Capacity of Photographic Emulsions," *J. Opt. Soc. Am.,* **48,** 926 (1958).

5.17 F. T. S. Yu, "Synthesis of an Optical Sound Spectrograph," *J. Acoust. Soc. Am.,* **51,** 433 (1972).

5.18 F. T. S. Yu, "Generating Speech Spectrograms Optically," *IEEE Spectrum,* **12,** 51 (1975).

5.19 F. T. S. Yu, "Optimal Linearization of Holography," *Appl. Opt.,* **12,** 2483 (1969).

5.20 F. T. S. Yu, "Linear Optimization in the Synthesis of Nonlinear Spatial Filters," *IEEE Trans. Inf. Theory,* **IT-17,** 524 (1971).

5.21 F. T. S. Yu and G. C. Kung, "Synthesis of Optimum Complex Spatial Filters," *J. Opt. Soc. Am.,* **61,** 147 (1972).

5.22 K. Biedermann, "A Function Characterizing Photographic Film that Directly Relates to Brightness of Holographic Images," *Optik,* **28,** 160 (1969).

5.23 A. Tai and F. T. S. Yu, "Classification of *T-E* Curve," *Appl. Opt.,* **17,** 2450 (1978).

5.24 E. L. O'Neill, *Introduction to Statistical Optics,* Addison-Wesley, Reading, Mass., 1963.

5.25 A. Tai, "Optical Spectrum Analysers and their Applications in Biomedical Instrumentations," Ph.D. Dissertation, Wayne State University, Detroit, Mich. 1978.

5.26 T. A. Shankoff, "Phase Holograms in Dichromated Gelatin," *Appl. Opt.,* **7,** 2101 (1968).

5.27 R. J. Collier, C. B. Burckhardt, and L. H. Lin, *Optical Holography,* Academic Press, New York, (1971).

5.28 R. G. Zech, "Data Storage in Volume Holograms," Ph.D. Dissertation, University of Michigan, Ann Arbor, Mich. (1974).

5.29 B. J. Chang and C. D. Leonard, "Dichromated Gelatin for the Fabrication of Holographic Optical Elements," *Appl. Opt.,* **18,** 2407 (1979).

5.30 H. Kogelnik, "Coupled Wave Theory for Thick Hologram Grating," *Bell Sys. Tech. J.,* **48** 2909 (1969).

5.31 A. Graube, "Holograms Recorded with Red Light in Dye-Sensitized Dichromated Gelatin," *Opt. Commun.*, **8,** 251 (1973).

5.32 B. J. Chang and K. Winick, "Silver-Halide Gelatin Holograms," *SPIE Proc.*, **215** (1980).

5.33 S. Iwasa and J. Feinleb, "The PROM Device in Optical Processing Systems," *Opt. Eng.*, **13,** 235 (1974).

5.34 W. S. Colburn, R. G. Zech, and L. M. Ralston, "Holographic Optical Elements," Technical Report AFAL-TR-72-409, Air Force Avionics Laboratory, (1973).

5.35 L. H. Lin and H. L. Beaucamp, "Write-Read-Erase in Situ Optical Memory Thermoplastic Hologram," *Appl. Opt.*, **9,** 2088 (1970).

5.36 T. D. Beard, W. P. Bleha, and S. Y. Wong, "AC Liquid-Crystal Light Valve," *Appl. Phys. Lett.*, **22,** 90 (1973).

5.37 J. Grinberg et al., "A New Real Time Non-Coherent to Coherent Light Image Converter: The Hybrid Field Liquid Crystal Light Valve," *Opt. Eng.*, **14,** 217 (1975).

5.38 B. D. Guenther, C. R. Christensen, and J. Upatnieks, "A Coherent Optical Processing: Another Approach," *IEEE J. Quant. Elec.*, **QE-15,** 1348 (1979).

5.39 A. D. Gara, "Real-Time Tracking of Moving Objects by Optical Correlation," *Appl. Opt.*, **18,** 172 (1979).

5.40 W. P. Bleha et al., "Application of the Liquid Crystal Light Valve to Real Time Optical Data Processing," *Opt. Eng.*, **17,** 371 (1978).

5.41 B. H. Soffer et al., "Optical Computing with Variable Grating Mode Liquid Crystal Devices," *SPIE,* **232** 47 (1980).

6

Fourier Transform Properties of Lenses and Optical Information Processing

We do not attempt to treat optical information processing in complete detail here. We do, however, cover a few examples to give a feeling for the principles involved. The reader who wishes to pursue the subject further is referred to the collections edited by Pollack et al. (ref. 6.1), by Tippett et al. (ref. 6.2), and by Caulfield (ref. 6.3). In these references many useful techniques not mentioned here will be found.

Mention must be made of a few important contributors to the field. It was in the early 1950s that the communication theory aspects of optical processing techniques became evident. The most important initial impact was provided by the classic papers "Fourier Treatment of Optical Processes," by Elias, Grey, and Robinson (ref. 6.4), and "Optics and Communication Theory," by Elias (ref. 6.5). However, the very first application of communication theory techniques to modern optics may be O'Neill's paper "Spatial Filtering in Optics" (ref. 6.6). From the broad interest in this new concept, a symposium, "Communication and Information Theory Aspects of Modern Optics," was held in 1960 (ref. 6.7). Since that time the application of communication theory to optics has commanded great interest. The potential applications of coherent spatial filtering were particularly evident in the field of radar signal processing, and it was in this field that Cutrona and his associates at the University of Michigan published "Optical Data Processing and Filtering Systems" in 1960 (ref. 6.8). This article stimulated additional interest in these techniques. The 1964 paper, "Signal Detection by Complex Spatial Filtering," by Vander Lugt (ref. 6.9) introduced the most interesting subject of optical character recognition. Since then, optical information processing has been shown to have a vast variety of applications. It is the combination of communication theory and optics that has stimulated interest in modern optical information processing, and has led to a large amount of interesting research.

Prior to discussing linear optical processing systems, we treat the important transform properties of lenses. A thorough discussion of the properties of lenses requires a lengthy presentation of the basic theory of geometrical optics, which is beyond the purpose of this text. However, we use a system theory point of view, which, although not coming directly from the principles of geometrical optics, is quite consistent with these principles in its results. At the beginning of this chapter, we discuss in detail the phase transforms of the lenses and their imaging properties.

6.1 PHASE TRANSFORMATION OF THIN LENSES

A lens is made of glass or some other transparent material. The index of refraction of a lens is usually greater than that of free space, in which case the velocity of wave propagation inside the lens is lower than the velocity in free space. Prior to going into the general discussion, we should specify what we mean by a *thin lens*. If a light ray entering at a point on one side of a lens emerges at about the same point on the other side of the lens, then the lens may be regarded as thin. That is, the transverse displacement of the ray of light inside a thin lens is negligible. Thus nothing more than a simple phase retardation takes place in a wave front passing through a thin lens. The amount of retardation is proportional to the thickness of the lens.

Referring to fig. 6.1, we may write the phase retardation across the wave front as

$$\phi(x, y) = k[z_0 + (\eta - 1)z(x, y)], \tag{6.1}$$

where $z(x, y)$ is the thickness variation of the lens, z_0 is the maximum thickness of the lens, η is the refractive index of the lens, and k is the wave number. It

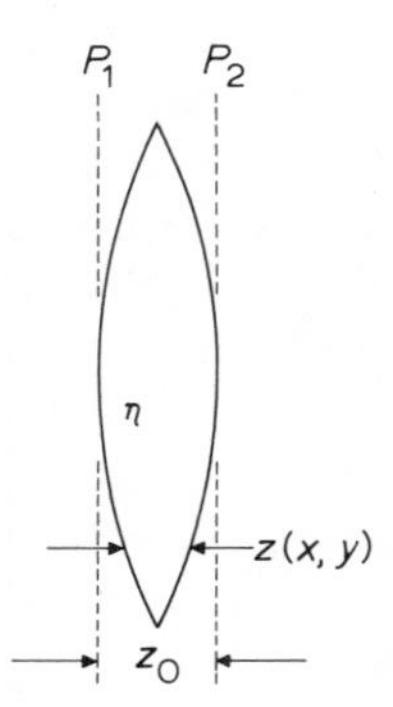

Fig. 6.1 Thickness variation of a convex lens.

is clear that $k\eta z(x, y)$ and $k[z_0 - z(x, y)]$ are the phase retardations due to the lens and free space, respectively. Consequently a thin lens can be represented by a general spatial phase transform.

$$T(x, y) = \exp[i\phi(x, y)] = \exp\{ik[z_0 + (\eta - 1)z(x, y)]\}. \tag{6.2}$$

Thus if the thin lens is illuminated by a monochromatic light source in which the complex light field distribution on the plane P_1 of fig. 6.1 is assumed to be $E(x, y)$, then the complex light field at the plane P_2, on the other side of the lens, can be written as

$$E'(x, y) = E(x, y)T(x, y). \tag{6.3}$$

In order to better understand the effects of the thickness variation in image formation, we can specify the form of the phase transformation for some practical lenses. The most widely used lenses are those with simple convex and concave surfaces. As an example, for a strictly convex lens (a positive lens, fig. 6.1) the phase transformation may be determined as follows. Let us assume that the radii of curvature of the two spherical surfaces are different and let us divide the lens into left and right halves, as shown in fig. 6.2. The thickness variation of the left half can be written

$$z_l(x, y) = z_{0l} - [R_l - (R_l^2 - \rho^2)^{1/2}] = z_{0l} - R_l\left\{1 - \left[1 - \left(\frac{\rho}{R_l}\right)^2\right]^{1/2}\right\}, \tag{6.4}$$

where z_{0l} is the maximum thickness of the left half, R_l is the radius of the curvature, and

$$\rho^2 = x^2 + y^2.$$

Similarly for the right half, the thickness variation can be written

$$\begin{aligned} z_r(x, y) &= z_{0r} - [R_r - (R_r^2 - \rho^2)^{1/2}] \\ &= z_{0r} - R_r\left\{1 - \left[1 - \left(\frac{\rho}{R_r}\right)^2\right]^{1/2}\right\}. \end{aligned} \tag{6.5}$$

The total thickness variation of the lens is the sum of eqs. (6.4) and (6.5):

$$\begin{aligned} z(x, y) &= z_l(x, y) + z_r(x, y) \\ &= z_0 - R_l\left\{1 - \left[1 - \left(\frac{\rho}{R_l}\right)^2\right]^{1/2}\right\} - R_r\left\{1 - \left[1 - \left(\frac{\rho}{R_r}\right)^2\right]^{1/2}\right\}, \end{aligned} \tag{6.6}$$

where $z_0 = z_{0l} + z_{0r}$.

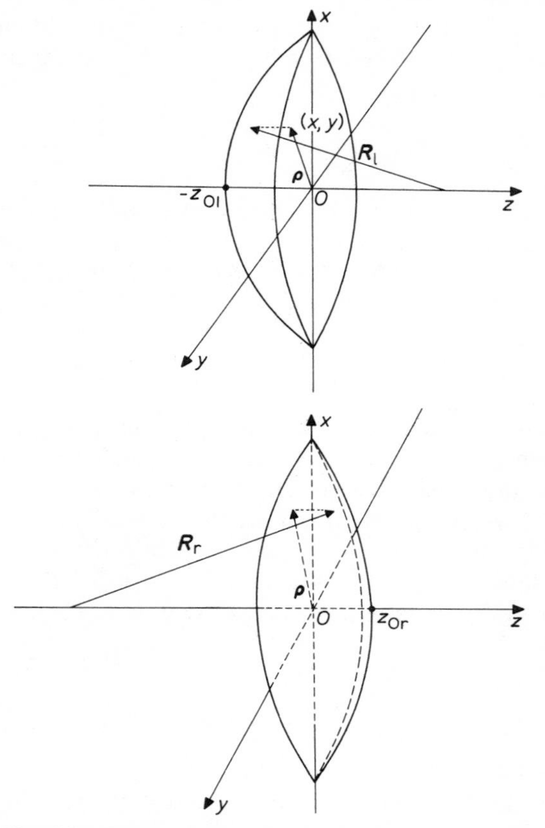

Fig. 6.2 Determination of the lens thickness variation.

Equation (6.6) may be simplified if it is confined to a relatively small region of the lens near the optical axis. With this restriction, the following approximations hold:

$$\left[1 - \left(\frac{\rho}{R_l}\right)^2\right]^{1/2} \simeq 1 - \frac{1}{2}\left(\frac{\rho}{R_l}\right)^2, \tag{6.7}$$

and

$$\left[1 - \left(\frac{\rho}{R_r}\right)^2\right]^{1/2} \simeq 1 - \frac{1}{2}\left(\frac{\rho}{R_r}\right)^2. \tag{6.8}$$

The thickness variation of the lens can then be approximated by

$$z(x, y) \simeq z_0 - \frac{\rho^2}{2}\left(\frac{1}{R_l} + \frac{1}{R_r}\right). \tag{6.9}$$

It may be noted that the paraxial approximations of eqs. (6.7) and (6.8) give approximately the same result as would be found if the spherical surfaces of the lens were replaced by parabolic surfaces.

From eqs. (6.9) and (6.2) the phase transformation of the convex lens is

$$T(x, y) = \exp(ik\eta z_0) \exp\left[-ik(\eta - 1)\frac{\rho^2}{2}\left(\frac{1}{R_1} + \frac{1}{R_r}\right)\right]. \tag{6.10}$$

Several of the quantities that pertain to the lens itself may be combined as follows:

$$f = \frac{R_1 R_r}{(\eta - 1)(R_1 + R_r)}. \tag{6.11}$$

Although it is not obvious at this point, the quantity f is actually the *focal length* of the lens, i.e., the distance from the lens center, on the axis, at which rays originally parallel to the axis converge to a point. Using eq. (6.11), eq. (6.10) may be written

$$T(x, y) = C_1 \exp\left(-i\frac{k}{2f}\rho^2\right), \tag{6.12}$$

where $C_1 = \exp(ik\eta z_0)$ is a complex constant.

Similarly, for a concave (i.e., negative) lens (fig. 6.3), the phase transformation can be shown to be

$$T(x, y) = C_2 \exp\left(i\frac{k}{2f}\rho^2\right), \tag{6.13}$$

where C_2 is a complex constant. Similar derivations may be carried out for the phase transformations for any other type of thin lens. Note that the essential

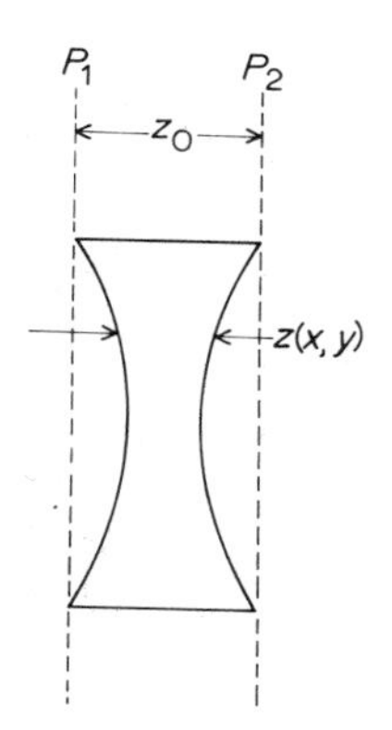

Fig. 6.3 Thickness variation of a concave lens.

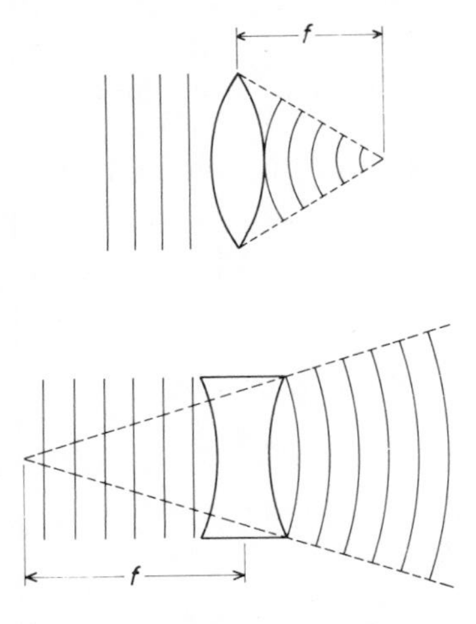

Fig. 6.4 Convergent and divergent effects of positive and negative lenses on an incident plane wave.

difference in the phase transformations of a convex and concave lens is in the positive and negative quadratic phase delays.

The significance of the phase transformation of lenses can be seen if it is assumed that a monochromatic plane wave field is incident normally on a convex lens. Then the complex light field immediately behind the lens (i.e., on P_2 of fig. 6.1) is

$$E'(x, y) = C \exp\left(-i\frac{k}{2f}\rho^2\right), \tag{6.14}$$

where C is a complex constant. This expression may be interpreted as a quadratic phase transform, which is the approximation to a spherical wave front. This spherical wave front converges toward a point of the optical axis at a distance f behind the lens. This confirms the previous statement that f is the focal length. On the other hand, if the lens is concave the wave front is diverging about a point on the optical axis at the distance f in front of the lens. These two cases of quadratic phase transformation can be illustrated in fig. 6.4. A convex lens is appropriately termed a *converging lens*, while a concave lens is a *diverging lens*.

The quadratic phase transformation has been derived under the paraxial approximation. Without paraxiality, the transformed wave front will depart from perfect sphericity, and various types of *aberration* will occur. Thus practical lenses are often corrected to be free from aberrations, at least to some degree. This correction may be accomplished by grinding the lens surfaces aspherically in order to improve the sphericity of the emerging wave front.

6.2 FOURIER TRANSFORM PROPERTIES OF LENSES

It is very useful that a two-dimensional Fourier transformation may be obtained from a positive lens. Fourier transform operations usually bring to mind complicated electronic spectrum analyzers or digital computers. However, this complicated transform can be performed extremely simply in a coherent optical system, and because the optical transform is two-dimensional, it has greater information capacity than transforms carried out by means of electronic circuitry.

Prior to the derivation of the Fourier transform properties of a lens, we will use Huygens' principle for the determination of the complex light field on a planar surface P_2 due to a planar light source P_1, as shown in fig. 6.5.

Let us assume that the complex light field at P_1 is $f(x, y)$. Then the complex light distribution at P_2 may be determined by means of Huygens' principle:

$$g(\alpha, \beta) = C \iint_S f(x, y) \exp(ikr)\, dx\, dy, \tag{6.15}$$

where S denotes the surface integral, C is an arbitrary complex constant, $k = 2\pi/\lambda$, and

$$r = [l^2 + (\alpha - x)^2 + (\beta - y)^2]^{1/2}. \tag{6.16}$$

If it is assumed that the separation of the (x, y) and (α, β) coordinate systems is large compared with the spatial dimensions of P_1 and P_2, then r can be paraxially approximated,

$$r \simeq l + \frac{1}{2l}[(\alpha - x)^2 + (\beta - y)^2]. \tag{6.17}$$

The complex light field at P_2 is therefore

$$g(\alpha, \beta) = C \iint_S f(x, y) \exp(ikl) \exp\left\{ i\frac{\pi}{\lambda l}[(\alpha - x)^2 + (\beta - y)^2] \right\} dx\, dy. \tag{6.18}$$

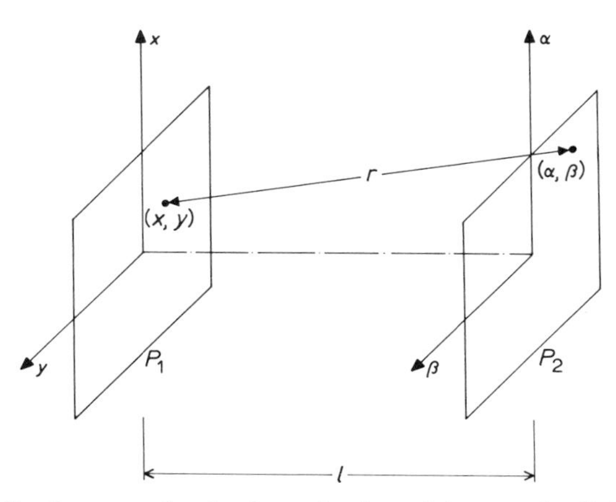

Fig. 6.5 Geometry for the determination of the complex light field.

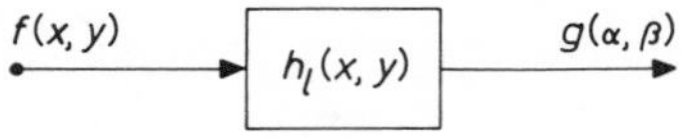

$f(x, y)$ → $h_l(x, y)$ → $g(\alpha, \beta)$

Fig. 6.6 Input-output linear system analog of the optical setup shown in fig. 6.5.

This can be written in the simplified form,

$$g(\alpha, \beta) = C_1 f(x, y) * h_l(x, y), \tag{6.19}$$

where C_1 is an arbitrary complex constant, $*$ denotes the spatial convolution, and

$$h_l(x, y) = \exp\left[i\frac{\pi}{\lambda l}(x^2 + y^2)\right]$$

is known as the spatial impulse response. Equations (6.18) or (6.19) define the input-output linear system analog of fig. 6.5, as shown in fig. 6.6.

An additional optical element is required to obtain Fourier transformation by a lens. If the lens is positive, and a point source of monochromatic light is placed at the front focal point, then the light passing through the lens is collimated into a monochromatic plane wave (fig. 6.7). This output wave is the spatial Fourier transform of the point source.

Thus in fig. 6.5, a point source in the plane P_1 is equivalent to a spatial delta function $\delta(x, y)$. A positive lens is placed in the plane P_2, and the separation of the coordinate system is made equal to the focal length of the lens. Then the complex light field in front of the lens is

$$g(\alpha, \beta) = C_1 \exp\left[i\frac{\pi}{\lambda f}(\alpha^2 + \beta^2)\right]. \tag{6.20}$$

The action of the lens is to make the spherical wave field into a plane wave field. This is merely to say that the lens must introduce a phase transformation

$$T(\alpha, \beta) = \exp\left[-i\frac{\pi}{\lambda f}(\alpha^2 + \beta^2)\right], \tag{6.21}$$

so that the complex light field behind the lens is

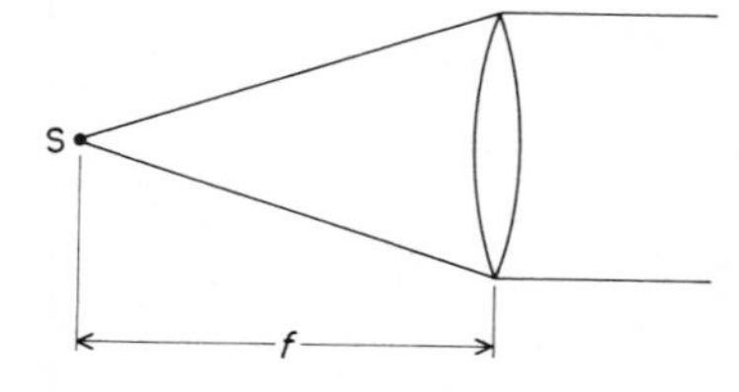

Fig. 6.7 Fourier transform of a monochromatic point source S.

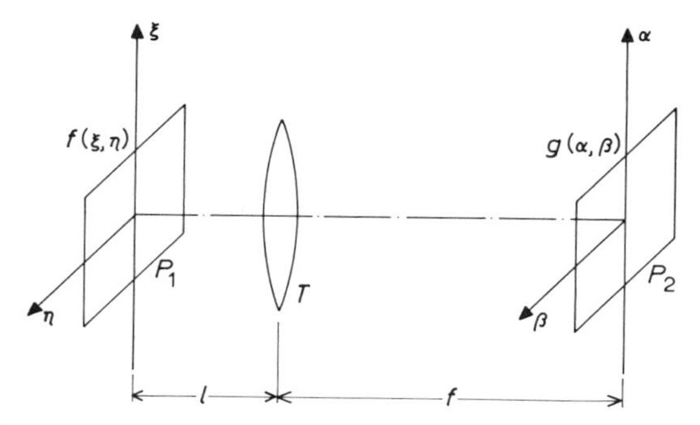

Fig. 6.8 Geometry for the determination of the optical Fourier transformation.

$$g_1(\alpha, \beta) = C_1, \tag{6.22}$$

a wave field parallel to the (α, β) plane.

Let us consider fig. 6.8, a simple optical system. If the complex light field at P_1 is $f(\xi, \eta)$, then the complex light distribution at P_2 can be written as

$$g(\alpha, \beta) = C\{[f(\xi, \eta) * h_l(\xi, \eta)]T(x, y)\} * h_f(x, y), \tag{6.23}$$

where C is again an arbitrary complex constant, $h_l(\xi, \eta)$ and $h_f(x, y)$ are the corresponding spatial impulse responses, and $T(x, y)$ is the phase transformation of the lens.

The linear system analog for eq. (6.23) is shown in fig. 6.9. Equation (6.23) may be written in the integral form

$$g(\alpha, \beta) = C \iint_{S_1} \left[\iint_{S_2} \exp\left(i \frac{k}{2} \Delta \right) dx\, dy \right] f(\xi, \eta)\, d\xi\, d\eta, \tag{6.24}$$

where S_1 and S_2 denote the surface integrals of the light field P_1 and the lens T, respectively, and

$$\Delta \triangleq \left\{ \frac{1}{l} \left[(x - \xi)^2 + (y - \eta)^2 \right] + \frac{1}{f} [(\alpha - x)^2 + (\beta - y)^2 - (x^2 + y^2)] \right\}. \tag{6.25}$$

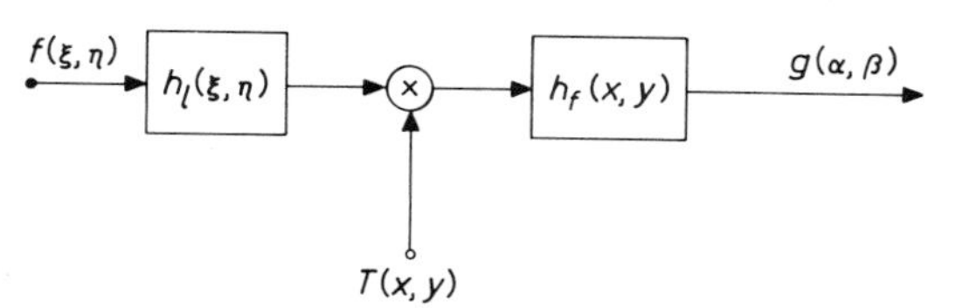

Fig. 6.9 Linear system analog of the optical setup shown in fig. 6.8.

It may be noted that the surface integral of the lens may be assumed to be of infinite extent, since the lens is very large compared to the spatial apertures at P_1 and P_2 (paraxiality).

Equation (6.25) may be written as

$$\Delta = \frac{1}{f}[v\xi^2 + vx^2 + \alpha^2 - 2v\xi x - 2x\alpha + v\eta^2 + vy^2 + \beta^2 - 2v\eta y - 2y\beta], \tag{6.26}$$

where $v = f/l$.

By completing the square, eq. (6.26) can be written as

$$\Delta = \frac{1}{f}\Bigg[(v^{1/2}x - v^{1/2}\xi - v^{-1/2}\alpha)^2 - \alpha^2\left(\frac{1-v}{v}\right) - 2\xi\alpha + (v^{1/2}y - v^{1/2}\eta - v^{-1/2}\beta)^2 - \beta^2\left(\frac{1-v}{v}\right) - 2\eta\beta\Bigg]. \tag{6.27}$$

By using this in eq. (6.24) we have

$$g(\alpha, \beta) = C \exp\left[-i\frac{k}{2f}\left(\frac{1-v}{v}\right)(\alpha^2 + \beta^2)\right] \times \iint_{S_1} f(\xi,\eta) \exp\left[-i\frac{k}{f}(\alpha\xi + \beta\eta)\right] d\xi\, d\eta \times \iint_{S_2} \exp\left\{i\frac{k}{2f}[(v^{1/2}x - v^{1/2}\xi - v^{-1/2}\alpha)^2 + (v^{1/2}y - v^{1/2}\eta - v^{-1/2}\beta)^2]\right\} \times dx\, dy. \tag{6.28}$$

Since the integrations over S_2 are assumed to be taken from $-\infty$ to $+\infty$, we obtain a complex constant that can be incorporated with C. Thus we have

$$g(\alpha, \beta) = C_1 \exp\left[-i\frac{k}{2f}\left(\frac{1-v}{v}\right)(\alpha^2 + \beta^2)\right] \times \iint_{S_1} f(\xi,\eta) \exp\left[-i\frac{k}{f}(\alpha\xi + \beta\eta)\right] d\xi\, d\eta. \tag{6.29}$$

From this it is clear that, except for a spatial quadratic phase variation, $g(\alpha, \beta)$ is the Fourier transform of $f(\xi,\eta)$. As a matter of fact, the quadratic phase factor vanishes if $l = f$. Evidently if the signal plane P_1 is placed at the front focal plane of the lens, the quadratic phase factor disappears, which leaves an exact Fourier transform relation. Thus eq. (6.29) may be written as

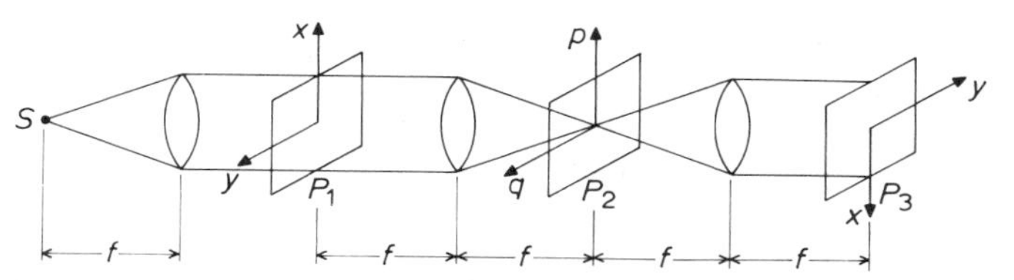

Fig. 6.10 Successive Fourier transformations of lenses. A monochromatic point source is located at S.

$$G(p, q) = C_1 \iint_{S_1} f(\xi, \eta) \exp\left[-i(p\xi + q\eta)\right] d\xi\, d\eta \qquad \text{for} \quad v = 1, \tag{6.30}$$

where $p = k\alpha/f$ and $q = k\beta/f$ are the spatial frequency coordinates.

It must be emphasized that the exact Fourier transform relation takes place under the condition $l = f$. Under the condition $l \neq f$, a quadratic phase factor will be included. Furthermore, it can easily be shown that a quadratic phase factor also results if the signal plane P_1 is placed behind the lens.

In conventional Fourier transform theory, the transformation from the spatial domain to the spatial frequency domain requires the kernel $\exp[-i(px + qy)]$ and the transformation from the spatial frequency domain to the spatial domain requires the conjugate kernel $\exp[i(px + qy)]$. Obviously, a positive lens always introduces the kernel $\exp[-i(px + qy)]$. Therefore, in an optical system we take only successive transforms, rather than a transform followed by its inverse, as shown in fig. 6.10.

6.3 OPTICAL IMAGE FORMATION

In this section we develop some equations for a general optical imaging system. Image formation is considered for the two extremes, i.e., for the completely incoherent and the completely coherent cases.

In fig. 6.11, a hypothetical optical imaging system is shown. Assume that the light emitted by the source Σ is monochromatic, and suppose that an image of the input signal is formed at the output plane of the optical system. In order to demonstrate the image formation of the optical system, we let $u(x, y)$ be the

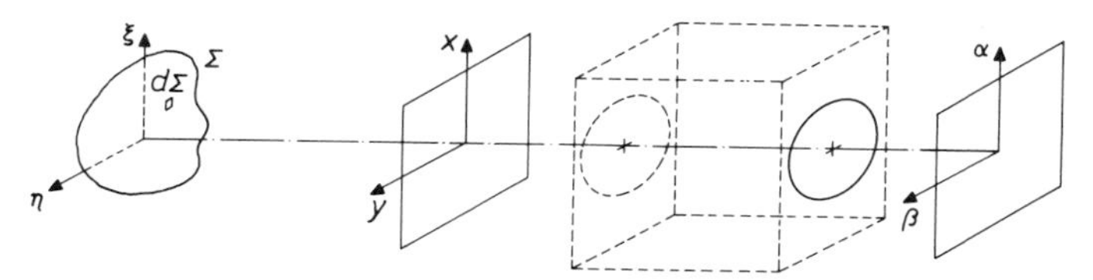

Fig. 6.11 Hypothetical optical imaging system. The black-box optical system is shown lying between the input or signal plane (x, y) and the output or image plane (α, β).

complex light amplitude distribution on the input signal plane due to an incremental light source $d\Sigma$. If the transmittance of the input plane is $f(x, y)$, then the complex light field immediately behind the signal plane is $u(x, y)\, f(x, y)$.

If it is assumed that the optical system in the black box is linearly spatially invariant with a spatial impulse response of $h(x, y)$, then the complex light field at the output plane of the system due to $d\Sigma$ can be determined by the convolution equation

$$g(\alpha, \beta) = u(x, y)f(x, y) * h(x, y). \tag{6.31}$$

At this point it may be emphasized that the assumption of linearity of the optical system is generally valid for small amplitude disturbances; however, the spatial-invariance condition may be applicable only over a small region of the signal plane.

From eq. (6.31), the irradiance in the image plane due to $d\Sigma$ is

$$dI(\alpha, \beta) = g(\alpha, \beta)g^*(\alpha, \beta)\, d\Sigma. \tag{6.32}$$

Therefore the total irradiance of the image due to the whole light source is

$$I(\alpha, \beta) = \iint_{\Sigma} |g(\alpha, \beta)|^2\, d\Sigma, \tag{6.33}$$

which can be written out as the convolution integral

$$I(\alpha, \beta) = \iiiint_{-\infty}^{\infty} \Gamma(x, y; x', y')h(\alpha - x, \beta - y)h^*(\alpha - x', \beta - y') \times f(x, y)f^*(x', y')\, dx\, dy\, dx'\, dy', \tag{6.34}$$

where

$$\Gamma(x, y; x', y') = \iint_{\Sigma} u(x, y)u^*(x', y')\, d\Sigma. \tag{6.35}$$

Now we choose two points Q_1 and Q_2 on the input signal plane; Q_1 is taken as the origin of the (x, y) plane, and Q_2 is arbitrary. If r_1 and r_2 are the respective distances from Q_1 and Q_2 to $d\Sigma$, then the complex disturbances at Q_1 and Q_2 due to $d\Sigma$ are, respectively,

$$u_1(x, y) = \frac{[I(\xi, \eta)]^{1/2}}{r_1} \exp(ikr_1) \tag{6.36}$$

and

$$u_2(x, y) = \frac{[I(\xi,\eta)]^{1/2}}{r_2} \exp(ikr_2), \tag{6.37}$$

where $I(\xi,\eta)$ is the irradiance across the light source. By substituting eqs. (6.36) and (6.37) in eq. (6.35) we have

$$\Gamma(x, y) = \iint_{\Sigma} \frac{I(\xi,\eta)}{r_1 r_2} \exp[ik(r_1 - r_2)]\, d\Sigma. \tag{6.38}$$

In the paraxial case, $r_1 - r_2$ may be approximated by

$$r_1 - r_2 \simeq \frac{2}{r_1 + r_2}(\xi x + \eta y) \simeq \frac{1}{r}(\xi x + \eta y), \tag{6.39}$$

where r is the separation between the light source plane and the signal plane. Then eq. (6.38) can be reduced to

$$\Gamma(x, y) = \frac{1}{r^2} \iint_{\Sigma} I(\xi,\eta) \exp\left[i\frac{k}{r}(\xi x + \eta y)\right] d\xi\, d\eta. \tag{6.40}$$

Equation (6.40) is the inverse Fourier transform of the source irradiance.

Now one of the two extreme cases of the hypothetical optical imaging system can be seen by letting the light source become infinitely large. If the irradiance of the source is relatively uniform, that is $I(\xi, \eta) \simeq K$, eq. (6.40) becomes

$$\Gamma(x, y) = K_1 \delta(x, y), \tag{6.41}$$

where K_1 is an appropriate positive constant. This equation describes a completely incoherent optical imaging system.

On the other hand, if the light source is vanishingly small, eq. (6.40) becomes

$$\Gamma(x, y) = K_2, \tag{6.42}$$

where K_2 is a positive constant. This equation in fact describes a completely coherent optical imaging system.

Referring to eq. (6.34), for the completely incoherent case $[\Gamma(x, y) = K_1\delta(x, y)]$, the irradiance at the output is

$$I(\alpha, \beta) = \iiiint_{-\infty}^{\infty} \delta(x' - x, y' - y) h(\alpha - x, \beta - y) \times h^*(\alpha - x', \beta - y') f(x, y) f^*(x', y')\, dx\, dy\, dx'\, dy', \tag{6.43}$$

which can be reduced to

$$I(\alpha, \beta) = \iint_{-\infty}^{\infty} |h(\alpha - x, \beta - y)|^2 |f(x, y)|^2 \, dx \, dy. \tag{6.44}$$

It is clear from eq. (6.44) that for the incoherent case the image irradiance is the convolution of the signal irradiance with respect to the impulse response irradiance. In other words, for the completely incoherent case, the optical system is linear in irradiance, i.e.,

$$I(\alpha, \beta) = |h(x, y)|^2 * |f(x, y)|^2. \tag{6.45}$$

By Fourier transformation, eq. (6.45) can be expressed in the spatial frequency domain:

$$I(p, q) = |H(p, q)|^2 |F(p, q,)|^2, \tag{6.46}$$

where $I(p, q)$, $H(p, q)$, and $F(p, q)$ are the Fourier transforms of $I(\alpha, \beta)$, $h(x, y)$, and $f(x, y)$, respectively, and p and q are the spatial frequency coordinates.

In a more convenient form, eq. (6.45) can be written as

$$I(\alpha, \beta) = h_i(x, y) * f_i(x, y), \tag{6.47}$$

where $h_i(x, y) = |h(x, y)|^2$ and $f_i(x, y) = |f(x, y)|^2$ are the irradiance impulse responses of the system. Then eq. (6.46) may be written as

$$I(p, q) = H_i(p, q) F_i(p, q), \tag{6.48}$$

where $H_i(p, q)$ and $F_i(p, q)$ are the Fourier transforms of $h_i(x, y)$ and $f_i(x, y)$, respectively. In fact, $H_i(p, q)$ is given by

$$H_i(p, q) = \iint_{-\infty}^{\infty} h(x, y) h^*(x, y) \exp[-i(px + qy)] \, dx \, dy. \tag{6.49}$$

Then by the Fourier multiplication theorem, eq. (6.49) becomes

$$H_i(p, q) = \frac{1}{4\pi^2} \iint_{-\infty}^{\infty} H(p', q') H^*(p' - p, q' - q) \, dp' \, dq', \tag{6.50}$$

which is the convolution of the complex transfer function with respect to its conjugate. On the other hand, for the completely coherent case $[\Gamma(x, y) = K_2]$, eq. (6.34) becomes

$$I(\alpha, \beta) = g(\alpha, \beta) g^*(\alpha, \beta) = \iint_{-\infty}^{\infty} h(\alpha - x, \beta - y) f(x, y) \, dx \, dy$$

$$\times \iint_{-\infty}^{\infty} h^*(\alpha - x', \beta - y') f^*(x', y') \, dx' \, dy'. \tag{6.51}$$

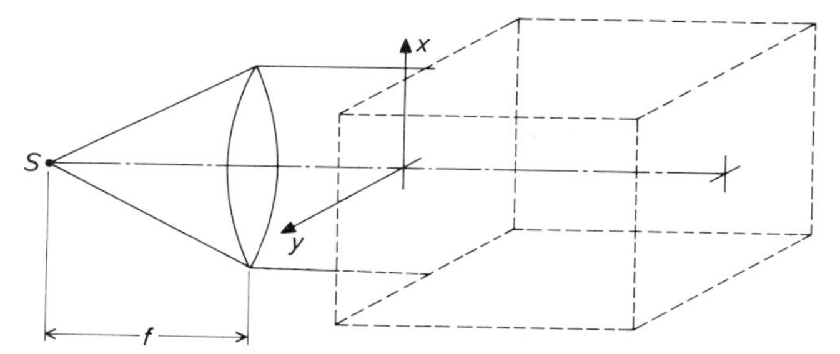

Fig. 6.12 A simple optical configuration to achieve $\Gamma(x, y) = K$. A monochromatic point source is located at S.

From eq. (6.51) it is obvious that the optical system is linear in complex amplitude, i.e., that

$$g(\alpha, \beta) = \iint_{-\infty}^{\infty} h(\alpha - x, \beta - y) f(x, y)\, dx\, dy. \tag{6.52}$$

Again by Fourier transformation eq. (6.52) becomes

$$G(p, q) = H(p, q) F(p, q), \tag{6.53}$$

where $G(p, q)$, $H(p, q)$, and $F(p, q)$ are the corresponding Fourier transforms of $g(\alpha, \beta)$, $h(x, y)$, and $f(x, y)$, respectively.

A coherence-preserving optical system that makes $\Gamma(x, y) = K$ (a constant) can be achieved by the configuration of fig. 6.12. In this figure, a source of monochromatic light is located at the front focal point of a positive lens. As a result, a collimated planewave illuminates the signal plane (x, y) of the system.

6.4 BASIC INCOHERENT OPTICAL INFORMATION PROCESSING SYSTEMS

In this section we discuss optical information processing of an incoherent source. Let a signal transparency $f_1(x, y)$ be located at the input signal plane P_1 of an optical system (fig. 6.13). If the transparency is illuminated by an incoherent light source so that the irradiance at P_1 is uniform within the region of interest, then the irradiance immediately behind the transparency is

$$I(x, y) = I_0 f_1(x, y), \tag{6.54}$$

where I_0 is the uniform irradiance that illuminates the signal transparency.

If the input signal transparency is imaged on a one-to-one basis into another signal transparency $f_2(x, y)$ at P_2, then the irradiance immediately behind the second transparency can be written as

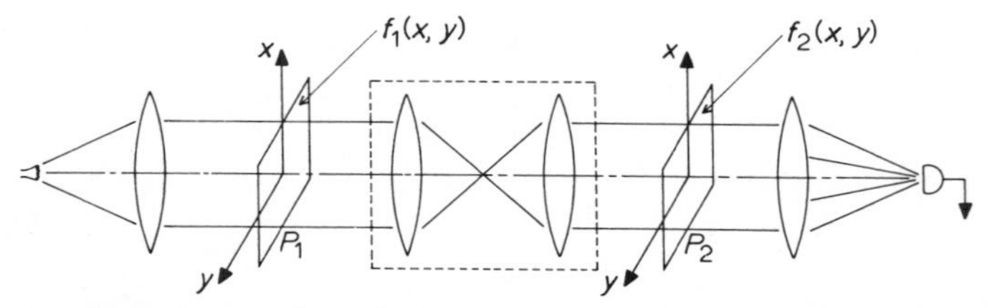

Fig. 6.13 Incoherent processing of signal multiplication and integration. The filament represents an incoherent source, and the imaging system is shown enclosed in dashed lines; a photodiode is shown at the right.

$$I(x, y) = I_0[f_1(x, y) * h_i(x, y)]f_2(x, y). \tag{6.55}$$

The irradiance impulse response of the optical system $h_i(x, y)$ is given, as usual, by $h_i(x, y) = |h(x, y)|^2$, where $h(x, y)$ is the complex amplitude spatial impulse response.

If the spatial frequency response of the signal $f_1(x, y)$ is within the limits of the spatial frequency response of the optical system, then the spatial impulse response of the system may be approximated by a Dirac delta function and the irradiance behind P_2 may be approximated by

$$I(x, y) = I_0 f_1(x, y) f_2(x, y), \tag{6.56}$$

i.e., the output irradiance is, in this approximation, proportional to the product $f_1(x, y)f_2(x, y)$.

The integration of the two processing signals can be accomplished incoherently by focusing the output light on a photodetector, as shown in fig. 6.13. The output voltage from the photodetector is

$$V = K \iint_S f_1(x, y) f_2(x, y)\, dx\, dy,$$

where K is a proportionality constant, and S denotes surface integration over the output plane P_2.

The integration process of fig. 6.13 can be modified to achieve a multichannel multiplication integration, as shown in fig. 6.14. Then the voltages at the outputs of the photodetectors are

$$V(x) = K \int_l f_1(x, y) f_2(x, y)\, dy, \tag{6.57}$$

where K is a proportionality constant, and l denotes the line integral in the y direction of the signal for a fixed x.

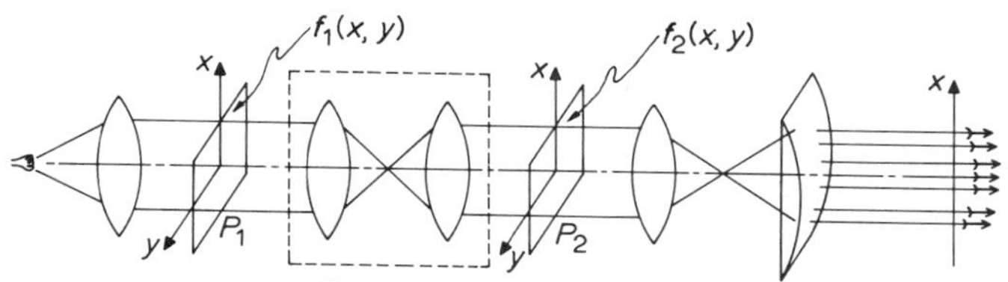

Fig. 6.14 Incoherent processing of multichannel signal multiplication and integration. An array of linear photodetectors is shown at the right.

It may be obvious that, if we translate one of the processing signals, for instance, $f_2(x, y)$, along the y axis, then the signal detected by the array of photodetectors may be written,

$$V(x, a) = K \int_l f_1(x, y) f_2(x, y - a)\, dy, \tag{6.58}$$

where a is the translation distance.

The processing signals we have considered so far are positive real. This is because a transparency is a passive device whose transmittance T is bounded:

$$0 \le T(x, y) \le 1. \tag{6.59}$$

However, if the processing signal contains positive and negative peaks, its realization can be accomplished by using an appropriate bias level and scaling factor. In this case, the transmittance of the signal is given by

$$T(x, y) = K_0 + K_1 f(x, y), \tag{6.60}$$

where K_0 and K_1 are the bias level and scaling factor, and $T(x, y)$ satisfies the transmittance condition of eq. (6.59).

If the processing signal satisfies eq. (6.60) under the constraint of eq. (6.59), then the output integral can be written as

$$V = K \iint_S [K_0 + K_1 f_1(x, y)][K_0' + K_1' f_2(x, y)]\, dx\, dy, \tag{6.61}$$

where K is an appropriate constant. It is apparent that error terms due to the bias levels are introduced into the evaluation of the integration. Even though the values of the bias levels and the scaling factor are known, the removal of these errors may involve variation of the processing signals $f_1(x, y)$ and $f_2(x, y)$. Often, however, a remedy will be obtained by use of the coherent processing to be described in the following section.

6.5 COHERENT OPTICAL INFORMATION PROCESSING SYSTEMS

Let us consider a general coherent optical information processing system (fig. 6.15). If a monochromatic point source is located at the front focal length of a collimating lens L_1, then a monochromatic plane wave will illuminate the input signal plane P_1. A processing signal $f(x, y)$ is inserted at P_1, and the complex light field at the output plane P_3 may be written as

$$g(\alpha,\beta) = Kf(x, y) * h(x, y), \tag{6.62}$$

where K is an appropriate constant and $h(x, y)$ is the spatial impulse response of the system.

If we assume that the spatial frequency of the processing signal $f(x, y)$ lies within the spatial frequency limit of the optical system, then the spatial impulse response $h(x, y)$ may be approximated by the Dirac delta function, so that the output light field becomes

$$g(\alpha, \beta) = Kf(x, y), \tag{6.63}$$

which is proportional to the input processing signal. Recalling the Fourier transform properties of lenses (see sec. 6.2), we see that the complex light amplitude distribution on the plane P_2 is proportional to $F(p, q)$, where $F(p, q)$ is the Fourier transform of the input processing signal $f(x, y)$. The irradiance on P_2 is therefore proportional to $|F(p, q)|^2$.

Several applications of elementary coherent optical systems are described on the following pages.

Coherent Optical Spectrum Analysis

If the coherent optical system shown in fig. 6.15 is termianted at P_2, as shown in fig. 6.16, then the system is essentially a two-dimensional sepctrum analyzer. The complex light field at P_2 is

$$E(p, q) = K\iint_S f(x, y) \exp[-i(xp + yq)]\, dx\, dy = KF(p, q), \tag{6.64}$$

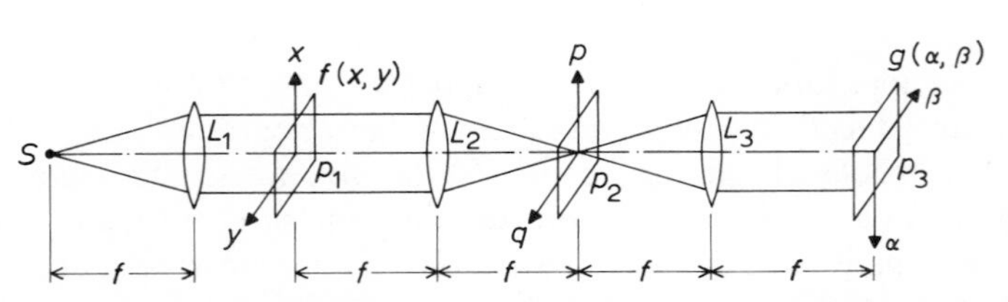

Fig. 6.15 A coherent optical information processing system. Here and in the remaining illustrations of this chapter, S is a monochromatic point source.

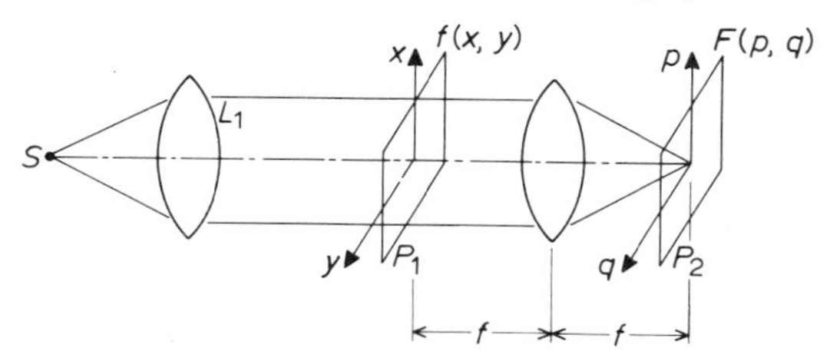

Fig. 6.16 A coherent spectrum analyzer.

where K is a proportionality constant, S denotes surface integration over the input plane P_1, the factor $F(p, q)$ is the two-dimensional Fourier transform of the input procession $f(x, y)$, and p and q are the spatial frequency coordinates. The corresponding irradiance is therefore

$$I(p, q) = E(p, q)E^*(p, q) = K^2|F(p, q)|^2, \tag{6.65}$$

which is proportional to the power spectrum of the processing signal.

The optical spectrum analyzer of fig. 6.16 can be modified to yield a multichannel one-dimensional spectrum analyzer by the addition of a cylindrical lens, as shown in fig. 6.17. If the input multichannel signal is $f(x_i, y)$, then the complex light field at the output plane of the analyzer is

$$E(x_i, q) = K \exp\left[-i\frac{k}{2f}\left(\frac{1-v}{v}\right)q^2\right]\int f(x_i, y)e^{-iqy}\,dy,$$
$$i = 1, 2, \ldots, n, \tag{6.66}$$

where $v = f/l$, and subscript i denotes the corresponding channel in the x axis.

But since $l = 3f$, i.e., $v = 1/3$, eq. (6.66) becomes

$$E(x_i, q) = K \exp\left(-i\frac{k}{f}q^2\right)F(x_i, q), \qquad n = 1, 2, \ldots, n, \tag{6.67}$$

where $F(x_i, q)$ is the one-dimensional Fourier transform of $f(x_i, y)$.

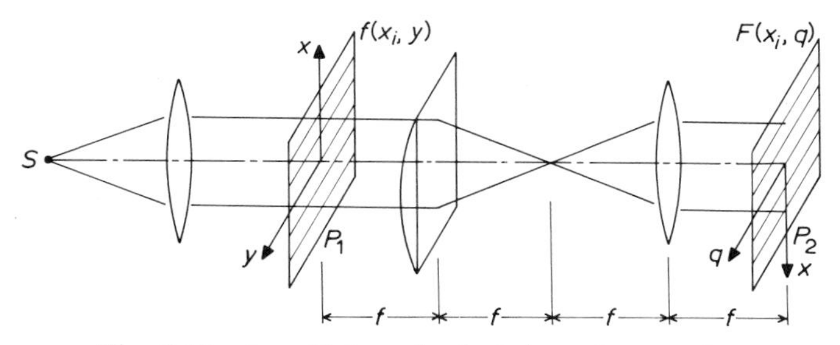

Fig. 6.17 A multichannel coherent spectrum analyzer.

The corresponding irradiance is therefore

$$I(p, q) = E(x_i, q)E^*(x_i, q) = K^2|F(x_i, q)|^2, \qquad i = 1, 2, \ldots, n. \tag{6.68}$$

Spatial Frequency Domain Synthesis

It is possible to synthesize a desired linear-filtering operation in the spatial frequency domain. Such a spatial frequency filter consists of a transparency located at the spatial frequency domain of a coherent optical system, as shown in fig. 6.18. In fact, the desired filter can be synthesized by direct manipulation of the complex amplitude transmittance across the frequency domain. If the processing signal $f(x, y)$ is inserted at the spatial domain P_1, then a Fourier transform of the input signal is distributed at the spatial frequency domain P_2. If a filter of transparency $H(p, q)$ is inserted at P_2, then the complex light field immediately behind P_2 is

$$E(p, q) = KF(p, q)H(p, q), \tag{6.69}$$

where K is a proportionality constant.

The filter has, in general, a complex amplitude transmittance

$$H(p, q) = |H(p, q)| \exp[i\phi(p, q)], \tag{6.70}$$

satisfying the physically realizable conditions,

$$|H(p, q)| \leq 1 \tag{6.71}$$

and

$$0 \leq \phi(p, q) \leq 2\pi. \tag{6.72}$$

Such a transmittance function may be represented by a set of points within or on a unit circle in the complex plane, as shown in fig. 6.19. The amplitude transmission of the filter changes with the optical density, and the phase delay varies with the thickness.

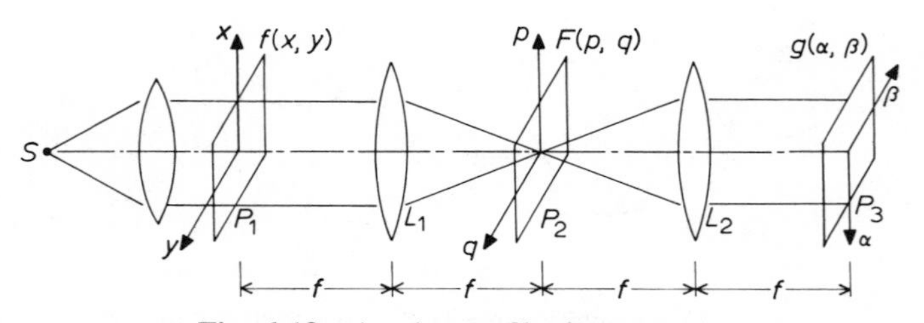

Fig. 6.18 A coherent filtering system.

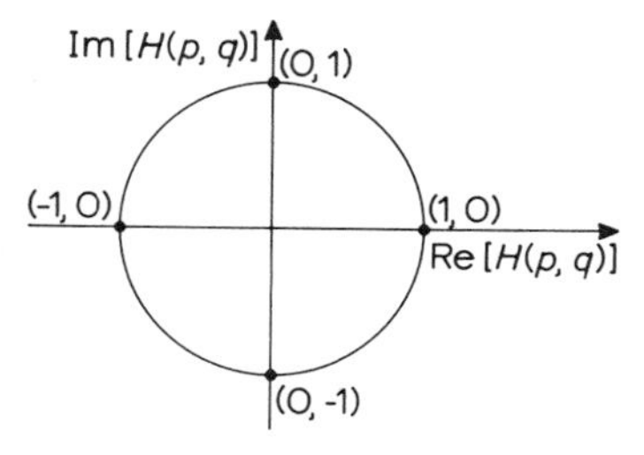

Fig. 6.19 A complex amplitude transmittance function.

The second lens L_2 of fig. 6.18 takes an inverse Fourier transformation of the complex light field $E(p, q)$ to the output plane P_3, such that the complex-amplitude light distribution across P_3 is

$$g(\alpha, \beta) = K\iint_S F(p, q)H(p, q) \exp[i(p\alpha + q\beta)]\, dp\, dq, \qquad (6.73)$$

where the surface integration is taken over the spatial frequency domain P_2.
Alternatively, by the Fourier multiplication theorem, eq. (6.73) can be written as

$$g(\alpha, \beta) = K\iint_S f(x, y)h(\alpha - x, \beta - y)\, dx\, dy = Kf(x, y) * h(x, y), \qquad (6.74)$$

where the integral is taken over the input spatial domain, and $h(x, y)$ is the spatial impulse response of the filter.

It may be pointed out that the frequency domain filter $H(p, q)$ can consist of apertures or slits of any shape. Depending on the arrangement of apertures, there results a low-pass, high-pass, or bandpass spatial filter. It is clear that any opaque portion in the filter represents a spatial frequency-band rejection. Note also that the inclusion of a phase plate causes a phase delay in the phase-plate portion of the filter. Complex filter synthesis is realizable since we have complete freedom to synthesize the amplitude and the phase responses independently. A technique introduced by Vander Lugt for the synthesis of complex spatial filters is given in sec. 6.6.

Spatial Domain Synthesis

Instead of inserting a complex filter $H(p, q)$ in the spatial frequency plane P_2 (fig. 6.18), put a complex transparency $h(x, y)$ in the input plane P_1. Then the complex light field at P_2 is

$$E(p, q) = K\iint_S f(x, y)h(x, y) \exp[-i(px + qy)]\, dx\, dy, \qquad (6.75)$$

where K is a proportionality constant.

If the processing signal is made to translate in P_1, then eq. (6.75) can be written

$$E(p, q) = K\iint_S f(x - \alpha, y - \beta)h(x, y) \exp[-i(px + qy)]\, dx\, dy, \tag{6.76}$$

where α and β are the variables corresponding to the x and y lateral displacements. It is clear now that, if the measurement of the light field takes place at $p = q = 0$, then the integral of eq. (6.76) can be written

$$g(\alpha, \beta) = K\iint_S f(x - \alpha, y - \beta)h(x, y)\, dx\, dy, \tag{6.77}$$

which we recognize as the crosscorrelation function of $f(x, y)$ and $h(x, y)$. Thus

$$g(\alpha, \beta) = Kf(x, y) \circledast h(x, y). \tag{6.78}$$

On the other hand, if the spatial coordinates of the input processing signal $f(x, y)$ are reversed, and then translated laterally in the input spatial domain, such that the input signal $f(x - \alpha, y - \beta)$ is replaced by $f(\alpha - x, \beta - y)$, then at the origin of P_2 the complex light field is

$$g(\alpha, \beta) = K\iint_S f(\alpha - x, \beta - y)h(x, y)\, dx\, dy. \tag{6.79}$$

This is the convolution integral of $f(x, y)$ and $h(x, y)$. It is identical to eq. (6.74).

In this section, we have discussed the two methods of synthesis that are available for coherent optical information processing, namely:

1 Frequency domain synthesis, in which a complex spatial filter is introduced in the frequency domain, and the filtering operation is performed directly on the complex frequency spectrum.

2 Spatial domain synthesis, in which a complex reference function is introduced in the input spatial domain, and the filtering operation is performed directly on the processing signal.

Theoretically, these two techniques of synthesis yield essentially the same result. However, in practice the spatial domain synthesis requires a scanning mechanism, and thus more instrumentation; operations also take longer than in scanner-free frequency domain synthesis. It may sometimes be advantageous to mix these two syntheses together; then the output signal may be written

$$g(\alpha, \beta) = \mathscr{F}^{-1}\{H(p, q)\mathscr{F}[f(x, y)h(x, y)]\}, \tag{6.80}$$

where $\mathscr{F}$ and $\mathscr{F}^{-1}$ denote the direct and inverse Fourier transforms. $H(p, q)$ is the complex filter function in the frequency domain, and $h(x, y)$ is the complex reference function in the input spatial domain.

Multichannel Frequency Domain Synthesis

One of the most important features of the coherent information processing system is that of two-dimensionality. This feature is useful when the processing signals are functions of two variables. The signal can be displayed in two spatial dimensions and then both variables can be processed simultaneously. On the other hand, if the same signal were to be processed electronically, some sort of scanning technique would be required.

Occasionally the processing signals are functions of one variable, and the additional capacity of the optical system may not be needed. In this case, the free coordinate of the optical system may be converted into a sequence of one-dimensional channels, such that a finite number of one-variable processing signals may be processed simultaneously. A multichannel system is shown in fig. 6.20. The input processing signal $f(x, y)$ is composed of a finite number of one-dimensional signals, such that

$$f(x, y) = f(x_i, y), \qquad i = 1, 2, \ldots, n, \tag{6.81}$$

where x_i represents the ith channel on the x axis, and n is the total number of channels. It may be emphasized that the practical limit of n is the highest number of stripe elements that the optical system can resolve. The complex light field at the front of cylindrical lens L_{c2} is

$$E(x_i, q) = K \exp\left(-i\frac{k}{2f}q^2\right)F(x_i, q), \qquad i = 1, 2, \ldots, n, \tag{6.82}$$

which is proportional to the Fourier transform $F(x_i, q)$ up to a quadratic phase factor.

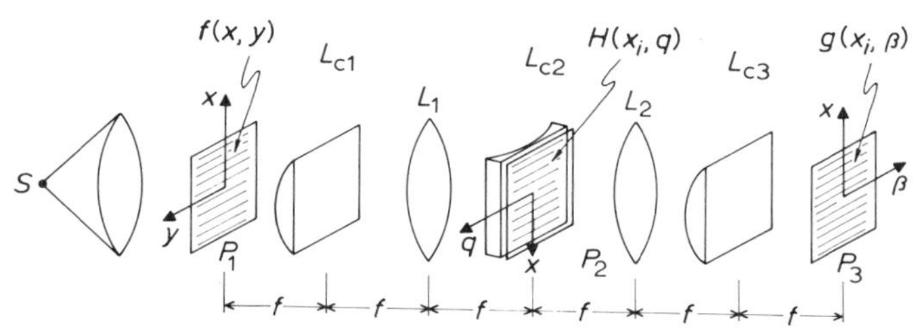

Fig. 6.20 A multichannel coherent filtering system.

If L_{c2} has a phase transform of

$$T(x, q) = \exp\left(i\frac{k}{f}q^2\right), \tag{6.83}$$

and if a multichannel filter $H(x_i, q)$ is inserted in the spatial frequency plane P_2, then the complex light field behind P_2 is

$$E'(x_i, q) = K \exp\left(i\frac{k}{2f}q^2\right)F(x_i, q)H(x_i, q), \qquad i = 1, 2, \ldots, n. \tag{6.84}$$

Because of the inverse transform operations of lenses L_2 and L_{c3}, the complex light amplitude distribution at the output plane P_3 is

$$g(x_i, \beta) = K\int f(x_i, \beta - y)h(x_i, y)\, dy = Kf(x_i, y) * h(x_i, y), \quad i = 1, 2, \ldots, n. \tag{6.85}$$

If the input signals are to undergo the same filtering in the spatial frequency domain, then separation into individual channels is not necessary. In this special case, the cylindrical lenses are not needed, and the optical system is the same as that of fig. 6.18. Thus the Fourier transform of the multichannel input signals on P_2 is $F(p, q)$. The filter transparency is $H(p, q)$, which is independent of p. Therefore the complex amplitude distribution across the output plane P_3 is

$$g(x, \beta) = K\int f(x, \beta - y)h(x, y)\, dy. \tag{6.86}$$

On the other hand, if the multichannel processing signals are required to operate at the spatial domain, then the input signals $f(x_i, y)$ and the reference function $h(x_i, y)$ can both be inserted in the input plane P_1 of fig. 6.20. The complex light field distribution at the spatial frequency domain in this case is

$$g(x_i, q) = K \exp\left(-i\frac{k}{2f}q^2\right)\int f(x_i, y)h(x_i, y)\exp[-iqy]\, dy, \quad i = 1, 2, \ldots, n. \tag{6.87}$$

If the input signals $f(x_i, y)$ are translated along the y axis, then the evaluation of eq. (6.87) along the line $q = 0$ is

$$g(x_i, \beta) = K\int f(x_i, \beta + y)h(x_i, y)\, dy, \qquad i = 1, 2, \ldots, n. \tag{6.88}$$

This can be accomplished by placing a slit along the line $q = 0$. It is clear that eq. (6.88) is the multichannel crosscorrelation, which can be expressed as

$$g(x_i, \beta) = Kf(x_i, y) \circledast h(x_i, y), \qquad i = 1, 2, \ldots, n. \tag{6.89}$$

On the other hand, for convolution we could simply reverse the input processing signals such that the output signal would be

$$g(x_i, \beta) = K\int f(x_i, \beta - y)h(x_i, y)\, dy = Kf(x_i, y) * h(x_i, y),$$
$$i = 1, 2, \ldots, n. \tag{6.90}$$

6.6 COHERENT OPTICAL COMPLEX SPATIAL FILTERING

Coherent optical information processing, described in the previous section, permits the synthesis of a vast variety of spatial filters. Such optical filters consist of transparencies that are inserted in the appropriate locations in the optical system. It was in 1963 that Vander Lugt proposed and successfully demonstrated a new method of synthesizing a complex spatial filter (refs. 6.9, 6.10). By means of this technique it is possible to uniquely combine amplitude and phase information in a single filter. Vander Lugt's technique has stimulated great interest in the areas of optical signal detection, and pattern and character recognition. It is because of these interesting applications that we treat this topic in an independent section of this chapter.

The basic concept of matched filtering was introduced in chap. 1. This matched filtering may be easily demonstrated in a simple coherent system diagramed in fig. 6.21. Let us assume that the input processing signal is obscured by some additive Gaussian noise; assume, further, that the matched complex filter $H(p, q)$, has a transmittance that is proportional to the complex conjugate

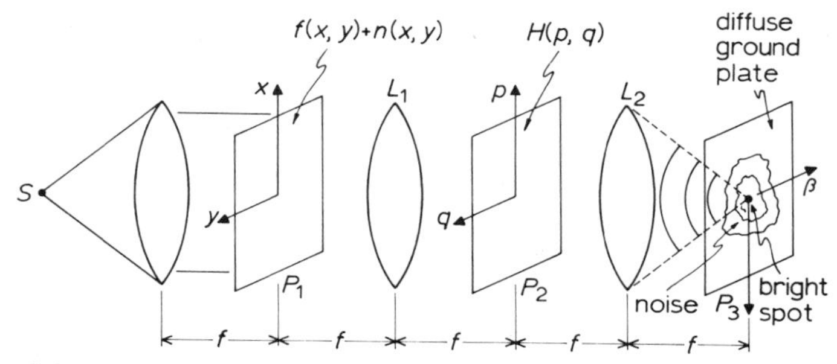

Fig. 6.21 Coherent matched filtering. The signal plus noise is $f(x, y) + n(x, y)$, and the matched filter is $H(p, q)$.

of the Fourier transform of the signal, that is,

$$H(p, q) = KF^*(p, q), \tag{6.91}$$

where $F(p, q)$ is the Fourier transform of the input signal $f(x, y)$.

The light amplitude distribution immediately behind the filter is

$$E(p, q) = K_1[|F(p, q)|^2 + N(p, q)\, F^*(p, q)], \tag{6.92}$$

where N is the Fourier transform of the random noise.

In eq. (6.92), the term $|F(p, q)|^2$ is, of course, real. It represents a plane wave with some amplitude weighting. Thus the complex light field due to the signal term will be imaged into a small region by L_2. It is clear that the effect of the matched filter is to eliminate the phase variation of the Fourier transform signal. It is not intended to restore the signal at the output plane, but rather to compress it into a small bright spot.

On the other hand, the noise term $NF^*(p, q)$ is differently affected. The filter has no compensating effect on the phase spectrum of $N(p, q)$, but it does alternate the amplitude of the noise spectrum in some regions where the signal amplitude spectrum is low. The net effect is that the noise is amplitude-weighted and reproduced at the output plane P_3 essentially unchanged, except that it is attenuated relative to the peak signal.

The complex light field at the output plane P_3 of the matched filter system is therefore

$$g(\alpha, \beta) = K_1\left[\iint f(x, y) f(\alpha + x, \beta + y)\, dx\, dy + \iint n(x, y) f(\alpha + x, \beta + y)\, dx\, dy)\right]. \tag{6.93}$$

Alternatively, eq. (6.93) can be written

$$g(\alpha, \beta) = K[f(x, y) \circledast f(x, y) + n(x, y) * f(-x, -y)]. \tag{6.94}$$

The first term of eq. (6.94) is the autocorrelation of the input signal, and the second term is the convolution of the input noise and the mirror image of the input signal. Nevertheless, it is clear that the effect of the matched filter is to compress most of the light diffracted by the signal into a small region; this is the autocorrelation function of the signal. In the case of noise, however, the matched filter does not compress the light into the small region, and no bright spot is observed at the output plane.

So far the basic concept of matched filtering has been discussed; however, to synthesize a two-dimensional matched filter may be difficult in practice. Nevertheless, there exists a stratagem well known to electrical engineers, based on the fact that a band-limited complex function may be represented by a real-valued function of twice the bandwidth of the complex function. In other words, the amplitude and the phase components of the complex function are both modulated by a carrier function whose frequency is greater than the highest frequency component of the complex function. Then by a suitable demodulation technique it is possible to reproduce the complex function from the modulated real function. Using this technique to synthesize a complex spatial filter, we can represent the transparency by means of the real function

$$H(p, q) = K_1 + K_2|S(p, q)| \cos[\alpha_0 p + \phi(p, q)], \tag{6.95}$$

where $S(p, q) = |S(p, q)| \exp[-i\phi(p, q)]$ is the complex function, K_1 and K_2 are the appropriate constants such that $0 \leq H(p, q) \leq 1$, p and q are the corresponding spatial frequency coordinates, and α_0 is the carrier frequency of the modulation function. It is clear that the amplitude and phase components of the complex function are properly modulated.

If this complex spatial filter is inserted in the spatial frequency plane of fig. 6.21, then the complex light distribution at the output plane P_3 is

$$g(\alpha, \beta) = K \iint F(p, q)\, H(p, q) \exp[-i(\alpha p + \beta q)]\, dp\, dq. \tag{6.96}$$

This can be written as

$$\begin{aligned} g(\alpha, \beta) = KK_1 &\iint F(p, q) \exp[-i(\alpha p + \beta q)]\, dp\, dq \\ &+ \frac{KK_2}{2} \iint F(p, q) S(p, q) \exp\{-[i(\alpha - \alpha_0) p + \beta q]\}\, dp\, dq \\ &+ \frac{KK_2}{2} \iint F(p, q) S^*(p, q) \exp\{-i[(\alpha + \alpha_0) p + \beta q]\}\, dp\, dq. \end{aligned} \tag{6.97}$$

The first term of eq. (6.97) represents a diffraction pattern around the optical axis of the output plane P_3, and the second and third terms represent diffraction around the points at $\alpha = \alpha_0$ and $\alpha = -\alpha_0$, respectively. Therefore the value of the spatial carrier frequency α_0 should be chosen sufficiently large to ensure that the three diffraction patterns will not overlap.

It may be emphasized that the bias level K_1 in eq. (6.95) must be chosen large enough to ensure the physical realization of the filter function. However, when $|S(p, q)|$ has a large dynamic range, the large value of K_1 often leads to poor utilization of the recording medium. This problem can be overcome by using interferometric techniques to synthesize the complex filter.

Interferometric Techniques in the Synthesis of Complex Spatial Filters

There are several methods of synethsizing complex spatial filters. In this section we mention some that make use of interferometric techniques. One interesting method centers around a modification of the Rayleigh interferometer (fig. 6.22; see also ref. 6.2, p. 128). A portion of the collimated monochromatic light illuminates the detecting singal $S(x, y)$, and a portion is converged by lens L_2 to form a reference point source. The complex light field at the back focal plane of L_3 is therefore

$$E(p, q) = K_1 \exp\left(-i\frac{\alpha_0}{2}p\right) + K_2 S(p, q) \exp\left(i\frac{\alpha_0}{2}p\right), \tag{6.98}$$

where $S(p, q)$ is the Fourier transform of the detecting signal.

The corresponding irradiance on the surface of the photographic plate is

$$I(p, q) = E(p, q) E^*(p, q) = K_1^2 + K_2^2 |S(p, q)|^2 + K_1 K_2 S(p, q) \exp(ip\alpha_0) + K_1 K_2 S^*(p, q) \exp(-ip\alpha_0), \tag{6.99}$$

where $S(p, q) = |S(p, q)| \exp[i\phi(p, q)]$. If the recording is in the linear region of the photographic emulsion (see chap. 5), then the transparency of the recorded complex spatial filter may be written

$$H(p, q) = K\{K_1^2 + K_2^2 |S(p, q)|^2 + 2K_1 K_2 |S(p, q)| \cos[\alpha_0 p + \phi(p, q)]\}. \tag{6.100}$$

Equation (6.100) is not only a real-valued function, but is non-negative as well; i.e., $H(p, q) \geq 0$. Filter synthesis by this interferometric technique offers these

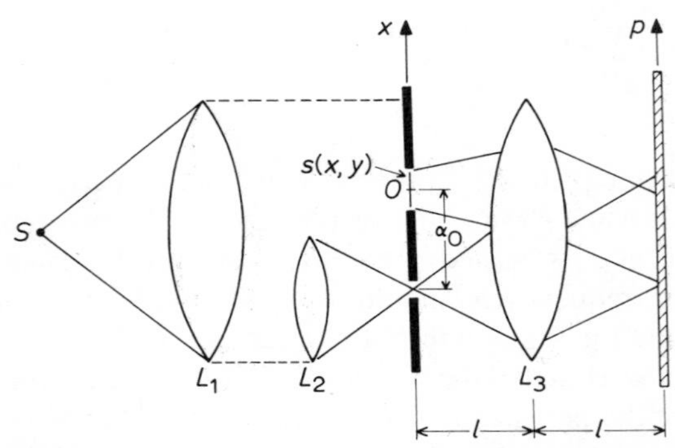

Fig. 6.22 Synthesis of a complex spatial filter; a modified Rayleigh interferometer. The detecting signal is $s(x, y)$, the pinhole behing L_2 provides a point reference source, and a photographic plate P is placed in the (p, q) plane.

advantages at the expense of increased spatial frequency bandwidth required by the amplitude and phase modulation.

If this complex spatial filter is inserted in the spatial frequency plane of a coherent optical processing system as shown in fig. 6.23, then the field immediately behind the spatial filter is

$$E(p, q) = F(p, q)\, H(p, q), \tag{6.101}$$

where $F(p, q)$ is the Fourier transform of the input signal $f(x, y)$. By substitution of eq. (6.100) into eq. (6.101), we have

$$E(p, q) = K[K_1^2 F(p, q) + K_2^2 F(p, q) |S(p, q)|^2 + K_1 K_2 F(p, q)\, S(p, q) \times \exp(ip\alpha_0) + K_1 K_2 F(p, q)\, S^*(p, q) \exp(-ip\alpha_0)]. \tag{6.102}$$

The field at the output plane P_3 is the inverse Fourier transform of eq. (6.102):

$$g(\alpha, \beta) = \iint E(p, q) \exp[-i(\alpha p + \beta q)]\, dp\, dq, \tag{6.103}$$

which can be written as the symbolic equation

$$g(\alpha, \beta) = K[K_1^2 f(x, y) + K_2^2 f(x, y) * s(x, y) * s^*(-x, -y) + K_1 K_2 f(x, y) * s(x + \alpha_0, y) + K_1 K_2 f(x, y) * s^*(-x + \alpha_0, -y)]. \tag{6.104}$$

The first and second terms of eq. (6.104) represent the zero-order diffraction, which appears at the origin of the output plane; the third and fourth terms are the convolution and crosscorrelation terms, which are diffracted in the neighborhood of $\alpha = \alpha_0$ and $\alpha = -\alpha_0$, respectively. The zero-order and the con-

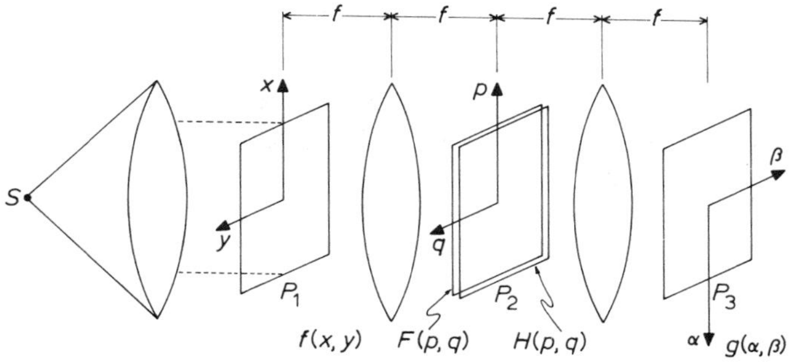

Fig. 6.23 Coherent detecting by complex spatial filtering. The input signal is $f(x, y)$, filter is $H(p, q)$, and the output signal is $g(\alpha, \beta)$.

volution terms are of no particular interest here; it is the crosscorrelation term that is used in signal detection.

Now if the input signal is assumed to be a detecting signal imbedded in an additive white Gaussian noise n,

$$f(x, y) = s(x, y) + n(x, y), \tag{6.105}$$

then the correlation term of eq. (6.104) can be written

$$R(\alpha, \beta) = K_1K_2[s(x, y) + n(x, y)] * s^*(-x + \alpha_0, -y). \tag{6.106}$$

Since the crosscorrelation with respect to $n(x, y)$ and to $s^*(-x + \alpha_0, -y)$ can be shown to be approximately equal to zero, this equation can be written as

$$R(\alpha, \beta) = K_1K_2 s(x, y) * s^*(-x + \alpha_0, -y), \tag{6.107}$$

which is proportional to the autocorrelation of $s(x, y)$.

In order to ensure that the zero-order and the first-order diffraction terms of eq. (6.104) will not overlap, the spatial carrier frequency α_0 may be approximated by the inequality

$$\alpha_0 > l_f + \tfrac{3}{2} l_s, \tag{6.108}$$

where l_f and l_s are the spatial lengths in the x direction of the input signal $f(x, y)$ and the detecting signal $s(x, y)$, respectively. To show that this is true, we consider the length of the various output terms of $g(\alpha, \beta)$, as illustrated in fig. 6.24. If l_f and l_s are the respective lengths of input signal $f(x, y)$ and detecting signal $s(x, y)$ in the x direction, then it is clear that lengths of first, second, third, and fourth terms of eq. (6.104) are l_f, $2l_s + l_f$, $l_f + l_s$, and $l_f + l_s$, respectively. From fig. 6.24, we conclude that to achieve complete separation the spatial carrier frequency α_0 must satisfy the inequality (6.108).

Note that the interferometric technique for the complex spatial filter synthesis is similar to the modulation technique in communication theory. The filter synthesis (such as shown in fig. 6.22) is equivalent to single sideband modulation of an electronic communication system. The detecting signal spectrum occupies a bandwidth l_s centered at the spatial carrier frequency α_0.

Following the analogy of spatial filter synthesis and basic communication theory, it is clear that a multiplexed spatial filter may be synthesized. The required complex spatial frequency Fourier spectrum of n detecting signals can be recorded by n carrier frequencies. The minimum value of the carrier frequency for the (i, j)th detecting signal can be approximated by the inequalities:

$$|\alpha_0(i, j + 1) - \alpha_0(i, j)| > l_{fx} + \tfrac{1}{2} l_{sx}(i, j + 1) + \tfrac{1}{2} l_{sx}(i, j), \tag{6.109}$$

and

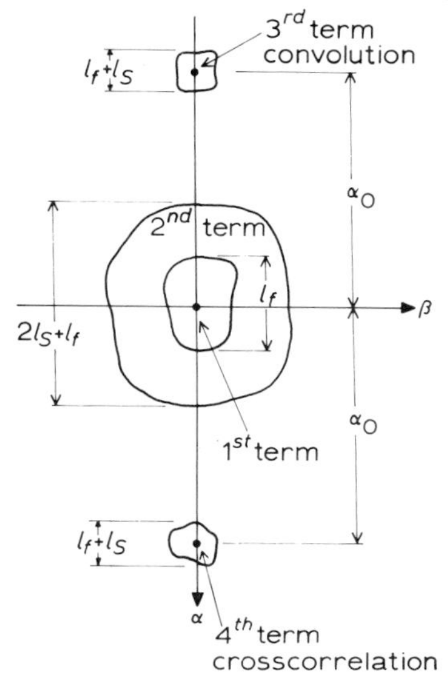

Fig. 6.24 Sketch of the output diffraction of a complex spatial filter.

$$|\alpha_0(i+1, j) - \alpha_0(i, j)| > l_{fy} + \tfrac{1}{2} l_{sy}(i+1, j) + \tfrac{1}{2} l_{sy}(i, j), \quad (6.110)$$

where l_{fx} and l_{fy} are the spatial lengths of the input signal in the x and y directions, l_{sx} *and* l_{sy} are the spatial lengths corresponding to the (i, j)th detecting signal, and $\alpha_0(i, j)$ is the vector representation of the spatial carrier frequency with respect to the (i, j)th detecting signal. A sketch of the detecting signal matrix and the reference point source is given in fig. 6.25.

Complex spatial filters are generally sensitive to rotation and to misscaling of the input detecting signal. That is, when the input signal is improperly matched to the angular orientation of the detecting signal or is given improper magnification, then the response of the filter is greatly decreased from that of a

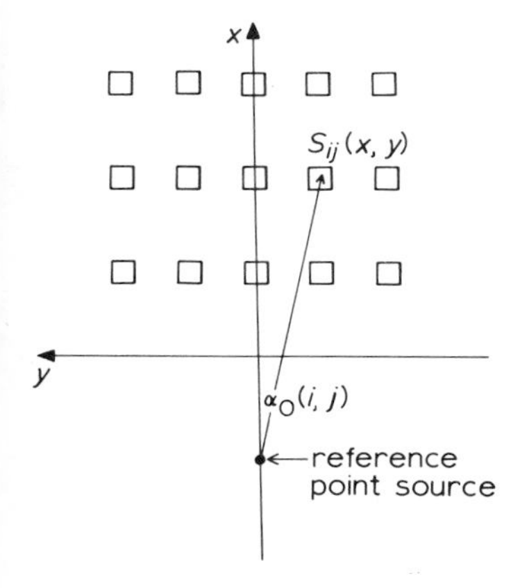

Fig. 6.25 Input plane for multiplex filter synthesis.

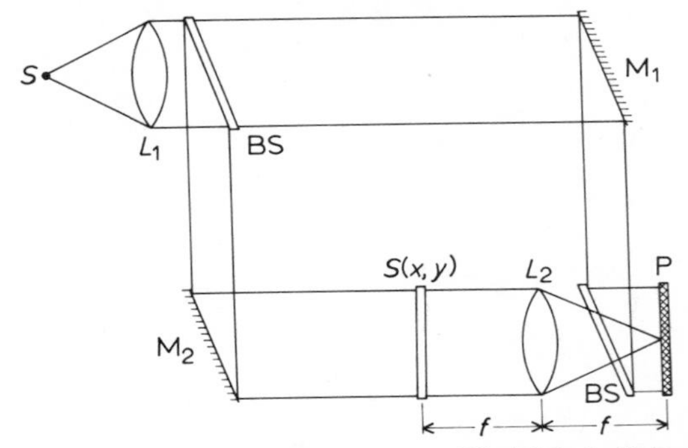

Fig. 6.26 Synthesis of a complex spatial filter; a modified Mach–Zehnder interferometer. BS, beam splitter; $s(x, y)$, detecting signal; P, photographic plate.

correctly matched one. The degree of sensitivity to the mis-orientation and mis-scaling depends to a large extent on characteristics of the detecting signal. For instance, a detecting signal of other than circular symmetry is more rotationally sensitive than a signal with circular symmetry. Rotational sensitivity may be reduced to some extent by additional filtering. In some cases, the only practical solution to the orientation and scale problems is a manual search operation.

At this point, let us give two alternative coherent optical systems for the synthesis of complex spatial filters (figs. 6.26, 6.27). The optical system shown in fig. 6.26 is a modified Mach–Zehnder interferometer. An oblique plane wave front on the photographic plate is produced by the tilted mirror M_1 and the second beam splitter. The lower part of the interferometer projects the Fourier transform of the detecting signal $s(x, y)$ onto the photographic plate. Hence an interference pattern of the two wave fronts results on the recording medium. The principle of the synthesis shown in fig. 6.27 is essentially the same as the modified Rayleigh interferometric technique, except that a smaller lens L_2 is used to produce the Fourier transformation of the detecting signal, and that a prism produces an appropriate oblique reference wave.

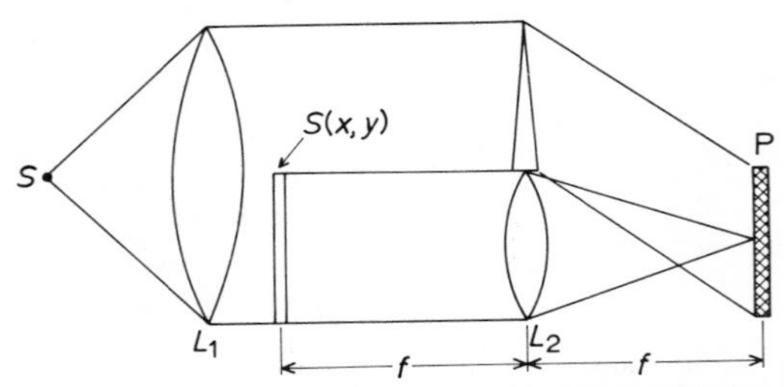

Fig. 6.27 Snythesis of a complex spatial filter. P, photographic plate; $s(x, y)$, detecting signal.

Complex Spatial Filtering in Character Recognition

A particularly interesting application of coherent spatial filtering is that of character recognition. The concept of matched filtering plays an important role in this application. Under the usual assumption of spatial invariance, a complex spatial filter is said to be matched with the detecting signal $s(x, y)$ if and only if the impulse response of the filter is

$$h(x, y) = Ks^*(-x, -y). \tag{6.111}$$

The transfer characteristic of the filter is given by

$$H(p, q) = Ks^*(p, q), \tag{6.112}$$

where $s^*(p, q)$ is the complex conjugate of the Fourier transform of the detecting signal $s(x, y)$.

To demonstrate that matched filtering provides a means for the recognition of characters, let us assume that the matched filter was designed to detect a particular character $s_i(x, y)$. Now if the input signal $\alpha(x, y)$ consists of a sequence of different nonoverlapping characters,

$$f(x, y) = s_1(x, y) + s_2(x, y) + \cdots + s_n(x, y), \tag{6.113}$$

then it can be shown that the magnitude of the autocorrelation is generally larger than that of the crosscorrelation. That is,

$$|s_i(x, y) \circledast s_i^*(x, y)| \geq |s(x, y) \circledast s_i^*(x, y)|, \tag{6.114}$$

To show that this is true, let us write down the autocorrelation function

$$R_{ii}(\alpha, \beta) = \iint s_i(x, y)\, s_i^*(x + \alpha, y + \beta)\, dx\, dy, \tag{6.115}$$

and the crosscorrelation function

$$R_{ji}(\alpha, \beta) = \iint s_j(x, y)\, s_i^*(x + \alpha, y + \beta)\, dx\, dy, \qquad i \neq j. \tag{6.116}$$

At the peak $\alpha = \beta = 0$ of the autocorrelation function we can write,

$$|R_{ii}(0, 0)| = \iint |s_i(x, y)|^2\, dx\, dy, \tag{6.117}$$

and, similarly,

$$|R_{ji}(0, 0)| = \frac{\left| \iint s_j(x, y) s_i^*(x, y)\, dx\, dy \right|^2}{\iint |s_j(x, y)|^2\, dx\, dy}, \qquad i \neq j. \tag{6.118}$$

But the Schwarz inequality states

$$\left| \iint s_j(x, y)\, s_i^*(x, y)\, dx\, dy \right|^2 \leq \iint |s_i(x, y)|^2\, dx\, dy \iint |s_j(x, y)|^2\, dx\, dy. \tag{6.119}$$

Therefore it follows that

$$|R_{ji}(0, 0)| \leq |R_{ii}(0, 0)|, \tag{6.120}$$

the equality holding only when

$$s_j(x, y) = K s_i(x, y). \tag{6.121}$$

It must be emphasized that character recognition by means of complex spatial filtering is not always a unique process without errors. Nevertheless in most cases it is a workable detection scheme. Examples are shown in figs. 6.28 and 6.29, by Vander Lugt. For a more detailed discussion of the complex spatial filtering problem the reader can refer to the excellent article, "Character Reading by Optical Spatial Filtering," by Vander Lugt, et al. (reprinted as chap. 7 or ref. 6.2).

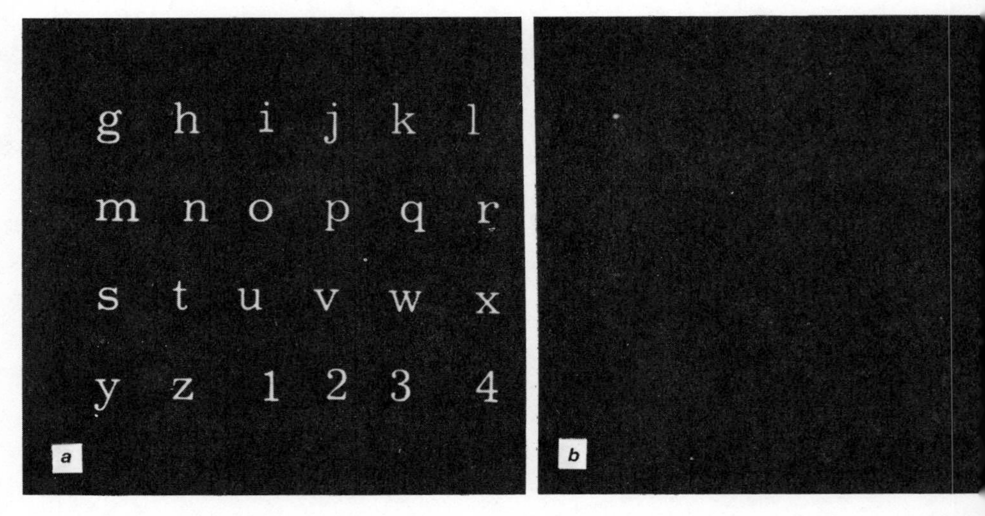

Fig. 6.28 Character recognition. (*a*) Alphanumerics. (*b*) Detecting of the letter "g." (By permission of A. B. Vander Lugt.)

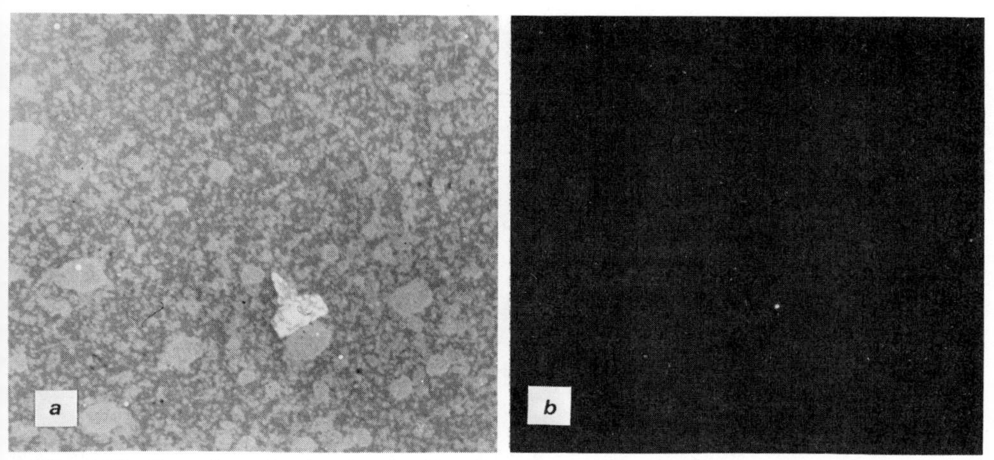

Fig. 6.29 Detecting of a signal embedded in random noise. (*a*) Input scene. (*b*) Detected signal. (By permission of A. B. Vander Lugt.)

6.7 COHERENT OPTICAL SPECTRUM ANALYSIS WITH AREA MODULATION

Area modulation has been used for sound track recording in motion pictures for some time (ref. 6.11). The use of area modulation for spectrum analysis was demonstrated by Dyachenko et al. (ref. 6.12) and Felstead (ref. 6.13). Speech spectrograms have also been generated optically by Aleksoff using sound tracks directly as input signals (ref. 6.14). Furthermore, the application of area-modulated signals for optical correlation was illustrated by Pernick (ref. 6.15).

In this section, we demonstrate the use of area modulation for coherent optical spectrum analysis.

Area modulation is a simple operation that produces a binary transmittance of two discrete tones. The basic forms of area modulation, *unilateral* and *bilateral*, are described in fig. 6.30. These two forms are chosen in our present application because of their simplicity. We note forms are chosen in our present application because of their simplicity. We note that other forms of area modulation may also be used. Since an area-modulated signal is in binary form (i.e., transparent and opaque), the problem of the nonlinearity of the recording material can be avoided. As a matter of fact, high-contrast films that are highly nonlinear are preferable to assure a good tonal separation. Besides allowing less stringent control in the production of the input transparency, the area modulation technique makes possible the use of a much larger class of recording materials. These materials may be unacceptable with the density modulation technique because of their nonlinear transmittance characteristics.

We can describe the two-dimensional amplitude transmittance function for a *bilateral* area-modulated signal by

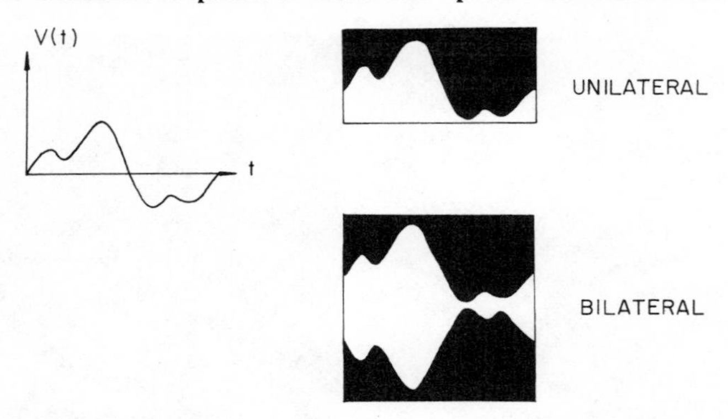

Fig. 6.30 Unilateral and bilateral area modulation formats.

$$T_1(x, y) = \operatorname{rect}\left(\frac{x}{L}\right)\operatorname{rect}\left\{\frac{y}{2[B + f(x)]}\right\}, \tag{6.122}$$

and a *unilateral* area-modulated signal by

$$T_2(x, y) = \operatorname{rect}\left(\frac{x}{L}\right)\operatorname{rect}\left\{\frac{y - [f(x) - B]/2}{B + f(x)}\right\}, \tag{6.123}$$

where (x, y) is the spatial coordinate system, L is length of the input transparency in the x direction, B is the spatial bias level, $f(x)$ is the input signal function, and

$$\operatorname{rect}\left(\frac{x}{L}\right) \triangleq \begin{cases} 1, & |x| \leq L/2, \\ 0, & \text{otherwise}. \end{cases}$$

We note that the bias level is chosen such that $B \geq f(x)$, for all x.

If the area-modulated transparency of eqs. (6.122) or (6.123) is inserted into the input plane of a coherent optical processor (fig. 6.31), the complex light distribution at the output spatial frequency plane will be

$$G(p, q) = C \iint T(x, y) \exp[-i(px + qy)]\, dx\, dy, \tag{6.124}$$

where (p, q) is the spatial frequency coordinate system, C is an arbitrary compex constant, and $T(x, y)$ is the input transparency. We note that eq. (6.124) is generally difficult to evaluate. However, if we restrict our evaluation to only the p-axis spatial frequency coordinate, then for the case of *bilateral* area-modulated signal we have

$$G_1(p, 0) = C \iint T_1(x, y) \exp(-ipx)\, dx\, dy, \tag{6.125}$$

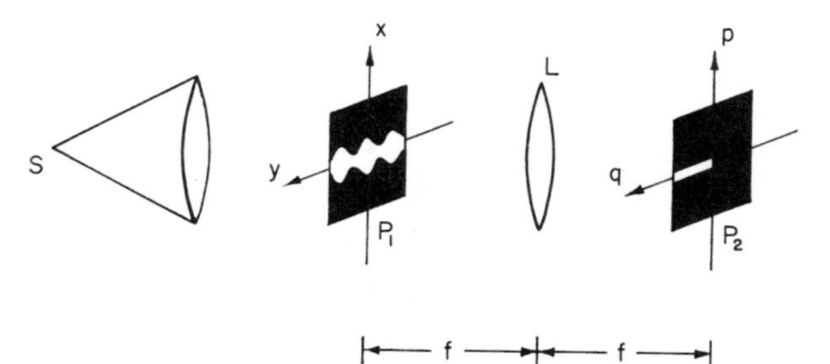

Fig. 6.31 Optical spectrum analysis with area modulation: S, monochromatic point source; L, transform lens.

which can be shown to be

$$G_1(p, 0) = 2C \int \operatorname{rect}\left(\frac{x}{L}\right)[B + f(x)] \exp(-ipx)\, dx$$

$$= 2C \int_{-L/2}^{L/2} B \exp(-ipx)\, dx + \int_{-L/2}^{L/2} f(x) \exp(-ipx)\, dx]. \quad (6.126)$$

Thus we see that eq. (6.126) represents a one-dimensional Fourier transformation of a constant B and an input signal function $f(x)$. We also note that eq. (6.126) is essentially the same form as for the density-modulated case.

We also note that, for a *unilateral* area-modulated signal, the complex amplitude distribution along the p-axis coordinate would be

$$G_2(p, 0) = \iint T_2(x, y) \exp(-ipx)\, dx$$

$$= C\left[\int_{-L/2}^{L/2} B \exp(-ipx)\, dx + \int_{-L/2}^{L/2} f(x) \exp(-ipx)\, dx\right]$$

$$= \tfrac{1}{2} G_1(p, 0), \quad (6.127)$$

which is identical with the result obtained for the bilateral case, except that the irradiance is reduced by a factor of four.

In analysis, we consider a bilateral sinusoidal area-modulated signal inserted in the input plane of a coherent optical processor, i.e.,

$$f(x) = A \sin(p_0 x + \theta), \quad (6.128)$$

where p_0 is some arbitrary angular spatial frequency, and θ is a constant phase factor. By substitution eq. (6.128) into eq. (6.126), we show that

$$G_1(p, 0) = K_1\, \delta(p) + K_2\, \delta(p - p_0) + K_3\, \delta(p + p_0), \quad (6.129)$$

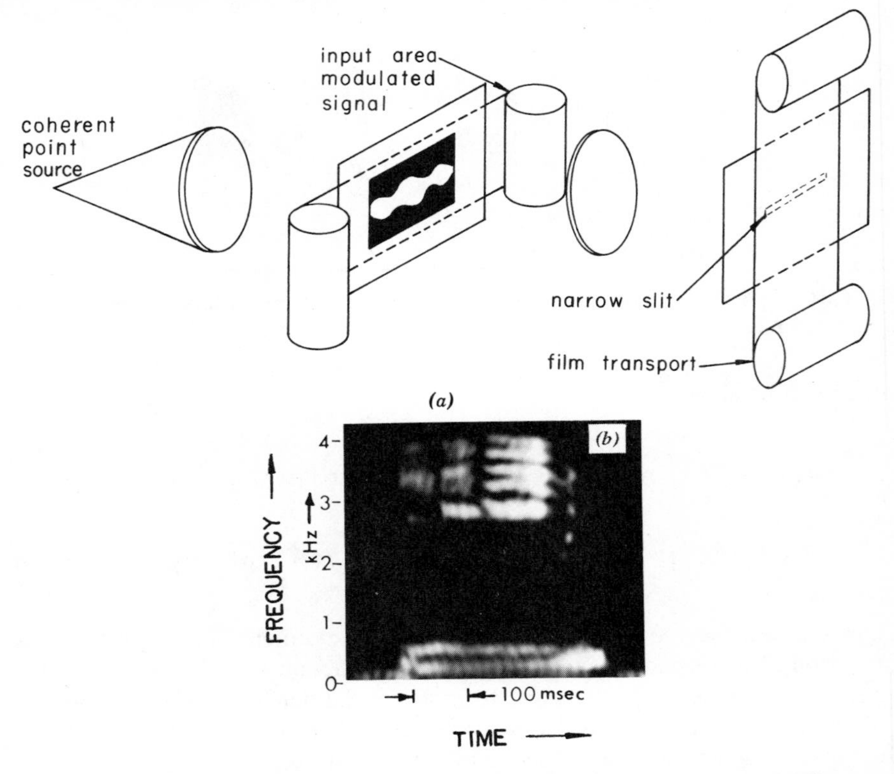

Fig. 6.32 Continuous generation of speech spectrogram by coherent optics. (*a*) Spectrum analysis with area modulation. (*b*) Speech spectrogram of a word "beet" generated with area modulation technique. (Courtesy of C. C. Aleksoff of the Environmental Research Institute of Michigan.)

where K is a proportional constant. From the above equation, except the zero-order diffraction, two spectra points are located at $p = p_0$ and $p = -p_0$, respectively. Thus we see that the coherent optical processor can indeed perform spectrum analysis with area modulation. The basic advantage of the area modulation technique is that a larger class of recording material can be used, since the recording is not restricted to the linear region of the recording transparency. To sum up this section we provide a section of a speech spectrogram generated by the area modulation technique, shown in fig. 6.32.

6.8 OPTICAL CORRELATION USING MELLIN TRANSFORMS

We have noted in sec. 6.6 that correlation detection by complex spatial filterings is generally sensitive to orientation and scale problems of the input signal. The scale of the input signal from which the complex spatial filter is formed must be precisely matched, otherwise a large decrease of correlation peak will result.

Although there are methods that have been proposed to overcome this mis-scaling problem, most of them fall short of solving it. In this section we illustrate a technique by which the mis-scaling problem may be alleviated, the Mellin transformation, as proposed by Casasent and Psaltis (ref. 6.16).

The Mellin transformation (ref. 6.17) was applied in time-varying circuit theory by Gerardi (ref. 6.18) in 1959, in space-variant image restoration by Sawchuk (ref. 6.19) in 1974, and by others. The two-dimensional Mellin transform along the imaginary axis is defined by

$$M(ip, iq) = \iint_{0}^{\infty} f(\mathscr{E}, \eta)\mathscr{E}^{-(ip+1)}\eta^{-(iq+1)}\, d\mathscr{E}\, d\eta, \tag{6.130}$$

where (p, q) is the transform plane coordinate system.

The basic interest of the Mellin transform for optical correlation detection lies in the optical synthesis of the transformation. Since the information processing operation relies on the Fourier transformation properties of lenses, the synthesis must involve the same transformation operation. It can be shown that, if we replace the variables $\mathscr{E} = e^x$ and $\eta = e^y$, then the Fourier transform of $f(e^x, e^y)$ yields the Mellin transform of $f(\mathscr{E}, \eta)$:

$$M(p, q) = \iint_{-\infty}^{\infty} f(e^x, e^y) \exp[-i(px + qy)]\, dx\, dy, \tag{6.131}$$

where we let $M(ip, iq) = M(p, q)$ for simple notation, and (p, q) is the spatial frequency coordinate system. An inverse Mellin transform can also be defined:

$$M^{-1}[M(p, q)] = f(\mathscr{E}, \eta) = \frac{1}{2\pi}\iint_{-\infty}^{\infty} M(p, q)\mathscr{E}^{ip}\eta^{iq}\, dp\, dq, \tag{6.132}$$

where M^{-1} denotes the inverse Mellin transformation. It can be shown that eq. (6.132) is equivalent to the inverse Fourier transform by replacing the variables $\mathscr{E} = e^x$ and $\eta = e^y$.

We note that the basic advantage of the Mellin transformation for application to coherent optical processing is its scale-invariant property. In other words, the Mellin transforms of two different-scale, but otherwise identical, spatial functions are scale-invariant, i.e.,

$$M_2(p, q) = a^{i(p+q)}M_1(p, q), \tag{6.133}$$

where a is an arbitrary scale factor, $M_1(p, q)$ and $M_2(p, q)$ are the Mellin transforms of $f_1(x, y)$ and $f_2(x, y)$, respectively and $f_1(x, y)$ and $f_2(x, y)$ are the two identical, except for different-scale, functions. From the above equation we see that the magnitudes of the Mellin transforms are

$$|M_2(p, q)| = |M_1(p, q)|, \tag{6.134}$$

which is essentially the same scaling. Thus we see that the Mellin transform of an object function is scale invariant, i.e.,

$$|M[f(\mathscr{E}, \eta)]| = |M[f(a\mathscr{E}, a\eta)]|, \tag{6.135}$$

when M denotes the Mellin transform.

Unlike the Fourier transform, the magnitude of the Mellin transform is, however, not equal to a shift of object function, i.e.,

$$|M[f(x, y)]| \neq |M[f(x - x_0, y - y_0)]|. \tag{6.136}$$

In optical correlation detection, we would assume that a complex spatial filter of $M[f(x, y)] = M(p, q)$ is synthesized by the interferometric technique as shown in fig. 6.33, i.e.,

$$H(p, q) = K_1 + K_2|M(p, q)|^2 + 2K_1K_2|M(p, q)| \cos[\alpha_0 p + \phi(p, q)], \tag{6.137}$$

where $M(p, q) = |M(p, q)| \exp[i\phi(p, q)]$. We now place this spatial filter in the spatial frequency plane P_2 of a coherent optical processor, as shown in fig. 6.23. If an object transparency of $f(e^x, e^y)$ is inserted in the input plane P_1 of the optical processor, the complex light distribution at the output plane P_3 can be shown to be

$$\begin{aligned} g(\alpha, \beta) = {} & K_1 f(e^x, e^y) + K_2 f(e^x, e^y) * f(e^x, e^y) * f^*(e^x, e^y) \\ & + K_1K_2 f(e^x, e^y) * f(e^x + e^{\alpha_0}, e^y) \\ & + K_1K_2 f(e^x, e^y) * f^*(-e^x + e^{\alpha_0}, -e^y), \end{aligned} \tag{6.138}$$

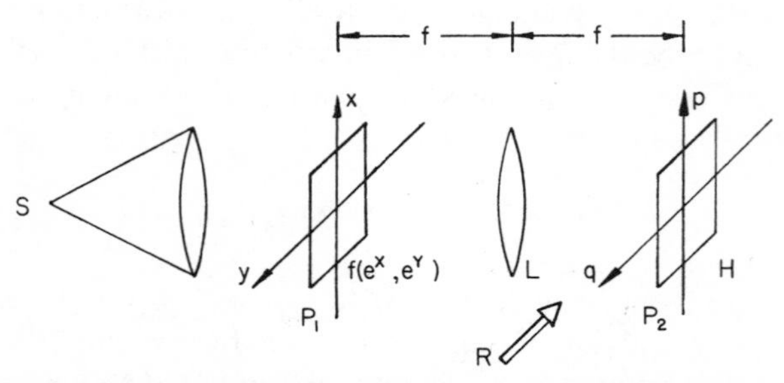

Fig. 6.33 Optical generation of a Mellin transform matched filter. S, monochromatic point source; $f(e^x, e^y)$ input signal; L, transform lens; R, reference beam; H, holographic plate.

where $*$ denotes the convolution operation and superscript* denotes the complex conjugate. The first and second terms of eq. (6.138) represent the zero-order diffraction, which appears at the origin of the output plane; the third and fourth terms are the convolution and correlation terms, which appear in the neighborhood of $\alpha = \alpha_0$ and $\alpha = \alpha_0$, respectively, as sketched in fig. 6.24. Thus we see that it is possible to optically correlate, through the Mellin transform technique, two object functions that differ greatly in scale or size. The importance of the Mellin transform technique as applied in correlation detection may lie in its role in the implementation of a coherent optical pattern recognition system. Methods of implementing Mellin transform in real-time mode on optically and electronically addressed devices have been reported by Casasent and Psaltis (ref. 6.20). The interested reader can refer to this reference.

PROBLEMS

6.1 By means of paraxial approximations, show that the phase transform of the concave lens of fig. 6.3 is indeed given by eq. (6.13).

6.2 Determine the complex light field and the corresponding irradiance at the back focal length of the lens shown in fig. 6.34 where $F(\xi, \eta)$ represents the amplitude transmittance of the transparency, and $l < f$. The incident wave is plane monochromatic.

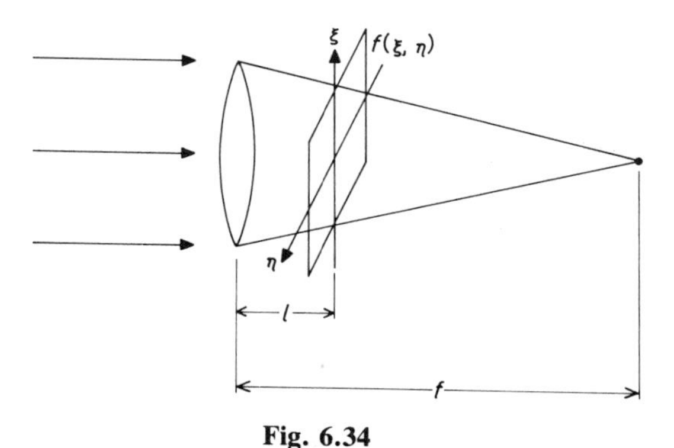

Fig. 6.34

6.3 Consider a lens cut from a thin section of a cone, as shown in fig. 6.35. By means of the paraxial approximation illustrated in sec. 6.1, determine the corresponding phase transformation. Explain the effect of the lens upon a normally incident monochromatic plane wave.

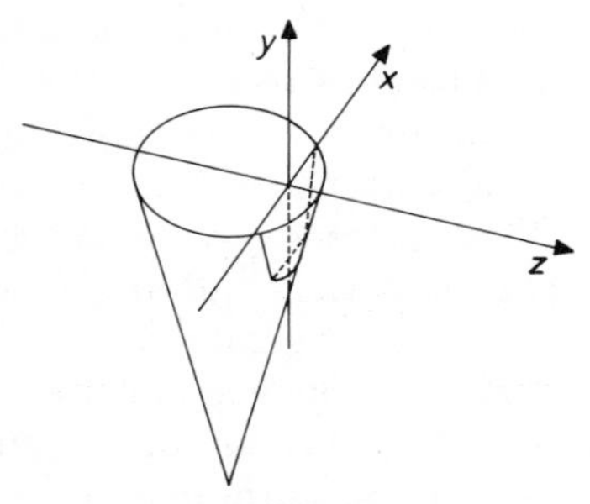

Fig. 6.35

6.4 The amplitude transparency of a zone-lens grating is given as

$$T(x, y) = \tfrac{1}{2}(1 + \cos ax^2) \qquad \text{for all} \quad y,$$

where a is an arbitrary constant.

(a) Show that this transparency may be decomposed as the combination of three lenses, i.e., flat, concave, and convex cylindrical lenses.

(b) Determine the corresponding focal lengths of this transparency.

(c) If this transparency is illuminated by a normally incident monochromatic plane wave, determine the effect cf the transmitted light field at the back focal plane of the transparency.

6.5 Consider a monochromatic plane wave that is obliquely illuminating a diffraction screen, which consists of an aperture in the shape of a large slit (fig. 6.36). A convex lens is placed behind the aperture as shown. Determine the diffraction pattern at the back focal plane of the lens.

6.6 A monochromatic plane wave is normally incident on an aberration-free objective lens. Due to some alignment error, the image is observed on a

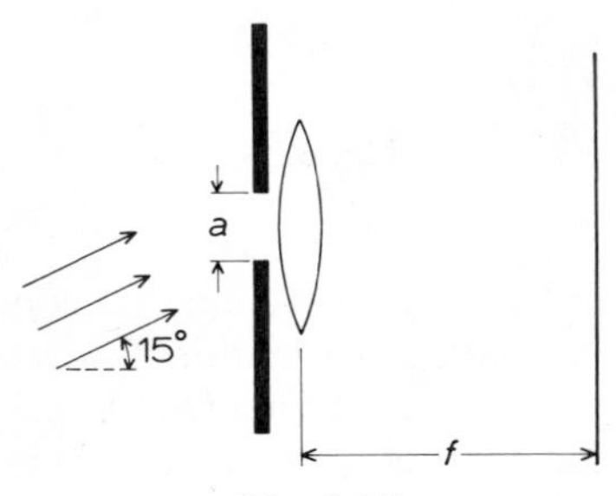

Fig. 6.36

plane displaced slightly from the focal plane of the lens. Determine the largest distance error for which the irradiance is still sufficiently accurate to represent the Fraunhofer diffraction pattern.

6.7 The Michelson stellar interferometer is shown in fig. 6.37. A diffraction screen of two parallel slits is placed at the front of an objective lens of a telescope. The light from a distant star enters one of the slits after reflections from mirrors M_{11} and M_{12}, and enters another slit after reflections from mirrors M_{21} and M_{22}. Assume that a filter selecting monochromatic light of wavelength λ is placed at the front of the objective lens. Let a and b be the separations of the slits and mirrors, respectively, and let f be the focal length of the lens. Determine the locations of the maximum and minimum irradiance when:

(a) The star lies on the axis of the telescope.

(b) The star lies at a small angle θ to the axis of the telescope, in the direction perpendicular to the slits.

6.8 A spatial filter with time-varying characteristics may be realized by means of the Fabry-Perot inteferometric technique. The filter consists of the thin transparent glass plates that are used to support a highly reflective

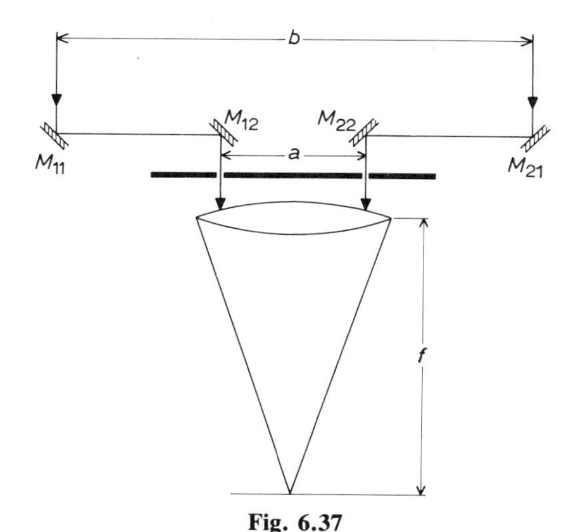

Fig. 6.37

substance coated on the inner surfaces (fig. 6.38). A monochromatic plane wave is normally incident upon the surface of one of the glass plates. If we denote by r and t the respective reflection and transmission coefficients of the reflective substance, and neglect the effect of the glass plates, show that the complex wave field transmitted thorugh the filter is

$$E(\omega) = K \frac{t^2 \exp(-i\omega d/c)}{1 - r^2 \exp(-i2\omega d/c)},$$

where ω is the circular frequency of the incident plane wave, c, is the velocity of the wave propagation, d is the separation of the glass plates, and K is an arbitrary constant.

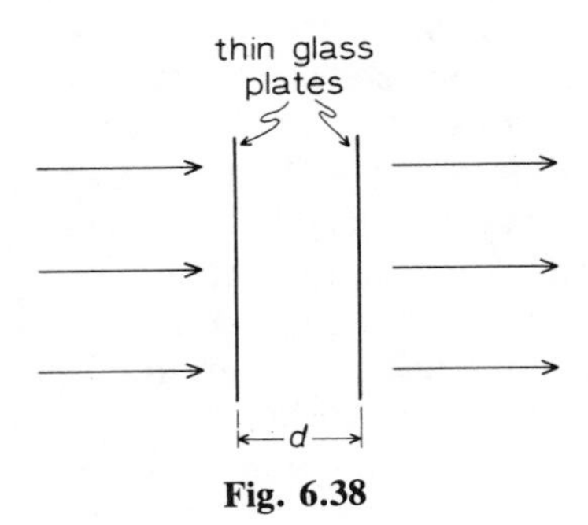

Fig. 6.38

6.9 Given two positive thin lenses of focal lengths f_1 and f_2.

(a) Calculate the resultant focal length of these two lenses in contact.

(b) If these two lenses are separated by a distance d, where $d < f_1$ and $d < f_2$, then evaluate the distance of the focal point behind the second lens.

6.10 **(a)** Calculate the temporal mutual coherence function of a light source whose spectrum is approximated by rect $[|\lambda - \lambda_0|/\Delta\lambda]$, where the center wavelength $\lambda_0 = 6000$ Å and spectral width $\Delta\lambda = 2$ Å.

(b) How many fringes can be formed in a Michelson interferometer before the visibility drops to 50 percent?

6.11 Let us consider a far distant star (i.e., a circular source at infinity); if the subtended angle at the Michelson stellar interferometer is $\theta = 10 \times 10^{-8}$ radian, then:

(a) Sketch the degree of mutual coherence γ_{12} at the interferometer.

(b) From part (a), determine the size of the star.

6.12 Given a monochromatic source consisting of two parallel slit sources; one has a finite width, the other is assumed infinitely small. These two sources may be described as $I_S(x) + a_0\delta(x - x_0)$, where a_0 is a positive constant.

(a) Sketch a Young's experiment to show that the degree of mutual coherence γ_{12} can be measured.

(b) Determine the portion of $I_S(x)$ that can be completely derived from $|\gamma_{12}|^2$.

6.13 Given the coherent optical processor shown in fig. 6.39. Denote the transmittance of the input transparency by $f(x, y)$ and the output complex light field by $F(p, q)$ [the Fourier transform of $f(x, y)$]. The output irradiance is recorded on a linear photographic plate. If this recorded transparency is inserted in the input plane, in place of the original transparency, show that the output light distribution is proportional to the autocorrelation function of $f(x, y)$. S is a monochromatic point source.

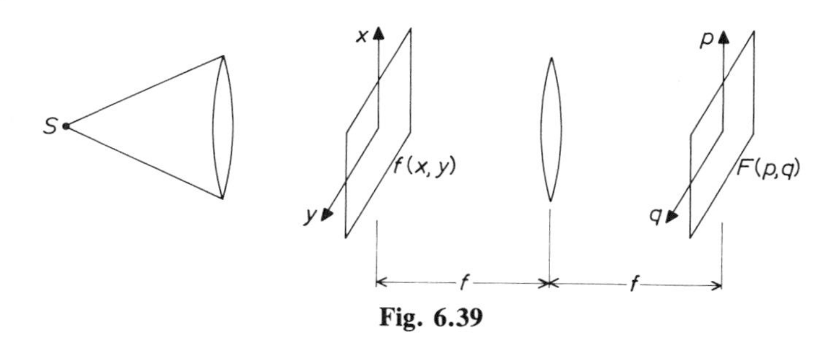

Fig. 6.39

6.14 Suppose the transmittance $T(x, y)$ of a recorded photographic plate is given as

$$T(x, y) = \tfrac{1}{2} + \tfrac{1}{4} \sin \sqrt{x^2 + y^2}.$$

Design a coherent optical processor and an appropriate spatial filter so that the output irradiance will be of the highest contrast.

6.15 A transparency consists of a signal embedded in an additive white Gaussian noise. Synthesize an appropriate coherent optical information processor for the signal filtering. If the signal is assumed spatially limited, i.e., if

$$f(x, y) = \begin{cases} A(x, y), & 0 \leqslant x^2 + y^2 \leqslant a^2 \\ 0, & \text{otherwise} \end{cases},$$

then construct a simple amplitude filter such that the output signal-to-noise ratio will be appreciably increased.

6.16 For the system of fig. 6.15, design an appropriate spatial filter such that the output signal is the spatial derivative of the input transparency.

6.17 Let the input transparency of the detector sketched in fig. 6.23 be given by

$$f(x, y) = \exp[-(x^2 + y^2)] \exp[i\beta(x^2 + y^2)^{1/2}],$$

where β is a positive real constant, and the additive spatial noise is assumed to be white Gaussian.

(a) Synthesize a matched filter such that the signal-to-noise ratio at the output plane will be optimum.

(b) Determine the corresponding light amplitude distribution on the output plane of the processor.

6.18 Certain relationships between the real and imaginary parts of a linear system function are necessary. These relationships may be expressed by means of Hilbert transform pairs,

$$f_r(x) = -\frac{1}{\pi}\int_{-\infty}^{\infty} \frac{f_i(\alpha)}{x - \alpha}\, d\alpha$$

and

$$f_i(x) = \frac{1}{\pi}\int_{-\infty}^{\infty} \frac{f_r(\alpha)}{x - \alpha}\, d\alpha,$$

where $f(x) = f_r(x) + if_i(x)$, a complex-valued function, with $f_r(x)$ and $f_i(x)$ the respective real and imaginary parts. Design a coherent optical system that is able to perform the respective Hilbert transformations.

6.19 The optical processing technique of prob. 6.18 may be extended to a two-dimensional Hilbert transform pair,

$$f_r(x, y) = \frac{1}{\pi^2}\int_{-\infty}^{\infty}\int_{-\infty}^{\infty} \frac{f_i(\alpha, \beta)}{(x - \alpha)(y - \beta)}\, d\alpha\, d\beta$$

and

$$f_i(x, y) = \frac{1}{\pi^2}\int_{-\infty}^{\infty}\int_{-\infty}^{\infty} \frac{f_r(\alpha, \beta)}{(x - \alpha)(y - \beta)}\, d\alpha\, d\beta,$$

where $f(x, y) = f_r(x, y) + if_i(x, y)$. Design an optical system that can perform the two-dimensional transformations.

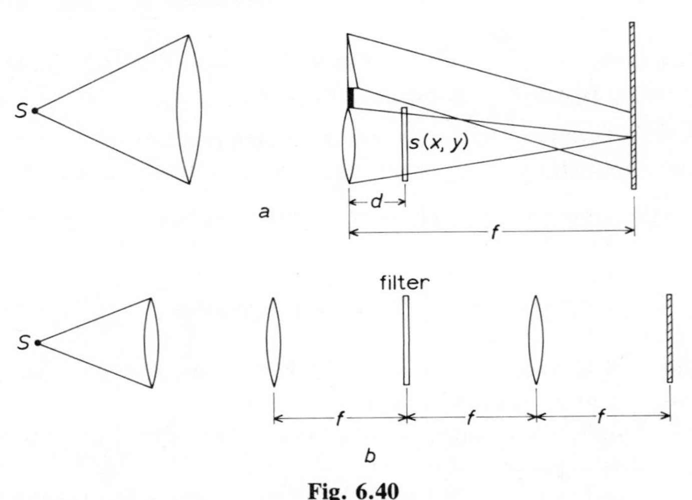

Fig. 6.40

6.20 Given a method of synthesizing a complex spatial filter as shown in fig. 6.40*a*, a signal transparency $s(x, y)$ is inserted at a distance d behind the transform lens. The amplitude transmittance of the recorded spatial filter is linearly proportional to the signal transparency. If this resulting spatial filter is used for signal detection, determine the appropriate location of the input plane of the system, shown in fig. 6.40*b*.

6.21 The complex spatial filter of the previous problem is inserted in the input plane of the optical processor shown in fig. 6.41. What is the complex light distribution at the output plane?

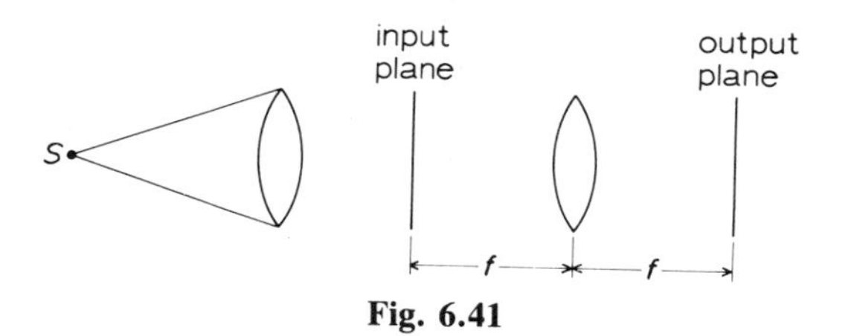

Fig. 6.41

6.22 Two input transparencies of amplitude transmittances $f_1(x, y)$ and $f_2(x, y)$ are placed in the front focal plane of the transform lens of an optical processor, as shown in fig. 6.42. The transparencies are centered in the (x, y) coordinate plane at $(a, 0)$ and $(-a, 0)$, respectively. Compute the irradiance at the back focal plane of the lens.

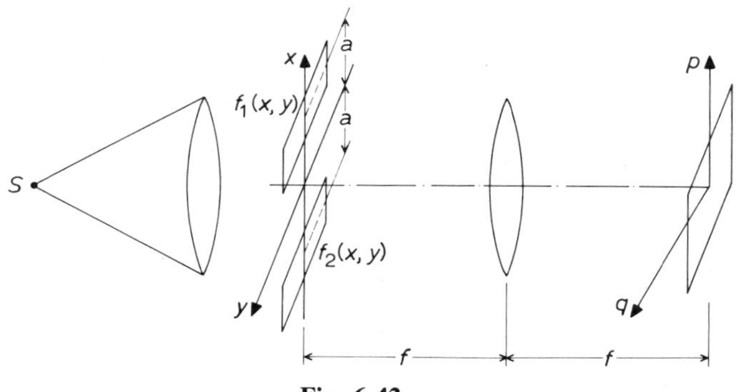

Fig. 6.42

6.23 The output irradiance of the system of fig. 6.42 is recorded on a linear photographic plate. If this transparency is placed at the input plane of the optical system, as shown in fig. 6.43:

(a) Determine the complex light distribution at the back focal plane of the transform lens.

(b) Discuss briefly the important functional operations, with respect to $f_1(x, y)$ and $f_2(x, y)$, at the output plane (α, β).

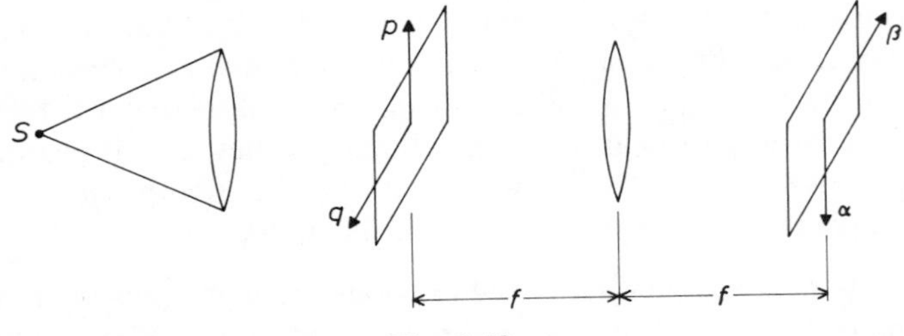

Fig. 6.43

6.24 Consider the coherent optical processing system with an incident monochromatic plane wave (fig. 6.44). Two signal transparencies of transmittance $f_1(x, y)$ and $f_2(x, y)$ are inserted in the front focal plane of the first transform lens, with centers at $(a, 0)$ and $(-a, 0)$. If there is a sinusoidal grating with amplitude transmittance

$$T(p, q) = \tfrac{1}{2} + \tfrac{1}{2}\cos ap \qquad \text{for every } q,$$

determine the light amplitude distribution at the output plane of the optical system.

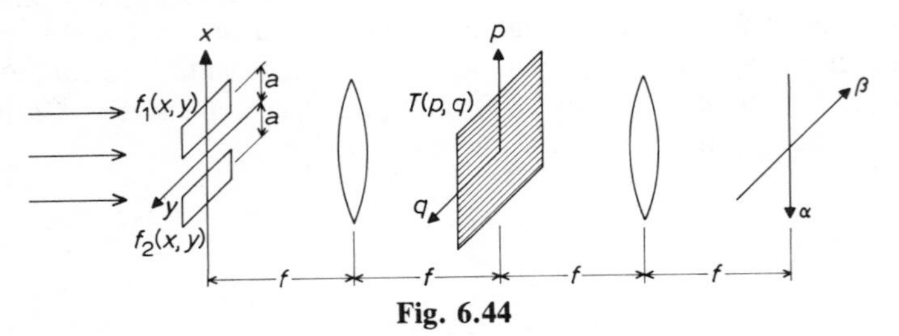

Fig. 6.44

6.25 Repeat prob. 6.24, if a sinusoidal grating of transmittance

$$T(p, q) = \tfrac{1}{2} + \tfrac{1}{2}\sin ap \qquad \text{for every } q$$

replaces the grating of the previous problem. Show the effect on output functional operations.

6.26 Consider a spatial-multiplex transparency, which is the sequential record of two images $A_1(x, y)$ and $A_2(x, y)$, made through a sinusoidal grating in contact with the film. The corresponding transmittance may be written

$$T(x, y) = K + A_1(x, y)\cos p_0 x + A_2(x, y)\cos q_0 y,$$

where K is an arbitrary constant, and p_0 and q_0 are the respective carrier frequencies in the (p, q) spatial frequency coordinates. The images $A_1(x, y)$ and $A_2(x, y)$ are assumed to be spatially band-limited within a circular region of radius $r = r_0$, where $r_0 < \frac{1}{2}p_0$, and $r_0 < \frac{1}{2}q_0$. This transparency is placed at the input plane P_1 of the optical processor in fig. 6.23.

(a) Determine the complex light distribution on the spatial frequency plane P_2.

(b) Design a spatial filter so that the output light field will reimage $A_1(x, y)$ and $A_2(x, y)$, one at a time.

6.27 We see that, with the optical arrangement shown in fig. 6.45, the light

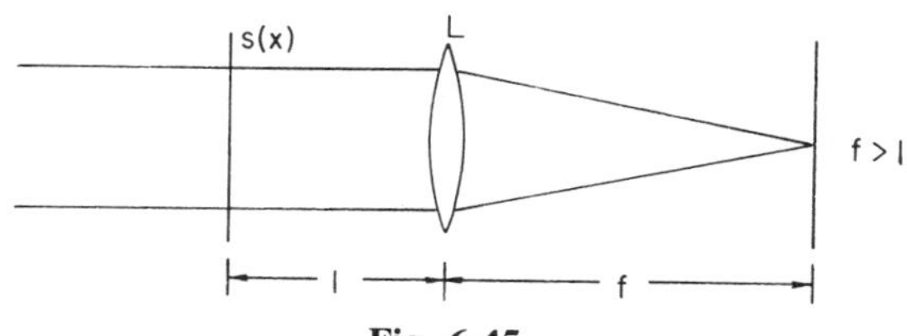

Fig. 6.45

distribution at the focal plane would not correspond to the *exact* Fourier transform of the input signal $s(x)$. Find the focal length of a lens that we can insert behind the lens L such that we would have the exact Fourier transform of the input at the focal plane.

6.28 Let $f(x)$ represent a sinusoidal signal recorded on a bilateral area modulation sound track. The amplitude transmittance function may be described by eq. (6.122). If this transparency is inserted in the input plane of a coherent optical processor, then:

(a) Calculate the complex light distribution at the back focal plane of the transform lens.

(b) If the signal spectrum of part (a) is filtered by a narrow slit filter as shown in fig. 6.31, then evaluate the corresponding inverse Fourier transform.

6.29 Given a cathode-ray-scanner (CRT) with a spatial resolution of about 0.5 line/mm. The CRT is used to generate a bilateral area-modulated speech signal, from 25 and 4,000 Hz. Determine the minimum modulating frequency and the velocity of the scanning electron beam.

6.30 **(a)** Identify the Fourier and Mellin transformations, which is either shift- or scale-invariant?

(b) Illustrate an optical arrangement so that the resultant Fourier transformation is a shift- and scale-invariant.

(c) Evaluate the optical setup of part (b); show that the resultant Fourier spectrum is indeed scale- and shift-invariant.

6.31 With reference to the Mellin transform of sec. 6.8, if the input signal transparency is a circular open aperture of radius r, show that the corresponding Mellin transform is independent of r.

REFERENCES

6.1 D. K. Pollack, C. J. Koester, and J. T. Tippett, Eds., *Optical Processing of Information*, Spartan Books, Baltimore, Md., 1963.

6.2 J. T. Tippett et al., Eds., *Optical and Electro-Optical Information Processing,* MIT Press, Cambridge, Mass., 1965.

6.3 H. J. Caulfield, *Handbook of Optical Holography*, Academic Press, New York, 1979.

6.4 P. Elias, D.S. Grey, and D. Z. Robinson, "Fourier Treatment of Optical Processes." *J. Opt. Soc. Am.*, **42**, 127 (1952).

6.5 P. Elias, "Optics and Communication Theory," *J. Opt. Soc. Am.*, **43**, 229 (1953).

6.6 E. L. O'Neill, "Spatial Filtering in Optics," *IRE Trans. Inform. Theory*, **IT-2**, 56 (1956).

6.7 E. L. O'Neill, Ed., *Communication and Information Theory Aspects of Modern Optics*, General Electric Co., Electronics Laboratory, Syracuse, N.Y., 1962.

6.8 L. J. Cutrona et. al., "Optical Data Processing and Filtering Systems," *IRE Trans. Inform. Theory*, **IT-6**, 386 (1960).

6.9 A. Vander Lugt, "Signal Detection by Complex Spatial Filtering," *IEEE Trans. Inform. Theory*, **IT-10**, 139 (1964).

6.10 A. Vander Lugt, *Signal Detection by Complex Spatial Filtering*, Radar Laboratory Report No. 4594-22-T, Institute of Science and Technology, University of Michigan, Ann Arbor, 1963.

6.11 R. Spottiswoode, *Film and Its Techniques*, University of California Press, Berkeley, Cal., 1964.

6.12 A. A. Dyachenko, M. V. Persikov, and O. E. Shushpaov, "Use of Shadowgraphs for Spectral Functional Analysis by the Methods of Coherent Optics," *Optics and Spectroscopy*, **31**, 249 (1971).

6.13 E. B. Felstead, "Optical Fourier Transformation of Area-Modulated Spatial Functions," *Appl. Opt.*, **10**, 2468 (1971).

6.14 F. T. S. Yu, "Generating Speech Spectrograms Optically," *IEEE Spectrum,* **12**, 51 (1975).

6.15 B. J. Pernick, "Area-Modulated Signal Recordings for Coherent Optical Correlations," *Appl. Opt.*, **11**, 1425 (1972).

6.16 D. Casasent and D. Psaltis, "Scale Invariant Optical Correlation Using Mellin Transforms," *Opt. Commun.*, **17**, 59 (1976).

6.17 R. Bracewell, *The Fourier Transform and Its Applications*, McGraw-Hill, New York, (1963), chap. 12.

6.18 F. R. Gerardi, "Application of Mellin and Hankel Transforms to Networks with Time-Varying Parameters," *IRE Trans. Circ. Theory*, 197 (1959).

6.19 A. A. Sawchuk, "Space-Variant Image Restoration by Coordinate Transformation," *J. Opt. Soc. Am.*, **64**, 138 (1974).

6.20 D. Casasent and D. Psaltis, "Scale Invariant Optical Transform," *Opt. Eng.*, **15**, 258 (1976).

6.21 H. H. Hopkins, "The Concept of Partial Coherence in Optics," *Proc. Roy. Soc.*, ser. A., **208**, 263 (1961).

6.22 H. H. Hopkins, "On the Diffraction Theory of Optical Images," *Proc. Roy. Soc.*, ser. A., **217**, 408 (1953).

6.23 J. Rhodes, "Analysis and Synthesis of Optical Images," *Am. J. Phys.*, **21**, 337 (1953).

7

Techniques and Applications of Coherent Optical Processing

Communication theory originated with a group of mathematically oriented electrical engineers whose interest was centered on electrical communication. However, from the very beginning interest in the optical standpoint has never been totally absent. Recent advances in electro-optical devices have made the relationship between optics and communication theory more profound than ever.

During the past two decades, optical information processing has received increased attention from a growing number of engineers and physicists. This increase of activity has stemmed in part from the discovery of a strong coherent source (i.e., laser), and in part from a realization that optical configurations can be used to perform a wide variety of processing operations on signals. Although both coherent and noncoherent processing techniques have been employed for various analog processings, the former has been generally accepted as more versatile and useful. In coherent optical processings the phenomenon of diffraction is employed. If a lens is placed at a focal distance away from a transparency, the light at a focal length behind the lens is distributed according to a two-dimensional spectral analysis of the object transparency. By simple insertion of complex spatial filters in this spectral plane, a wide variety of information processing operations can be performed. With a rather simple arrangement of the optical configurations, the output light distribution, affected by the spatial filter function, can be imaged.

We note that optical configuraitons are capable of performing essentially any linear operation on a function of two variables. With the addition of cylindrical lenses to the configuration, the optical system can be converted to multichannel one-dimensional processing, as described in the previous chapter. In this chapter we describe some techniques and applications of coherent optical information processing. In view of the broad area of interest, we confine ourselves to a few cases that we consider of general interest.

7.1 RESTORATION OF BLURRED PHOTOGRAPHIC IMAGES

One of the interesting applications of coherent optical information processing is the restoration of blurred photographic images (refs. 7.1–7.7).

Some of the physical constraints discoverable from the standpoint of information theory have also been discussed in previous articles (refs. 7.8, 7.9). In this section we discuss the synthesis of a complex spatial filter to reduce the effect of blurring. As we pointed out in chap. 6, a complex spatial filter may be realized by the combination of an amplitude filter and a thin-film phase filter. Such a synthesis, however, is generally difficult to achieve in practice (ref. 7.10). By means of the holographic techniques, which are considered in chap. 9, the desired phase filter may be easy to realize. The preparation of such a phase filter has been studied by Stroke and Zech (ref. 7.2) for the restoration of blurred images, and by Lohmann and Paris (ref. 7.11) for optical data processing.

In this section we consider the synthesis of a phase filter that, when combined with an amplitude filter, can be used for the restoration of an image that has been blurred. The complex filtering process that we discuss may be able to correct some of the blurred images, but it is by no means optimum.

The Fourier transform expression of a linearly distorted (blurred) image can be expressed by

$$G(p) = S(p)D(p), \tag{7.1}$$

where $G(p)$ is the distorted-image function, $S(p)$ is the nondistorted-image function, $D(p)$ is the distorting function of the imaging system, and p is the spatial frequency. Then the corresponding inverse filter transfer function for the restoration is

$$H(p) = \frac{1}{D(p)}. \tag{7.2}$$

We note that (ref. 7.8) the inverse filter function is generally not physically realizable, particularly for blurred images due to linear motion or defocusing. If we are willing, however, to accept a certain degree of error, then an approximate inverse filter might be possible to realize. For example, let the transmission function of a linear smeared point image be

$$f(\xi) = \begin{cases} 1, & -1/2\Delta\xi \leq \xi \leq 1/2\Delta\xi \\ 0, & \text{otherwise} \end{cases}, \tag{7.3}$$

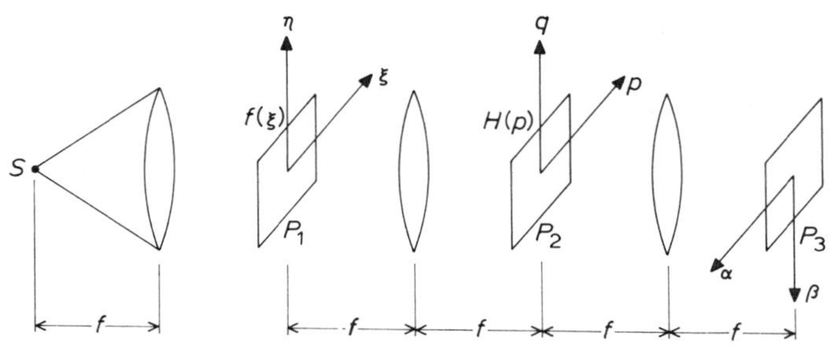

Fig. 7.1 Coherent data processing system.

where $\Delta\xi$ is the smear length. If a transparency satisfying eq. (7.3) is inserted in the input plane P_1 of a coherent optical data processor, as shown in fig. 7.1, the resultant complex light field on the spatial frequency plane is

$$F(p) = \Delta\xi \frac{\sin(p\Delta\xi/2)}{p\Delta\xi/2}, \tag{7.4}$$

which is essentially the Fourier transform of the smeared point image. A plot of the Fourier spectrum given by eq. (7.4) is shown in fig. 7.2. It can be seen that the Fourier spectrum is bipolar.

In principle, the smeared image may be corrected by means of inverse filtering (ref. 7.8). A suitable inverse filter function is

$$H(p) = \frac{p\Delta\xi/2}{\sin(p\Delta\xi/2)}. \tag{7.5}$$

It is noted that the inverse filter function itself is not only a bipolar but also an infinite poles function. Thus the filter is not physically realizable. However, if we are willing to sacrifice some of the resolution, then an approximate filter function may be realized. In order to do so, we will combine the amplitude filter

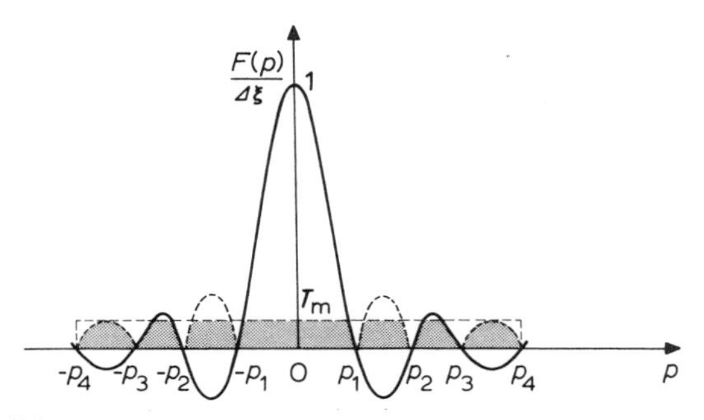

Fig. 7.2 The solid curve represents the Fourier spectrum of a linear smeared point image. The shaded area represents the corresponding restored Fourier spectrum of a point image. The zeroes are given by $p_n = 2n\pi/\Delta\xi$, $n = 1, 2, 3, \ldots$.

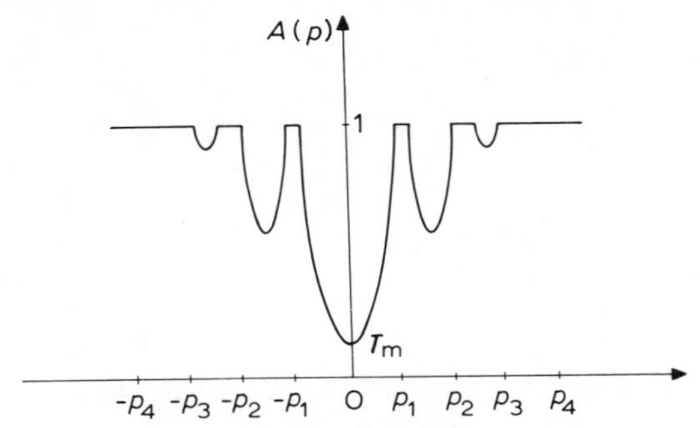

Fig. 7.3 Amplitude filter function.

of fig. 7.3 with the independent phase filter of fig. 7.4. The transfer function of this combination is

$$H(p) = A(p)e^{i\phi(p)}. \tag{7.6}$$

If this approximated inverse filter is inserted in the spatial frequency plane of the data processor of fig. 7.1, the restored Fourier transfer function will be

$$F_1(p) = F(p)H(p). \tag{7.7}$$

If we let T_m be the minimum transmittance of the amplitude filter, then the restored Fourier spectrum of the point image is the shaded spectrum shown in fig. 7.2. We can now define the relative degree of image restoration (ref. 7.6):

$$\mathcal{D}(T_{\mathrm{m}})(\text{percent}) = \frac{1}{T_{\mathrm{m}}\Delta p}\int_{\Delta p} \frac{F(p)H(p)}{\Delta\xi}\,dp \times 100, \tag{7.8}$$

where Δp is the spatial bandwidth of interest. In fig. 7.2, for example, $\Delta p = 2p_4$. From eq. (7.8) we can plot degree of image restoration as a function of T_{m} (fig. 7.5). We see the perfect restoration is approached as T_{m} approaches zero. However, at the same time the restored Fourier spectrum is also vanishing, and so no image will be reconstructed. Thus perfect restoration cannot be achieved in practice. These considerations aside, it seems that noise (film granularity and speckling) is the major limiting factor in the image restoration. To

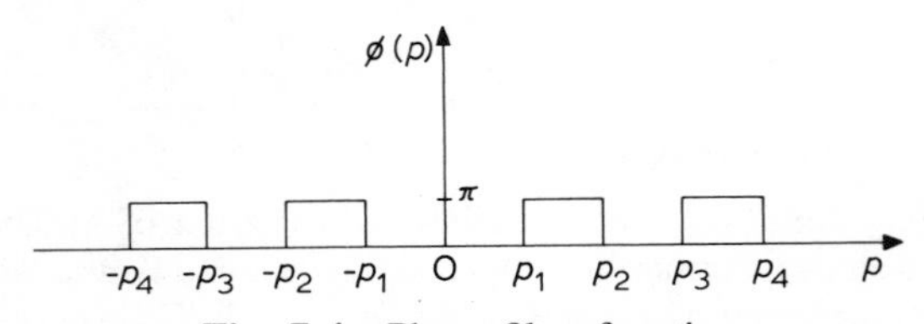

Fig. 7.4 Phase filter function.

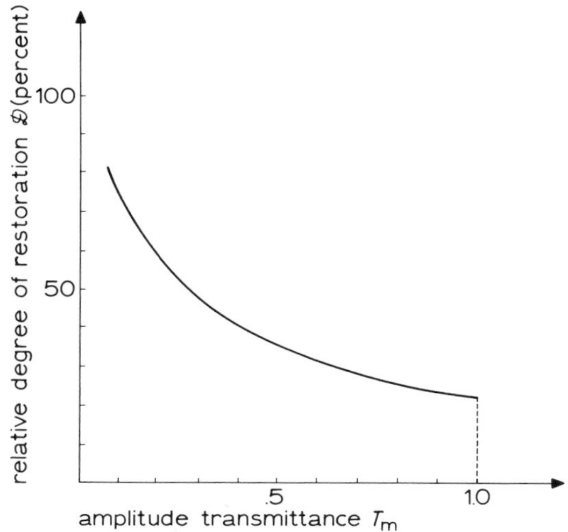

Fig. 7.5 Relative degree of image restoration as a function of T_m for linear image motion.

achieve a high degree of restoration, a lower transmittance T_m is required, and the restored Fourier spectrum is therefore weaker. In turn, a lower signal-to-noise ratio of the restored image will result. Therefore, considering the noise problem, it is clear that our optimum value T_m must be obtained, at least in practice, for optimum image restoration.

As mentioned earlier in this section, a phase filter may be synthesized by means of a holographic technique (perhaps by a computer-generated hologram). We now see how such a phase filter works in image restoration. Let us assume that the transmittance of a holographic phase filter is

$$T(p) = \tfrac{1}{2}\{1 + \cos[\phi(p) + \alpha_0 p]\}, \tag{7.9}$$

where α_0 is an arbitrarily chosen constant, and

$$\phi(p) = \begin{cases} \pi, & p_n \leq p \leq p_{n+1},\ n = \pm 1, \pm 3, \pm 5, \ldots \\ 0, & \text{otherwise} \end{cases}. \tag{7.10}$$

With the amplitude filter added, the complex filter function can be written as

$$H_1(p) = A(p)T(p) = \tfrac{1}{2}A(p) + \tfrac{1}{4}[H(p)\exp(i\alpha_0 p) + H^*(p)\exp(-i\alpha_0 p)]. \tag{7.11}$$

where $H(p)$ is the approximate inverse filter function of eq. (7.6). Note also that $H(p) = H^*(p)$, because of eq. (7.10).

Now, if this complex filter $H_1(\mathrm{p})$ is inserted in the spatial frequency plane P_2

of fig. 7.1, then the complex light field behind P_2 is

$$F_2(p) = \tfrac{1}{2}F(p)A(p) + \tfrac{1}{4}[F(p)H(p)\exp(i\alpha_0 p) + F(p)H^*(p)\exp(-i\alpha_0 p)]. \tag{7.12}$$

It is clear that the first term of eq. (7.12) is the restored Fourier spectrum due to the amplitude filter alone, which will be diffracted on the optical axis at the output plane P_3 of fig. 7.1. The second and third terms are the restored Fourier spectra of the smeared image, in which the restored images due to these terms will be diffracted away from the optical axis at the output plane, and centered at $\alpha = \alpha_0$ and $\alpha = -\alpha_0$, respectively. As an illustration, fig. 7.6 shows the calculated irradiance of a restored point image blurred by linear motion, after transmission through an amplitude filter ($T_m = 0.1$), a phase filter ($p \leq p_4$), and the combination of the amplitude and phase filters. The results of an experiment using such filters are shown in fig. 7.7. It may be worth pointing out that the restoration of defocused images can be accomplished by a similar procedure, except that the complex spatial filter in this latter case must have rotational symmetry.

It should be emphasized that the relative degree of restoration is with respect to the spatial bandwidth of interest. It is clear that the ultimate limit of Δp is restricted by the diffraction limits of the optical imaging and processing systems, whichever comes first (ref 7.12). Therefore, it does not follow that a high degree of restoration implies restoration beyond the diffraction limit.

7.2 SYNTHETIC-APERTURE RADAR

One of the interesting and unique applications of coherent optical techniques is the processing of synthetic-aperture antenna data. The discussion in this section follows for the most part that of Cutrona et al. (ref. 7.13).

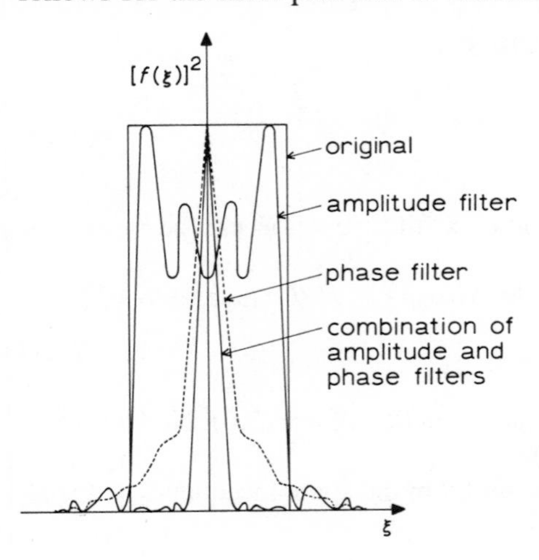

Fig. 7.6 Calculated irradiance of a restored point image blurred by linear motion (By permission of J. Tsujiuchi.)

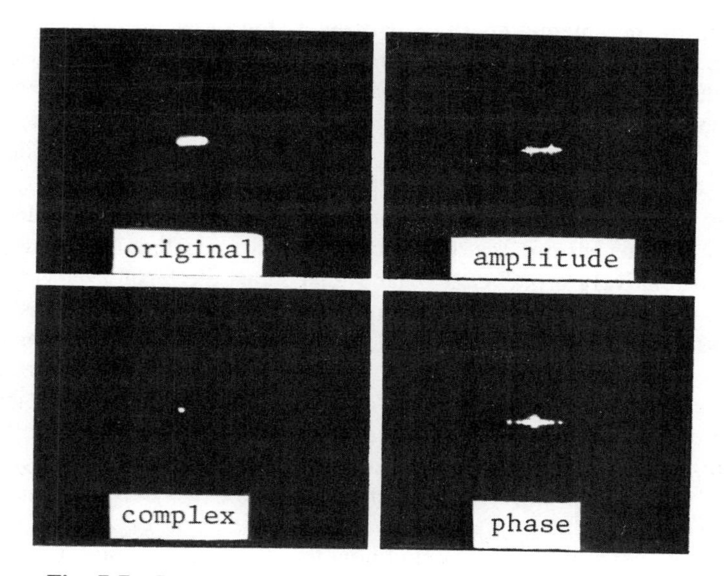

Fig. 7.7 Restored images. (By permission of J. Tsujiuchi.)

Let us consider a sidelooking radar system carried by an aircraft in level flight, as shown in fig. 7.8. Suppose that a sequence of pulsed radar signals is directed onto the terrain from the radar system in the plane and that return signals depending on the reflectivity of the terrain across an area adjacent to the flight path are received. Let us define the cross-track coordinate of the radar image as the "ground range" coordinate and the along-track coordinate as the "azimuth" coordinate. It may be convenient, also, to define the coordinate joining the radar trajectory of the plane and any target under consideration as "slant range" coordinate. If a conventional type of radar system is used, then the azimuth resolution will be (at least in principle) of the order of $\lambda r_1/D$, where λ is the wavelength of the radar signals, r_1 is the slant range, and D is the along-track

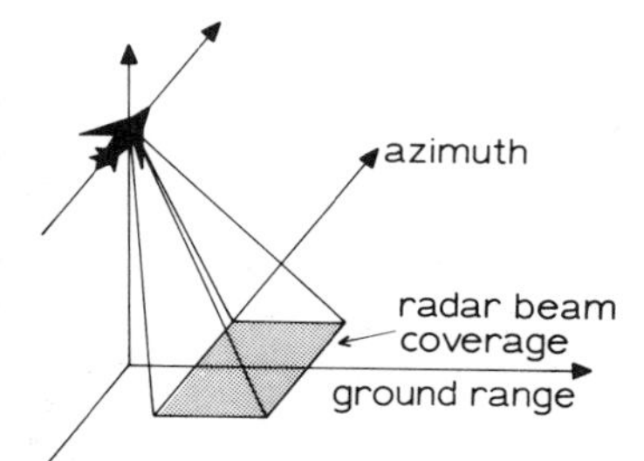

Fig. 7.8 Sidelooking radar.

dimension of the antenna aperture. However, the radar wavelength is several orders of magnitude larger than the optical wavelength, and therefore a very large value of antenna aperture D is required in order to have an angular resolution comparable to that of a photoreconnaissance system. The required antenna length may be hundreds, or even thousands, of feet; obviously it is impractical to realize this in an aircraft.

However, this difficulty can be resolved by means of the synthetic-aperture technique. Let us assume that the aircraft carries a small sidelooking antenna, and that a relatively broadbeam radar signal scans the terrain by virtue of the aircraft motion. The radar pulses are emitted in a sequence of positions along the flight path, which can be treated as if they were the positions occupied by the elements of a linear antenna array. The return radar signal at each of the respective positions is then recorded "coherently" as a function of time; that is, the radar receiver has available a reference signal such that it is able to record the amplitude and phase information simultaneously. The various recorded complex wave forms are then properly combined to synthesize an effective aperture.

In order to examine in more detail how this synthetic-aperture technique can be accomplished, let us consider first a point-target problem and then extend the results, by superposition, to a more complicated case. We assume a point target located at x_1 in fig. 7.9. The radar pulse is produced by periodic rectangular modulation of a sinusoidal signal of radian frequency ω. This periodic pulsing provides the range information and the fine azimuth resolution, provided that the distance traveled by the aircraft between sample pulses is smaller than $\pi/\Delta p$, where Δp is the spatial bandwidth of the terrain reflections. Now the radar signals returned to the aircraft from a point object may be written as

$$S_1(t) = A_1 \exp\left[i\omega\left(t - \frac{2r}{c}\right)\right], \tag{7.13}$$

where A_1 is an appropriate complex constant. The complex quantity A_1 has contained in it such factors as transmitted power, target reflectivity, phase shift, and the inverse fourth-power propagation law. By the use of the paraxial approx-

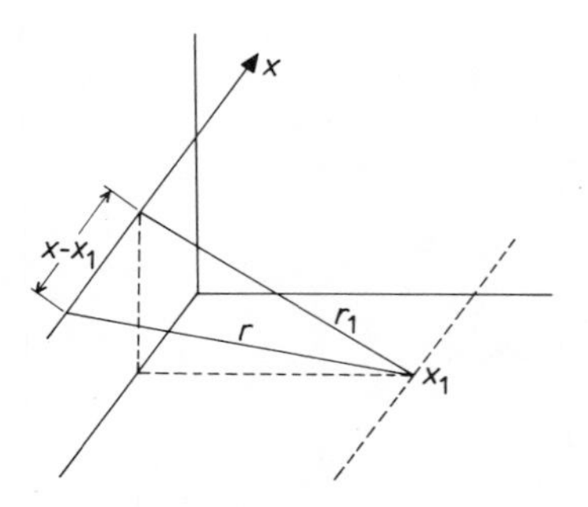

Fig. 7.9 Geometry of sidelooking radar.

imation, the range r may be approximated by

$$r \simeq r_1 + \frac{(x - x_1)^2}{2r_1}. \tag{7.14}$$

Substitution of eq. (7.14) into eq. (7.13) yields

$$S_1(t) = A_1(x_1, r_1) \exp\left\{i\left[\omega t - 2kr_1 - \frac{k}{r_1}(x - x_1)^2\right]\right\}, \tag{7.15}$$

where $k = 2\pi/\lambda$. Equation (7.15) is a function of t and x, in which the aircraft motion links the space variation to the time variable by the relation

$$x = vt, \tag{7.16}$$

where v is the velocity of the aircraft. Now if we assume that the terrain at range r_1 consists of a collection of n points targets, then by superposition the total returned radar signal may be written

$$S(t) = \sum_{n=1}^{N} S_n(t)$$
$$= \sum_{n=1}^{N} A_n(x_n, r_1) \exp\left\{i\left[\omega t - 2kr_1 - \frac{k}{r_1}(vt - x_n)^2\right]\right\}. \tag{7.17}$$

If the returned radar signal of eq. (7.17) is synchronously demodulated, then the demodulation signal may be written as

$$S(t) = \sum_{n=1}^{N} |A_n(x_n, r_1)| \cos\left[\omega_c t - 2kr_1 - \frac{k}{r_1}(vt - x_n)^2 + \phi_n\right], \tag{7.18}$$

where ω_c is the arbitrary carrier frequency and ϕ_n is the arbitrary phase angle.

In order to store the return radar signal of eq. (7.18), a cathode ray tube is used. The demodulated signal is fed in to modulate the intensity of the electron beam, which is swept vertically across the cathode ray tube in synchronism with return radar pulses (fig. 7.10). If this modulated cathode ray display is imaged on a strip of photographic film, which is drawn at a constant horizontal velocity, then the successive range traces will be recorded side by side, producing a two-dimensional format (fig. 7.11). The vertical lines represent the successive range sweeps and the horizontal dimension represents the azimuthal position, which is the sample version of $S(t)$. This sampling is carried out in such a way that, by the time the samples have been recorded on the film, the sampled version is essentially indistinguishable from the unsampled version. In this recording, it is clear that the time variabler is converted to a space variable defined in terms of the distance along the recorded film. With the proper reading exposure, the

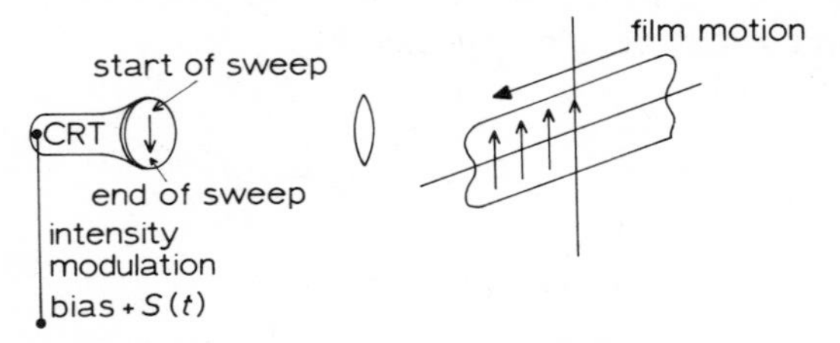

Fig. 7.10 Recording of the return radar signal for subsequent optical processing.

transparency of the recorded film represents the azimuthal history of the returned radar signal. Thus considering only the data recorded along a line $y = y_1$ on the film, the transmittance can be written as

$$T(x, y_1) = K_1 + K_2 \sum_{n=1}^{N} |A_n(x_n, r_1)| \cos\left[\omega_x x - 2kr_1 - \frac{k}{r_1}\left(\frac{v}{v_f}x - x_n\right)^2 + \phi_n\right], \tag{7.19}$$

where K_1 and K_2 are bias and proportionality constants, $x = v_f t$ is the film coordinate, v_f is the velocity of the film motion, and $\omega_x = \omega_c/v_f$.

Since the cosine can be written into the sum of two exponential conjugate forms, the summation of eq. (7.19) can be written as the sum of two terms T_1 and T_2, which are given by

$$T_1(x, y_1) = \frac{K_2}{2} \sum_{n=1}^{N} |A_n(x_n, r_1)| \times \exp\left\{i\left[\omega_x x - 2kr_1 - \frac{k}{r_1}\left(\frac{v}{v_f}\right)^2\left(x - \frac{v_f}{v}x_n\right)^2 + \phi_n\right]\right\}, \tag{7.20}$$

and

$$T_2(x, y_1) = \frac{K_2}{2} \sum_{n=1}^{N} |A_n(x_n, r_1)| \times \exp\left\{-i\left[\omega_x x - 2kr_1 - \frac{k}{r_1}\left(\frac{v}{v_f}\right)^2\left(x - \frac{v_f}{v}x_n\right)^2 + \phi_n\right]\right\}. \tag{7.21}$$

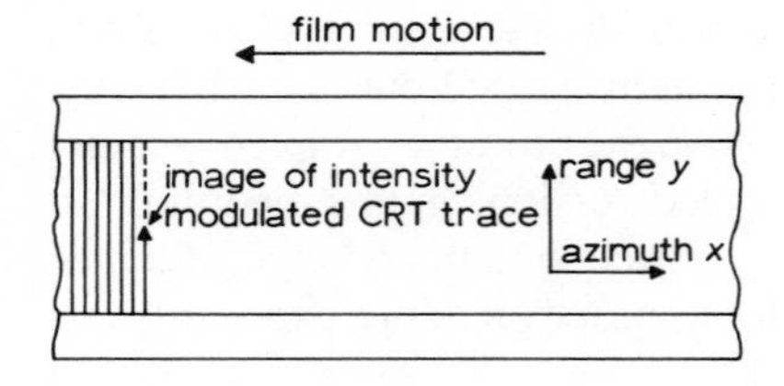

Fig. 7.11 Radar signal recording format.

For simplicity in illustration, we restrict ourselves to the one-target problem; thus for $n = j$ eq. (7.20) can be written as

$$T_1(x, y_1) = C \exp(i\omega_x x) \exp\left\{-i\frac{k}{r_1}\left(\frac{v}{v_f}\right)^2\left(x - \frac{v_f}{v}x_j\right)^2\right], \tag{7.22}$$

where C is an appropriate complex constant. The first exponent of eq. (7.22) introduces a linear phase function, i.e., a simple tilt of the transmitted wave front. The oblique angle of this tilt from the plane of the film is given by

$$\sin\theta = \frac{\omega_x}{k_1}, \tag{7.23}$$

where $k_1 = 2\pi/\lambda_1$, with λ_1 the wavelength of the illuminating light source. From the second exponent of eq. (7.22), it can be seen that the transmitted function is that of a positive cylindrical lens centered at:

$$x = \frac{v_f}{v}x_j, \tag{7.24}$$

with a focal length of

$$f = \frac{1}{2}\frac{\lambda}{\lambda_1}\left(\frac{v_f}{v}\right)^2 r_1, \tag{7.25}$$

where λ_1 is the wavelength of the illuminating light source.

Therefore it can be seen that eq. (7.20), except for a linear phase function, is a superposition of N positive cylindrical lenses, centered at locations given by

$$x = \frac{v_f}{v}x_n, \qquad n = 1, 2, \ldots, N. \tag{7.26}$$

Similarly, eq. (7.21) contains a linear phase factor $-\theta$, and represents a superposition of N negative cylindrical lenses, with centers given by eq. (7.26) and with focal lengths given by eq. (7.25).

In order to reconstruct the image, the transparency yielding eq. (7.19) is illuminated by a monochromatic plane wave, as shown in fig. 7.12. Then by the Fresnel–Kirchhoff theory we can show that the real images produced by $T_1(x, y_1)$ and the virtual images produced by $T_2(x, y_1)$ will be reconstructed at the front and back focal planes of the film. The relative positions of the images of the point scatterers are preserved along the line foci because the multiple centers of the lenslike structure of the film are determined by the positions of the point scatterers. However, the reconstructed image will be spread in the y direction; this is because the film, in fact, is a realization of a one-dimensional function along $y = y_1$, and hence exerts no focal power in this direction.

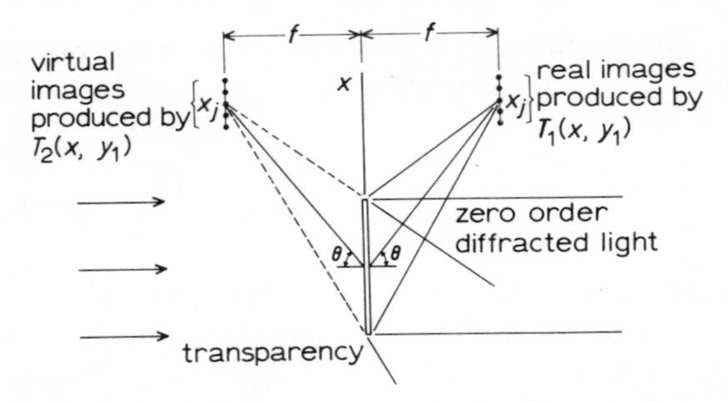

Fig. 7.12 Image reconstruction produced by the film transparency for $y = y_1$.

Since it is our aim to reconstruct an image not only in the azimuthal direction but also in the range direction, it is necessary to image the y coordinate directly onto the focal plane of the azimuthal image. In order to accomplish this we recall that from eq. (7.25) the focal length of the azimuthal image distribution is linearly proportional to the range r_1. In turn, the focal distance is linearly related to the y coordinate under consideration. Thus, to construct the terrain map, we must image the y coordinate of the transmitted signal onto a tilted plane determined by the focal distances of the azimuthal direction. This imaging procedure may be carried out easily by inserting a positive conical lens immediately behind the recording film, as shown in fig. 7.13. It is clear that, if the transmittance of the conical lens is

$$T_l(x, y_1) = \exp\left(-i\frac{k_1}{2f}x^2\right), \tag{7.27}$$

where f is a linear function of r_1, as shown in eq. (7.25), then it is possible to remove all the virtual diffraction from the entire titlted plane to the point at infinity, while leaving the transmittance in the y direction unchanged. Thus, if the cylindrical lens is placed at the focal distance from the film transparency, it will create a virtual image of the y direction at infinity. Now the azimuthal and the range images (i.e., the x and y directions) coincide, but at the point of

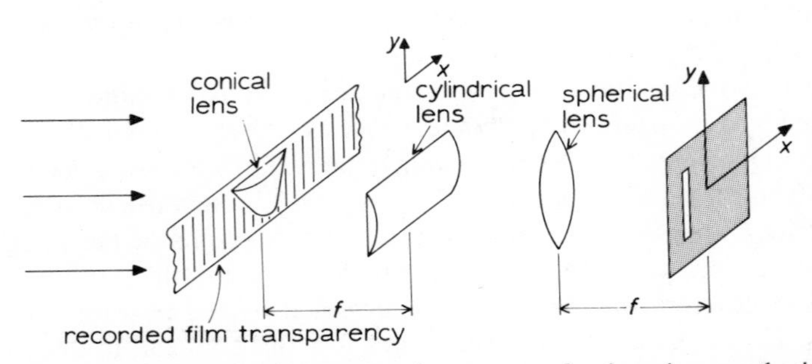

Fig. 7.13 An optical information processing system for imaging synthetic-aperture antenna data.

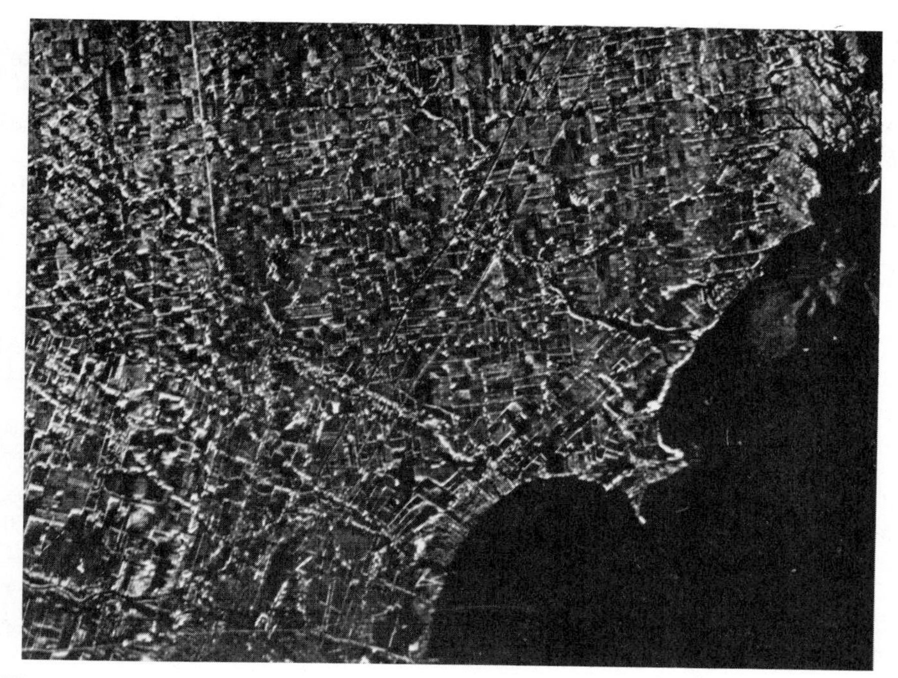

Fig. 7.14 Synthetic-aperture radar image of the Monroe, Mich. area. (By permission of L. J. Cutrona.)

infinity. They can be brought back to a finite distance by a spherical lens. In this final operation, a real image of the azimuthal and range coordinates of the terrain will be mapped in focus at the output plane of the system. In practice, however, the desired image may be recorded through a slit in the output plane.

Figure 7.14 is an example of the images obtainable by optical processing of synthetic-aperture radar data. The radar image was obtained in the Monroe area, located south of Detroit, Michigan. This photograph shows a variety of scatterers including city streets, wooded areas, and farmland. Lake Erie, with some broken ice floes, can be seen on the right.

7.3 OPTICAL PROCESSING OF BROAD-BAND SIGNALS

An important application of coherent optical systems is in the spectrum analysis of broad-band signals. The optical processing technique suggested by Thomas (ref. 7.14) was found to be capable of generating a near realtime spectrum analysis of a large space-bandwidth signal. The application of this technique to the processing of wide-band radio signals has been reported by Markevitch (ref. 7.15).

In this section, we consider an optical processing technique that is capable of handling broad-band signals. Before going into this discussion, we must recognize a basic limitation of the multichannel optical spectrum analyzer (see sec. 6.5): the resolution is limited by the length of the input channel (i.e., the width of the input aperture). However, this basic limitation may be easily overcome by using a two-dimensional optical spectrum analyzer such as sketched in fig. 6.16.

If an input transparency $f(x, y)$ is inserted in the input plane of the analyzer, then the complex light field distributed on the (p, q) plane will be recalled to be

$$F(p, q) = C \iint f(x, y) \exp[-i(xp + yq)]\, dx\, dy, \tag{7.28}$$

which is the Fourier transform of the input signal $f(x, y)$, where C is a complex constant.

Now we see, by means of this basic two-dimensional Fourier transformation, that a large space-bandwidth signal may be processed by means of a slightly different technique. The broad-band signal may be recorded, for example, by photographing a cathode ray tube raster pattern without interleaving, as shown in fig. 7.15. The broad-band signal is made to modulate the intensity of the CRT beam. The scan rate of the CRT beam is adjusted so that the maximum frequency content of the broad-band signal will be properly recorded, without loss of resolution. The required scan velocity of the electron beam is therefore

$$v \geq \frac{f_m}{R}, \tag{7.29}$$

where v is the scan velocity, f_m is the maximum frequency content of the broad-band signal, and R is the resolution limit of the optical system. Of course, the return sweep of the CRT beam is assumed to be much faster than $1/f_m$.

The transmittance of the recorded format may be represented by

$$f(x, y) = \sum_{n=1}^{N} f(x) f(y), \tag{7.30}$$

where $N = h/b$. Here

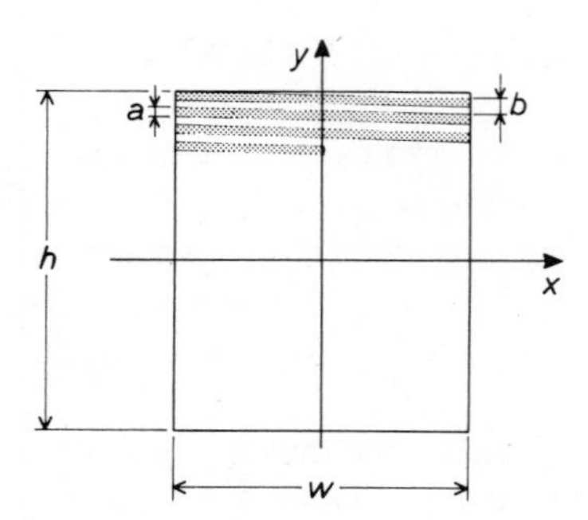

Fig. 7.15 Raster scan input signal transparency.

$$f(x) = \begin{cases} f\left[x + (2n - 1)\dfrac{w}{2}\right], & -\dfrac{w}{2} \leq x \leq \dfrac{w}{2} \\ 0, & \text{otherwise} \end{cases} \tag{7.31}$$

and

$$f(y) = \begin{cases} 1, & \dfrac{h}{2} - (n - 1)b - a \leq y \leq \dfrac{h}{2} - (n - 1)b \\ 0, & \text{otherwise} \end{cases} \tag{7.32}$$

If this format is inserted in the input plane of the analyzer, then the complex light field distributed on the spatial frequency domain may be written as

$$F(p, q) = C \iint \sum_{n=1}^{N} f(x)f(y) \exp[-i(xp + yq)]\, dx\, dy, \tag{7.33}$$

where C is a complex constant.

Equivalently, eq. (7.33) can be written as

$$F(p, q) = C \sum_{n=1}^{N} \int f(x) \exp(-ixp)\, dx \int f(y) \exp(-iyq)\, dy. \tag{7.34}$$

By substituting eqs. (7.31) and (7.32) in eq. (7.34), we obtain

$$F(p, q) = C_1 \sum_{n=1}^{N} \operatorname{sinc}\left(\frac{qa}{2}\right)\left\{\operatorname{sinc}\left(\frac{pw}{2}\right) * F(p) \exp\left[i\frac{pw}{2}(2n - 1)\right]\right\} \times \exp\left\{-i\frac{q}{2}[h - 2(n - 1)b - a]\right\}, \tag{7.35}$$

where C_1 is a complex constant, and

$$F(p) = \int f(x') \exp(-ipx')\, dx',$$

$$x' = x + (2n - 1)\frac{w}{2}, \qquad \text{and} \qquad \operatorname{sinc} X \triangleq \frac{\sin X}{X}.$$

For simplicity, let us assume that the broad-band signal is a simple sinusoid [i.e., $f(x') = \sin p_0 x$], $w \simeq h$, and $b \simeq a$. Then eq. (7.35) may be written as

$$F(p, q) = C_1 \operatorname{sinc}\left(\frac{qa}{2}\right) \sum_{n=1}^{N} \left\{\operatorname{sinc}\left(\frac{pw}{2}\right) * \frac{1}{2}[\delta(p - p_0) + \delta(p + p_0)] \times \exp\left[i\frac{wp_0}{2}(2n - 1)\right]\right\} \exp\left\{-i\frac{q}{2}[w - (2n - 1)b]\right\}. \tag{7.36}$$

To further simplify the analysis, let us consider only one of the components, say $\delta(p - p_0)$, thus

$$F_1(p, q) = C_1 \operatorname{sinc}\left[\frac{w}{2}(p - p_0)\right]\operatorname{sinc}\left(\frac{qa}{2}\right)\exp\left(-i\frac{qw}{2}\right) \times \sum_{n=1}^{N} \exp\left[i\frac{1}{2}(2n - 1)(wp_0 + bq)\right]. \tag{7.37}$$

The corresponding irradiance may be written as

$$I_1(p, q) = |F_1(p, q)|^2 = C_1^2 \operatorname{sinc}^2\left[\frac{w}{2}(p - p_0)\right]\operatorname{sinc}^2\left(\frac{qa}{2}\right) \times \sum_{n=1}^{N} \exp[i2n\theta] \sum_{n=1}^{N} \exp[-i2n\theta], \tag{7.38}$$

where

$$\theta = \frac{1}{2}(wp_0 + bq). \tag{7.39}$$

But (ref. 7.16)

$$\sum_{n=1}^{N} e^{i2n\theta} \sum_{n=1}^{N} e^{-2n\theta} = \left(\frac{\sin N\theta}{\sin \theta}\right)^2. \tag{7.40}$$

Therefore, $I_1(p, q)$ may be written as

$$I_1(p, q) = C_1^2 \operatorname{sinc}^2\left[\frac{w}{2}(p - p_0)\right]\operatorname{sinc}^2\left(\frac{qa}{2}\right)\left(\frac{\sin N\theta}{\sin\theta}\right)^2. \tag{7.41}$$

It may be seen from eq. (7.41) that the first sinc factor represents a relatively narrow spectral line in the p direction, located at $p = p_0$, which is derived from the Fourier transform of a pure sinusoid truncated within the width w of the input transparency. The second sinc factor represents a relatively broad spectral band in the q direction, which is derived from the Fourier transform of a rectangular pulse of width a (i.e., the channel width). The last factor deserves special mention, in that for large values of N it approaches a sequence of narrow pulses. The locations of the pulses (i.e., the peaks), may be obtained by letting $\theta = n\pi$, which gives

$$q = \frac{1}{b}(2\pi n - wp_0), \qquad n = 1, 2, \ldots. \tag{7.42}$$

Thus this factor yields the fine spectral resolution in the q direction.

To continue the interpretation of eq. (7.41), we note that irradiance in the p direction is confined within a relatively narrow region, which essentially depends on w. The half-width of the spectral spread (i.e., from the center to the first zero) is seen to be

$$\Delta p = \frac{2\pi}{w}, \tag{7.43}$$

which is the resolution limit of an ideal transform lens. In the q direction, the irradiance is first confined within a relatively broad spectral band (which primarily depends on the channel width a) centered at $q = 0$, and then modulated by a sequence of narrow periodic pulses. The half-width of the broad spectral band is

$$\Delta q = \frac{2\pi}{a}. \tag{7.44}$$

The separation of the narrow pulses is obtained from a similar equation, i.e.,

$$\Delta q_1 = \frac{2\pi}{b}. \tag{7.45}$$

It may be seen from eqs. (7.44) and (7.45) that there will be only a few pulses located within the spread of the broad spectral band for each p_0, as shown in fig. 7.16.

The actual location of any of the pulses is determined by the signal frequency. Thus if the signal frequency changes, the position of the pulses also changes, in accordance with

$$dq = \frac{w}{b} dp_0. \tag{7.46}$$

The displacement in the q direction is proportional to the displacement in the p direction. Since the pulse width decreases as the number of the scan lines N

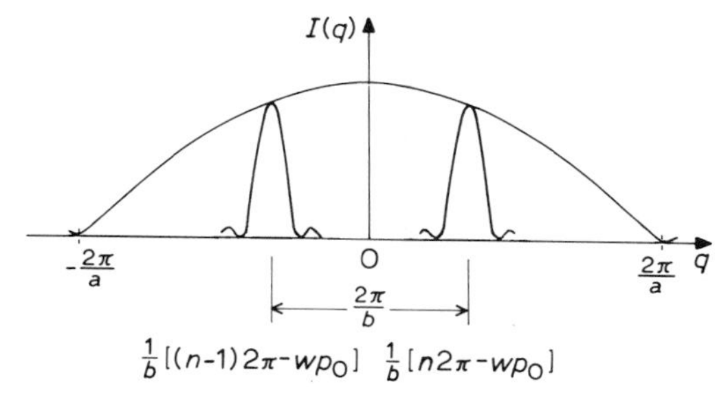

Fig. 7.16 A broad spectral band modulated by a sequence of narrow pulses.

increase, the output spectrum would yield a frequency discrimination equivalent to that obtained with a one-dimensional processor for a continuous signal, which is *NW* long.

In order to avoid the ambiguity in reading the output plane, all but one of the periodic pulses should be ignored. This may be accomplished by masking all of the output plane except the region

$$-\frac{\pi}{b} \leq q \leq \frac{\pi}{b}, \qquad 0 \leq p \leq \infty, \tag{7.47}$$

as shown in fig. 7.17. The periodic pulses are $2\pi/b$ apart; therefore, as the input signal frequency advances, one pulse leaves the open region defined by eq. (7.47) at $q = -\pi/b$ while another pulse enters the region at $q = \pi/b$. As a result of eq. (7.46), a single bright spot would be scanned out diagonally on the output plane. The frequency locus of the input signal is also determined in cycles per second. Finally, to remove the nonuniform characteristic of the second sinc factor of eq.(7.41), a weighting transparency may be placed at the output plane of the processor.

So far we have assumed that the input signal has been recorded on a linear photographic film (a subject upon which we elaborate in the next chapter). The analysis has been carried out on a one-frame basis. However, the operation could also be performed on a continuously running tape basis. Thus it is possible to synthesize a near real-time optical spectrum analyzer.

A convenient mode of near real-time operation would be to use a continuous strip of photographic film and a rapid film developer, as diagramed in fig. 7.18. A modulating signal is fed into the z-axis of a cathode ray tube. The modulated electron beam is then swept across the CRT by means of an internal sawtooth

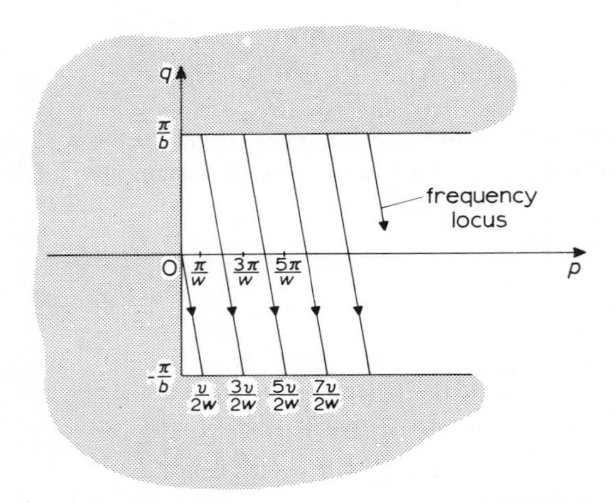

Fig. 7.17 Frequency locus at the output spectral plane.

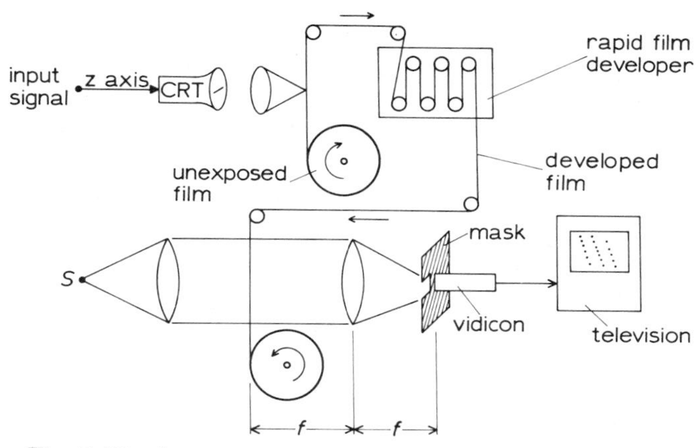

Fig. 7.18 Synthesis of a near real-time spectrum analyzer.

generator. The output of the scanned CRT is then imaged on a strip of slow-moving photographic film. The exposed film then proceeds to a rapid film developer. This rapid film developer is commercially available (refs. 7.14, 7.15); developing and fixing take a few seconds to a minute. After developing, the film is transported to the input plane spectrum analyzer. The output spectrum may be displayed on a television screen as shown in fig. 7.18. The continuous time-varying spectrum of the signal will be observed. This output spectrum can also be photographed on a continuous strip of film (by means of a movie camera) for later use.

The access time from the input processing signal to the output spectrum display is generally limited by the transit time of the film traveling from the CRT to the optical processor. This access time depends somewhat on the physical configuration of the system; however, the most important limiting factor is the film development time. It may be feasible to reduce the access time to something between a few seconds and a minute.

The wide-band optical spectrum analyzer offers considerable flexibility in data handling. To name just one application, the output time-varying spectrum may be directly connected to an appropriate analog-digital converter and then to a digital computer for signal processing. This application may be important to the future of automatic signal detection, synthesis, and recognition. A near real-time continuous optical spectrum analyzer with a space-bandwidth product capability greater than 10^7 is within the current state of the art (refs. 7.14, 7.15).

7.4 SYNCHRONOUS DUAL-CHANNEL OPTICAL SPECTRUM ANALYZER

In comparing two time signals, it is often necessary to perform synchronous Fourier transformation on them. For example, in order to measure the propagation time delay of a signal embedded in noise, we may simultaneously transform the signals of the transmitter and receiver to produce two synchronous spectrograms. By measuring the time difference of the characteristic spectrum between the two spectrograms, the propagation time of the signal can be obtained. To achieve this synchronous dual-channel processing capability with electronic spectrum analyzers, we may just record the two signal synchronously on a magnetic disc or tape and then process them sequentially. Alternatively, the signals may be processed through two independent processors simultaneously. With the first appraoch, the processing time would be doubled while, with the second approach, the equipment cost would be doubled. In this section, we demonstrate a technique by which a single-channel optical spectrum analyzer can be modified to perform dual-channel synchronous processing with no sacrifice in processing time and very little addition in cost (ref. 7.17).

The concept of an optical spectrum analyzer has already been discussed in chap. 6. Basically, the input voltage-time signal is used to modulate the intensity of a light source (e.g., CRT, LED) and the time-varying light intensity is mapped into a spatial signal of varying amplitude transmittance by recording the intensity pattern linearly on a moving film strip. The developed transparency is then scanned across the input aperture of the optical processor while the output spectrum is being recorded on another moving film strip at the output plane through a narrow slit. To modify this basic system for dual-channel synchronous processing, a dual-beam (with independent intensity controls) CRT is used in the temporal-to-spatial signal conversion. The intensity of each beam is modulated by one of the input signals, as shown in fig. 7.19. The time-varying intensity pattern is then recorded as two parallel spatial signals on a moving film strip. If we simply transform this input transparency with a positive lens, we would have at the output the spectra of the two input signals superimposed on each other. In order to obtain each spectrum individually, the two spectra must be separated without ambiguity. This can be achieved with the use of a half-wave retardation plate.

A linearly polarized laser is used as the illuminating source, as shown in fig. 7.20. One of the channels at the input plane is covered with a half-wave

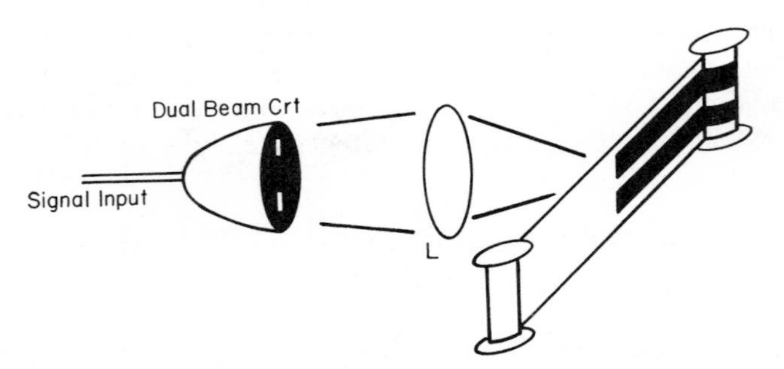

Fig. 7.19 Generation of input transparency for dual-channel synchronous processing.

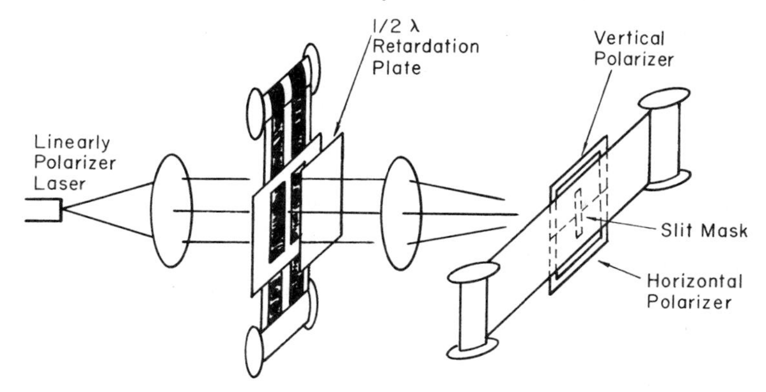

Fig. 7.20 Dual-channel synchronous processing.

retardation plate. The retardation plate rotates the angle of polarization of the laser light by 90°. The modulated light field of one channel would therefore have a polarization perpendicular to that of the second channel. At the output plane, we would once again have the spectra of the two channels superimposed on each other. However, the spectra would have a polarization perpendicular to each other. By covering the top sideband with a horizontal polarizing sheet, one of the sidebands of each spectrum would be filtered out. The two polarizing sheets are made to overlap at the middle, thus blocking out entirely the undesirable zero-order diffraction. With this simple modification, the spectra of the two input signals can be displayed separately and simultaneously on the output plane.

The dual-channel operation is achieved with little additional cost and with virtually no sacrifice in performance. The one-dimensional processing format of the basic optical spectrum analyzer is not very efficient in the use of the system capacity. In the extension into a dual-channel system, the additional capability is achieved by simply utilizing the system capacity more efficiently.

We may also point out that multichannel processing can also be performed with astigmatic processors. However, with the astigmatic processors, the spectrum of the channels are displayed side by side in parallel. They cannot be used to generate spectrograms (frequency versus time display) of two channels continuously and synchronously as required in time delay measurements. The major disadvantage of this technique is that cross-talk cannot be completely eliminated because cross-polarized light is inadvertently passed through the polarizing sheets. The SNR is dependent mainly on the quality of the polarizers. With off-the-shelf commercial polarizers, the SNR would be in the range of 20–23 db.

7.5 WIDE-BAND SPECTRUM ANALYSIS WITH AREA MODULATION

The concept of optical spectrum analysis for temporal signals has been illustrated in sec. 6.5. Essentially, the one-dimensional temporal input signal is converted into a one-dimensional spatial signal that is then processed with an optical spectrum analyzer. However, such a one-dimensional signal conversion is very inefficient in utilizing the capacity of the two-dimensional optical analyzer. In order to process a signal with a large space bandwidth, a very large optical system would be required. In 1966 Thomas (ref. 7.14) introduced a wide-band processing technique that would fully make use of the two-dimensional property of the optical system for the processing of one-dimensional temporal signals (sec. 7.3). With this technique, signals with very large space bandwidth can be processed without the need for an unacceptably large optical system. It was shown that the space-bandwidth product available with such a technique could be as high as16×10^6.

The most commonly used technique for the temporal-to-spatial signal conversion is density modulation. The time-varying input signal is first converted into a spatial signal of varying amplitude transmittance. However, a very linear conversion is difficult to achieve with density modulation due to the nonlinear characteristic of the recording medium (e.g., photographic film) used in the signal conversion. Area modulation was proposed in sec. 6.9 as an alternate signal conversion format. With its two-tone recording characteristic, area modulation is not affected by the nonlinearity of the recording medium. We note that area modulation has been used for sound track recording in motion pictures for some time (ref. 7.18). The use of area modulation for spectrum analysis was demonstrated by Dyachenko et al. (ref. 7.19) and Felstead (ref. 7.20). The application of area-modulated signals for optical correlation was illustrated by Pernick (ref. 7.21). In this section, we demonstrate the feasibility of using area modulation for wide-band signal processing (ref. 7.22).

The area-modulated input for wide-band signal processing can be obtained with a CRT scanner. The input signal is first propely biased and then modulated with a very high frequency carrier sinusoid. The modulated signal is then fed into the y-input of the CRT. At the same time, a ramp function is used to produce a vertical sweep by modulating the x-input. The CRT scan is recorded on a film strip driven along the y direction. With the proper scan rate, vertical channels of area-modulated signals would be recorded side by side as shown in fig. 7.21. The vertical scan rate of the CRT beam should be fast enough to satisfy the criterion

$$v > \frac{f_m}{R}, \tag{7.48}$$

where v is the scan velocity of the CRT beam, f_m is the highest signal frequency, and R is the resolution of the CRT. The film drive scan rate should be fast enough to assure that there is no overlapping between adjacent channels. The frequency of the modulating sinusoid should be beyond the resolution of the CRT in order to produce a uniform transmittance. We note that the area-modulated signal obtained with the method we described would be in bilateral form (sec. 6.7). The unilateral form can be obtained by simply masking half of the CRT scan.

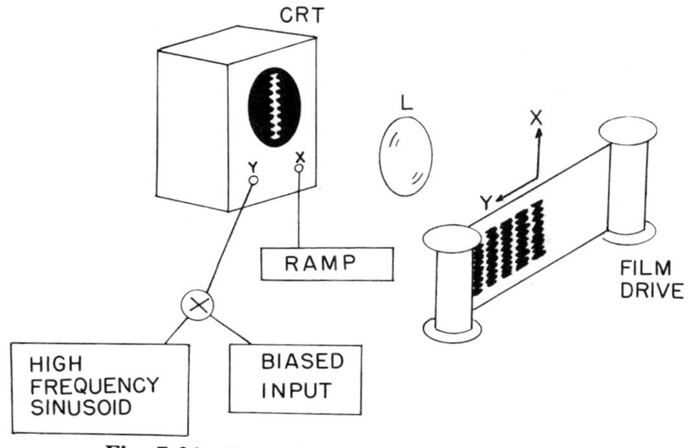

Fig. 7.21 Recording of area-modulated input.

From eq. (6.123) the transmittance of the resulting 2-D input transparency can be written as

$$f(x, y) = \sum_{n=1}^{n} \operatorname{rect}\left(\frac{x}{L}\right) \operatorname{rect} \frac{y - nD}{2[B + f_n(x)]}, \tag{7.49}$$

where $N = H/D$ is the number of scan lines in the input aperture, $f_n(x) = f(x + nD)$ is the transmittance function of the nth line, B is the bias level, D is the separation between the scan lines, and H and L are the height and width of the input format, as shown in fig. 7.22. If we insert the transparency of eq. (7.49) in the input plane of a coherent optical processor, then the complex light field distributed on the spatial frequency plane (p, q) can be written

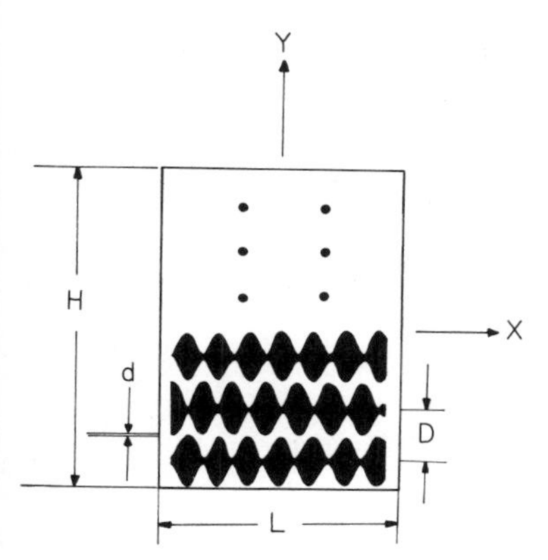

Fig. 7.22 Geometry of area-modulated input for wide-band processing.

$$F(p, q) = \sum_{n=-N}^{N} C \int_{-L/2}^{L/2} [B + f_n(x)] \operatorname{sinc}\{q[B + f_n(x)]\} \exp[-i(px - nDq)]\, dx. \tag{7.50}$$

In order to avoid ambiguity in reading the higher order diffractions, one of the first-order diffractions at the spatial frequency plane is chosen for the output reading. This can be accomplished by masking all the output plane except the region

$$-\frac{\pi}{D} \le q \le \frac{\pi}{D}, \qquad 0 \le p \le \infty. \tag{7.51}$$

If $D > 2B$, over the open aperture of eq. (7.51), the sinc factor of eq. (7.50) can be approximated by

$$\operatorname{sinc}\{q[B + f_n(x)]\} \simeq 1. \tag{7.52}$$

We note that the condition $D > 2B$ is also necessary to assure no overlapping between the adjacent channels of the input transparency. With reference to eq. (7.52), the complex light distribution behind the open aperture of the mask can be written as

$$F(p, q) = H(p, q) \sum_{n=1}^{N} C \int_{-L/2}^{L/2} [B + f_n(x)] \exp[-i(px + nDq)]\, dx, \tag{7.53}$$

where

$$H(p, q) = \begin{cases} 1, & -\dfrac{\pi}{D} \le q \le \dfrac{\pi}{D}, \quad 0 \le p \\ 0, & \text{otherwise} \end{cases} \tag{7.54}$$

For illustration, we assume that the wide-band signal is a single sinusoid, that is,

$$f_n(x) = \sin[p_0(x + nL)], \qquad n = 1, 2, \ldots, N. \tag{7.55}$$

The complex light distribution over the open aperture of the mask at the spatial frequency plane would be

$$F(p, q) = C_1 \operatorname{sinc}\left(\frac{LP}{2}\right) + C_2 \operatorname{sinc}\left[\frac{L}{2}(p - p_0)\right] \sum_{n=1}^{N} \exp[in(Lp_0 + Dq)], \tag{7.56}$$

where C_1 and C_2 are the proportionally complex constants. The corresponding irradiance can be written as

$$I(p, q) = |F(p, q)|^2$$
$$= K_1\left[\operatorname{sinc}\left(\frac{LP}{2}\right)\right]^2 + K_2\{\operatorname{sinc}[\frac{L}{2}(p - p_0)]\}^2$$
$$\cdot \sum_{n=1}^{N} \exp\,[in(Lp_0 + Dq)] \sum_{n=1}^{N} \exp\,[-in[Lp_0 + Dq)], \tag{7.57}$$

where K_1 and K_2 are the proportionality constants. By the following well-known identity,

$$\sum_{n=1}^{N} \exp\,[i2n\theta] \sum_{n=1}^{N} \exp\,[-i2n\theta] = \left(\frac{\sin N\theta}{\sin\,\theta}\right)^2, \tag{7.58}$$

$I\,(p, q)$ can be expressed as

$$I(p, q) = \left[K_1 \operatorname{sinc}\left(\frac{LP}{2}\right)\right]^2 + K_2\left\{\operatorname{sinc}\left[\frac{L}{2}(p - p_0)\right]\right\}^2\left(\frac{\sin N\theta}{\sin\,\theta}\right)^2, \tag{7.59}$$

where $\theta = \frac{1}{2}(Lp_0 + Dq)$. We note that eq. (7.59) is similar to the result obtained with the density modulation of eq. (7.41).

It may be seen from eq. (7.59) that the first sinc factor represents a narrow zero-order spectral line diffracted at the origin of the output plane. The second sinc factor represents a narrow spectral line in the p direction, located at $p = p_0$, which is derived from the Fourier transform of a sinusoid over the width L of the input format. For a large value of N, the quantity of $[\sin(N\theta/\sin\,\theta]^2$ converges to a sequence of narrow periodic pulses. The locations of the pulses can be obtained by letting $\theta = n\pi$, which is

$$q = \frac{1}{D}(2n\pi - Lp_0), \qquad n = 1, 2, \ldots. \tag{7.60}$$

Thus the last sinc factor yields the fine spectral resolution in the q direction. From eq (7.60) we also see that the location of any of the pulses is determined by the signal frequency p_0. As the signal frequency changes, the location of the pulse would change correspondingly. That is,

$$dq = \frac{L}{D}dp_0. \tag{7.61}$$

The displacement in the q direction is porportional to the displacement in the p direction. Since the pulse width decreases as the number of the scan lines N

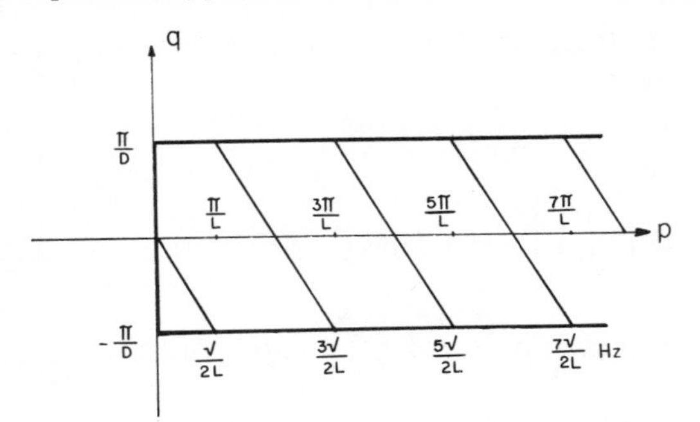

Fig. 7.23 Output spectrum frequency locus.

increases, the output spectrum would yield a frequency discrimination equivalent to that obtained with a one-dimensional processor for a continuous signal that is NL long.

We further note that the periodic pulses of $(\sin N\theta/\theta)^2$ are $2\pi/D$ apart. Therefore, as the input signal frequency advances, one pulse leaves the open region at $q = -\pi/D$, resulting in a diagonal frequency locus in the open region of the output mask, as shown in fig. 7.23.

For illustration, we show in fig. 7.24 the spectrum of a sinusoidal input. The spectrum is obtained with the wide-band processing technique using the area modulation format. The inserted white lines indicate the boundary of the slit aperture. We note that the spectrum within the aperture is identical to that

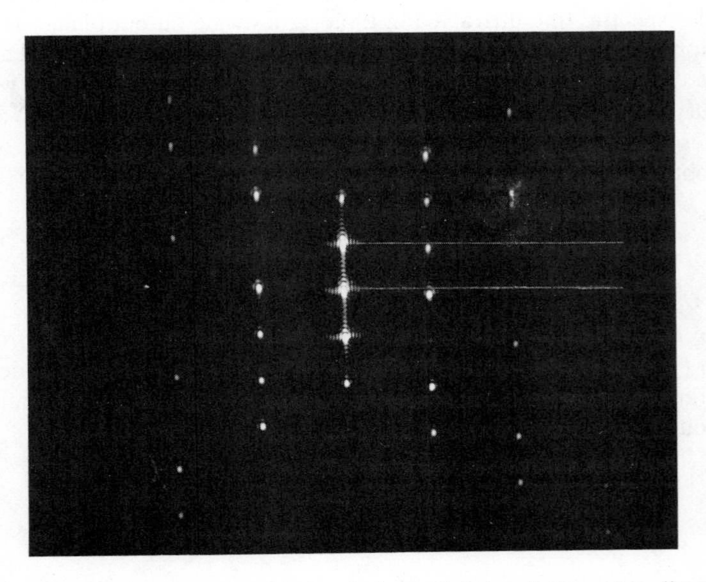

Fig. 7.24 Output spectrum of a wide-band sinusoidal signal. The white lines indicate slit boundaries.

obtained with density modulation (sec. 7.3) and the lack of spurious spectrum indicates that it does not suffer the adverse effect of film nonlinearity. However, the use of area modulation in wide-band signal processing has one drawback. The two-dimensional format of area modulation occupies a much larger system space. Thus, if area modulation is utilized to convert a signal with a given time-bandwidth product, a larger system space would be required as compared with the density modulation.

We have demonstrated the feasibility of utilizing area modulation for wide-band signal processing. The use of area modulation would eliminate the problem of nonlinearity with the recording medium. And since area modulation does not require the precise control of the exposure and film development to ensure linearity, its implementation is much easier. Furthermore, without the linearity restriciton, a much larger class of recording materials may be used. However, the available space-bandwidth product for a given system size is smaller with area modulation than with density modulation. Nevertheless, due to the efficient use of the two-dimensional format of the optical analyzer, the space-bandwidth product obtainable with area modulation is still sufficiently large to be useful.

7.6 NONLINEAR PROCESSING THROUGH HALFTONE SCREEN

In previous sections we have discussed techniques that relied on linear spatial-invariant operations. There are, however, techniques available for nonlinear processing operations (refs. 7.23–7.25). One such approach is an optical homomorphic processing system, as shown in fig. 7.25. We note that such an approach has been successfully applied with digital processing technique (ref. 7.26). Recently, Kato and Goodman (ref. 7.27) proposed a nonlinear processing technique with coherent optical system through halftone screen processes. The extension of Kato and Goodman's halftone screen technique to nonmonotonic nonlinear image processing was subsequently pursued by Dashiell and Sawchuk (ref. 7.28), Strand (ref. 7.29), and Liu et al. (ref. 7.30). The nonlinear processing technique can also be achieved with the Fabry–Perot interferometric method reported by Bartholomew and Lee (ref. 7.31), and with the utilization of film nonlinearity reported by Tai et al. (ref. 7.32). Mention must be made that realtime nonlinear processing of photographic images with liquid crystal valves is currently being pursued by Soffer et al. (ref. 7.33). It would be beyond the scope of this book to discuss all the various techniques available for nonlinear optical information processing; we refer the interested reader to those cited

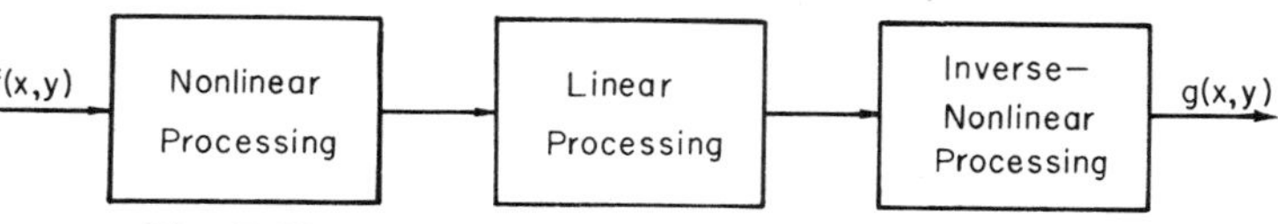

Fig. 7.25 Optical homomorphic processing system.

references. We do however, in this section take the time to discuss a nonlinear processing technique using a halftone screen.

The halftone screen process is a photographic process used widely in the publishing industry. A halftone screen is generally composed of a periodic array of vignetted dots whose transmittance function may be described by fig. 7.26*a*. If the halftone screen is used in conjunction with a high-contrast film of continuous tone light distribution, then the light modulated due to the halftone screen produces a periodic array of area-modulated dots over the film, as depicted in Fig. 7.26*d*. If the exposed film is inserted in the input plane P_1 of a coherent optical processor such as shown in fig. 7.1, then different orders of spectra distribution can be observed at the Fourier plane P_2. By properly controlling the exposure distribution, different sizes of dots can be produced through the use of this technique. In other words, if a slowly varying object transparency is contact printed onto a high-contrast film through a halftone screen, an image consisting of dot arrays of different sizes would be recorded. Thus by properly controlling the dot profiles of the halftone screen, the size of the dots in the halftone image will be nonlinearly related to the transmittance of the object transparency. Then by spatial filtering with a coherent optical processing sys-

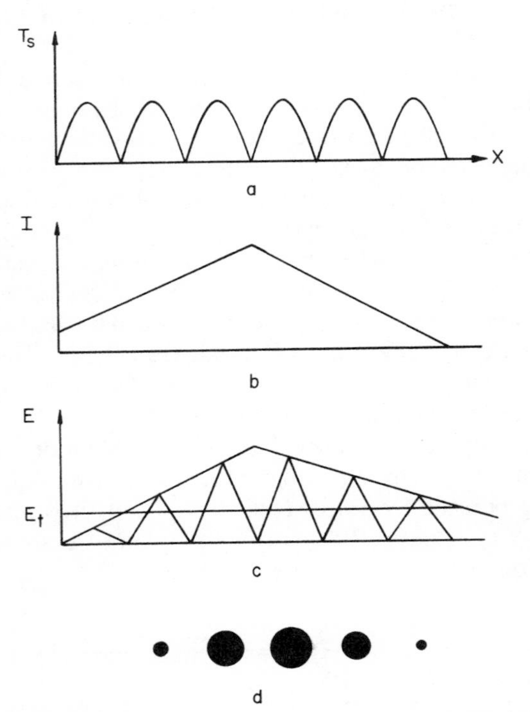

Fig. 7.26 The halftone screen process. (*a*) T_s, intensity transmittance of a halftone screen. (*b*) I, light irradiance distribution. (*c*) E, exposure distribution over high-contrast film; E_t, the threshold level of the high-contrast film. (*d*) Periodic array of area-modulated dots over the film.

tem, the halftone image transparency will yield a filtered image that is nonlinearly related to object transparency.

However, for nonlinear optical information processing, only one-dimensional halftone screens are utilized. The fundamental difficulty in nonlinear optical processing is the design of the halftone screens. Although there are commercial screens available for logarithmic transformer operation, the technique is generally suitable only for a small number of gray levels. We now describe a technique of genertaing one-dimensional halftone screens suitable for coherent optical processing.

Let us assume that we have available a one-dimensional mask that has a periodic intensity transmittance funtion $T(x)$ of period d, that is,

$$T(x) = \begin{cases} 1, & 0 \le x \le \dfrac{d}{D} \\ 0, & \dfrac{d}{D} < x \le d \end{cases}, \tag{7.62}$$

where d is the spatial period, and $D \geq 2$ is a positive constant. To construct a halftone screen, a low-contrast film is placed below in contact with the mask. We now adjust the setup so as to translate the film in discrete d/D steps in the x direction so that the exposures of the film would be

$$E(x) = It_n, \qquad \frac{(n-1)d}{D} \le x \le \frac{nd}{D}, \tag{7.63}$$

where $n = 1, 2, \ldots, N$, I is the localized irradiance, and t_n is the exposure time in the ith step.

If we assume that the recordings are in the linear region of the Hurter-Driffield (H and D) curve, then the density distribution of the recorded film would be

$$D(x) = \gamma \log E(x) - D_0, \tag{7.64}$$

where γ is the gamma of the film and D_0 is the linear intercept of the H and D curve. Because different values of t_n can be used, the previously described values of the periodic density distribution of the halftone screen can be obtained. We note that, if the periodic mask is a Ronchi type, then the density distribution of the recorded halftone screen will be an N-level screen, as shown in fig. 7.27.

We stress that, even in the linear region of the H and D curve, the discrete density steps are generally not proportional to the exposures. In order to obtain the prescribed density steps, a precalibrated exposure should be used. By Ronchi ruling of 19.68 lines/cm a 10-gray-level halftone screen can be generated (ref. 7.30).

Let us illustrate an example of photographic image equidensity contouring with halftone screen techniques. The first step is to produce a halftone trans-

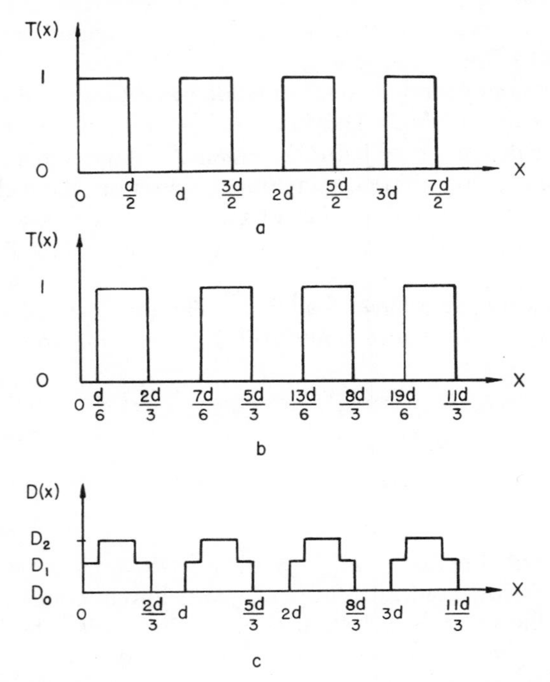

Fig. 7.27 Producing a three-level halftone screen (*a*)Transmittance function of the Ronchi ruling with respect to the first exposure. (*b*) Transmittance function with respect to the second exposure. (*c*) Density function of a three-level halftone screen.

parency of the photographic image that is suitable for coherent information processing. This can be done by contact printing the image transparency through a halftone screen onto a high-contrast film with incoherent or white-light illumination. The exposure of the high-contrast film (i.e., the halftone transparency) can be written by

$$E(x, y) = It\,10^{[-D(x) - D_1(x,y)]}, \tag{7.65}$$

where I is the irradiance, t is the exposure time, $D(x)$ and $d_1(x, y)$ are the density distributions of the halftone screen and the object transparency, respectively, and (x, y) is the spatial coordinate system. By properly recording the high-contrast film, the resultant transmittance function is

$$T(x, y) = \begin{cases} 1, & E(x, y) < E_t \\ 0, & E(x, y) > E_t \end{cases}, \tag{7.66}$$

where E_t is the threshold level of the light-contrast film. From the above equation we see that the image transparency is converted into a spatially modulated binary transparency, which is called halftone image transparency. We further note that by controlling the exposure It, a variety of mappings of photographic image densities into different halftone pulse widths can be achieved. However, the maximum number of different pulse widths in the halftone transparency cannot be higher than the number of gray levels of the halftone screen. The amplitude transmittance of the halftone transparency can be described as

$$f(x) = \sum_{n=-\infty}^{\infty} \delta(x - nd) * \text{rect}\left(\frac{x}{a}\right) \tag{7.67}$$

where $\delta(x)$ is the Dirac delta function, d is the pulse period, a is the pulse width, and

$$\text{rect}\left(\frac{x}{a}\right) \triangleq \begin{cases} 1, & \left|\frac{x}{a}\right| \leq \frac{1}{2} \\ 0, & \text{otherwise} \end{cases}.$$

If the halftone transparency of eq. (7.67) is inserted at the input plane P_1 of the coherent optical processing system of fig. 7.28, then at the spatial frequency plane P_2 the complex light distribution is

$$E(p) = \delta(p) + \sum_{n=-\infty}^{\infty} \frac{\sin(npa/2)}{npd/2} \delta\left(p - \frac{2\pi n}{d}\right), \tag{7.68}$$

where p is the angular spatial frequency coordinate. Thus we see that, except for the zero-order term, there are sequences of higher order diffractions distributed along the p-axis in the spatial frequency plane, each centered at $p = (2\pi n)/d$. In principle, it is possible to process these diffraction orders with complex spatial filters. If we single out the nth-order diffraction in the Fourier plane (assuming

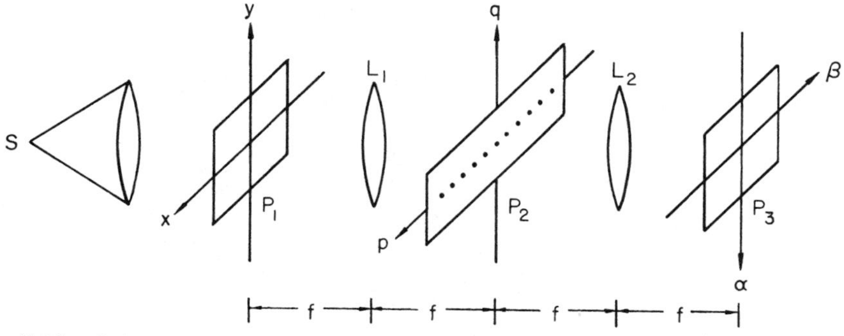

Fig. 7.28 Coherent optical information processor. S, monochromatic point source; L, transform lens.

that the sampling condition is fulfilled to guarantee the separation of the diffraction orders in the Fourier plane), then the output inadiance at plane P_3 would be

$$I_n = \left(\frac{\lambda f}{n\pi}\right)^2 \sin^2\left(\frac{n\pi a}{d}\right), \tag{7.69}$$

where f is the focal length of the transform lens. From this equation we see that there are N number of equidensity contours that can be sequentially extracted. In other words, by spatially singling out each of the diffraction orders in the spatial frequency plane P_2, a sequence of equidensity contour images can be obtained, one at a time, in the output image plane.

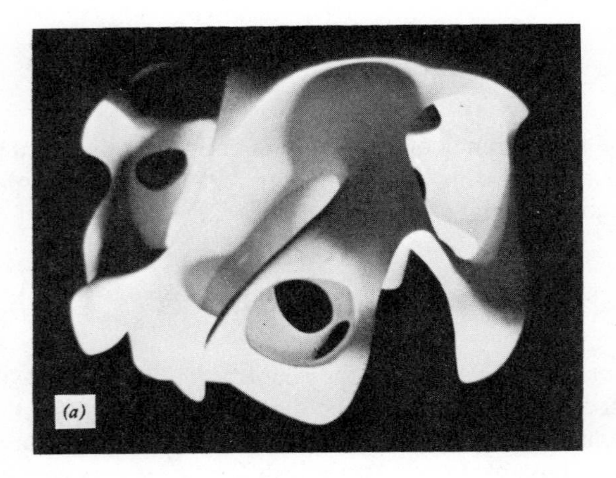

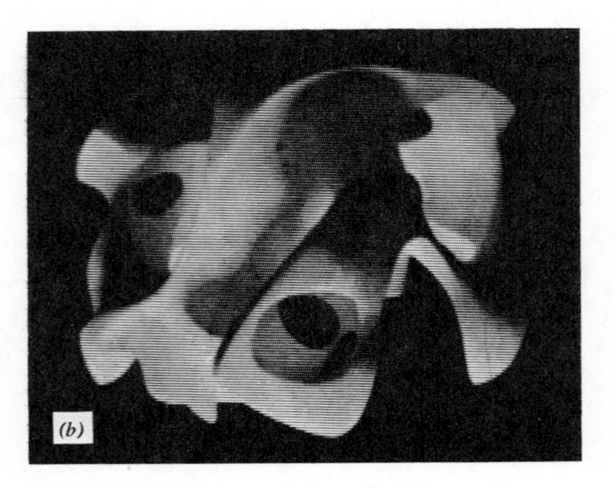

Fig. 7.29 Equidensity imaging through halftone screen technique. (*a*) Original object transparency. (*b*) Halftone image transparency.

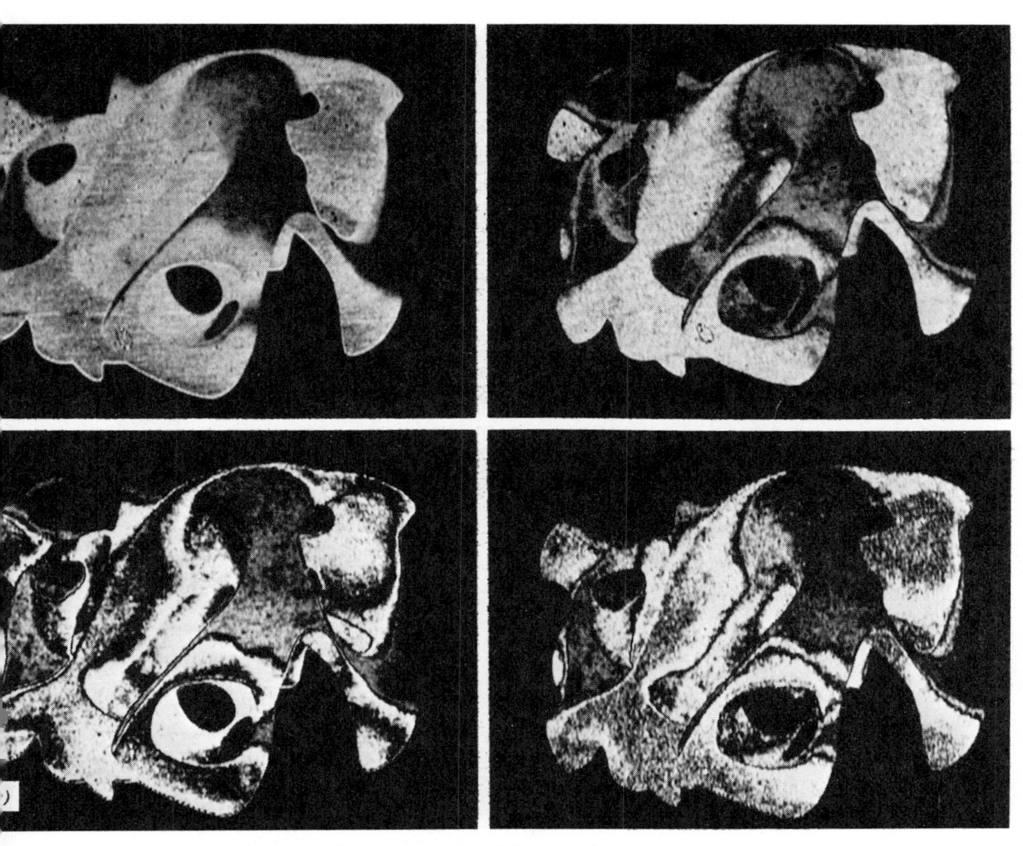

(*c*) Four output equidensity images.

In ending this section, we provide a set of equidensity images obtained by the halftone screen technique. Figure 7.29*a* shows the original photographic image, fig. 7.29*b* is the produced halftone image transparency, and fig. 7.29*c* shows the four equidensity images obtained by the halftone image transparency with a coherent optical processing technique.

7.7 LOGARITHMIC FILTERING USING FILM NONLINEARITY

In this section we demonstrate a technique that makes use of the inherent nonlinearity of the photographic film to perform the logarithmic transformation (ref. 7.32). Unlike the halftone screen, the photographic film itself can easily achieve a resolution of several hundred lines per millimeter and, therefore, a much greater space-bandwidth product. This capability of transforming images with very high spatial resolution maintains the basic advantage of the optical method over the digital techniques.

To produce a photopositive image, a two-step contact printing process is used. A linear amplitude transmittance of the input irradiance, a negative transparency, is first produced. The intensity transmittance of the negative transparency as a function of the input exposure can be described as $K_1E^{-\gamma}$, where K_1 is a proportionality constant, E is the exposure, and γ is the gamma of the film used. Using this negative transparency in contact printing, the relative exposure for the final positive transparency would be equal to the intensity transmittance of the first negative transparency. We wish to find the amplitude transmittance characteristics required for the second exposure so that the amplitude transmittance of the final transparency would be proportional to the logarithm of the original exposures, that is, $T_A = K_2 \log E$. We find that this amplitude transmittance characteristic is dependent on both the gamma of the film used in the first exposure and the proportionality constant K_2 (i.e., exposure range) desired for the logarithmic output. We first examine the amplitude transmittance characteristics required in the second exposure for different ranges with a given gamma for the first exposure. We show in fig. 7.30 the different

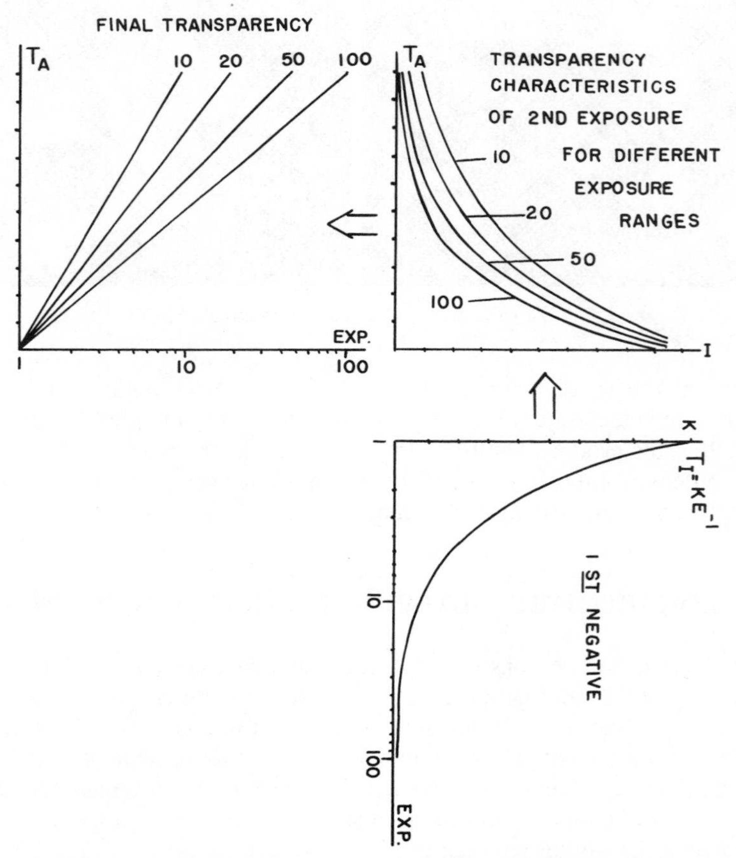

Fig. 7.30 A mapping technique to obtain a logarithmic transformation.

characteristics required to produce logarithmic transformation for exposure ranges of 10, 20, 50, and 100 with the gamma of the film used in the first exposure equal to 1. To produce a logarithmic transformation of different exposure ranges, it is not necessary to find different films for the second exposure because any one of these characteristics can be used with any gamma value in the first exposure to obtain a logarithmic transformation. That is, we only have to find one film whose characteristic matches any one of the curves in this family of curves. We can then use it in the contact printing process to produce a logarithmic transformation for any exposure range by varying the gamma of the film used in the first exposure, as demonstrated in fig. 7.31. One characteristic (the one that produces a range of 10 for $\gamma = 1$ in fig. 7.30) is used with gamma values of 0.5, 0.6, 0.7, and 1. Logarithmic transformation for ranges 100, 50, 25, and 10 are obtained.

One possible emulsion we have found is the Kodak Ortho High-Contrast film developed for 5 minutes in DK 50. It matches quite closely one of the curves

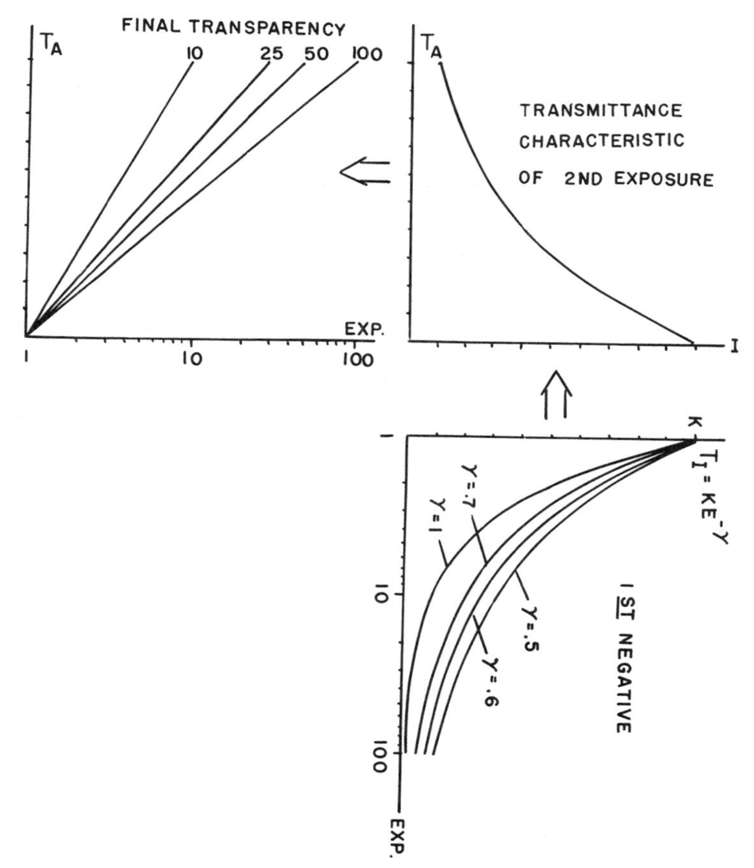

Fig. 7.31 A mapping technique to obtain a logarithmic transformation with various film gammas γ.

within a fairly large range of amplitude transmittance as shown in fig. 7.31. Using a gamma of 0.6 for the first exposure, a logarithmic transformation with an exposure range of 50 can be obtained. Some experimental results are shown in fig. 7.32. A Panatomic X film developed to a gamma of 0.6 is used in the first exposure. The exposure is confined to within the linear region of the H and D curve by pre-exposing the film to above the toe region. In the contact printing, the exposure is confined to that region by pre-exposing the Ortho film and confining the maximum exposure. The measured amplitude transmittance is very close to the ideal logarithmic transformation for the exposure range of about 50. As we have pointed out earlier, a higher exposure range can be obtained if a lower gamma film is used for the first exposure. The output amplitude transmittance range is about 0.15–0.9, which is compatible with the result obtained with the halftone screen method (ref. 7.27).

To test the performance of the logarithmic transformation, a set of sinusoidal gratings multiplied perpendicular to each other is used as the input signal transparency. That is,

$$f(x, y) = \sin p_0 x \cdot \sin q_0 y, \tag{7.70}$$

where p_0 and q_0 are the spatial frequencies of the gratings. Such an input is used because it provides a convenient continuous tone signal and the terms in its Fourier spectrum are well separated. In fig. 7.33*a* we have the spectrum of a linearly recorded transparency of these multiplied gratings. The density range of the object transparency is about 0–1.7. Besides the four first-order diffraction terms, four intermodulation terms are clearly visible. The multiplied signal is

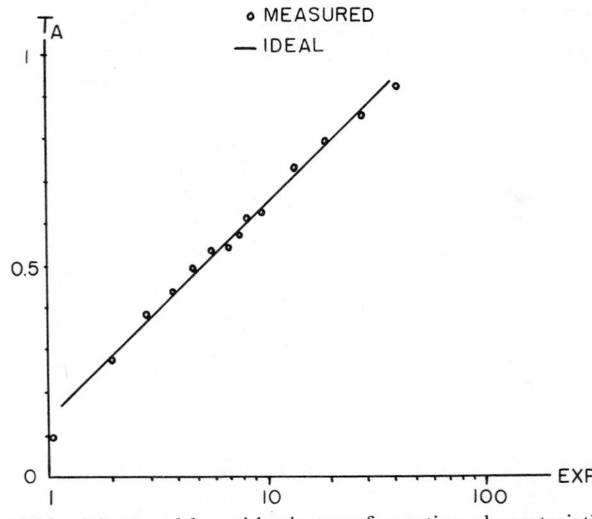

Fig. 7.32 Measured logarithmic transformation characteristic.

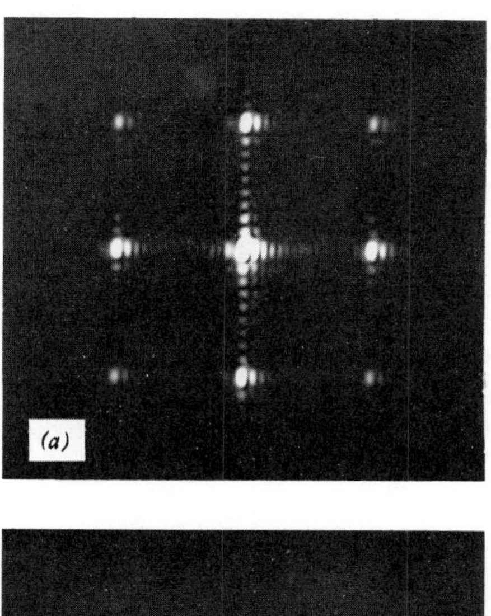

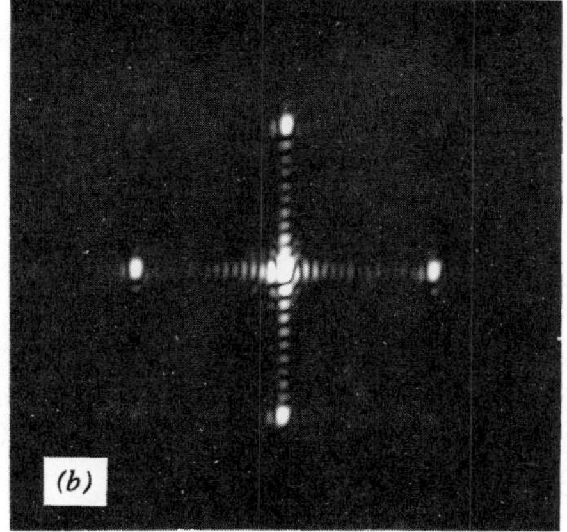

Fig. 7.33 Fourier spectra of two multiplied sinusoidal gratings. (*a*) Linearly transformed signal. (*b*) Logarithmically transformed signal.

then logarithmically transformed with the technique using film nonlinearity as described earlier. The signal function can now be written as

$$\log f(x, y) = \log \sin p_0 x + \log \sin q_0 y . \qquad (7.71)$$

The $\log \sin p_0 x$ and $\log \sin q_0 y$ functions are also periodic with the same fundamental frequencies p_0 and q_0, respectively. The spectrum of this logarithmically transformed signal is shown in fig. 7.33*b*. It can be seen that the intermodulation terms are significantly suppressed.

This technique is then applied to the filtering of a signal embedded in multiplicative noise. A picture of a girl multiplied with a Ronchi-type grating acting as noise is used as the input signal, as shown in fig. 7.34*a*. The multiplied signal transparency has a density range of about 1.8. We then try to filter out this multiplied grating by using an absorptive filter at the spatial frequency plane. In fig. 7.34*b* we show the result obtained by linear spatial filtering; the grating cannot be completely removed since the two signals are convolved in the spatial frequency domain. A better result is obtained with the logarithmic transformation, as shown in fig. 7.34*c*, using the same absorptive filter. Strictly speaking, with the logarithmic transformation, an exponential transformation should be performed at the end to get back the original signal. Such an operation, however, is not imperative in many applications since it affects only the gray scale distribution and not the resolution of the object. Such an inverse operation is not illustrated here; we refer the interested reader to the work of Dymek et al. (ref. 7.34).

By comparing a linear function and a logarithmic function, we see that the slope of the logarithmic function is larger at low input levels but smaller at high input levels. Thus the tonal differences between low-intensity areas with the logarithmically transformed signal are enhanced. Comparing figs. 7.34*b* and 7.34*c*, we see that the ears and eyes of the object stand out more distinctively with the logarithmically transformed case. The facial features, however, become less distinct because the logarithmic transformation decreases the tonal differences between high-intensity areas.

An important application of the logarithmic transformation is the detection of a signal embedded in multiplicative noise by means of complex spatial matched filtering. It is well known that, at the output plane of a complex matched filtering system, there is a correlation function between the input and the function used in the generation of the complex spatial matched filter. That is, if the input function is $f_1(x, y)$ and the function used for the matched filter is $f_2(x, y)$, then the correlation term of the matched filter output is

$$g(x, y) = Kf_1(x, y) * f_2^*(-x + b, -y), \tag{7.72}$$

centering at $x = -b$ of the output plane, where $*$ denotes convolution, * denotes complex conjugate, and K is a proportionality constant.

For a signal $s(x, y)$ embedded in additive noise $n(x, y)$, the input function becomes $s(x, y) + n(x, y)$. Using $s(x, y)$ to generate the matched filter, the corresponding filtered output signal can, therefore, be written as

$$g(x, y) = Ks(x, y) * s^*(-x + b, -y) + Kn(x, y) * s^*(-x + b, -y), \tag{7.73}$$

which is the autocorrelation function of the signal plus the crosscorrelation function of the signal and the noise.

If the signal is embedded in multiplicative noise, however, we would not have

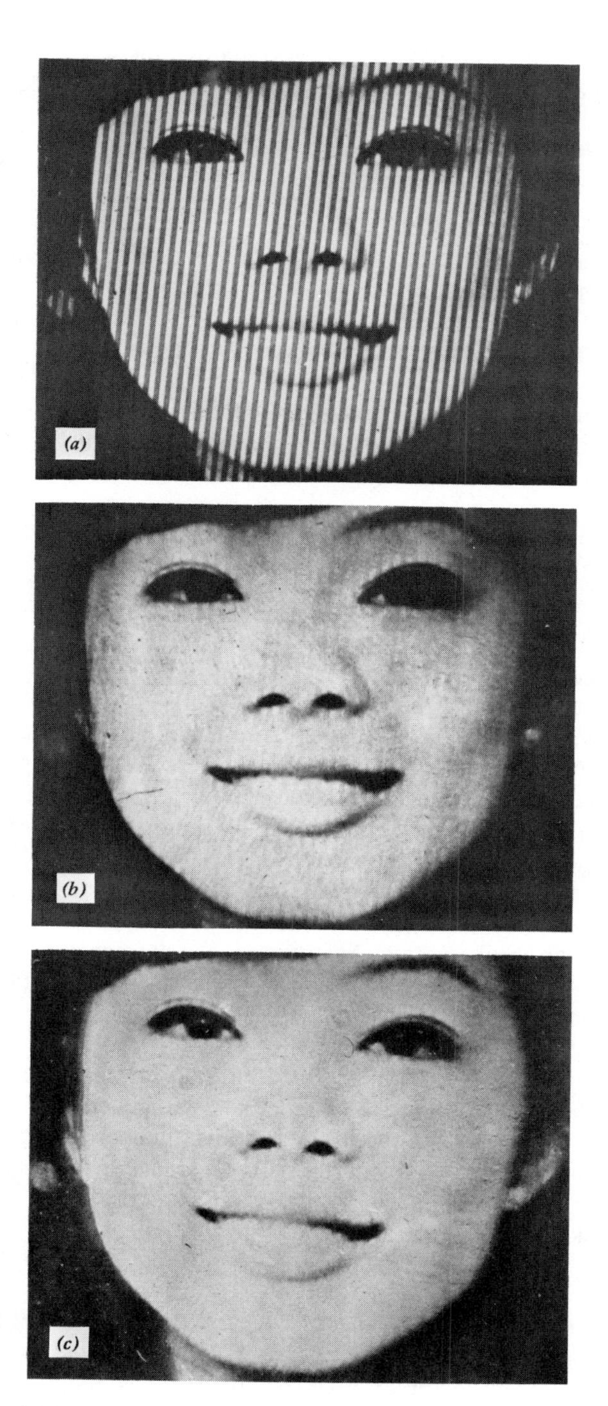

Fig. 7.34 Spatial filtering of a multiplied signal. (*a*) Original input signal. (*b*) Linear filtering. (*c*) Logarithmic filtering.

an autocorrelation function, but the crosscorrelation function between the signal and the product of the signal and the noise functions. That is, if the input function is $s(x, y)n(x, y)$, the output signal of the matched filter is

$$g(x, y) = ks(x, y)n(x, y) * s^*(-x + b, -y). \tag{7.74}$$

Now, if the input multiplicative signal is logarithmically transformed, the input signal function then becomes

$$\log s(x, y)n(x, y) = \log s(x, y) + \log n(x, y). \tag{7.75}$$

We therefore obtain a logarithmic signal embedded in an additive noise function. In the generation of the spatial matched filter, the logarithmically transformed signal $\log s(x, y)$ is used instead of $s(x, y)$. The corresponding correlation output is

$$g(x, y) = k[\log s(x, y)] * [\log s(-x + b, -y)]^* + k[\log n(x, y)] * [\log s(-x + b, -y)]^*, \tag{7.76}$$

which is the autocorrelation function of the logarithm of the signal function plus the crosscorrelation between the logarithm of the signal and the logarithm of the noise functions. Thus we see that in this case we obtain an autocorrelation function in the output plane that can be used for detection of a signal embedded in multiplicative noise. In an experimental demonstration the picture of the girl is once again used as the signal and the Ronchi grating used as the noise. The optimum complex matched filter was constructed with the well-known Vander Lugt method (sec. 6.6). For comparison, the autocorrelation function obtained by putting the original signal function back in the input plane of the optical processor is shown in fig. 7.35*a*. At its side is the corresponding densitometer scan of the photographic recording. In fig. 7.35*b* the result obtained with the usual linear spatial filtering scheme for a signal embedded in multiplicative noise is shown. The correlation peak is much lower and broader, and it may not be suitable for signal detection applications. In fig. 7.35*c* we show the result obtained with logarithmic filtering. A higher and sharper correlation peak can be observed. Since $\log n(x, y)$ is not a white Gaussian function, the crosscorrelation term between $\log s(x, y)$ and $\log n(x, y)$ is not expected to be arbitrarily small. The correlation peak, therefore, is somewhat broader than that of the autocorrelation peak, as shown in fig. 7.35*a*.

7.8 OPTIMUM SIGNAL DETECTION BY PREWHITENING

Correlation detection by complex matched filtering has been applied successfully in a large variety of detection systems. Substantial efforts have been

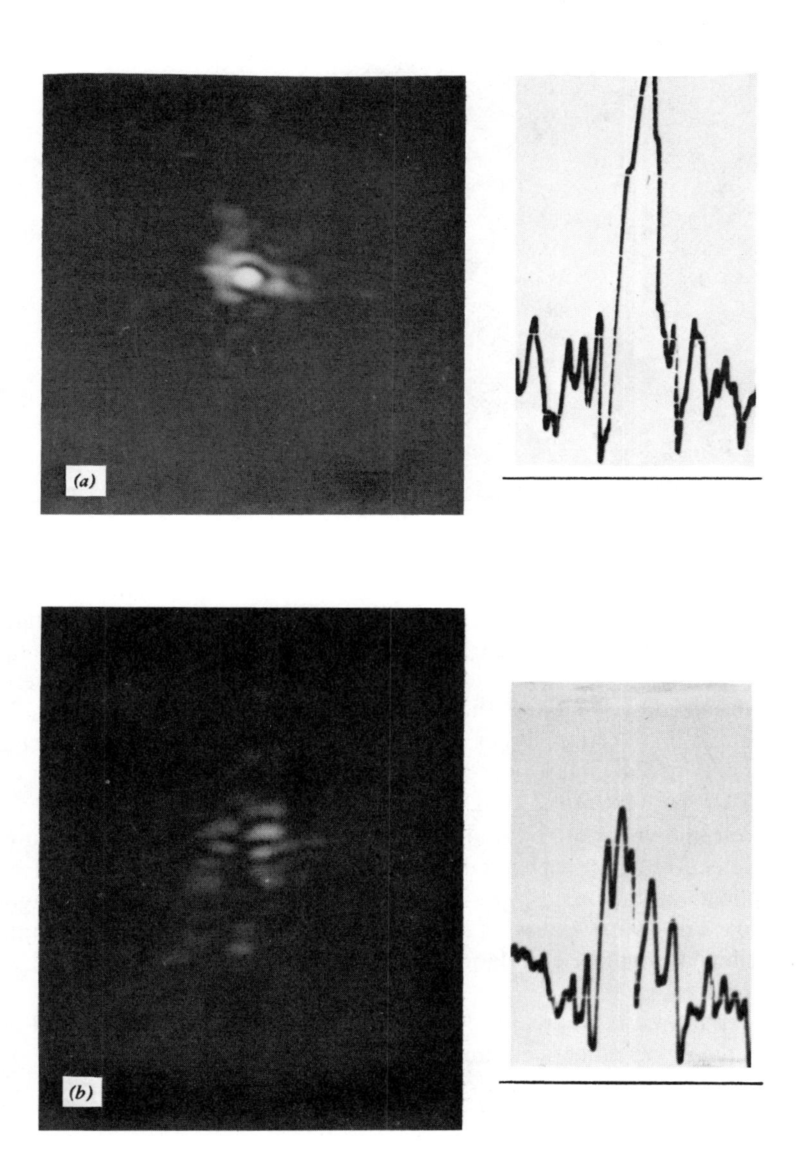

Fig. 7.35 Output signals of complex spatial matched filtering system. (a) Autocorrelation. (b) Linear filtering. (c) Logarithmic filtering.

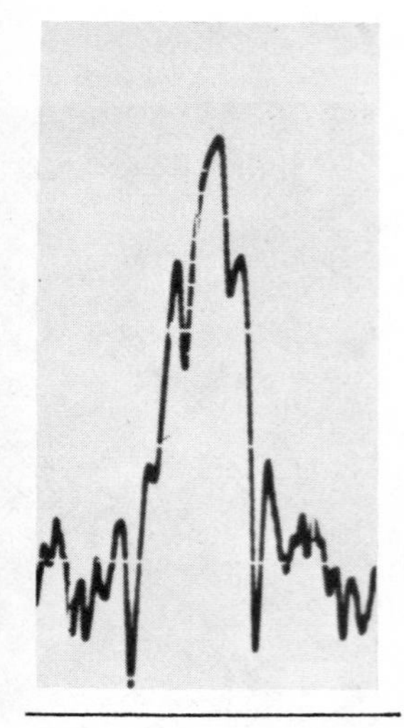

undertaken to improve the signal-to-noise ratio (SNR) of the process. Many of the proposed techniques, such as high-pass filtering, improve the SNR by increasing the selectivity of the correlation process. However, such techniques decrease the system tolerance for signal size and orientation variations. In this section we demonstrate a simple technique by Chao et al. (ref. 7.36), where the SNR is optimized by whitening the noise spectrum. Although the technique can only be applied to certain types of signal detection problems, a substantial improvement in SNR may be achieved without overly affecting the system sensitivity to size and orientation variations.

It is well known from communication theory (ref. 7.35) that optimum correlation detection can be achieved if the noise spectrum is white (sec. 1.5). In most applications the noise (i.e., the background cluster in which the signal is embedded) is not white Gaussian in nature. However, if the noise spectrum can be prewhitened, then the correlation detection process can still be optimized by match filtering.

Consider the case where we have an input signal $S(x, y)$ with a Fourier spectrum of $S(p, q)$ embedded in an additive colored noise $n(x, y)$ with a power spectral density of $N(p, q)$. Passing the noisy input through a filter with a transfer function of $H(p, q)$, the signal output is

$$S_0(x, y) = \iint S(p, q) H(p, q) \exp[i(px + qy)]\, dp\, dq, \tag{7.77}$$

and the noise output would have a power spectral density of

$$N_0(p, q) = N(p, q)|H(p, q)|^2. \tag{7.78}$$

The peak signal power at (x, y) is determined by

$$|S_0(x, y)|^2 = \left|\iint S(p, q)H(p, q) \exp[i(px + qy)]\, dp\, dq\right|^2, \tag{7.79}$$

and the average noise power is equal to

$$\bar{N}_0 = \int N_0(p, q)\, dp\, dq = \iint |H(p, q)|^2 N(p, q)\, dp\, dq. \tag{7.80}$$

To optimize the filtering process, we would like to maximize $|S_0(x, y)|^2/\bar{N}_0$, the output SNR. Using the Schwarz inequality (sec. 1.5), the SNR is found to be maximized when

$$\frac{|S_0(x, y)|^2}{\bar{N}_0} = \iint \frac{|S(p, q)|^2}{N(p, q)}\, dp\, dq. \tag{7.81}$$

The optimum filter function is

$$H(p, q) = \frac{KS^*(p, q)}{N(p, q)},$$

where K is a complex constant.

For the special case where the input noise is white [i.e., $N(p, q) = N$], the optimum filter would simply be

$$H(p, q) = \frac{KS^*(p, q)}{N} = K'S^*(p, q), \tag{7.82}$$

where K' is a proportionality constant.

For the general case that we are considering where $N(p, q)$ is not a constant, it would be necessary to generate the optimum filter $KS^*(p, q)/N(p, q)$. This can be achieved with two sequential filters, $K_1/\sqrt{N(p, q)}$ and $K_2S^*(p, q)/\sqrt{N(p, q)}$, as illustrated in fig. 7.36. The first filter is often referred to as the *whitening filter*.

In order to generate the spatial whitening filter $K_1/\sqrt{N(p, q)}$, we require a prior knowledge of the noise spectrum $N(p, q)$. However, in most practical applications, it is not feasible to isolate the noise from the signal for the generation of the whitening filter. While we may not be able to isolate the noise from the signal, we may be able to generate before hand a spatial noise that exhibits similar spectral distribution as the noise spectrum of the input. The speckle noise is a good example, since the statistical distribution of the speckle noise is

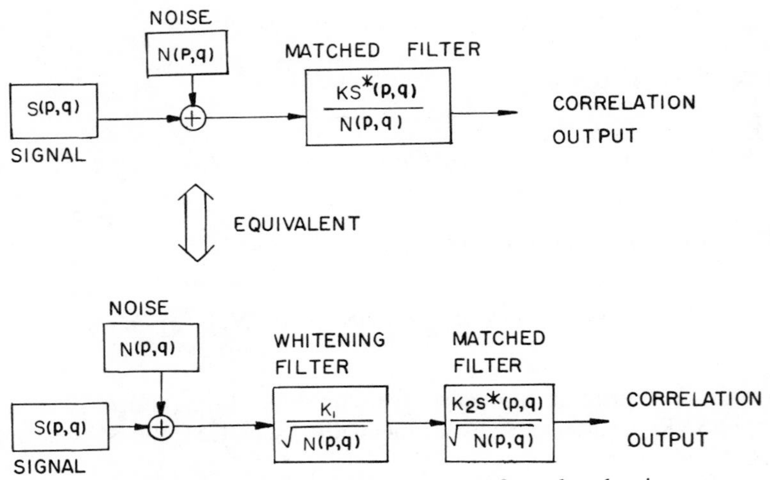

Fig. 7.36 Optimum filtering process for colored noise.

dependent upon the recording geometry (ref. 7.37). In other words, the same statistical distribution of speckle noise can be generated at any time by simply using the same recording geometry. There are also other possible applications of this technique. One example is the detection of targets from a particular background of terrain (e.g., corn fields, pine forest, etc.), since the background clusters of photographs taken from the same altitude would exhibit similar spectral distribution.

Before we describe the technique for generating such an averaged whitening filter and its use in the correlation detection process, we should note that the speckle noise is generally multiplicative and optimum correlation detection by matched filtering is effective only with additive noise. We have, however, in previous sections shown that effective correlation detection can also be applied to the multiplicative signal if the signal input is first logarithmically transformed to produce an additive form. Here we assume that such a logarithmic transformation has been performed with either the use of film nonlinearity (ref. 7.32) or a halftone screen (ref. 7.27).

For each individual input sample, the noise possesses a distinctive spectral distribution. In order to generate the whitening filter, we need the statistical power spectral density of the input noise. If the noise is ergodic, one common method of estimating the spectral density is to take M samples with a size of X and Y and obtain an average for their spectral distribution (ref. 7.38). That is,

$$N(p, q) \cong \frac{1}{M} \sum_{m=0}^{m=M} \left| \int_{-x/2}^{x/2} \int_{-y/2}^{y/2} n(x, y) \exp[-i(px + qy)] \, dx \, dy \right|^2. \quad (7.83)$$

The optimal arrangement for the generation of the whitening filter is illustrated in fig. 7.37. A roll of transparency containing spatial noise with the same

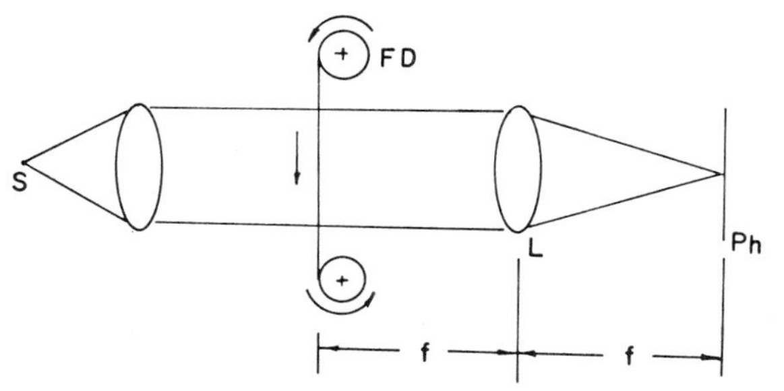

Fig. 7.37 A coherent optical system for the generation of a whitening spatial filter. *S*, monochromatic point source; *FD*, film drive; *L*, transform lens; *Ph*, photographic plate.

statistical spectral distribution as the noise spectrum is scanned across the input plane of an optical processor. At the same time, the Fourier spectrum is recorded on a photographic plate with a contrast of $\gamma = 1$. If the recording is confined to the linear region of the H and D curve, then the output amplitude transmittance is equal to

$$T_A = E^{-\gamma/2}, \tag{7.84}$$

where E is the input exposure. After exposure to the Fourier spectrum of the scanning input noise for a time interval of T seconds, the amplitude transmittance of the developed plate can be written as

$$W(p, q) = \frac{K}{\left[\int_0^T |N(p, q; t)|^2 \, dt\right]^{-\gamma/2}} \cong \frac{K}{\sqrt{N(p, q)}}, \qquad \text{for } \gamma = 1, \tag{7.85}$$

where $N(p, q)$ is the power spectral density of the input noise, and $K \ll 1$ is a proportionality constant.

The complex matched filter is constructed as shown in fig. 7.38. The signal input $s(x, y)$ is placed in the input plane P_1 and its Fourier spectrum is filtered

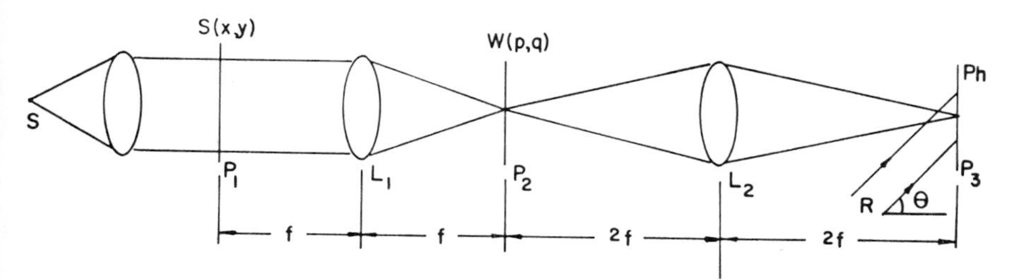

Fig. 7.38 Generation of a complex matched filter. *S*, monochromatic point source; L_1 and L_2, transform lenses; $s(x, y)$, input signal; $W(p, q)$, whitening filter; *Ph*, photographic plate; *R*, reference beam.

by the whitening filter at P_2. The filtered signal spectrum is then imaged onto P_3 with the relay lens. An offset reference beam is used to modulate the signal spectrum and the light field is then recorded on a holographic plate. The complex light distribution at P_3 can be written as

$$E(p, q) = \frac{KS(p, q)}{\sqrt{N(p, q)}} + \exp[ikp \sin \theta]. \tag{7.86}$$

If the recording is linear, then the amplitude transmittance of the resulting transparency would be

$$H(p, q) = K_1 + \left|\frac{KS(p, q)}{\sqrt{N(p, q)}}\right|^2 + \frac{KS(p, q)}{\sqrt{N(p, q)}} \exp[-ikp \sin \theta] + \frac{KS^*(p, q)}{\sqrt{N(p, q)}} \exp[ikp \sin \theta], \tag{7.87}$$

where K_1 is an appropriate instant. Thus we see that a matched filter is generated for the filtered signal function $KS(p, q)/\sqrt{N(p, q)}$.

The optical correlation detection process is shown in fig. 7.39. The input signal $s(x, y)$ in additive noise $n(x, y)$ is inserted in the input plane P_1 of the processor; then the correlation outputs at P_4 can be written

$$\Omega(x, y) = \mathscr{F}^{-1}\left[\frac{|KS(p, q)|^2}{N(p, q)}\right] + \mathscr{F}^{-1}\left[\frac{KS^*(p, q)n(p, q)}{N(p, q)}\right]. \tag{7.88}$$

Comparing the optimum filtering process depicted in fig. 7.36 with the optical implementation in fig. 7.39, we can see that they are equivalent. The optical correlation detection system is, therefore, optimum.

We would stress that the whitening process was designed to suppress the cross-correlation output of the random background noise $n(x, y)$. It is, therefore,

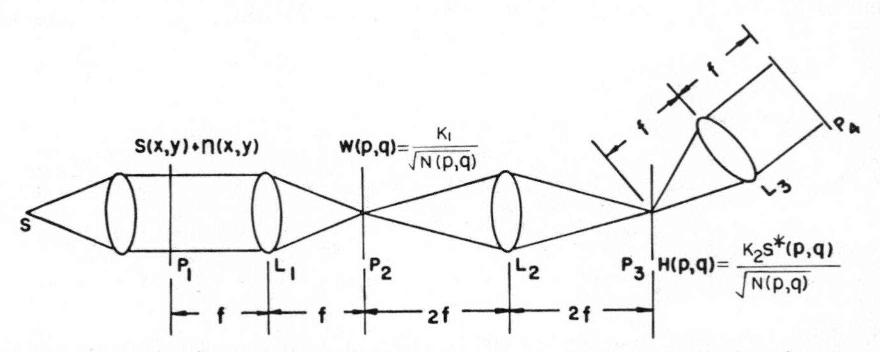

Fig. 7.39 An optimized correlation detection system. S, monochromatic point source; L_1, L_2, and L_3, transform lenses; $W(p, q)$ whitening filter; $H(p, q)$, matched filter.

independent of the input signal $s(x, y)$, and the improvement of the output correlation SNR can be obtained without severely affecting the system's sensitivity to the size of signal and the orientation variation.

Let us now demonstrate that the whitening of the input noise spectrum can indeed improve the SNR of a correlation detection system. In our experiment we choose an input signal (i.e., a Chinese character) to be embedded in an additive speckle noise, as shown in fig. 7.40. We note that the additive speckle noise is similar to a multiplicative speckle after the logarithmic transformation. A similar set of speckle noise (i.e., the same spectral distribution) is recorded on a long roll of film for the generation of a whitening filter, as we have discussed. We now show the experimental results that we have obtained from this prewhitening process. For comparison, we first perform the correlation detection *without* the use of whitening filter. In fig. 7.41*a* we show the result of the autocorrelation output that is obtained by the insertion of the input signal alone without noise. Below this we show the result obtained from a photometer scan of the correlation peak. In fig. 7.41*b* we show the output correlation of the input signal embedded in additive random speckle noise. Below this part is the result obtained from the photometer scan. We then repeat the experiments with the use of a whitening filter. In fig. 7.42*a* we show the autocorrelation output that is obtained by the insertion of the input signal alone (without input noise) and the corresponding photometer scan. Figure 7.42*b* shows the corresponding correlation detection for the signal embedded in a random speckle noise. From the results of figs. 7.41*b* and 7.42*b*, we see that the crosscorrelation of the input noise is significantly suppressed with the prewhitening process. Thus a higher SNR can indeed be obtained by this prewhitening technique.

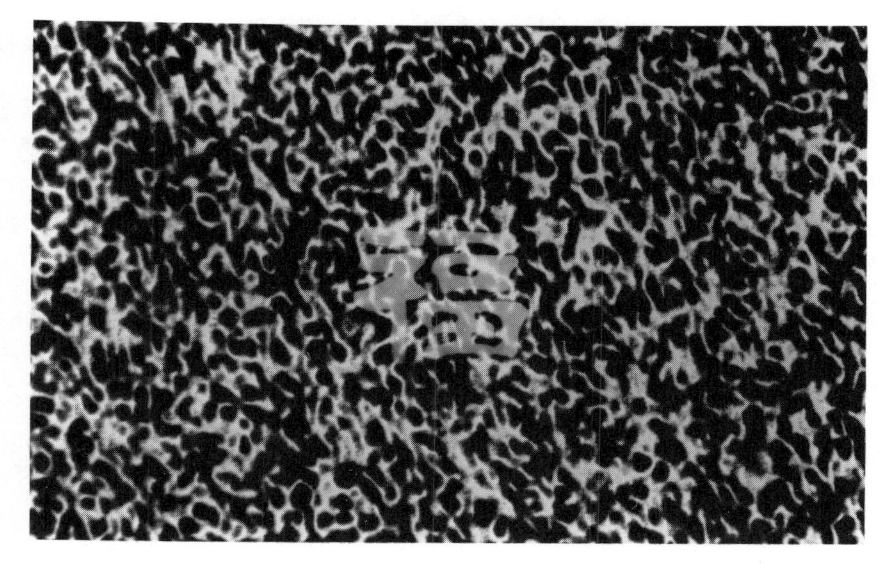

Fig. 7.40 An input Chinese character embedded in an additive speckle noise.

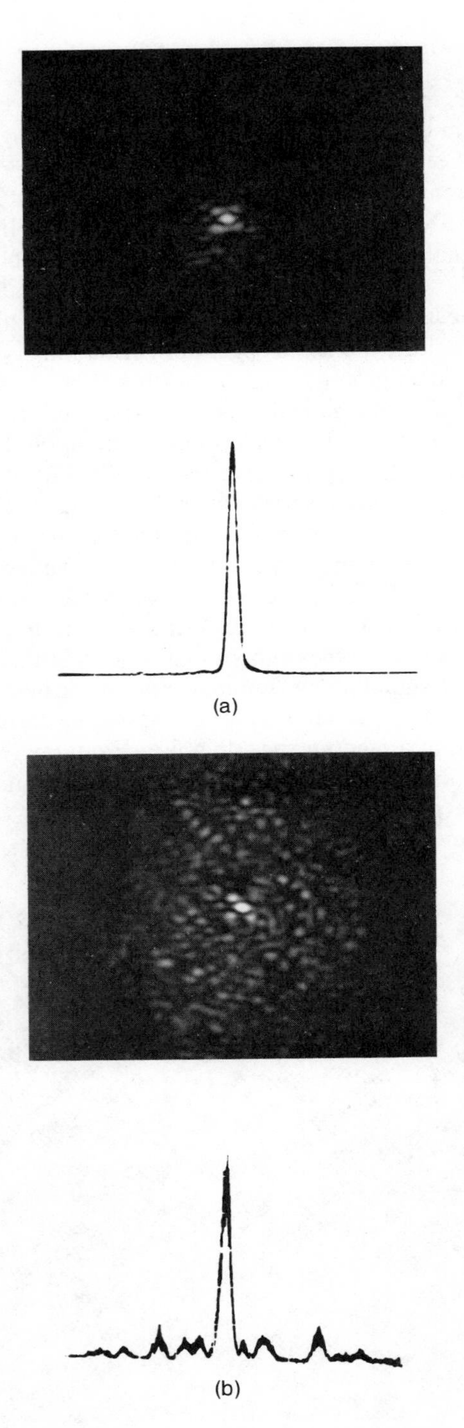

Fig. 7.41 Output correlation detection without the use of whitening filter. (*a*) Autocorrelation of the input signal. (*b*) Output correlation of the signal embedded in additive speckle noise.

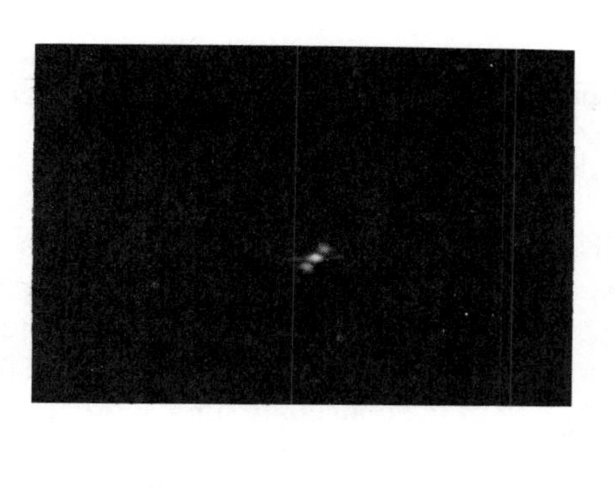

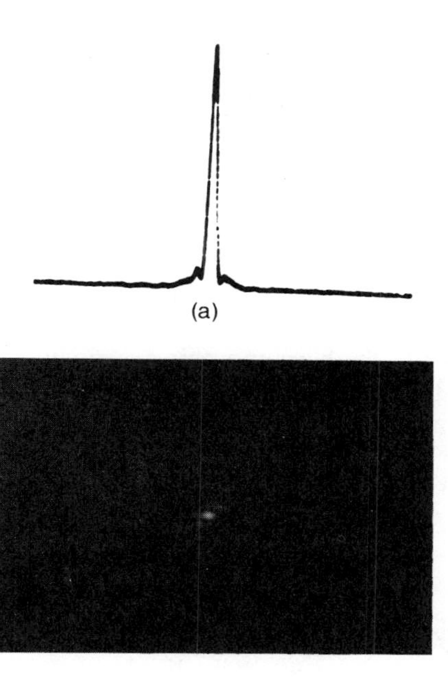

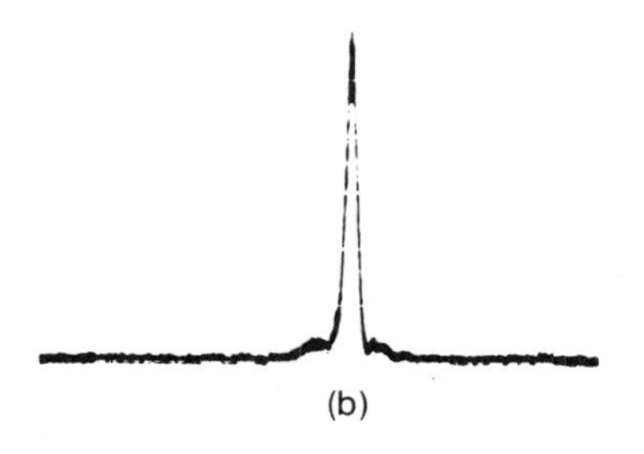

Fig. 7.42 Output correlation detection with the use of whitening filter. (*a*) Autocorrelation of the input signal. (*b*) Output correlation of the signal embedded in additive speckle noise.

7.9 SPACE-VARIANT PROCESSING

So far we have discussed only the space-invariant processing operation of images. In other words, every image point to be processed is affected by the same processing operation. Although the space-variant processing concept has been known and successfully applied in system theory and digital processing for some time, the application of the concept to optical processing is relatively new. In 1965 Cutrona (ref. 7.39) proposed a general optical space-variant processing concept parallel with the Fourier transform technique in terms of system eigenfunctions. This technique is, however, only of theoretical interest, since it is not known how to find the set of eigenfunctions. Attempts have been made, by Heinz et al. (ref. 7.40), to use optical space-variant processing in matrix multiplication; some calculated results are reported in their paper. In a recent article, Deen et al. (ref. 7.41), reported a space-variant operation using volume holograms. They have shown some positive results (ref. 7.42). The most interesting example of space-variant operation is the geometrical transformation by Bryngdahl (refs. 7.43, 7.44). He uses a computer-generated hologram for the operation. Mention must also be made of Sawchuk's work (ref. 7.45), the computer technique of space-variant image restoration by coordinate transformation.

We now describe a basic concept for achieving optical space-variant processing similar to that of the homomorphic processing system, as illustrated in sec. 7.6. In general, a space-variant system may be described by a block diagram, as shown in fig. 7.43*a*. The input-output relation may be described by the equation

$$g(x, y) = \iint h(x, y; x', y') f(x', y') \, dx' \, dy', \tag{7.89}$$

where $h(x, y; x', y')$ is space-variant impulse response. Needless to say, if $h(x, y; x', y')$ is described as a function of the coordinate difference, i.e.,

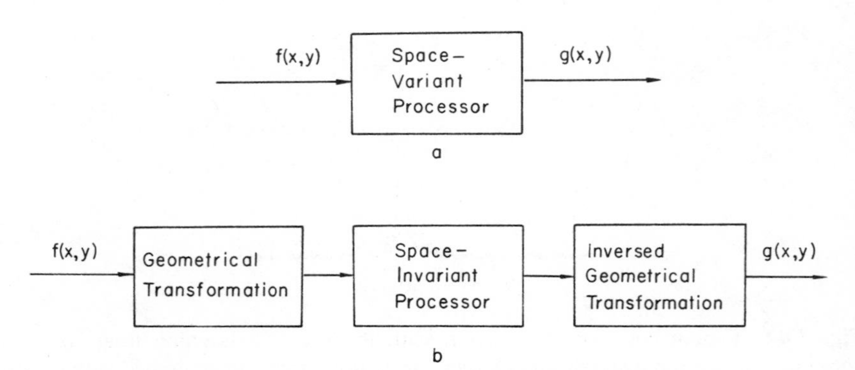

Fig. 7.43 Techniques of space-variant processing.

Fig. 7.44 Space-variant processing. (*a*) Space-variant operation utilizing a computer-generated hologram. *S*, monochromatic point source; $s(x, y)$, object transparency; *H*, computer-generated hologram; *L*, transform lens. (*b*) Experimental results. (By permission of O. Bryngdahl.)

$h(x - x', y - y')$, then the system is space-invariant. For example, a technique of optical space-variant processing may be accomplished by the system shown in the block diagram of fig. 7.43*b*. In other words, whenever a space-variant processing can be decomposed as in fig. 7.43*b*, first the necessary geometrical transformation may be performed. The geometrical transform signal will then be processed by a space-invariant processor, and the required inversed geometrical transformation will be performed last.

We now provide an example of space-variant operation due to Bryngdahl (ref. 7.43), as shown in fig. 7.44. He uses a computer-generated hologram by Lee (ref. 7.46) in a coherent optical processor, as shown in fig. 7.44*a*, and obtains the experimental results shown in fig. 7.44*b*. The principle behind the computer-generated hologram is a generalized grating such that the grating frequency varies as a function of spatial coordinates (x, y). In other words, the local angular spatial frequency of grating (p_x, q_y) at a given localized object point (x, y) will carry the point object to

$$\alpha = \frac{\lambda f}{2\pi} p_x, \qquad \beta = \frac{\lambda f}{2\pi} q_y, \tag{7.90}$$

where λ is the wavelength of the light source, f is the focal length of the lens, and (α, β) is the spatial coordinate system. In a more recent article, Casasent and Psaltis (ref. 7.47) have shown that the geometrical transformations can also be accomplished by means of a nonlinear beam scanning device that writes the data onto an optical light value. Although this method sacrifices the parallel processing capability, it can process in a near real-time mode. Interest in space-variant processing of photographic images is somewhat recent, and further progress can be expected from future research.

7.10 SPECTRAL ANALYSIS OF ULTRASONIC BLOOD FLOW

In this section we present an application of coherent optical processing technique for displaying and analyzing a blood-flow-generated ultrasonic Doppler spectrum. The system is cost-effective and, when coupled with a real-time recording material, can produce spectrograms on-line. Other advantages include a large, continuously variable bandwidth, an instantaneous display of velocity profile, and simultaneous display of temporal spectra (ref. 7.48).

The system makes use of the Fourier transformation property of lenses in processing time signals. To process time signals, the signal must first be converted into a spatial format that may then be used as the input transparency for the optical spectrum analyzer. The output is a Fourier transform in the spatial frequency domain. It should be noted that the temporal frequency spectrum is a direct mapping of the spatial frequency spectrum whose zero frequency appears exactly on the optical axis of the output plane. The advantage of such a system is that the signal is processed simultaneously at the speed of light, therefore, in real-time.

Figure 7.45 shows a technique that may be employed to accomplish the required temporal to spatial signal conversion. The sinusoidal time signal at the upper left-hand corner is used to modulate the intensity of a cathode ray tube (CRT). Because intensity is a positive real quantity, the signal has to be biased so that the intensity displayed on the CRT is given as $B + I(x)$, where B is the bias and $I(x)$ is proportional to the input voltage. Some sort of a recording material is then used to produce a transparency for use as the input in the optical processor. It should be noted that the effect of bias is simply to introduce an additional zero-order diffraction in the spatial frequency domain that can be easily eliminated by using a stop-band filter in the output plane.

The most commonly used recording material is photographic film. A typical characteristic of a silver halide photographic emulsion is shown at the lower left-hand corner in fig. 7.45. If exposure is confined within the linear region, the amplitude transmittance of the resulting transparency will be inversely proportional to the input intensity. The inversion occurs because the silver halide emulsion produces a negative image of the input signal. This inversion has no effect on the spectrum of the signal.

Although photographic film is the most commonly employed recording material, many new materials have appeared in recent years. Some can develop an

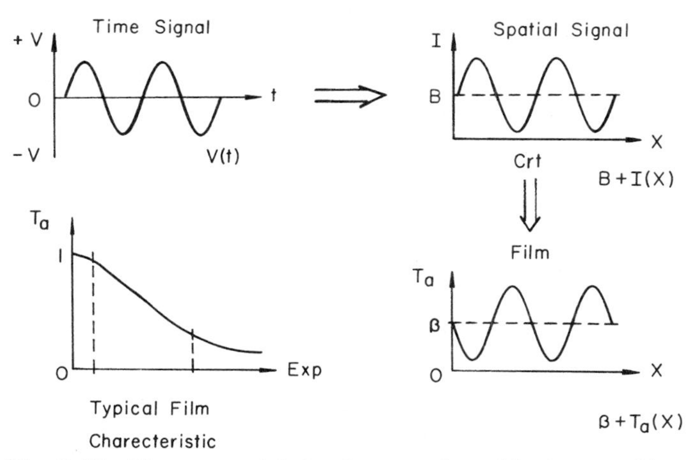

Fig. 7.45 Time to spatial signal conversion with photographic film.

Table 7.1
A Comparison of Photographic Recording Materials

Material	Resolution, lines/mm	Sensitivity, J/cm^2	Recycling Speed
Pockel effect light modulator ($Bi_4Ti_3O_{12}$)	100	5×10^{-6}	30 msec
Liquid crystal	50	1×10^{-4}	0.1 second development; 2 seconds erasure
Photographic film	1000	1×10^{-7}	10 seconds development; not reusable

image in near real-time and can record and erase in rapid succession. Table 7.1 lists two such recording materials and their characteristics compared with photographic film. The pockel effect light modulator is probably the most suitable recording media for use in a real-time optical spectrum analyzer. It is very fast and has good resolution and sensitivity; its disadvantage is high cost. Liquid crystal is slower, requiring about 2 seconds to record and erase. The delay is mainly associated with the erase operation as it takes only 100 msec for the image to develop. Another disadvantage of the liquid crystal medium is its relatively poor contrast (10:1). Photographic film has extremely high resolution and sensitivity. Using a rapid developer, the image can be developed in under 10 seconds. Photographic materials, however, are not reusable, yet their relatively low cost and high resolution more than make up the difference in many applications.

It should be emphasized that the delay time associated with photographic film is caused by the film development processes. The actual optical processing is still performed in real-time at the speed of light. This means that the frequency spectrum can be obtained within 10 seconds after the time signal is fed into a CRT to be converted into a spatial signal and produce a transparency. There will not be any delay due to accumulation caused by nonreal-time processing as typically experienced with electronic spectrum analyzers. The output frequency spectrum is obtained at the same rate as the time signal being fed into the optical spectrum analyzer.

The key elements of the optical spectrum analyzer described are shown in fig. 7.46. The addition of a cylindrical lens to the basic spectrum analyzer yields a one-dimensional Fourier transform processor. The action of the spherical lens in combination with the cylindrical lens is to perform a one-dimensional transform of the input spatial signal in the y direction and to magnify the spectrum into an easily visible size (d_2/d_1). The system simply images the input signal in the x direction, but magnifies it by a factor of f_2/f_1 as shown in the figure. Therefore, by appropriate selection of focal lengths and distances, we can obtain a spectrum of any desired size.

The integrated system for processing ultrasonic Doppler blood flow signals is

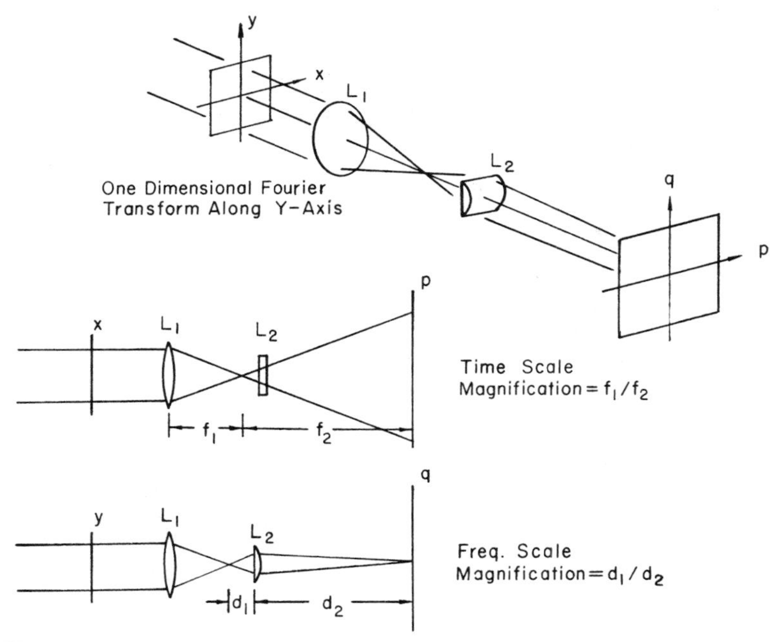

Fig. 7.46 Arrangement for a one-dimensional optical spectrum analyzer.

shown in fig. 7.47. An audio signal associated with blood flow obtained through the Doppler flowmeter is amplified and is used to modulate the intensity axis of a CRT (scope). A high frequency (1 MHz) sawtooth voltage is connected to the vertical input to produce a band, and a ramp (or low-frequency sawtooth) is used to make the beam scan across the CRT. Since the scanning is done electronically, this approach allows the recording material to remain stationary. This technique also enables the output spectrum over a given time interval to be displayed at once without additional storage or recording devices. If a real-time recording material is used, it will record the intensity variations on the CRT through the beam splitter. The spectral wave form will appear instantaneously on the ground glass as depicted in fig. 7.47. The ground glass is replaced by photographic film when a permanent record is desired. The time-scale calibration is given by the period selected for each scan. The frequency calibration is accomplished by feeding a single-frequency audio signal to the intensity modulation producing the first band on the recording material. This gives a single spot in the spectrogram corresponding to the input audio frequency.

The efficiency of this system has been demonstrated experimentally with humans having no known vascular disease and with controlled laboratory flow facilities. In the human experiments, blood flow in the brachial artery was sensed using a Parker 806 continuous-wave directional Doppler instrument. Figure 7.48 shows a typical spatial signal photographed on film by leaving the camera aperture open during the entire scanning interval. The intensity variation is

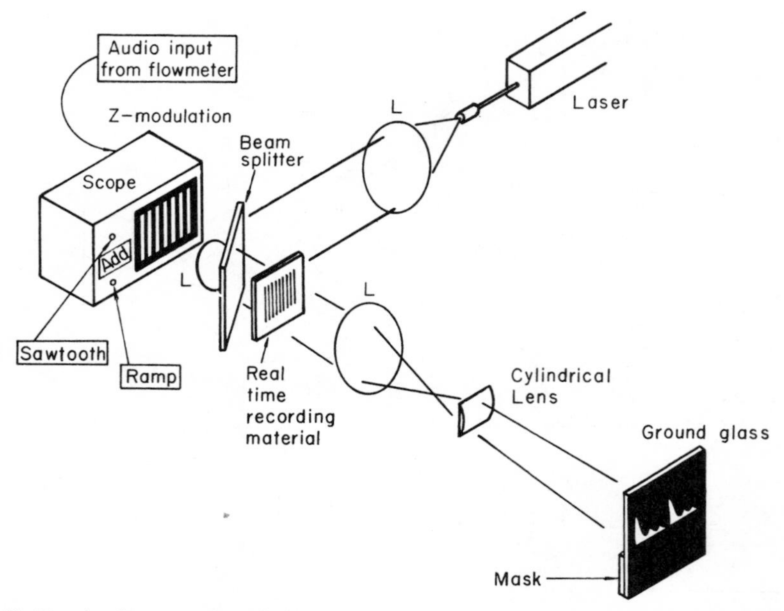

Fig. 7.47 An integrated real-time optical spectrum analyzer for Doppler flowmeter signals.

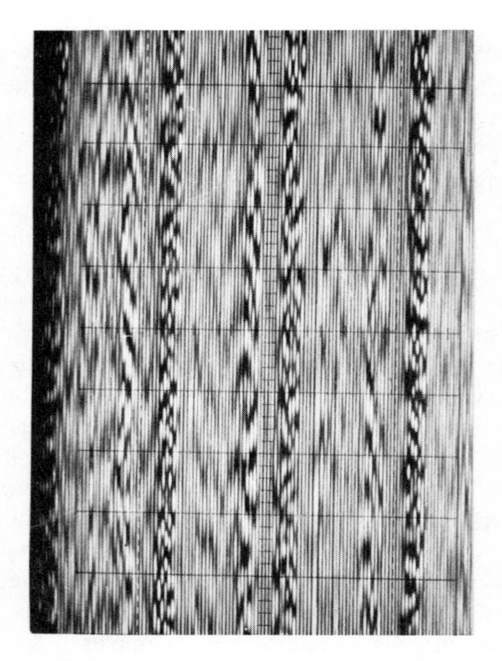

Fig. 7.48 A typical spatial signal obtained during ultrasonic Doppler blood flow sensing.

produced by the input audio signal, which is derived from the ultrasonic Doppler flowmeter. Although both horizontal and vertical axes are commonly used in optical transformation, we operate in one direction only, by taking the Fourier transformation of the discrete time samples in parallel.

Figure 7.49 is the corresponding blood flow spectrogram recorded using the optical spectrum analyzer, where the Doppler frequency is displayed versus time. The brightness of the plot corresponds to the intensity of the backscattered ultrasonic energy. The contour of the waveform is generally referred to as the high-frequency envelope. It should be noted that the results shown in fig. 7.49 compare closely with spectrograms obtained using specialized electronic spectrum analyzers (ref. 7.49). Moreover, the resolutions displayed in fig. 7.49 are even better than the electronically obtained spectrograms. It has been shown that the high-frequency envelope corresponds exactly to the blood flow velocity wave form sensed by electromagnetic flow meters (ref. 7.48).

In concluding this section, we would like to point out that the combination of a coherent optical spectrum analyzer and an ultrasonic blood flow transducer may provide a unique solution to the problem of analyzing Doppler blood flow spectra. It is feasible to construct an integrated coherent optical spectrum analyzer that is instantaneous, has large, continuously variable bandwidth, and is capable of simultaneous analysis and display of frequency spectra. This system may offer the additional advantages of being able to process the signal at any instant of time and to produce a frequency distribution that is useful in situations where velocity (frequency) profiles are important. The sensitivity and resolution of the coherent optical system are also superior to most conventional electronic counterparts.

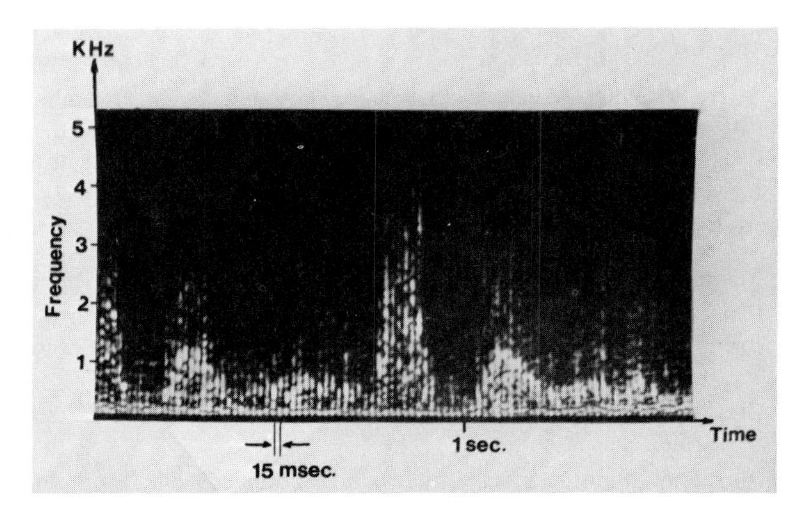

Fig. 7.49 Spectrogram of blood flow in brachian artery obtained with optical spectrum analyzer.

PROBLEMS

7.1 An image of stationary irradiance from a distant object is projected on the recording medium of a camera. The object moves at a variable speed during the time the shutter is open, so that the image is severely smeared on the recording medium. The transmission function of the recorded film is given by

$$T(x, y) = A(x - x', y) \exp\left(\frac{-x^2}{2}\right), \qquad -2 \leq x' \leq 2,$$

where $A(x, y)$ is the unsmeared image of the object. Design a coherent optical processor and an appropriate inverse spatial filter to deblur the image.

7.2 If the object of prob. 7.1 is moving at a constant acceleration, the transmission function of the recorded film may be written

$$T(x, y) = A(x - x', y)\left(1 - \frac{x}{2a}\right), \qquad 0 \leq x' \leq a,$$

where a is an arbitrary constant. Synthesize a complex inverse filter to deblur this image.

7.3 Consider the transmission function of a linear smeared image to be

$$T(x, y) = \begin{cases} A(x - x', y), & -\dfrac{\Delta x}{2} \leq x' \leq \dfrac{\Delta x}{2} \\ 0, & \text{otherwise} \end{cases},$$

where $A(x, y)$ is the unsmeared image, and Δx is the corresponding smear distance. For some reason, the minimum transmittance T_m of the filter is not to go below 25 percent. Using the techniques described in sec. 7.1:

(a) Design an optimum complex spatial filter.

(b) Compute the relative degree of image restoration.

7.4 In image deblurring, we consider synthesizing an inverse filter with a combination of an amplitude filter and a holographic filter.

(a) Sketch an optical arrangement that can obtain both the holographic and the amplitude filters.

(b) By properly controlling the film gamma, show that the amplitude filter can be obtained.

(c) Show that the inverse filter can be obtained with a combination of the amplitude and the holographic filters.

(d) If the inverse filter of part (c) is inserted in the spatial frequency plane of the coherent optical processor of fig. 7.1, evaluate the output deblurred image irradiance.

7.5 Consider the complex image subtraction of prob. 6.21. If we let one of the input transparencies be an open aperture, say $f_2(x, y) = 1$, then determine the output irradiance distribution around the optical axis.

7.6 With reference to prob. 7.5, if $f_1(x, y) = \exp[i\phi(x, y)]$ is a pure phase object, show that the phase variation of $f_1(x, y)$ can be observed at the output plane of the optical processor.

7.7 Consider a multisource coherent optical processor for image subtraction, as shown in fig. 7.50. We assume that the two input object transparencies are separated by a distance of $2h_0$.

(a) Determine the grating spatial frequency and the spacing of these coherent point sources.

(b) Evaluate the intensity distribution of the subtracted image at the output plane.

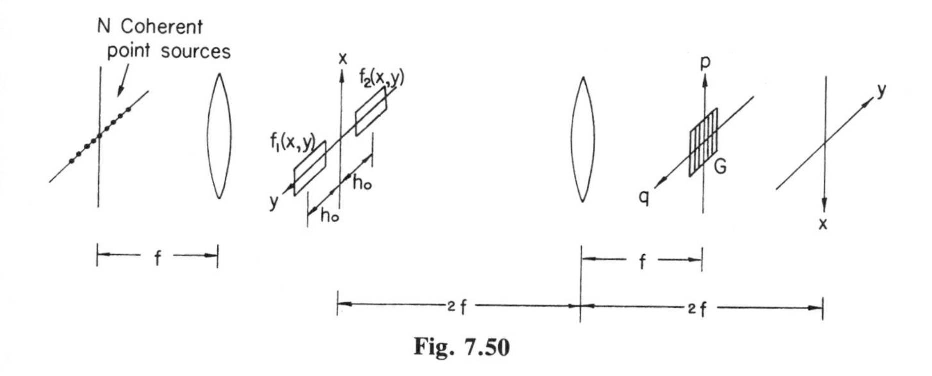

Fig. 7.50

7.8 With reference to the synthetic aperture radar of sec. 7.2, if the aircraft collects data from a point object over a travel distance L, then:

(a) Determine the resulting angular resolution.

(b) If all the available data is used, show that the resolution in a synthetic aperture radar system is equal to half the size of the actual antenna carried by the aircraft. (Note, antenna of size A has a radiated beamwidth $\theta = \lambda/A$).

7.9 Referring to prob. 7.8, if a radar pulse function is

$$s(t) = \operatorname{rect}\left(\frac{t}{T}\right)\exp[i\omega t],$$

where ω is in radians per second, then:

(a) Calculate the corresponding ambiguity function.

(b) Sketch the result of part (a).

7.10 With reference to the sidelooking radar of fig. 7.9, we assume the wavelength of the radar signal is 2 cm, the slant range distance is 20,000 m, and the velocity ratio of the aircraft to the film motion is 10^5. If the recorded format is illuminated by a 6000 Å coherent source, then:

(a) What is the focal length of the recorded Fresnel zone lens?

(b) What is the f-number of the recorded zone lens? Note that the f-number of the signal impinging on the aircraft flight path is r_1/A, where A is the beamwidth at range r_1.

7.11 With reference to eq. (7.40) and the broad-band optical spectrum analyzer described in sec. 7.3:

(a) Determine the corresponding frequency resolution at the output spectral plane. (Hint: the frequency resolution, that is the 3-db point, in the q direction may be determined by setting eq. (7.40) equal to half the comb function peak amplitude, $N^2/2$.)

(b) If the number of scan lines N becomes very large, show that the number of resolution elements at the output plane is linearly proportional to the number of scan lines.

7.12 The spatial resolution of a CRT scanner is 0.2 line/mm and the scanner has an area of 10 × 13 cm. If the CRT is used for wide-band intensity modulation, then:

(a) Determine the limit of the space-bandwidth product.

(b) If the time duration of the processing signal is 2 seconds and if we assume that the separation between the scanned lines is the resolution limit of the CRT, then determine the highest allowable frequency limit.

7.13 Let us consider that the area modulation is a unilateral form and that channels are separated by a distance equal to the resolution element (or spot size) of the electron beam.

(a) Evaluate the available space-bandwidth product for a given system size.

(b) Compare the result obtained from part (a) with the density modulation case.

7.14 Repeat prob. 7.13 for bilateral area modulation.

7.15 We refer to the wide-band spectrum analysis with area modulation of sec. 7.5. Assume that the input is a periodic triangular signal as shown in fig. 7.51.

(a) Determine the minimum bias level required for the wide-band area modulation.

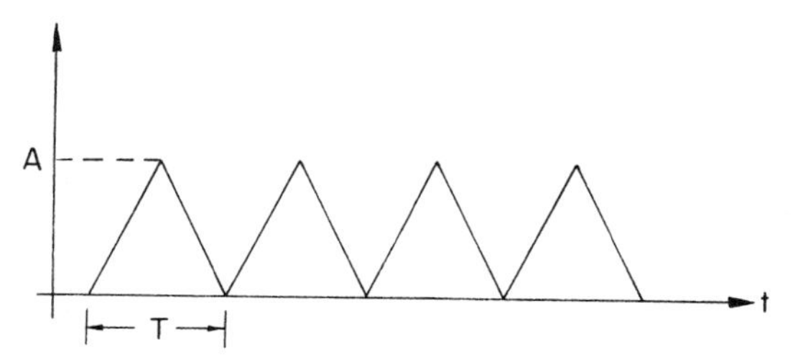

Fig. 7.51

(b) Sketch the corresponding wide-band modulated format.

(c) If the recorded format of part (b) is insertd in a coherent spectrum analyzer, evaluate the spectrum within a narrow slit along the q axis of the spatial frequency plane, i.e.,

$$H(p, q) = \begin{cases} 1, & \dfrac{-\pi}{D} \leq q \leq \dfrac{\pi}{D} \\ 0, & \text{otherwise} \end{cases},$$

where D is the separation between the channels, as shown in fig. 7.22.

(d) Compare part (c) with direct Fourier transformation of the triangular signal of fig. 7.51.

7.16 Given a scanning optical correlator, as shown in fig. 7.52. The reference function $f(x, y)$ at P_1 is imaged onto the input function $g(x, y)$ at P_2 through an x-y scanner. The resultant signal is then Fourier transformed by L_3 onto a photo detector, which is placed at the Fourier plane P_3. The detected signal is then displayed on a CRT screen, as shown in the figure.

(a) Describe the complex light field immediately behind P_2.

(b) Determine the irradiance distribution at P_3.

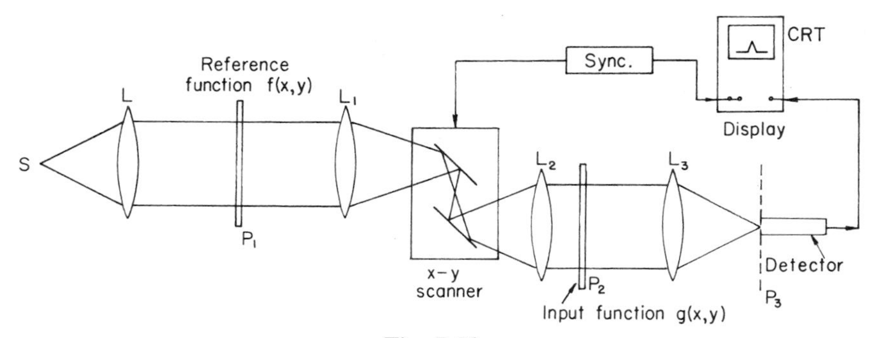

Fig. 7.52

7.17 With reference to prob. 7.16, if the reference signal and the input signal transparencies are given in fig. 7.53, then:

(a) Sketch the correlation detection signal that is displayed on the CRT screen.

(b) Describe a simple method to improve the correlation peak.

(c) If the reference signal transparency is preprocessed with high-pass filtering, then would the autocorrelation peak be affected?

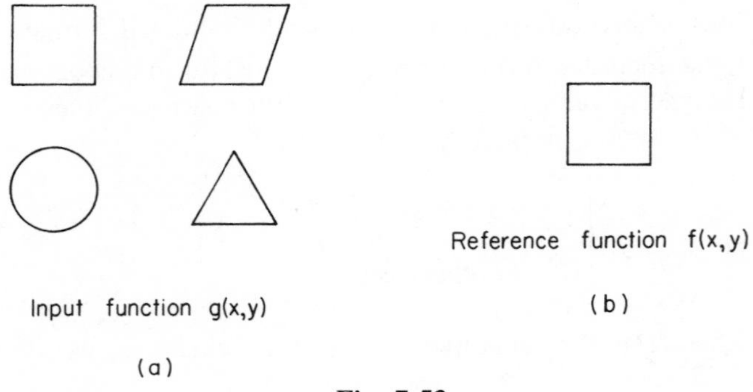

Fig. 7.53

7.18 Figure 7.54 shows the intensity transmittance distribution of a halftone screen. The screen is used for contact printing on a high-contrast film with a collimated white light. We assume that this thresholding exposure is set at $T_{in} = 0.6$.

(a) Determine the transmittance distribution of the recorded film.

(b) If the recorded transparency is inserted at the input plane of a coherent optical processor, then evaluate the irradiance distribution at the spatial frequency plane.

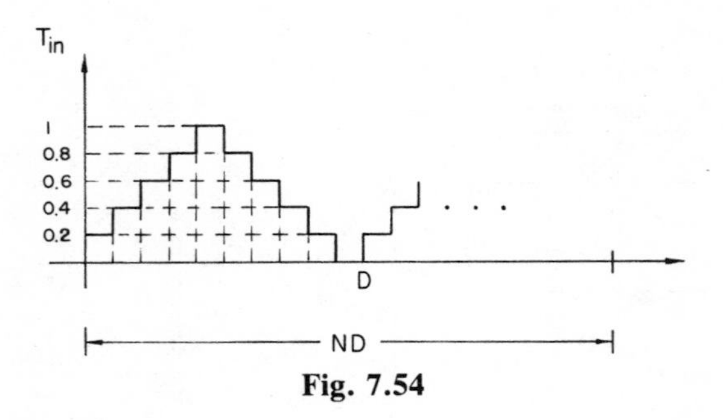

Fig. 7.54

7.19 **(a)** Describe a method of applying a halftone screen processing technique to achieve a logarithmic transformation.

(b) Is it possible to utilize a halftone screen processing technique to obtain an exponential transformation? State the major difficulties.

7.20 With reference to the nonlinear processing system of fig. 7.25, show that a higher correlation peak may be obtained through the homomorphic filtering system.

7.21 As compared with the halftone processing technique, state the major advantages and disadvantages of obtaining a logarithmic transformation utilizing the film nonlinearity.

7.22 **(a)** Show that equidensitometry images of sec. 7.6 can also be obtained with an incoherent source (e.g., white-light source), instead of a coherent source.

(b) What are the advantages of utilizing a white-light source?

7.23 Matched filtering by the prewhitening technique has been applied to both temporal and spatial signals. The spatial signal is generally processed with a coherent optical processor while the temporal signal is generally processed with an electronic counterpart.

(a) The proposed optical prewhitening technique of sec. 7.8 is limited to ergodic noise. Can the concept also be applied to a temporal signal? Explain.

(b) It is well known that the processing of a temporal signal is limited by the causality of the processing system. Explain how the causality constraint can be removed for spatial signal processing.

(c) Differentiate the dynamic constraints for the temporal and spatial prewhitening filters.

7.24 Given an intensity-modulated optical spectrum analyzer, as shown in fig. 7.55. The input signal represents a continuous density-modulated speech signal, from 25 to 4000 Hz.

(a) Determine the width Δx of the optical window required for narrow- and wide-band spectrum analysis. (Note: a 300-Hz bandwidth is used for wide-band analysis and 25 Hz is for narrow-band.)

(b) Show that the time and frequency resolutions cannot be resolved simultaneously.

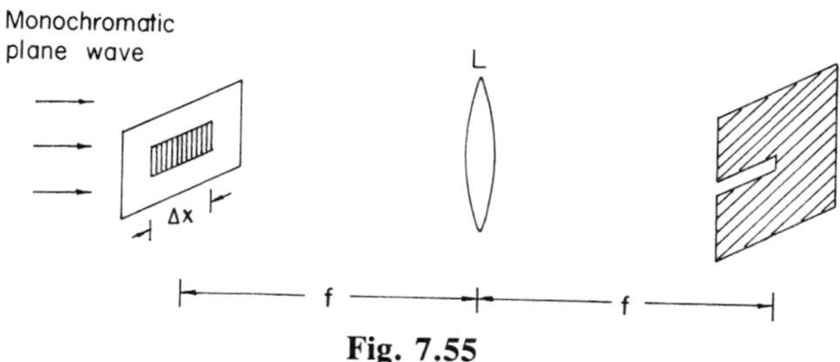

Fig. 7.55

7.25 A coherent optical processor for image subtraction is shown in fig. 7.56, where $O_1(x, y)$ and $O_2(x, y)$ are the input signal transparencies.

(a) Evaluate the output irradiance recorded onto the moving film.

(b) What would be the result if the π-phase filter were removed from the processor?

(c) Compare this technique with the image subtraction of prob. 6.25.

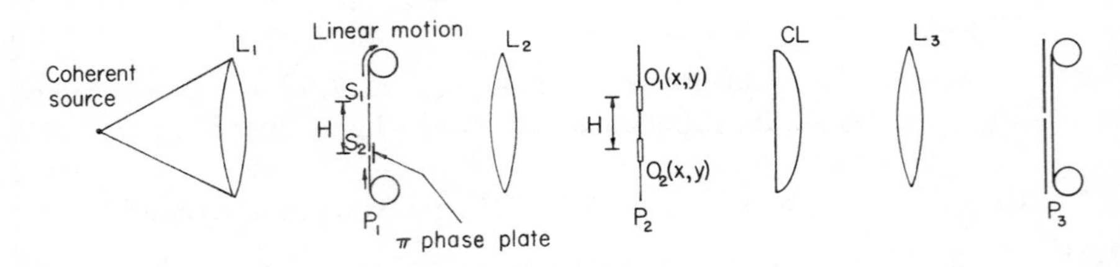

Fig. 7.56

REFERENCES

7.1 J. Tsujiuchi, "Correction of Optical Images by Compensation of Aberrations and Spatial Frequency Filtering," in E. Wolf, Ed., *Progress in Optics*, Vol. II, North-Holland Publishing Company, Amsterdam, 1963.

7.2 G. W. Stroke and R. G. Zech, "A Posteriori Image-Correcting Deconvolution by Holographic Fourier-Transform Division," *Phys. Lett.*, ser. A, **25,** 89 (1967).

7.3 G. W. Stroke, F. Furrer, and D. R. Lamberty, "Deblurring of Motion-Blurred Photographs Using Extended-Range Holographic Fourier-Transform Division," *Opt. Commun.*, **1,** 141 (1969).

7.4 J. L. Horner, "Optical Spatial Filtering with the Least Mean-Square-Error Filter," *J. Opt. Soc. Am.*, **59,** 553 (1969).

7.5 J. L. Horner, "Optical Restoration of Images Blurred by Turbulence Using Optimum Filter Theory," *Appl. Opt.*, **8,** 167 (1970).

7.6 J. Tsujiuchi, T. Honda, and T. Fukaya, "Restoration of Blurred Photographic Images by Holography," *Opt. Commun.*, **1,** 379 (1970).

7.7 E. N. Leith, "Image Deblurring Using Diffraction Gratings," *Opt. Lett.*, **5,** 70 (1980).

7.8 F. T. S. Yu, "Image Restoration, Uncertainty, and Information," *J. Opt. Soc. Am.*, **58,** 742 (1968); *Appl. Opt.* **8,** 53 (1969).

7.9 F. T. S. Yu, "Optical Resolving Power and Physical Realizability," *J. Opt. Soc. Am.*, **59,** 497 (1969); *Opt. Commun.*, **1,** 319 (1970).

7.10 T. M. Halladay and J. D. Gallatin, "Phase Control by Polarization in Coherent Spatial Filtering," *J. Opt. Soc. Am.*, **56,** 869 (1966).

7.11 A. W. Lohmann and D. P. Paris, "Computer Generated Spatial Filters for Coherent Optical Data Processing," *Appl. Opt.*, **7,** 651 (1968).

7.12 F. T. S. Yu, "Coherent and Digital Image Enhancement, Their Basic Differences and Constraints," *Opt. Commun.*, **3,** 440 (1971).

7.13 L. J. Cutrona et al., "On the Application of Coherent Optical Processing Techniques to Synthetic-Aperture Radar," *Proc. IEEE*, **54,** 1026 (1966).

7.14 C. E. Thomas, "Optical Spectrum Analysis of Large Space-Bandwidth Signals," *Appl. Opt.*, **5,** 1782 (1966).

7.15 B. V .Markevitch, Optical Processing of Wideband Signals, 3rd Annual Wideband Recording Symposium, Rome Air Development Center, April, 1969.

7.16 R. V. Churchill, *Fourier Series and Boundary Value Problems*, McGraw-Hill, New York, 1941, p. 32.

7.17 A. Tai and F. T. S. Yu, "Synchronous Dual-Channel Optical Spectrum Analyser," *Appl. Opt.*, **18,** 1297 (1979).

7.18 R. Spattiswoode, *Film and Its Techniques*, University of California Press, Berkeley, Cal., 1964.

7.19 A. A. Dyachenko, M. V. Persikov, and O. E. Shushpaav, "Use of Shadowgraph for Spectral Functional Analysis by the Methods of Coherent Optics," *Opt. Spectrosc.*, **31,** 249 (1971).

7.20 E. B. Felstead, "Optical Fourier Transformation of Area-Modulated Spatial Functions," *Appl. Opt.*, **10,** 2468 (1971).

7.21 B. J. Pernick, "Area-Modulated Signal Recordings for Coherent Optical Correlations," *Appl. Opt.*, **11,** 1425 (1972).

7.22 A. Tai and F. T. S. Yu, "Wide-Band Spectrum Analysis with Area Modulation," *Appl. Opt.*, **18,** 460 (1979).

7.23 J. D. Armitage and A. W. Lohmann, "Theta Modulation in Optics," *Appl. Opt.*, **4,** 399 (1965).

7.24 K. T. Stalker and S. H. Lee, "Use of Nonlinear Optical Elements in Optical Information Processing," *J. Opt. Soc. Am.*, **64,** 545 (1974).

7.25 M. Marquest and J. Tsujiuchi, "Interpretation des Aspects Particuliers des Images Obtenues dans une Expérience de Déliamage," *Opt. Acta,* **8,** 267 (1961).

7.26 A. V. Oppenheim, R. W. Schafer, and T. G. Stockham, Jr., "Nonlinear Filtering of Multiplied and Convolved Signals," *Proc. IEEE*, **56,** 1264 (1968).

7.27 H. Kato and J. W. Goodman, "Nonlinear Filtering in Coherent Optical Systems Through Halftone Screen Processes," *Appl. Opt.*, **14,** 1813 (1975).

7.28 S. R. Dashiell and A. A. Sawchuk, "Nonlinear Optical Processing: Analysis and Synthesis," *Appl. Opt.*, **16,** 1009 (1977).

7.29 T. C. Strand, "Non-Monotonic Non-Linear Image Processing Using Halftone Techniques," *Opt. Commun.*, **15,** 60 (1975).

7.30 H. K. Liu, J. W. Goodman, and J. Chan, "Equidensitometry by Coherent Optical Filtering," *Appl. Opt.*, **15,** 2394 (1976).

7.31 B. J. Bartholomew and S. H. Lee, "Nonlinear Optical Processing with Fabry–Perot Interferometers Containing Phase Recording Media," *Appl. Opt.*, **19,** 201 (1980).

7.32 A. Tai, T. Cheng, and F. T. S. Yu, "Optical Logarithmic Filtering Using Inherent Film Nonlinearity," *Appl. Opt.*, **16,** 2559 (1977).

7.33 B. H. Soffer, D. Boswell, A. M. Lackner, P. Chavel, A. A. Sawchuk, T. C. Strand, and A. R. Tanguay, Jr., "Optical Computing with Variable Grating Mode Liquid Crystal Devices," *Proc. SPIE*, **218,** 26 (1980).

7.34 M. S. Dymek, A. Tai, T. H. Chao, and F. T. S. Yu, "Exponential Transformation Using Film Nonlinearity for Optical Homomographic Filtering," *Appl. Opt.*, **19,** 829 (1980).

7.35 J. Thomas, *Introduction to Statistical Communication Theory*, John Wiley, New York, 1969, chap. 5.

7.36 T. H. Chao, A. M. Tai, M. S. Dymek, and F. T. S. Yu, "Optimum Correlation Detection by Prewhitening," *Appl. Opt.*, **19,** 2461 (1980).

7.37 J. W. Goodman, *Laser Speckle and Related Phenomena*, Vol. 9 in J. C. Dainty, Ed., *Topics in Applied Physics*, Springer, New York, 1975.

7.38 M. Schwartz, *Information, Transmission, Modulation and Noise*, McGraw-Hill, New York, 1970, chap. 6.

7.39 L. J. Cutrona, "Recent Development in Coherent Optical Technology," in J. T. Tippett et al., Eds., *Optical and Electrooptical Information Processing*, Cambridge, Mass., 1965.

7.40 R. A. Heinz, J. O. Artman, and S. H. Lee, "Matrix Multiplication by Optical Method," *Appl. Opt.*, **9,** 2161 (1970).

7.41 L. M. Deen, J. F. Walkup, and M. O. Hagler, "Representations of Space-Variant Optical Systems Using Volume Holograms," *Appl. Opt.*, **14,** 2438 (1975).

7.42 R. J. Marks, II, J. F. Walkup, M. O. Hagler, and T. F. Knife, "Space-Variant Processing of 1-D Signal," *Appl. Opt.*, **16,** 739 (1977).

7.43 O. Bryngdahl, "Optical Map Transformations," *Opt. Commun.*, **10,** 164 (1974).

7.44 O. Bryngdahl, "Geometrical Transformation in Optics," *J. Opt. Soc. Am.*, **64,** 1092 (1974).

7.45 A. A. Sawchuk, "Space-Variant Image Restoration by Coordinate Transformation," *J. Opt. Soc. Am.*, **64,** 138 (1974).

7.46 W. H. Lee, "Binary Synthetic Holograms," *Appl. Opt.*, **13,** 1644 (1974).

7.47 D. Casasent and D. Psaltis, "Scale Invariant Optical Transform," *Opt. Eng.*, **15,** 258 (1976).

7.48 J. C. Lin, F. T. S. Yu, and A. M. Tai, "Ultrasonic Blood Flow Spectral Analysis Using Coherent Optics," *IEEE Trans. Biomed. Eng.*, **BME-25,** 243 (1978).

7.49 D. W. Baker, "Pulse Ultrasonic Doppler Blood-Flow Sensing," *IEEE Trans. Sonic Ultrasonic*, **17,** 171 (1970).

8

Optical Processing with Incoherent Source

The use of coherent light enables optical systems to carry out many sophisticated information processing operations (ref. 8.1). However, coherent optical processing systems are plagued with coherent artifact noise, which frequently limits their processing capability. Although many optical information processing operations can be implemented by systems that use incoherent light (refs. 8.2–8.5), there are other severe drawbacks. The incoherent processing system is capable of reducing the inevitable artifact noise, but it generally introduces a dc-bias buildup problem, which results in poor noise performance. Techniques have been developed for coherent operation with light of reduced coherence (refs. 8.6, 8.7); however, these techniques also possess severe limitations.

Attempts at reducing the temporal coherence requirements on the light source in optical information processing fall into two general categories: one, the use of incoherent instead of coherent optical processing has been pursued by Lowenthal and Chavel (ref. 8.8) and Lohmann (ref. 8.9), among others. The other, the reduction of coherence while still operating in the linear-in-amplitude, has been pursued by Leith and Roth (ref. 8.10) and by Morris and George (ref. 8.11). The latter is the one that we believe to hold the most promise, and it is the concept of complex amplitude processing that we are pursuing.

Since its invention as a strong coherent source, the laser has become a fashionable device for many applications, particularly for coherent optical information processing. This trend has been largely due to its complex amplitude processing capability. However, coherent processing systems are also contaminated with coherent artifact noise, which Gabor (ref. 8.12) has noted to be the number one enemy in optical information processing systems. Moreover, the coherent source is generally expensive and the optical processing environment is usually very critical. As we look at the optical information processing technique from a different standpoint, a question arises: is it necessarily true that all information processing operations require a coherent source? The answer to this question is that there are many optical information processing operations that can be easily carried out with reduced coherence requirements. In other words, if the coherence requirement for certain information processing operations is not too high, the information processing may be carried out by incoherent or white-light sources.

In this chapter, we describe a technique that permits certain information processing (refs. 8.13, 8.14) operations to be carried out by a spectrally broad-band light source (i.e., white light). This method is capable of performing information processing that obeys the concept of coherent light rather than incoherent optics. That is, we use a white (temporally incoherent) light source for our processing system and the optical system is linear in complex amplitude rather than in intensity as in the conventional incoherent processing system. In other words, the system is operating in partially coherent mode in the Fourier plane, so that the input signal can be processed in a complex amplitude.

8.1 A WHITE-LIGHT OPTICAL PROCESSING TECHNIQUE

We now describe an optical processing technique that can be carried out by a white-light source, as shown in fig. 8.1. Note that the white-light processing system is similar to a coherent processing system except for the use of a white-light source and a high diffraction efficiency grating inserted in the input plane P_1. If we place a signal transparency $s(x, y)$ in contact with the diffraction grating, the complex light field for every wavelength λ behind the achromatic transform lens L_1 is

$$E(p, q; \lambda) = C \iint s(x, y)[1 + \cos(p_0 x)] \exp[-i(px + qy)]\, dx\, dy, \qquad (8.1)$$

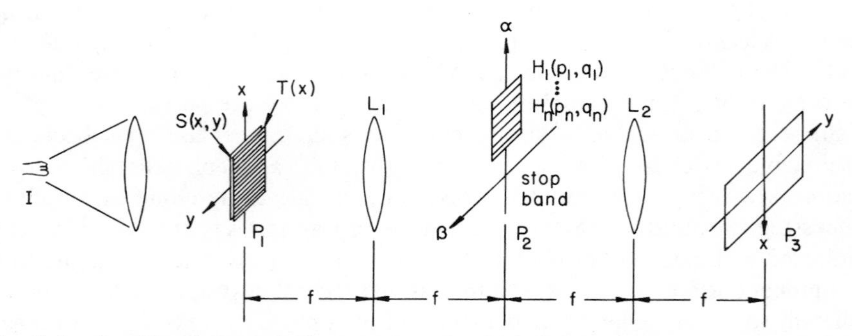

Fig. 8.1 A white-light optical processor. I, white-light point source; $s(x, y)$, input signal transparency; $T(x)$, high efficiency diffraction grating; $H_n(p_n, q_n)$, complex spatial filters.

where the integral is over the spatial domain of the input plane P_1, (p, q) denotes the angular spatial frequency coordinate system, and C is a complex constant.

For simplicity of analysis, we drop the proportionality constant and eq. (8.1) becomes

$$E(p, q; \lambda) = S(p, q) + S(p - p_0, q) + S(p + p_0, q), \tag{8.2}$$

where $S(p, q)$ is the Fourier spectrum of $s(x, y)$, $p = (2\pi/\lambda f)\alpha$, and $q = (2\pi/\lambda f)\beta$, (α, β) is the linear spatial coordinates system of (p, q), and f is the focal length of the achromatic transform lens. In terms of the spatial coordinates of α and β, eq. (8.2) can be written

$$E(\alpha, \beta; \lambda) = C_1 S(\alpha, \beta) + C_2 S\left(\alpha - \frac{\lambda f}{2\pi} p_0, \beta\right) + C_3 S\left(\alpha + \frac{\lambda f}{2\pi} p_0, \beta\right). \tag{8.3}$$

From the above equation, we see that two first-order signal spectra bands (i.e., second and third terms) are dispersed into rainbow color along the α axis, and each spectrum is centered at $\alpha = \pm(\lambda f/2\pi)p_0$.

In the analysis we assume that a sequence of complex spatial filters for various λ_n are available, i.e., $H(p_n; q_n)$, where $p_n = (2\pi/\lambda_n f)\alpha$, $q_n = (2\pi/\lambda_n f)\beta$. If we place these complex spatial filters in the spatial frequency plane with each centered at $\alpha = (\lambda_n f/2\pi)p_0$, then the complex light field behind the spatial frequency plane is

$$E(p, q; \lambda) = S(p - p_0, q) \sum_{n=1}^{N} H(p_n - p_0, q_n). \tag{8.4}$$

The corresponding complex light distribution at the output plane P_3 of the processor for each λ is

$$g(x, y; \lambda) = \sum_{n=1}^{N} \iint S(p - p_0, q) H(p_n - p_0, q_n) \exp[i(p_n x + q_n y)]\, dp_n\, dq_n, \tag{8.5}$$

where the integration is over the spatial domain. We assume that the signal spectrum is spatial frequency limited and that the bandwidth of $H(p_n, q_n)$ is extended to this limit, i.e.,

$$H(p_n, q_n) = \begin{cases} H(p_n, q_n), & \alpha_1 < \alpha < \alpha_2 \\ 0, & \text{otherwise} \end{cases} \tag{8.6}$$

where $\alpha_1 = (\lambda_n f/2\pi)(p_0 + \Delta p)$, and $\alpha_2 = (\lambda_n f/2\pi)(p_0 - \Delta p)$ are the upper and the lower spatial limits of $H(p_n, q_n)$, and Δp is the bandwidth of the input signal. The limiting wavelengths of the dispersed spectrums at the upper and the lower edges of the filters are

$$\lambda_l = \lambda_n \frac{p_0 + \Delta p}{p_0 - \Delta p}, \qquad \text{and} \qquad \lambda_h = \lambda_n \frac{p_0 - \Delta p}{p_0 + \Delta p}, \tag{8.7}$$

and the corresponding wavelength spread over the filters is

$$\Delta\lambda_n = \lambda_n \frac{4p_o \Delta p}{p_0^2 - (\Delta p)^2}. \tag{8.8}$$

If the spatial frequency p_0 of the grating is high, then the wavelength spreads over the filters can be approximated by

$$\Delta\lambda_n \simeq \frac{4\Delta p}{p_0} \lambda_n, \qquad p_0 \gg \Delta p. \tag{8.9}$$

Since the complex spatial filterings take place in discrete Fourier spectral bands of the light source, the filtered signals are *mutually* incoherent. The output light intensity distribution is

$$I(x, y) \simeq \sum_{n=1}^{N} \Delta\lambda_n |g(x, y; \lambda_n)|^2 = \sum_{n=1}^{N} \Delta\lambda_n |s(x, y; \lambda_n) * h(x, y; \lambda_n)|^2, \tag{8.10}$$

where $h(x, y; \lambda_n)$ is the spatial impulse response of the filter $H(p_n, q_n)$, and $*$ denotes the convolution operation. From the above equation, we see that the white-light processing technique is indeed capable of processing the signal in a complex amplitude. Since the output intensity is the sum of the mutually incoherent narrow-band irradiances, the annoying coherent artifact noise can be suppressed.

In addition, the white-light source contains all the color wavelengths of the visible light. As we see in the next few sections, the technique is particularly suitable for color image processing. We further note that the white-light processor of fig. 8.1 can also be used for coherent and partially coherent light.

In application to complex signal detection, we assume that a sequence of complex spatial filters of various λ_n's are available, i.e.,

$$H(p_n, q_n) = K_1 + K_2 + K|S(p_n, q_n)| \cos|\beta_0 q_n + \phi(p_n, q_n)|, \qquad n = 1, 2, \ldots, N, \tag{8.11}$$

where $S(p_n, q_n) = |S(p_n, q_n)| \exp[i\phi(p_n, q_n)]$ is the signal spectrum for wavelength λ_n.

If we place these complex spatial filters in the spatial frequency plane with each centered at $\alpha = (\lambda f/2\pi)p_0$, then the complex light field at the output image plane P_3 would be

$$g(x, y) \simeq \sum_{n=1}^{N} \Delta\lambda_n |s_n(x, y)\, e^{ip_0x} + s_n(x, y)\, e^{ip_0x} * s_n(x, y)\, e^{ip_0x}$$

$$* s_n^*(x, y)\, e^{ip_0x} + s_n(x, y)\, e^{ip_0x} * s_n(x, y + \beta_0)\, e^{ip_0x} + s_n(x, y)\, e^{ip_0x}$$

$$* s_n^*(-x, -y + \beta_0)\, e^{ip_0x}, \qquad (8.12)$$

where $*$ denotes the convolution operation and the superscript $*$ denotes the complex conjugate. For simplicity the proportionality constants K were dropped from the above equation. From eq. (8.12) we see that the first and second terms represent the zero-order terms, which are diffracted in the neighborhood of (0, 0) in the output plane, and the third and fourth terms are the convolution and correlation terms, which are diffracted in the neighborhood of $(0, -\beta_0)$ and $(0, \beta_0)$, respectively, as shown in fig. 8.2. Furthermore, we note that the diffracted output signal takes place by *incoherent* addition of the discrete spectral bands; therefore, the annoying coherent artifact noise can be suppressed.

Let us now discuss the correlation term of eq. (8.12), i.e.,

$$R(x, y) = \sum_{n=1}^{N} \Delta\lambda_n \iint S\left(\alpha - \frac{\lambda f}{2\pi}p_0, \beta\right) S_n^*\left(\alpha - \frac{\lambda_n f}{2\pi}p_0, \beta\right) \exp(-i\beta_0 y)$$

$$\exp\left| i\frac{2\pi}{\lambda f_n}(\alpha x + \beta y)\right| d\alpha\, d\beta. \qquad (8.13)$$

From this equation we see that there is a *mismatch* in location and in scaling of the incoming signal spectrum with respect to the filter funciton. In other words, if the spatial carrier frequency p_0 of the diffraction grating is high, a narrower spread in the x direction of the correlation peak can be obtained. Thus the accuracy of the complex filtering in the x direction is somewhat lower than that in the y direction. In other words, this white-light processing technique is

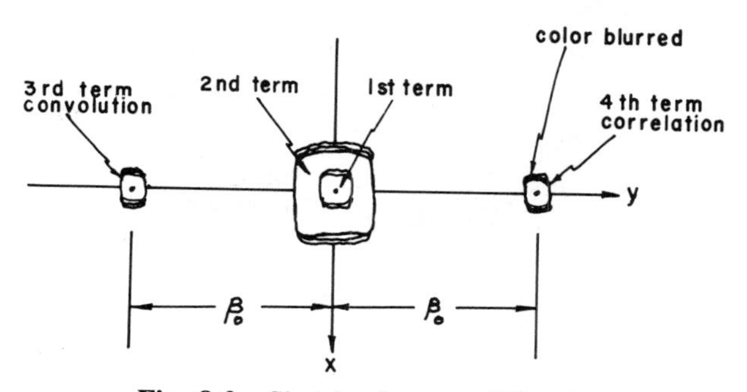

Fig. 8.2 Sketch of output diffraction.

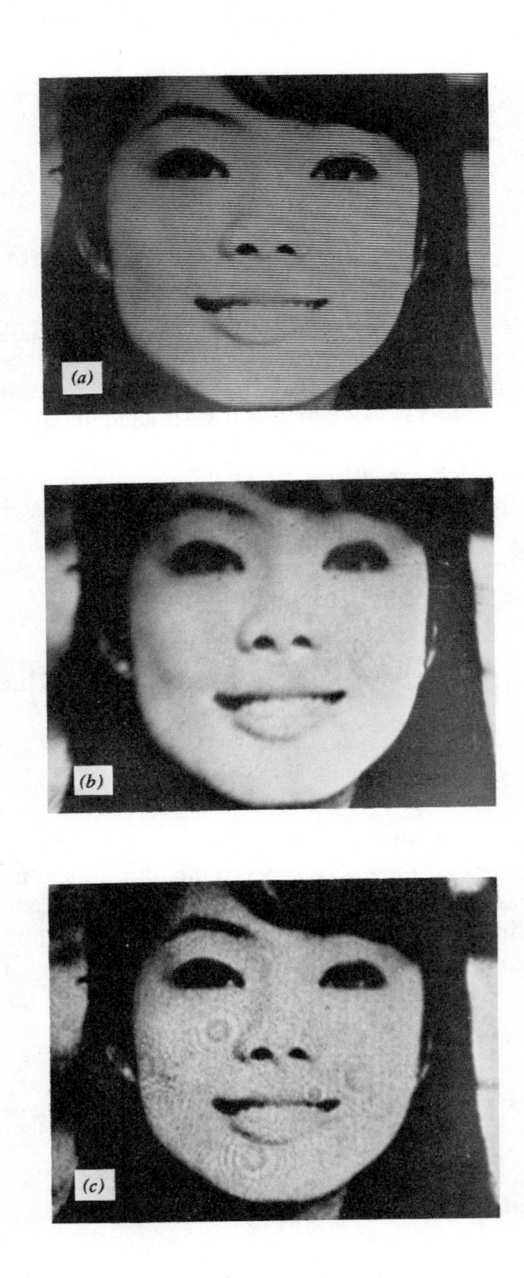

Fig. 8.3 Spatial filtering of an additive signal. (*a*) Original input signal, (*b*) Obtained by white-light processing, (*c*) Obtained by coherent processing.

effective only in one direction, and for some two-dimensional processing operations, this technique may pose certain limitations. Strictly speaking, this technique offers no coherent artifact noise and it can be applied to various problems in optical information processing. We emphasize that, although the technique uses a *temporal* incoherent source, the signal spectrum is displayed in partially coherent mode so that the signal can be processed in complex amplitude.

8.2 IMAGE FILTERING WITH WHITE-LIGHT SOURCE

In previous sections we have introduced a technique of white-light optical processing utilizing a diffraction grating. We have shown the feasibility of applying this white-light processing technique to complex signal filtering. Although the operation of this white-light processing technique is effective in one dimension, it can be applied for two-dimensional signals. We stress that the basic advantage of the white-light processing is the suppression of coherent artifact noise that plagues the coherent optical processing system. In this section we demonstrate an experimental result, showing that the white-light processing technique indeed suppresses the coherent artifact noise (ref. 8.15). We show that the white-light processing technique is a simple, economical, versatile technique, that may be used in many of the same applications as a coherent optical processor.

In our experimental illustration, we insert an input signal with an additive Ronchi-type grating (as shown in fig. 8.3*a*) in the input plane P_1 of the white-light processor of fig. 8.4. A tungsten arc lamp is used as a white-light point source in our experiment. With reference to the white-light processor (of fig. 8.4), we see that the signal spectrum is dispersed into rainbow color of spectral in the p direction of the spatial frequency plane P_2. Since the Ronchi grating is in the y direction with respect to the input spatial plane, a set of slanted spectra in rainbow color can be seen, as shown in fig. 8.5. If the carrier spectral lines are filtered out by an absorptive-type filter, then the additive Ronchi grating can be removed from output image plane P_3.

To illustrate the artifact noise effect, a thin dust plate is inserted between the collimating lens and the input plane P_1 of the optical processor. Figure 8.3*b*

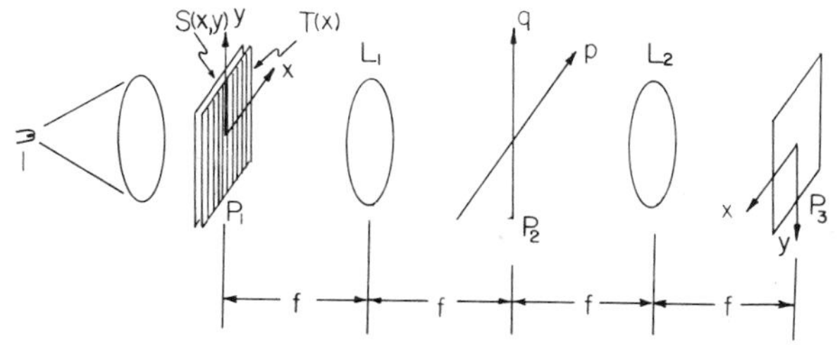

Fig. 8.4 A white-light optical processor utilizing a diffraction grating method. $s(x, y)$, input signal transparency; $T(x)$, diffraction grating; I, white-light source.

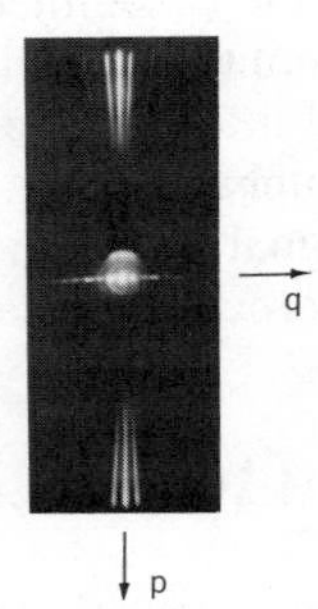

Fig. 8.5 Smeared rainbow color spectra.

shows the result obtained by the white-light processing technique, while fig. 8.3*c* is the result obtained with a He-Ne laser as light source. From these two experimental results, we see that the artifact noise effect is substantially suppressed with the white-light processing technique.

Thus compared with the conventional coherent optical processing technique, the white-light processing technique offers the following advantages:

1 The cost of the processing system is significantly lowered with the elimination of the laser source.

2 The alignment procedure of the white-light processing system is generally simpler than the coherent technique.

3 The output result is free from coherent artifact noise that usually plagues the coherent processing systems.

Finally, we stress that this white-light processing system is capable of performing complex information processing in a manner that obeys the concept of coherent light rather than incoherent optics. This white-light processing system is linear and spatially invariant in complex amplitude rather than intensity. In other words, the system operates in a partially coherent mode in the Fourier plane so that the signal can be processed in complex amplitude rather than in intensity.

8.3 MULTI-IMAGE REGENERATION

In this section we demonstrate that multi-image regeneration can be easily obtained by the white-light optical processing technique (ref. 8.16). We note that the simplicity and economical nature of the white-light processing technique make it particularly attractive for many practical applications, for example, multi-image regeneration for reproduction of integrated circuit masks.

For illustration, a signal transparency $s(x, y)$ is superimposed with a high-efficiency diffraction grating $T(x)$ at the input plane P_1 of the white-light optical processor of fig. 8.1. For simplicity, we assume that this diffraction grating is holographic type, and the complex light distribution at the spatial frequency plane P_2 for every λ can be described as

$$E(p, q; \lambda) = \iint s(x, y)T(x) \exp[-i(xp + yq)] \, dx \, dy, \tag{8.14}$$

where $T(x) = \frac{1}{2}[1 + \cos p_0 x]$, p_0 is the angular spatial frequency of the grating, (p, q) is the angular spatial frequency coordinate system, and the integration is over the spatial domain of the signal transparency. The corresponding complex light field in terms of (α, β) spatial coordinate system of P_2 can be shown to be

$$E(p, q; \lambda) = S(\alpha, \beta) + \tfrac{1}{2}S\left(\alpha - \frac{\lambda f}{2\pi}p_0, \beta\right) + \tfrac{1}{2}S\left(\alpha + \frac{\lambda f}{2\pi}p_0, \beta\right), \tag{8.15}$$

where f is the focal length of the achromatic transform lens. From this expression we see that two first-order signal spectra disperse in rainbow color along the α axis with respect to the wavelength of the broad-band light source.

In multi-image regeneration we place a sequence of diffraction gratings of various frequencies in some narrow Fourier spectral bands in the spatial frequency plane P_2, as illustrated in fig. 8.6. For simplicity, we evaluate only for the upper smeared Fourier spectral bands.

Now let us assume that the ampltiude transmittances of the gratings are

$$H_n(q_n) = \tfrac{1}{2}[1 + \cos(\beta_n q_n)], \qquad n = 1, 2, \ldots, N, \tag{8.16}$$

where β_n is an arbitrarily chosen spatial frequency, $q_n = (2\pi/\lambda_n f)\beta$, and λ_n is the nth center wavelength of the narrow spectral band. Alternatively, eq. (8.16) can be written as

$$H_n(\beta) = \tfrac{1}{2}[1 + \cos(q_0 \beta)], \qquad n = 1, 2, \ldots, N, \tag{8.17}$$

where $q_0 = (2\pi/\lambda_n f)\beta_n$.

In practice, the image spectrum is assumed to be spatially frequency limited and

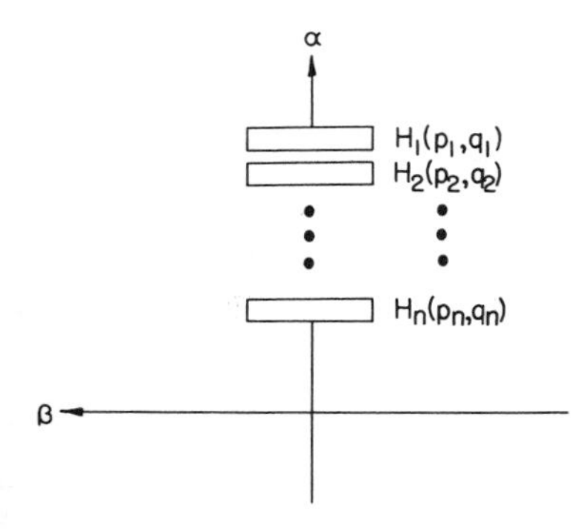

Fig. 8.6 Sketch of multi-image regeneration filterings for different spectral bands.

each of the gratings is designed to cover the entire image bandwidth Δp. The wavelength spread $\Delta\lambda$ over each of the gratings can be shown $\Delta\lambda_n \simeq (4\Delta p/p_0)\lambda_n$, $P_0 \gg \Delta p$. We see that $\Delta\lambda_n$ can be made very small if the spatial frequency p_0 of the diffraction grating is high. In other words, wavelength smear of regenerated images can be minimized. Thus for sufficiently large p_0, the complex light distribution behind the gratings $H_n(q_n)$ can be approximated by

$$E(p, q; \lambda) \simeq \frac{\Delta\lambda_n}{2}[S(p_n - p_0, q_n) + \tfrac{1}{2}S(p_n - p_0, q_n)e^{-i\beta_n q_n} + \tfrac{1}{2}S(p_n - p_0, q_n)e^{i\beta_n q_n}], \qquad n = 1, 2, \ldots, N, \tag{8.18}$$

where $\Delta\lambda_n$ is the corresponding narrow spectral band of the light source over the diffraction grating $H_n(q_n)$. Then at the back focal length of the achromatic transform lens L_2, we have the following output complex light distribution:

$$g(x, y; \lambda) = \frac{\Delta\lambda_n}{2}[s_n(x, y) + \tfrac{1}{2}s_n(x, y + \beta_n) + \tfrac{1}{2}s_n(x, y - \beta_n)]e^{iP_0 x}, \qquad n = 1, 2, \ldots, N. \tag{8.19}$$

Since each nth term of eq. (8.19) is derived from each narrow spectral band of the light source, they are mutually *incoherent*. The corresponding output irradiance is, therefore,

$$I(x, y) = \sum_{n=1}^{N}\left(\frac{\Delta\lambda_n}{2}\right)^2 [s_n^2(x, y) + \tfrac{1}{4}s_n^2(x, y + \beta_n) + \tfrac{1}{4}s^2{}_n(x, y - \beta_n)]. \tag{8.20}$$

From this equation we see that multi-images in different colors are regenerated at the output plane of the white-light processor. If the spatial frequencies of the gratings are $\beta_n = 2n\beta$ and $\beta_1 \geqq 2\Delta y$ at least two times higher than the size of the image, then we would have a sequence of nonoverlapping regenerated images displayed in different colors at the output plant, as sketched in fig. 8.7. We note that the intensity of the zero-order image can be suppressed by a neutral

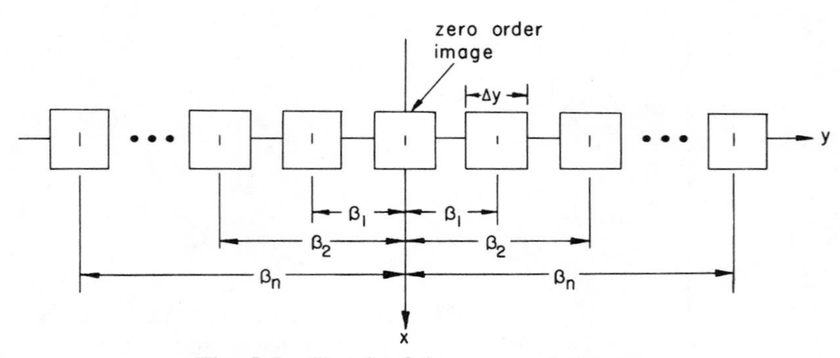

Fig. 8.7 Sketch of the regenerated images.

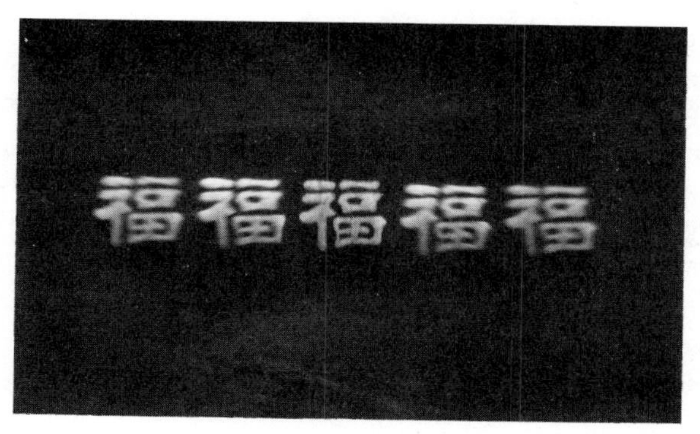

Fig. 8.8 A picture of the regenerated images obtained by the white-light processing technique. Note that the zero-order image irradiance has been suppressed.

density filter, so that uniform-intensity images can be obtained. Furthermore, from eq. (8.20), we see that the regenerated images take place by the spectral band $\Delta\lambda_n$, thus the annoying coherent artifact noise can be suppressed.

For simplicity of experimental demonstration, we use only two narrow gratings for the image regeneration; one is placed in the upper Fourier smeared spectrum and the other is inserted in the lower smeared spectrum. The spatial frequencies of these two gratings are chosen so that the regenerated images could be spatially separated. In our experiments, a xenon arc was used as the light source, and a high spatial frequency grating was used for the image spectra dispersion. An experimental result of this white-light image regeneration technique is shown in fig. 8.8. We would note that the zero-order image irradiance has been suppressed to obtain an uniform image intensity throughout the regenerated images. Although the optical parameters are not optimally adjusted by juggling the result that we have obtained, the quality of the regenerated images is reasonably good. We believe that with more careful and elaborate experimentation, higher quality regenerated images can be obtained.

8.4 SMEARED PHOTOGRAPHIC IMAGE DEBLURRING

An interesting application of coherent optical information processing is the restoration of smeared photographic images (sec. 7.1). A problem associated with image deblurring is the unavoidable coherent artifact noise in the coherent processing system, which frequently degrades the quality of the restored image. Thus if blurred photographic images can be restored by incoherent sources, higher quality restored images may be obtained. We have, in sec. 8.1, demonstrated a technique of complex spatial filtering with white-light sources. We have shown that the correlation peak in the direction of light dispersions is somewhat broader than in other directions. In other words, the white-light processor that we have proposed has one shortcoming; it is more effective in one direction. In this section we demonstrate that the same white-light processing technique can be applied to the problem of smeared image deblurring. We show that the deblurring process can indeed take place in the spectrally broad band of the light source.

Since the image deblurring by inverse spatial filtering takes the effect on every blurred image point, we consider the deblurring processing with an isolated image point. We now let the amplitude transmittance of a linear smeared point image be

$$f(y) = \begin{cases} 1, & \text{for } -\dfrac{\Delta y}{2} \leq y \leq \dfrac{\Delta y}{2}, \\ 0, & \text{otherwise} \end{cases} \tag{8.21}$$

where Δy is the smeared length. The coresponding Fourier spectrum is

$$F(q) = \Delta y \frac{\sin(q\Delta y/2)}{q\Delta y/2}. \tag{8.22}$$

We note that this blurred image can be corrected with an inverse spatial filtering process (sec. 7.1) for which the inverse spatial filter function is

$$H(q) = \frac{q\Delta y/2}{\sin(q\Delta y/2)}. \tag{8.23}$$

We see that the inverse filter function is not only a bipolar but also an infinite-pole function for which is not physically realizable (refs. 8.17, 8.18). However, we have shown in sec. 7.1 that the inverse spatial filter can be approximated by combining an amplitude and a phase filter.

Now we show that image deblurring can be carried out by a white-light processing technique. Let us assume that a linear smeared image $f(x, y)$ is inserted with a diffraction grating at the input plane P_1 of a white-light processor, as shown in fig. 8.9. We assume that the smearing is in the y direction of the spatial coordinate. The overall transmittance function of the input plane is

$$f(x, y)T(x) = f(x, y)[1 + \cos p_0 x], \tag{8.24}$$

where $T(x)$ is the diffraction grating and p_0 is the angular spatial frequency of the grating.

Since the input plane is illuminated by a collimated white-light source, the complex light distribution at the back focal plane P_2 of the achromatic transform lens L_1, for a given wavelength λ_0, can be shown to be

$$E(p, q; \lambda_0) = C_1 F(p, q) + C_2 F(p - p_0, q) + C_2 F(p + p_0, q) \tag{8.25}$$

where C_1, C_2, and C_3 are the appropriate complex constants, $F(p, q)$ is the Fourier spectrum of the smeared image transparency $f(x, y)$, $p = (2\pi/\lambda_0 f)\alpha$

and $q = (2\pi/\lambda_0 f)\beta$, (α, β) denotes the linear spatial coordinate system of the spatial frequency plane (p, q), and f is the focal length of the achromatic transform lens. Alternatively, eq. (8.25) can be written as

$$E(\alpha, \beta; \lambda_0) = C_1 F(\alpha, \beta) + C_2 F\left(\alpha - \frac{\lambda_0 f}{2\pi} p_0, \beta\right) + C_2 F\left(\alpha + \frac{\lambda_0 f}{2\pi} p_0, \beta\right). \tag{8.26}$$

From the above equations we see that, for a continuous spectral band of light sources, the second and third terms disperse linearly into rainbow colors along the α axis, and center at

$$\alpha = \alpha_0 = \pm \frac{\lambda_0 f}{2\pi} p_0, \tag{8.27}$$

where p_0 is the angular spatial frequency of the grating. In practice, the signal spectrum of the smeared image transparency is assumed to be spatial frequency limited, that is,

$$F(\alpha, \beta) = \begin{cases} F(\alpha, \beta), & (\alpha^2 + \beta^2)^{1/2} \leq \gamma, \\ 0, & \text{otherwise}, \end{cases} \tag{8.28}$$

where $\gamma = (\lambda f/2\pi)\Delta p$, and Δp denotes the angular spatial frequency limit.

Let us assume that a narrow-band deblurring filter $H(\beta)$, which is designed for wavelength λ_0, is inserted in the spatial frequency plane centered at $\alpha = \alpha_0$ of eq. (8.27); then the light intensity distribution immediately behind the filter is

$$I(\alpha, \beta) = \int_{\Delta\lambda} \left| F\left(\alpha - \frac{\lambda f}{2\pi} p_0, \beta\right) H(\beta) \right|^2 d\lambda, \tag{8.29}$$

where the integration is over the spectral band $\Delta\lambda$ near λ_0 and $\Delta\lambda$ is the narrow spectral bandwidth of the filter. For simplicity, the complex constant has been dropped from analysis. From the above equation the deblurred image irradiance at the output plane P_3 can be approximated by

$$\begin{aligned} I((x, y) &\simeq \Delta\lambda \left| \iint F\left(\alpha - \frac{\lambda_0 f}{2\pi} p_0, \beta\right) H(\beta) \exp\left[-i \frac{2\pi}{\lambda_0 f}(\alpha x + \beta y)\right] d\alpha\, d\beta \right|^2 \\ &\simeq \Delta\lambda \left| s(x, y) e^{ip_0 x} * h(y) \right|^2, \end{aligned} \tag{8.30}$$

where $h(y)$ is the impulse response of the deblurring filter and $*$ denotes the convolution operation. Thus we see that the deblurred image is diffracted at the optical axis of output plane P_3. We note further that the smeared image deblurring can take place with the entire broad spectral band of the white-light source. For example, a *fan-shape*-type filter (as shown in fig. 8.9) can be used to compensate for the size variation of the signal spectra so that the output deblurred image irradiance takes place with the entire spectral bandwidth of the light source. Therefore, a brighter incoherent deblurred image can be obtained at the output image plane.

We now estimate the minimum bandwidth requirement for the deblurring filter:

$$\alpha_1 = \frac{\lambda_0 f}{2\pi}(p_0 + \Delta p), \tag{8.31}$$

and

$$\alpha_2 = \frac{\lambda_0 f}{2\pi}(p_0 - \Delta p), \tag{8.32}$$

where α_1 and α_2 are the upper and lower spatial limits of the filter, as shown in fig. 8.10, λ_0 is the wavelength of the filter, f is the focal length of the achromatic transform lens, p_0 is the spatial frequency of the diffraction grating, and Δp is the spatial frequency limit of input transparency.

By the band-limited nature of the input transparency and the spatial limit of the filter, the two limiting wavelengths of the spectral lines arriving at the upper and the lower edges of the deblurring filter are

$$\lambda_1 = \lambda_0 \frac{p_0 + \Delta p}{p_0 - \Delta p}, \tag{8.33}$$

and

$$\lambda_2 = \lambda_0 \frac{p_0 - \Delta p}{p_0 + \Delta p}. \tag{8.34}$$

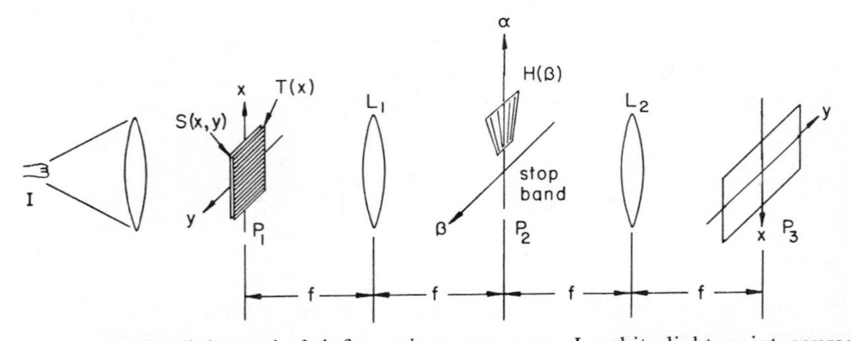

Fig. 8.9 A white-light optical information processor. I, white-light point source; $s(x, y)$, smeared image transparency; $T(x)$, diffraction grating; L_1 and L_2, achromatic transform lenses; $H(\beta)$, deblurring spatial filter.

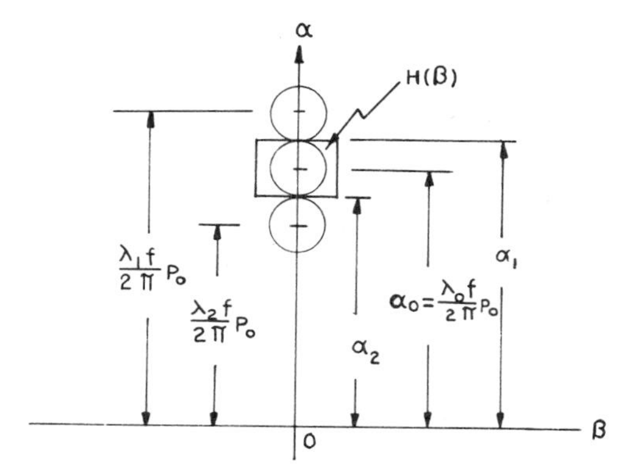

Fig. 8.10 Determination of wavelength spread over the spatial filter.

The spectral band over the deblurring filter can be determined by

$$\Delta\lambda = \lambda_1 - \lambda_2 = \lambda_0 \frac{4p_0 \Delta p}{p_0^2 - (\Delta p)^2}. \tag{8.35}$$

If the spatial carrier frequency p_0 is adequately high as compared with Δp, eq. (8.35) can be approximated by

$$\Delta\lambda \simeq \lambda_0 \frac{4\Delta p}{p_0}, \qquad p_0 \gg \Delta p. \tag{8.36}$$

With reference to fig. 8.11, the degree of restoration error due to different wavelength spreads may be estimated by

$$\epsilon = \frac{\Sigma \text{ shaded area}}{\Sigma \text{ solid area}} \times 100\%. \tag{8.37}$$

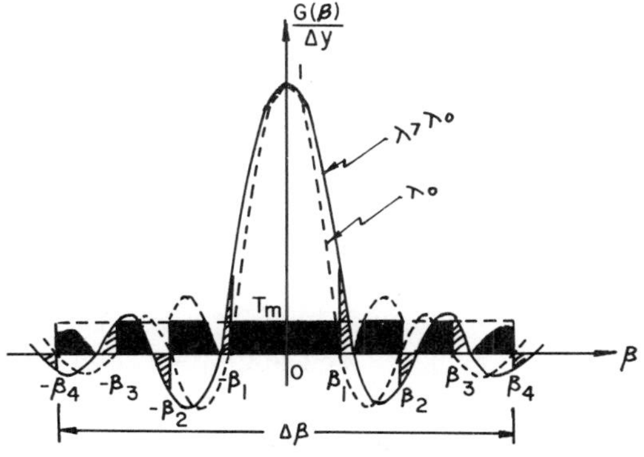

Fig. 8.11 Dashed and solid curves represent the Fourier spectrum of a linear smeared point image for λ_0 and for $\lambda > \lambda_0$, respectively, where λ_0 is the wavelength of the spatial filter, $q_n = 2\pi n/\Delta y$, $n = 1, 2, 3, \ldots, \Delta q$, the deblurred spatial band-width; T_m, maximum transmittance of the deblurred spectrum.

Thus from eq. (8.36) we see that a lower restoration error is achieveable if the spatial frequency p_0 of the diffraction grating is high.

In order to gain a feeling of magnitude, we provide a numerical example. Let us assume that the highest spatial frequency limit of the smeared image transparency is 25 lines/mm, the wavelength of the deblurring filter is 6000Å, the focal length of the transform lens is 250 mm, and the spatial carrier frequency of the diffraction grating is 1000 lines/mm. With reference to these data, the upper and lower spatial limits required for the filter are

$$\alpha_1 = (600 \times 10^{-6})(250)(1000 + 25) = 153.75 \text{ mm},$$

and

$$\alpha_2 = (600 \times 10^{-6})(250)(1000 - 25) = 146.25 \text{ mm}.$$

By the band-limited nature of the smeared image, the longest and the shortest wavelength limits are

$$\lambda_1 = 6000 \frac{1000 + 25}{1000 - 25} = 6307.7 \text{ Å},$$

and

$$\lambda_2 = 6000 \frac{1000 - 25}{1000 + 25} = 5707.3 \text{ Å}.$$

The spectral band over the spatial filter is, therefore,

$$\Delta\lambda = \lambda_1 - \lambda_2 = 600.4 \text{ Å}.$$

We note that λ_1 and λ_2 are about 300 Å off the deblurring filter wavelength $\lambda_0 = 6000$ Å. Thus the error due to wavelength spread (i.e., $\lambda_2 \leq \lambda \leq \lambda_1$) can be considered very small. Because the restoration takes place in spectrally broad-band $\Delta\lambda$ of an incoherent source, a brighter restored image with less artifact noise can be produced.

As a result of artifact noise reduction, we predict that a higher degree of image deblurring can be obtained with this technique than with the coherent technique.

We now describe a simple technique of synthesizing a deblurring filter in which the complex filter is synthesized as the product of an amplitude and a phase filter, as shown in fig. 8.12. The amplitude filter can be generated by recording the one-dimensional signal spectrum of a slit aperture on a photographic plate. The width of the slit corresponds to the smearing length of the blurred image. A one-dimensional cylindrical transform lens can be used to obtain an appropriate one-dimensional Fourier spectrum, as shown in fig. 8.13. the result obtained with the coherent processing technique. From these results we see that the artifact noise is substantially reduced with the white-light processing

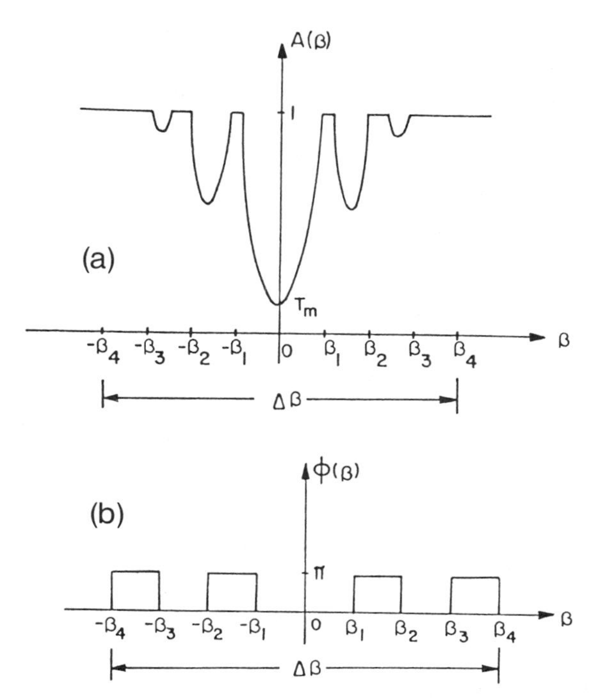

Fig. 8.12 Deblurring filter function. (*a*) Amplitude filter function. (*b*) Phase filter function.

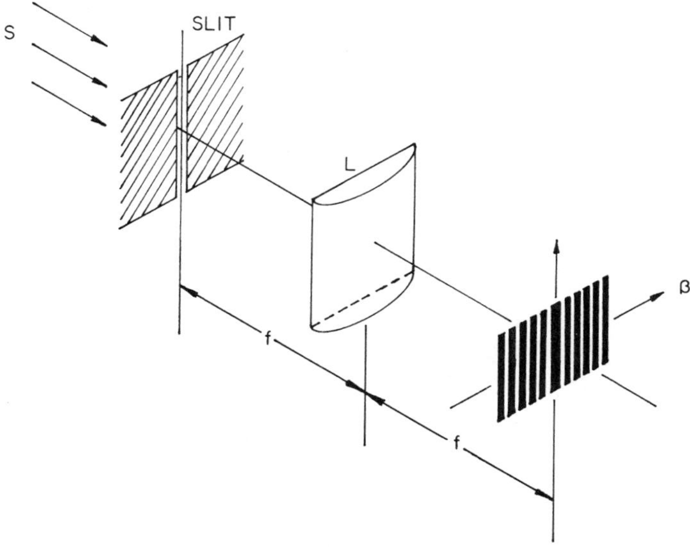

Fig. 8.13 Generation of amplitude filter. *S*, monochromatic plane wave; *L*, cylindrical transform lens.

To obtain the required amplitude-transmittance function, we control the film gamma (sec. 5.1) of the recorded film equal to one.

In the generation of a deblurring phase filter, a black-and-white bar pattern on a high-contrast film, as shown in fig. 8.14, is recorded. The width of the bar pattern is determined by the wavelength and the smearing length of the image. We also add a transparent reference point in an area near the recorded bar pattern. The recorded binary bar pattern is used as a mask to reproduce a number of gray-level bar patterns on a low-contrast photographic plate. If the recorded photographic plate is bleached, a set of spatial phase filters can be obtained. To search for an appropriate π-phase filter, we use a Ronchi-type grating as an input object in a coherent optical processor and place a bleached reference point over the zero-order spectrum at the spatial-frequency plane, as shown in fig. 8.15. If a contrast-reversed image is observed at the output plane, the corresponding bleached bar pattern must be a π-phase filter with respect to the wavelength of the coherent source.

In an experimental demonstration by Zhuang et al. (ref. 8.19), we first simulated a linear smeared photographic image. The simulation was accomplished by recording a linear object in motion on a photographic film and then contact printing to obtain a positive smeared-image transparency. To ensure linearity in amplitude transmittance, we controlled the overall film gamma such that it was about two. The white-light source that we used for our experiments was a 75-W xenon arc lamp with a 100-μm pinhole, which acted as a point source. Since we used a white-light source, index-matching liquid gates were not used in our experiments. The spectral width of the deblurring complex filter used

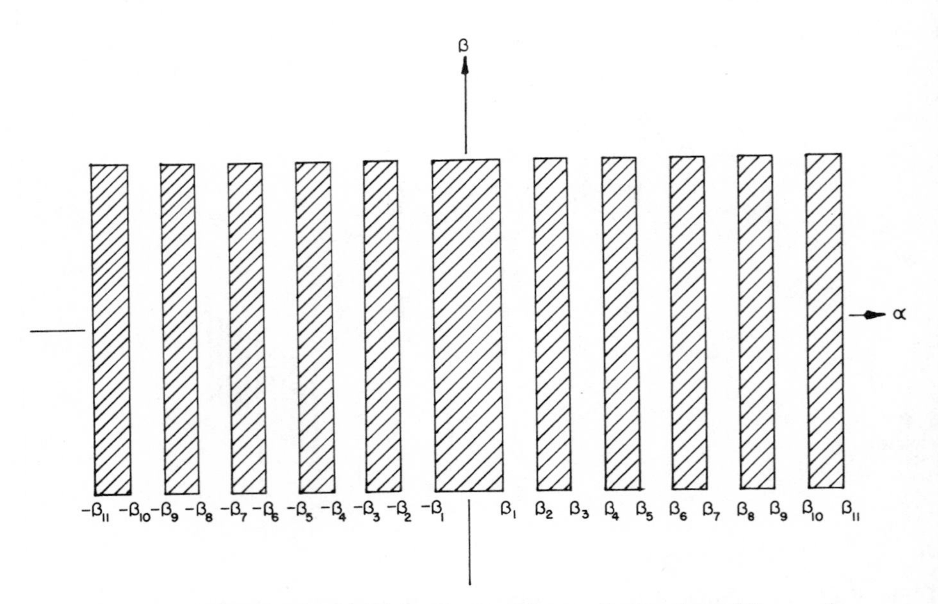

Fig. 8.14 Black-and-white bar pattern to be used as a phase filter mask.

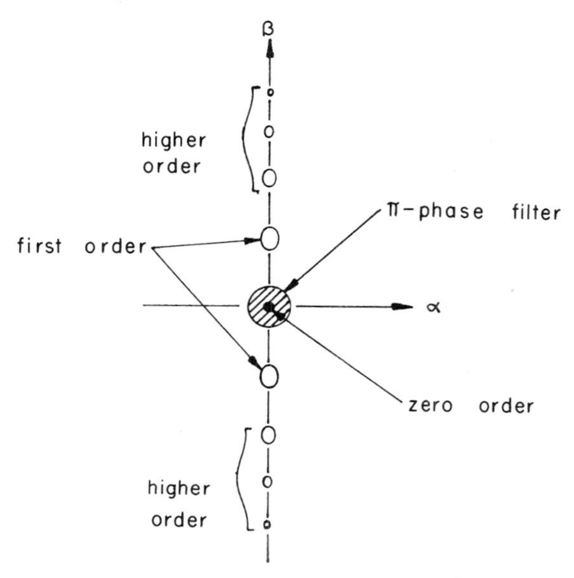

Fig. 8.15 A technique of determining a π-phase filter through contrast reversal.

was about 100 Å, the center wavelength λ_0 was 5154 Å, and the spatial bandwidth of the filter contained five main lobes. The corresponding Fourier spectrum of the deblurring filter is shown in fig. 8.16. We note that this Fourier spectrum is compatible with the result obtained by Swindell (ref. 8.20). As an experimental illustration, fig. 8.17*a* shows the linear smeared photograpahic image of the word "*optics*" as a blurred object. Figure 8.17*b* shows the deblurred image obtained with the white-light processing technique, and fig. 8.17*c* shows the result obtained with the coherent processing technique. From these results we see that the artifact noise is substantially reduced with the white-light processing

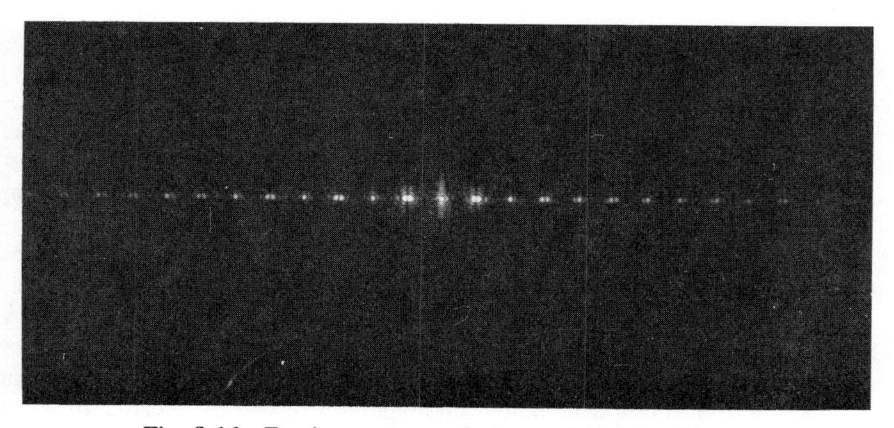

Fig. 8.16 Fourier spectrum of a linear deblurring filter.

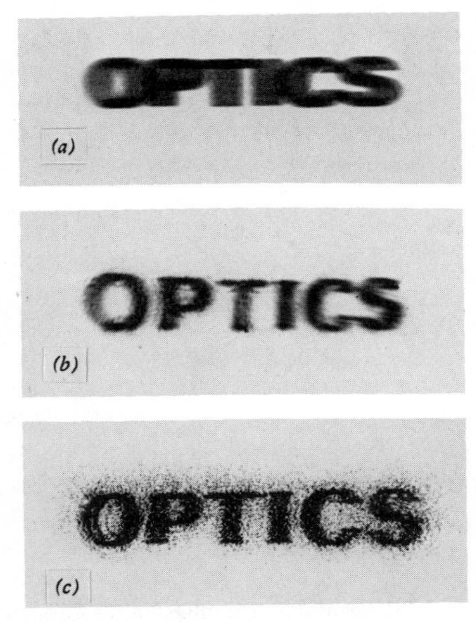

Fig. 8.17 Smeared image restoration of the word "OPTICS." (*a*) Original smeared image. (*b*) Deblurred with white-light technique. (*c*) Deblurred with coherent technique.

technique. The deblurred image obtained with the coherent technique appears to be sharper than the one obtained with the white-light technique because of the high spatial coherence of the light source. However, this drawback can be overcome if we use a smaller white-light source (i.e., a pinhole) and a broad spectral band deblurring filter to cover the entire smeared signal spectrum. Design of such a broad-band fan-shaped filter is under investigation.

8.5 IMAGE SUBTRACTION

Another interesting application of optical information processing is image subtraction. Image subtraction may be of value in many applications, such as urban development, highway planning, earth resources studies, remote sensing, meteorology, automatic surveillance, inspection, and so on. Optical image subtraction may also apply to communications as a means of bandwidth compression; for example, it would be necessary to transmit only the differrences between images in successive cycles, rather than the entire image in each cycle.

Optical image synthesis by complex amplitude subtraction was described by Gabor et al. (ref. 8.21). The technique involves successive recordings of two or more complex diffraction patterns on a holographic plate and the subsequent reproduction of the composite hologram images. A few years later, Bromley et al. (ref. 8.22) described a holographic Fourier subtraction technique, by which a real-time image and a previously recorded hologram image can be subtracted.

Although good image subtraction in their experiments was reported, it appears that the illumination for the hologram image reconstruction had to be arranged carefully. In their paper Lee et al. (ref. 8.23) proposed a technique by which image subtraction and addition can also be achieved using a diffraction grating. This technique involves the insertion of a diffraction grating in the spatial frequency domain of the coherent optical processor. We note that this technique offers the advantage of a real-time subtraction capability.

Since it would be an exhaustive effort to review various techniques of image subtraction, we refer the reader to the review paper by Ebersole (ref. 8.24). Most of the optical image synthesis techniques involve a coherent source to carry out the image subtraction. These sources also introduce artifact noise, which frequently limits the processing capability. In this section we apply the white-light processing technique, described in the previous sections, to image subtraction (ref. 8.25).

We insert two photographic image transparencies in contact with a phase grating at the input plant P_1 of the white-light optical processor shown in fig. 8.18. At the spatial frequency plane P_2, the complex light distribution for each wavelength λ of the light source may be described as

$$E(p, q; \lambda) = S_1(p - p_0, q)e^{-i\beta_0 q} + S_2(p - p_0, q)e^{i\beta_0 q}, \qquad (8.38)$$

where $S_1(p, q)$ and $S_2(p, q)$ are the Fourier spectra of the input signals $S_1(x, y)$ and $S_2(x, y)$, respectively, and β_0 is an arbitrary constant. Again we see that two input signal spectra disperse into rainbow colors along the α axis of the spatial frequency plane.

In image subtraction we would use a diffraction grating in the spatial frequency plane. Since the dispersed Fourier spectra vary with respect to the wavelength of the light source, we insert a fan-shaped grating to compensate for the wavelength variation. Let this fan-shaped grating be

$$H(q) = [1 + \sin \beta_0 q]; \qquad (8.39)$$

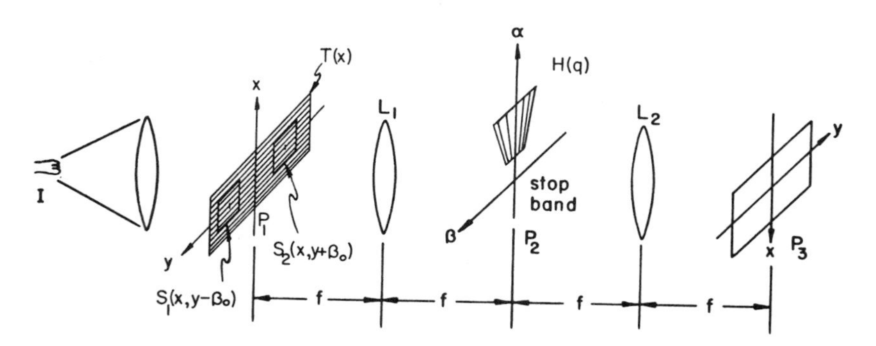

Fig. 8.18 A white-light optical processor. I, white-light point source; $T(x)$, high-efficiency phase grating; $H(q)$, slanted grating.

the output intensity distribution can be shown to be

$$I(x, y) \simeq \Delta\lambda\{|S_1(x, y - \beta_0)|^2 + |S_2(x, y - \beta_0)|^2 + \tfrac{1}{4}|S_1(x, y) - S_2(x, y)|^2 + |S_1(x, y - 2\beta_0)|^2 + |S_2(x, y + 2\beta_0)|^2\}. \qquad (8.40)$$

Thus the subtraction of the two input signals $[S_1(x, y) - S_2(x, y)]$ can be seen at the optical axis of the output plane. A difficulty arises in that, in practice, it is difficult to obtain a white-light point source. This shortcoming can be alleviated with the following source encoding technique (ref. 8.26).

Basically, optical image subtraction is a one-dimensional processing operation. Instead of utilizing a point source of light, a line source of light is used for the subtraction operation. Since the spatial coherence requirement for the subtraction operation is a point-pair concept, a strict coherence requirement is not needed. In other words, it is possible to encode an extended incoherent source in order to obtain a point-pair coherence requirement for image subtraction.

With reference to the incoherent image subtraction processing system of fig. 8.19, the processor is similar to a coherent optical processor except for an extended incoherent source and an encoding mask. Since image subtraction is a one-dimensional operation, for simplicity we adopt a one-dimensional notation for our analysis.

In evaluating the spatial coherence requirement, we would apply the partially coherent imaging theory (ref. 8.27). The mutual coherence function at the spatial frequency plane P_3 is

$$\Gamma_3(x_3, x_3') = \iint \Gamma_2(x_2, x_2') f(x_2) f^*(x_2') K_2(x_2, x_3) K_2^*(x_2', x_3')\, dx_2\, dx_2', \qquad (8.41)$$

where the integration is over the input plane P_2, x_2 and x_3 are the spatial coordinate systems of P_2 and P_3, $\Gamma_2(x_2, x_2')$ is the complex coherence function at the input plane P_2, $f(x_2)$ is the input function at P_2, which can be expressed as

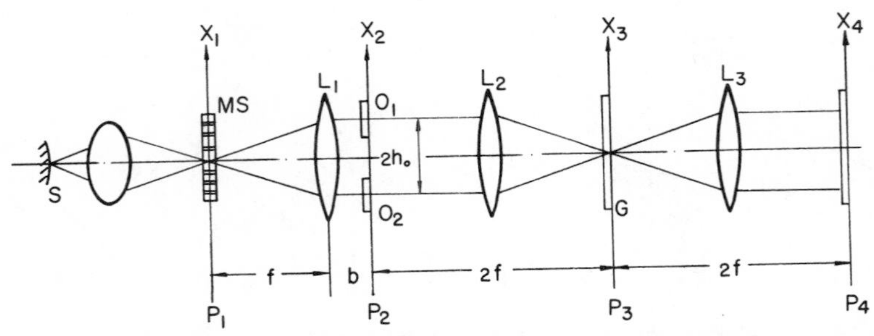

Fig. 8.19 Image subtraction with encoded extended incoherent source. S, mercury arc lamp; MS, multislit mask; O_1 and O_2, object transparencies; G, diffraction grating; L_1, L_2, and L_3 transform lenses.

$$f(x_2) = O_1(x_2 - h_0) + O_2(x_2 + h_0), \tag{8.42}$$

where $O_1(x_2)$ and $O_2(x_2)$ are the two input object transparencies, and

$$K_2(x_2, x_3) = \exp\left(i2\pi\frac{x_2x_3}{\lambda f}\right), \tag{8.43}$$

is the transmittance function between planes P_2 and P_3, λ is the wavelength of the light source, and f is the focal length of the transform lens L_2. Equation (8.41) can also be written

$$\Gamma_3(x_3, x_3') = \iint \Gamma_2(x_2, x_2')[O_1(x_2 - h_0) + O_2(x_2 + h_0)] \times [O_1^*(x_2' - h_0) + O_2^*(x_2' + h_0)] \exp\left[i2\pi\frac{x_3x_2 - x_3'x_2'}{\lambda f}\right] dx_2\, dx_2', \tag{8.44}$$

where the superscript * denotes the complex conjugate.

It is clear that the mutual coherence function immediately following the diffraction grating G, with a spacing period $d = (\lambda f)/h_0$, is

$$\Gamma_3'(x_3, x_3') = \left[\exp\left(i2\pi\frac{h_0}{\lambda f}x_3\right) - \exp\left(-i2\pi\frac{h_0}{\lambda f}x_3\right)\right] \times \left[\exp\left(-i2\pi\frac{h_0}{\lambda f}x_3'\right) - \exp\left(i2\pi\frac{h_0}{\lambda f}x_3'\right)\right]\Gamma_3(x_3, x_3'), \tag{8.45}$$

where we ignore the dc term of the diffraction grating. The image intensity at the output plane P_4 is

$$I(x_4) = \iint \Gamma_3'(x_3, x_3') \exp\left(i2\pi\frac{x_3 - x_3'}{\lambda f}x_4\right) dx_3\, dx_3', \tag{8.46}$$

where the integration is over the spatial frequency plane and x_4 is the output spatial coordinate system.

By substituting eqs. (8.44) and (8.45) into eq. (8.46) and integrating over the spatial frequency plane, we have

$$I(x_4) = \iint \Gamma(x_2, x_2')[O_1(x_2 - h_0) + O_2(x_2 + h_0)][O_1^*(x_2' - h_0) + O_2^*(x_2 + h_0)] \cdot [\delta(x_2 + x_4 + h_0)\delta(x_2' + x_4 + h_0)\delta(x_2' + x_4 - h_0)\delta(x_2' + x_4 - h_0) - \delta(x_2 + x_4 + h_0)\delta(x_2' + x_4 - h_0) - \delta(x_2 + x_4 - h_0) \times \delta(x_2' + x_4 + h_0)]\, dx_2\, dx_2', \tag{8.47}$$

where $\delta(x)$ is the Dirac delta function. Let us assume that $\Gamma(x_2, x_2')$ takes the form of $\Gamma(x_2 - x_2')$ and $\Gamma(x) = \Gamma^*(-x)$. If we evaluate eq. (8.47) by its terms and note that the input images are spatially limited, then we have

$$I(x_4) = \Gamma(0)[|O_1(-x_4)|^2 + |O_2(-x_4)|^2] - \Gamma(2h_0)O_1(-x_4)O_2^*(-x_4)$$
$$- \Gamma^*(2h_0)O_1^*(-x_4)O_2(-x_4) + \Gamma(0)[|O_1(-x_4 - 2h_0)|^2 + |O_2(-x_4 + 2h_0)|^2], \tag{8.48}$$

where $\Gamma(2h_0) = |\Gamma(2h_0)|\exp(i\phi)$, a complex quantity. We stress that the phase factor ϕ can be avoided by adjusting the grating position of G.

We now consider only the image terms around the origin of the output plane P_4:

$$I_o(-x_4) = |\Gamma(2h_0)||O_1(x_4) - O_2(x_4)|^2 + (1 - |\Gamma(2h_0)|)[|O_1(x_4)|^2 + |O_2(x_4)|^2], \qquad \text{for } \phi = 0. \tag{8.49}$$

From eq. (8.49) we see that the first term is proportional to the intensity of the subtracted image and the second term is proportional to the sum of the image irradiances, where $|\Gamma(2h_0)|$ is the degree of spatial coherence. If the degree of coherence $|\Gamma(2h_0)|$ is high, i.e., $\mu(2h_0) \simeq 1$, then eq. (8.49) reduces to

$$I_0(-x_4) \simeq |O_1(x_4) - O_2(x_4)|^2, \qquad \text{for } |\Gamma(2h_0)| \simeq 1. \tag{8.50}$$

Thus we see that the spatial coherence is only needed for every pair of points $|x_2 - x_2'| = 2h_0$. In other words, only a point-pair spatial coherence is required for the subtraction operation.

We now search for a source encoding such that a point-pair spatial coherence function can be found. We insert a mask transparency for the source encoding at the front focal plane P_1 of the collimator L_1 as shown in fig. 8.20. The spatial coherent function $\Gamma(x_2, x_2')$ over the input plane P_2 can be written (ref. 8.27) as

$$\Gamma(x_2, x_2') = \int S(x_1)K_1(x_1, x_2)K_1(x_1, x_2')\, dx_1, \tag{8.51}$$

where $S(x_1)$ is the intensity transmittance function of the mask, and $K_1(x_1, x_2)$ is the transmittance function between planes P_1 and P_2. We assume the mask is located within an isoplanatic patch and $K_1(x_1, x_2)$ can be written (ref. 8.27) as

$$K_1(x_1, x_2) = \exp i\left[2\pi\frac{x_1x_2}{\lambda f} + \epsilon\left(x_2 - x_1\frac{h}{f}\right)\right], \tag{8.52}$$

where $\epsilon(x)$ is the wave aberration of the collimator, and b is the distance between the collimator and the input plane P_2.

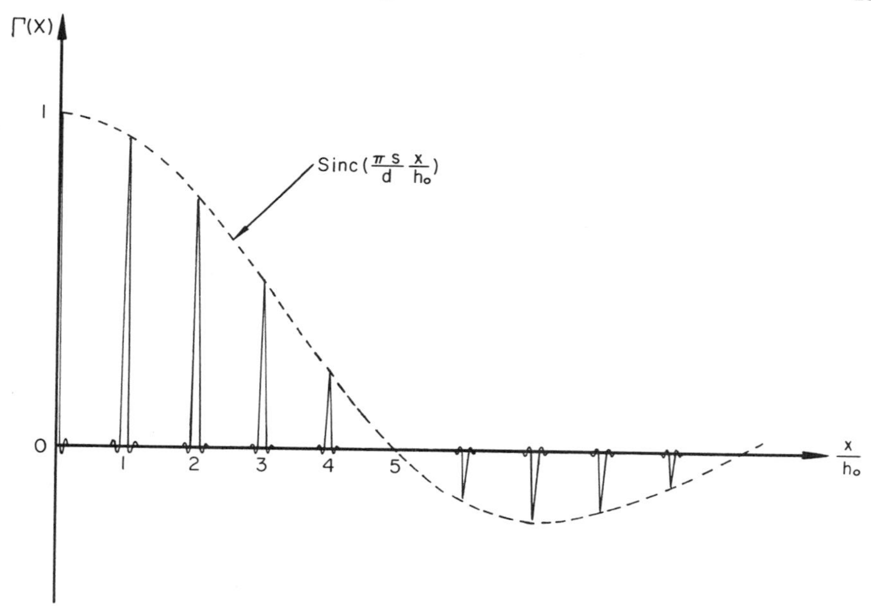

Fig. 8.20 Coherence function obtained with multislit source encoding, where $x = |x_2 - x_2'|$, and $s/d = \frac{1}{5}$.

We note that, if b is sufficiently small, i.e., if $[(b/f)x_1]_{max} \ll x_2$, the transmittance function of eq. (8.52) can be reduced to

$$K_1(x_1, x_2) \simeq \exp\left\{i\left[2\pi\frac{x_1 x_2}{\lambda f} + \epsilon(x_2)\right]\right\}. \tag{8.53}$$

By substituting eq. (8.53) into eq. (8.51), the spatial coherence function becomes

$$\Gamma(x_2, x_2') = \exp i[\epsilon(x_2) - \epsilon(x_2')] \int S(x_1) \exp\left[i2\pi\frac{x_1}{\lambda f}(x_2 - x_2')\right] dx_1. \tag{8.54}$$

From the above equation we see that the spatial coherence function is the Fourier transform of the mask transmittance function modulated by a phase (wave aberration) factor. The phase aberration will not affect the degree of mutual coherence $|\Gamma(x_2, x_2')|$. In the remaining analysis we ignore the phase aberration and assume that $\Gamma(x_2, x_2')$ takes the form $\Gamma(x_2 - x_2')$.

Now we evaluate the degree of coherence $|\Gamma(2h_0)|$ for two different cases. We first evaluate a single slit encoding, i.e.,

$$S(x_1) = \text{rect}\left(\frac{x_1}{s}\right), \tag{8.55}$$

where s is the slit width. By subtituting eq. (8.55) into eq. (8.54), we have

$$\Gamma(x_2 - x_2') = \operatorname{sinc} \frac{\pi s}{\lambda f}(x_2 - x_2'), \tag{8.56}$$

where the phase factor was ignored. If we let $x_2 - x_2' = 2h_0 = 2f\lambda/d$, then eq. (8.56) becomes

$$\Gamma(2h_0) = \operatorname{sinc}\left(2\pi\frac{s}{d}\right), \tag{8.57}$$

where d is the spacing period of the diffraction grating G. Thus we see that the degree of spatial coherence $|\Gamma(2h_0)|$ depends upon the ratio of the slit width s to the spacing d.

In order to gain a feeling of magnitude, we provide several values of $\Gamma(2h_0)$ in Table 8.1.

Table 8.1
Spatial Coherence Requirement for Single-Slit Mask

$\frac{s}{d}$	$\frac{1}{2}$	$\frac{1}{5}$	$\frac{1}{10}$	$\frac{1}{20}$
$\Gamma(2h_0)$	0	0.756	0.936	0.988

From Table 8.1 we see that a high degree of spatial coherence can be attained only through a very narrow slit. For example, if the spacing of the grating $d = 25\ \mu\text{m}$, to achieve a high degree of spatial coherence a slit width $s \leq 2.5\ \mu\text{m}$ should be used. Thus it would make the source too weak for practical processing operation.

As noted earlier, the spatial coherence requirement for image subtraction is a point-pair problem. It is possible to encode the extended source with N number of narrow slits. Thus with a multislit source encoding, an N-fold light power can be used for the image subtraction operation.

We now let the intensity transmittance of the encoding mask be

$$S(x_1) = \sum_{n=1}^{N} \operatorname{rect}\left(\frac{x_1 - nd'}{s}\right), \tag{8.58}$$

where s is the slit width and d' is the spacing between slits.

By substituting eq. (8.58) into eq. (8.54), the spatial coherence function becomes

$$\mu(x) = \frac{\sin\left(N\pi\dfrac{d'x}{\lambda f}\right)}{N\sin\left(\pi\dfrac{d'x}{\lambda f}\right)}\,\mathrm{sinc}\left(\frac{\pi s}{\lambda f}x\right), \tag{8.59}$$

where $x = x_2 - x_2'$. From the above equation we see that the last sinc factor is identical to the single-slit case of eq. (8.56), which represents a broad spread of coherence over x. However, the first factor, for large values of N, converges to a sequence of narrow pulses. The locations of the pulses (i.e., the peaks) occur at every $x = x_2 - x_2' = n(\lambda f/d')$. Thus this factor yields the fine spatial coherence discrimination at every point-pair separated at distance $\lambda f/d'$ over the input plane P_2.

We let the spacing of d' be equal to the spacing d of the diffraction grating G (i.e., $d' = d$); then the spatial coherence of eq. (8.59) becomes

$$\mu(x) = \frac{\sin\,[N\pi(x/h_0)}{N\sin[\pi(x/h_0)]}\,\mathrm{sinc}\left(\pi\frac{sx}{dh_0}\right), \tag{8.60}$$

where we substituted $d = \lambda f/h_0$. From eq. (8.60) we see that a sequence of narrow pulses occurs at $x = x_2 - x_2' = nh_0$, where n is an integer and their peak values are weighed by a broader sinc factor, as shown in fig. 8.20. It can be shown that the width of the pulses is inversely proportional to the number of slits N. Thus the multislit source encoding not only provides a point-pair coherence requirement for image subtraction, but also provides a higher available light power for the operation. In other words, the multislit encoding utilizes the light source more efficiently so that the inherent difficulty of acquiring a small incoherent source can be alleviated.

So far we have considered only the quasi-monochromatic light, but the effect of the temporal coherence has not been discussed. Since the scale of the Fourier spectrum varies with the wavelength, there is a temporal coherence requirement for every processing operation. With this consideration we must limit the temporal bandwidth $\Delta\lambda$ of the source so that the dispersed Fourier spectra will not spread beyond the allowable limit. In the image subtraction operation we should limit the spectrum spread within a very small fraction of the grating spacing d, i.e.,

$$\frac{P_m f\Delta\lambda}{2\pi} \ll d, \tag{8.61}$$

where p_m is the highest angular spatial frequency of the input objects, f is the focal length of the transform lens, and $\Delta\lambda$ is the spatial bandwidth of the source. Therefore, the temporal bandwidth of the source should be limited by the following inequality:

$$\frac{\Delta\lambda}{\lambda} \ll \frac{2\pi}{h_0 p_m}, \tag{8.62}$$

where λ is the center wavelength of the light source and $2h_0$ is separation of the input images.

In order to have some feeling of magnitude, we would let $h_0 = 6.6$ mm, $\lambda = 5461$ Å, and take a factor of 10 for eq. (8.62). The temporal bandwidth requirement $\Delta\lambda$ for various values of spatial frequencies p_m are tabulated in Table 8.2.

Table 8.2
Temporal Coherence Requirement

$\frac{p_m}{2\pi}$, lines/mm	0.5	1	5	20	100
$\Delta\lambda$, Å	166	83	16.5	4.1	0.8

From Table 8.2 we see that, if the spatial frequency of the input objects is low, a broader temporal bandwidth of the light source can be used. In other words, the higher the spatial frequency of the input objects, the narrower the temporal bandwidth required. We assume that all the lenses are archromatic.

In an experiment a mercury arc lamp with a green filter is used as an extended incoherent source. A multislit mask is used to encode the light source. The slit width s is 2.5 μm, the spacing of slits d' is 25 μm, and the overall size of the mask is about 2.5 × 2.5 mm^2. Thus it contains about 100 slits. The focal lengths of the transform lenses are 300 mm. A liquid gate containing two objects transparencies of sizes about 6 × 8 mm^2 is inserted immediately behind the collimator. A sinusoidal phase grating with a spacing period of 25 μ is used in the spatial frequency plane P_2 of fig. 8.19. The separation between the two input images to P_2 is 13.2 mm.

For experimental demonstration we provide two continuous tone images as input object transparencies, as shown figs. 8.21*a* and 8.21*b*. By comparing these two figure parts, we see that a liquid gate is withdrawn from the optical bench in fig. 8.21*b*. Figure 8.21*c* shows the subtracted image obtained with this incoherent processing technique, while fig. 8.21*d* is obtained with the coherent processing technique. From the result obtained with the incoherent technique, a profile of subtracted liquid gate can be seen, while from the result obtained with the coherent technique, the subtracted image is severely damaged by the coherent artifact noise.

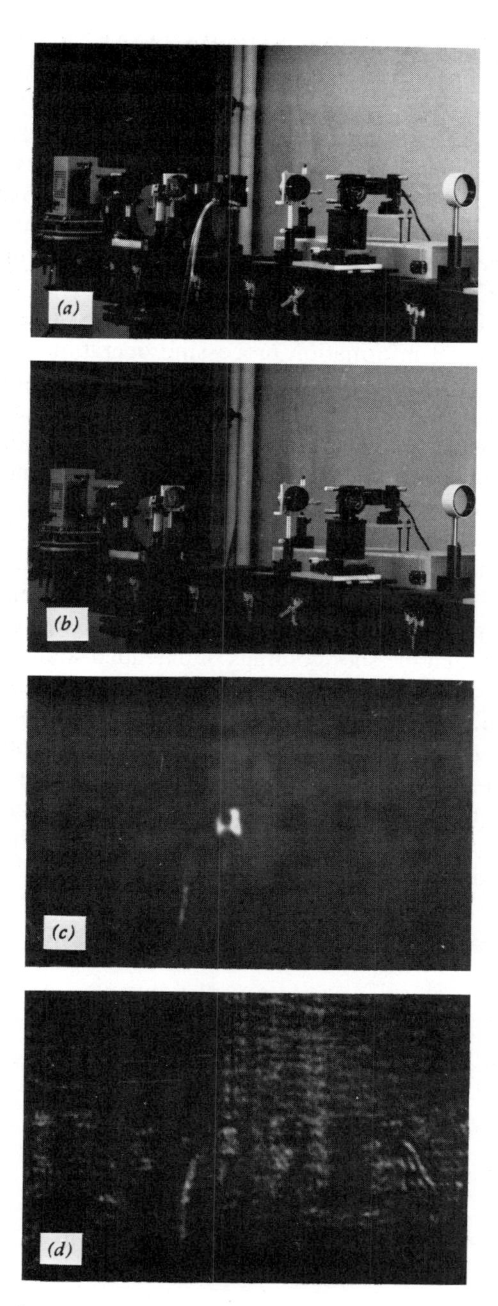

Fig. 8.21 Image subtraction: continuous tone object. (*a*) and (*b*) Input object transparencies. (*c*) Subtracted image obtained with incoherent technique. (*d*) Subtracted image obtained with coherent technique.

8.6 SOURCE ENCODING FOR PARTIALLY COHERENT PROCESSING

Among the basic limitations inherent in using an incoherent source for partially coherent processing is the extended source size. To achieve a broad spatial coherence requirement at the input plane of an optical information processor, a very small source size is needed. However, such a small light source is difficult to obtain in practice. We have shown in previous sections that an image subtraction processing operation can be carried out with an encoded extended source. In other words, a strictly broad coherence requirement may not be needed for some optical information processing operations.

In this section we describe a linear transformation relationship between spatial coherence function and source encoding intensity transmittance function (ref. 8.28). Since the spatial coherence requirement depends upon the information processing operation, a more relaxed coherence requirement may be used for specific processing operations.

The purpose of source encoding is to reduce the coherence requirement, so that an extended incoherent source may be used for certain information processing operations. In other words, the source encoding technique is capable of generating an appropriate coherence function for a specific information processing operation so that the shortcomings of extended source can be overcome.

We begin our discussion with Young's experiment under extended incoherent source illumination, as shown in fig. 8.22. First, we assume that a narrow slit is placed at plane P_1 behind an extended source. To maintain a high spatial coherence between slits Q_1 and Q_2 at P_2, it is known that the source size should be very narrow. If the separation between Q_1 and Q_2 is large, then a narrower slit size S_1 is required. In order to maintain a high spatial coherence between Q_1 and Q_2, the slit width should be (Sec. 4.1)

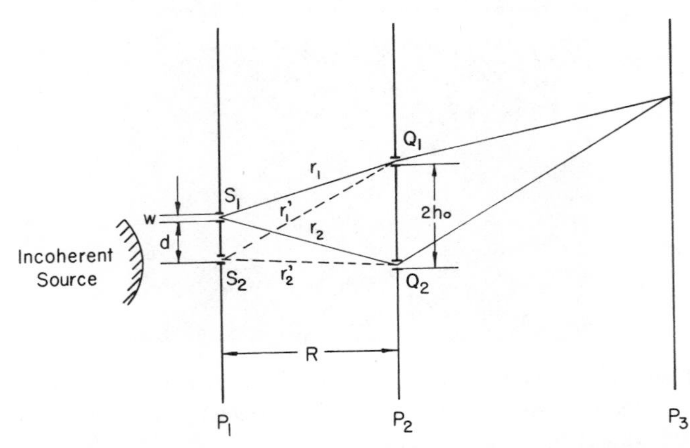

Fig. 8.22 Young's experiment with extended source illumination.

$$w \leq \frac{\lambda R}{2h_0}, \tag{8.63}$$

where R is the distance between planes P_1 and P_2, and $2h_0$ is the separation between Q_1 and Q_2.

Let us now consider two narrow slits S_1 and S_2 located in source plane P_1. We assume that the separation between S_1 and S_2 satisfies the following path length relation:

$$r_1' - r_2' = (r_1 - r_2) + m\lambda, \tag{8.64}$$

where the r's are the distances from S_1 and S_2 to Q_1 and Q_2, as shown in the figure, m is an arbitrary integer, and λ is the wavelength of the extended source. The interference fringes due to each of the two source slits S_1 and S_2 are in phase. A brighter fringe pattern can be seen at plane P_3. To further increase the intensity of the fringe pattern, we simply increase the number of source slits in appropriate locations in P_1 such that every separation between slits satisfies the coherence or fringe condition of eq. (8.64). If separation R is large, i.e., $R \gg d$ and $R \gg 2h_0$, then the spacing d between the source slits becomes

$$d = m\frac{\lambda R}{2h_0}. \tag{8.65}$$

From the above illustration, we see that, by properly encoding an extended source, it is possible to maintain the spatial coherence between Q_1 and Q_2, as well as increase the intensity of illumination. Thus, with a specific source encoding technique, an efficient utilization of the source power may be achieved.

In order to encode an extended source, a spatial coherence function for an information processing operation must first be found. With reference to the partially coherent optical processor of fig. 8.23, the spatial coherence function at input plane P_2 can be written as (ref. 8.27)

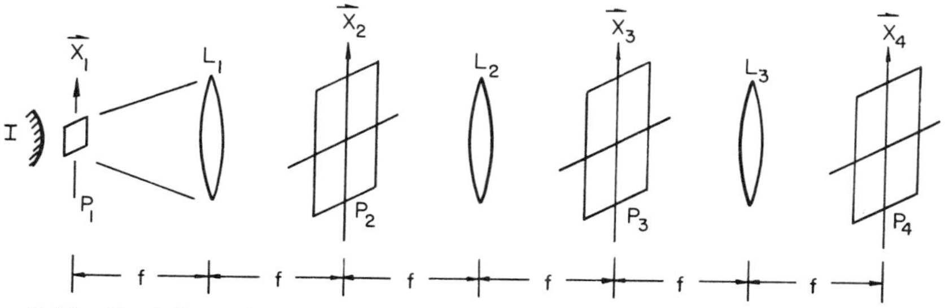

Fig. 8.23 Partially coherent optical processing with encoder extended incoherent source. I, Extended incoherent source; L_1, collimation lens; L_2 and L_3, transform lenses. Variables with arrows above are bold in text.

$$\Gamma(\mathbf{x}_2, \mathbf{x}_2') = \iint S(\mathbf{x}_1) K_1(\mathbf{x}_1, \mathbf{x}_2) K_1(\mathbf{x}_1, \mathbf{x}_2')\, \mathbf{dx}_1, \tag{8.66}$$

where $S(\mathbf{x}_1)$ is the intensity transmittance function of a source encoding mask, $K_1(\mathbf{x}_1, \mathbf{x}_2)$ is the transmittance function between source plane P_1 and the input plane P_2, and the integration is over the source plane P_1, which can be written as

$$K_1(\mathbf{x}_1, \mathbf{x}_2) \simeq \exp\left[i\left(2\pi \frac{\mathbf{x}_1 \mathbf{x}_2}{\lambda f} \right) \right]. \tag{8.67}$$

By substituting $K_1(\mathbf{x}_1, \mathbf{x}_2)$ into eq. (8.66), we have

$$\Gamma(\mathbf{x}_2 - \mathbf{x}_2') = \iint s(\mathbf{x}_1) \exp\left[i2\pi \frac{\mathbf{x}_1}{\lambda f}(\mathbf{x}_2 - \mathbf{x}_2') \right] \mathbf{dx}_1. \tag{8.68}$$

From the above equation we see that the spatial coherence function and source encoding intensity transmittance function form a Fourier transform pair

$$s(\mathbf{x}_1) = \mathscr{F}[\Gamma(\mathbf{x}_2 - \mathbf{x}_2')], \tag{8.69}$$

and

$$\Gamma(\mathbf{x}_2 - \mathbf{x}_2') = \mathscr{F}^{-1}[s(\mathbf{x}_1)], \tag{8.70}$$

where $\mathscr{F}$ denotes the Fourier transformation operation. If a spatial coherence function for an information processing operation is provided, the source encoding intensity transmittance function can be determined through Fourier transformation of eq. (8.69). We note that the source encoding function $S(\mathbf{x}_1)$ can consist of apertures or slits of any shape. We further note that in practice $S(\mathbf{x}_1)$ should be a positive real function that satisfies the following physical realizable condition:

$$0 \le S(\mathbf{x}_1) \le 1. \tag{8.71}$$

For example, if a spatial coherence function for an information processing operation is

$$\Gamma(x_2 - x_2') = \operatorname{rect}\left(\frac{|x_2 - x_2'|}{A} \right), \tag{8.72}$$

where A is an arbitrary positive constant, and

$$\operatorname{rect}\left(\frac{x}{A}\right) = \begin{cases} 1, & |x| \le A \\ 0, & \text{otherwise} \end{cases}$$

then the source encoding intensity transmittance would be

$$S(x_1) = \operatorname{sinc}\left(\frac{\pi A x_1}{\lambda f}\right). \tag{8.73}$$

Because $S(x_1)$ is a bipolar function, it is not physically realizable.

There is, however, a temporal coherence requirement for the incoherent source. In optical information processing operations, the scale of the Fourier spectrum varies with the wavelength of the light source. Thus a temporal coherence requirement should be imposed on every processing operation. If we restrict the Fourier spectra, due to wavelength spread, within a small fraction of the fringe spacing d of a complex spatial filter (e.g., deblurring filter), we have

$$\frac{P_m f \Delta\lambda}{2\pi} \ll d, \tag{8.74}$$

where $1/d$ is the highest spatial frequency of the filter, P_m is the angular spatial frequency limit of the input object transparency, f is the focal length of the transform lens, and $\Delta\lambda$ is the spectral bandwidth of the light source. The spectral width or the temporal coherence requirement of the light source is then

$$\frac{\Delta\lambda}{\lambda} \ll \frac{\pi}{h_0 P_m}, \tag{8.75}$$

where λ is the center wavelength of the light source, $2h_0$ is the size of the input object transparency, and $2h_0 = (\lambda f)/d$.

We now provide a numerical example. Let us assume that the size of the object is $2h_0 = 5$ mm, the wavelength of the light source is $\lambda = 5461$ Å, and we take the factor 10 for eq. (8.75) for consideration, that is;

$$\Delta\lambda = \frac{10\pi\lambda}{h_0 P_m}. \tag{8.76}$$

Several values of spectral width requirement $\Delta\lambda$ for various spatial frequencies P_m are tabulated in Table 8.3.

Table 8.3
SOURCE SPECTRAL REQUIREMENT

$\frac{P_m}{2\pi}$ lines/mm	0.5	1	5	20	100
$\Delta\lambda$, Å	218.4	109.2	21.8	5.46	1.09

From Table 8.3 we see that, if the spatial frequency of the input object transparency is low, a broader spectral width of light source can be used. In other words, the higher the spatial frequency required for an information processing operation, the narrower the spectral width of the light source needed.

We now illustrate examples of source encoding for partially coherent processing operations. Consider first the correlation detection operation (sec. 6.6).

In correlation detection the spatial coherence requirement is determined by the size of the detecting object (i.e., signal). In order to insure a physically realizable encoded source transmittance function, we assume that the spatial coherence function over the input plane P_2 is

$$\Gamma(|\mathbf{x}_2 - \mathbf{x}_1|) = \frac{J_1[(\pi/h_0)|\mathbf{x}_2 - \mathbf{x}_2'|]}{(\pi/h_0)|\mathbf{x}_2 - \mathbf{x}_2'|}, \tag{8.77}$$

where J_1 is a first-order Bessell function of first kind, and h_0 is the size of the detecting signal. A sketch of the spatial coherence as a function $|x_2 - x_2'|$ is shown in fig. 8.24*a*. By taking the Fourier transform of eq. (8.77), we obtain the following source encoding intensity transmittance function:

$$S(|\mathbf{x}_1|) = \text{cir}\left(\frac{|\mathbf{x}_1|}{w}\right), \tag{8.78}$$

where $w = f\lambda/h_0$ is the diameter of a circular aperture, as shown in fig. 8.24*a*,

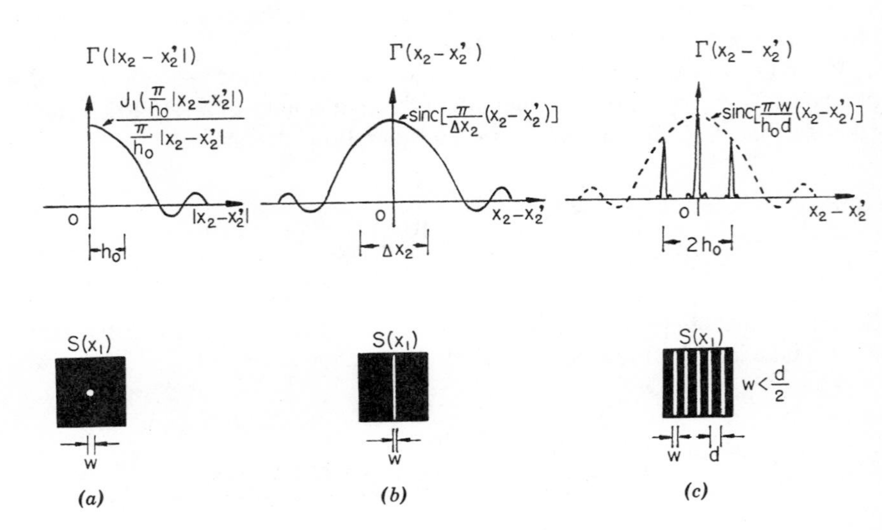

Fig. 8.24 Examples of spatial coherence requirements and source encodings. $\Gamma(x_2 - x_2')$, spatial coherence function; $S(x_1)$, source encoding transmittance. (*a*) For correlation detection. (*b*) For smeared image deblurring. (*c*) For image subtraction.

$$\text{cir}\left(\frac{|\mathbf{x}_1|}{w}\right) \triangleq \begin{cases} 1, & 0 \leq |\mathbf{x}_1| \leq w, \\ 0, & \text{otherwise} \end{cases}$$

f is the focal length of the collimating lens, and λ is the wavelength of the extended source. As a numerical example, we assume that the signal size is $h_0 = 5$ mm, the wavelength is $\lambda = 5461$ Å, the focal length is $f = 300$ mm. In this case the diameter D of the source encoding source should be about 32.8 μm or smaller.

Now consider the smeared image deblurring operation as the second example. We note that the smeared image deblurring is a one-dimensional processing operation and that the inverse filtering is a point-by-point processing concept such that the operation is taking place on the smearing length of the blurred object. Thus the spatial coherence requirement depends upon the smearing length of the blurred object. To obtain a physically realizable source encoding function, we let the spatial coherence function at the input plane P_2 be

$$\Gamma(|x_2 - x_2'|) = \text{sinc}\left(\frac{\pi}{\Delta x_2}|x_2 - x_2'|\right), \tag{8.79}$$

where Δx_2 is the smearing length. A sketch of eq. (8.79) is shown in fig. 8.24*b*. By taking the Fourier transform of eq. (8.79), we obtain

$$S(x_1) = \text{rect}\left(\frac{|x_1|}{w}\right), \tag{8.80}$$

where $w = (f\lambda)/(\Delta x_2)$ is the slit width of the source encoding aperture, as shown in fig. 8.24*b*, and

$$\text{rect}\left(\frac{|x_1|}{w}\right) = \begin{cases} 1, & 0 \leq |x_1| \leq w \\ 0, & \text{otherwise} \end{cases}.$$

For a numerical illustration, if the smearing length is $\Delta x_2 = 1$ mm, the wavelength is $\lambda = 5461°$ Å, and the focal length is $f = 300$ mm, then the slit width w should be about 163.8 μm or smaller.

We now consider image subtraction for our third illustration. Since image subtraction is a one-dimensional processing operation and since the spatial coherence requirement is dependent upon the corresponding point-pair of the images, a strictly broad spatial coherence function is not required. In other words, if we can maintain the spatial coherence between the corresponding image points to be subtracted, the subtraction operation can take place at the output image plane. Thus, instead of utilizing a strictly broad coherence function over the input plane P_2, we use a point-pair spatial coherence function. Again, to insure a physically realizable source-encoding transmittance, we let the point-pair spatial coherence function be (sec. 8.5)

$$\Gamma(|x_2 - x_2'|) = \frac{\sin\left(\frac{N\pi}{h_0}|x_2 - x_2'|\right)}{N \sin\left(\frac{\pi}{h_0}|x_2 - x_2'|\right)} \operatorname{sinc}\left(\frac{\pi w}{h_0 d}|x_2 - x_2'|\right), \tag{8.81}$$

where $2h_0$ is the main separation of the two input object transparencies at plane P_2, $N \gg 1$ a positive integer, and $w \ll d$. Equation (8.81) represents a sequence of narrow pulses that occur at $|x_2 - x_2'| = nh_0$, where n is a positive integer. Their peak values are weighted by a broader sinc factor, as shown in fig. 8.24*c*. Thus we see that a high degree of spatial coherence is maintained at every point-pair between the two input object transparencies. By taking the Fourier transformation of eq. (8.81), we obtain the following source encoding intensity transmittance:

$$S(|x_1|) = \sum_{n=1}^{N} \operatorname{rect}\left(\frac{|x_1 - nd|}{w}\right), \tag{8.82}$$

where w is the slit width, and $d = (\lambda f)/h_0$ is the separation between the slits. It is clear that eq. (8.82) represents N number of narrow slits with equal spacing d, as shown in fig. 8.24*c*. As a numerical example, we let the separation of the input objects $h_0 = 10$ mm, the wavelength $\lambda = 5461$ Å, the focal length of the collimator $f = 300$ mm. In such a setup the spacing d between the slits is 16.4 μm. The slit width w should be smaller than $d/2$, or about 1.5 μm. If the size of the encoding mask is 2 mm square, the number N of slits is about 122. Thus we see that with the source encoding technique it is possible to increase the intensity of the illumination N-fold, and at the same time maintain the point-pair spatial coherence requirement for image subtraction operation.

We now illustrate two examples obtained from the source encoding technique. The first experimental illustration is the result obtained for smeared photographic image deblurring with an encoded incoherent source, as shown in fig. 8.25. In this experiment a xenon arc lamp with a green interference filter is used as an extended incoherent source. A single-slit mask of about 100 μm is used as a source encoding mask. The smeared length of the blurred image is about 1 mm.

Figure 8.26 shows an experimental result obtained from an image subtraction operation with an encoded incoherent source. In this experiment a mercury arc lamp with a green filter is used as an extended incoherent source. A multislit mask is used to encode the light source. The slit width w is 2.5 μm and the spacing between slits is 25 μm. The overall size of the source encoding mask is about 2.5×2.5 mm^2. The mask contains about 100 slits.

From these experimental results, we see that the constraint of strictly broad spatial coherence requirement may be alleviated with a source encoding technique, so that it allows the optical information processing operation to be carried out with an extended incoherent source.

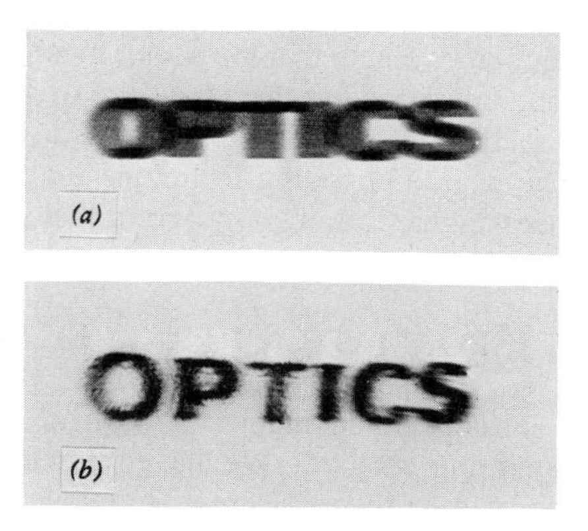

Fig. 8.25 Photographic image deblurring with an encoded extended incoherent source. (*a*) Input blurred object. (*b*) Deblurred image.

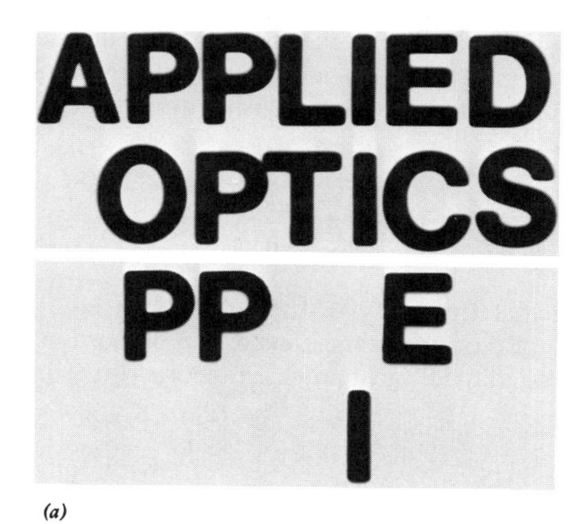

Fig. 8.26 Image subtraction with an encoded extended incoherent source. (*a*) Input object transparencies. (*b*) Subtracted image.

In concluding this section, we note that the use of incoherent sources to carry out the optical processing operation has the advantage of suppressing the coherent artifact noise. The incoherent processing system is usually simple and economical to operate. The source encoding technique may also be extended to white-light optical processing operation, a program that is being investigated.

8.7 OPTICAL INFORMATION PARALLEL PROCESSING

In sec. 8.1 we have introduced an incoherent optical processing technique utilizing a diffraction grating method. We have shown that complex signal processing can be performed by the incoherent technique. Here we describe an optical parallel processing technique that utilizes a multidiffraction grating method (ref. 8.29). We show that parallel complex spatial filtering can be realized with this technique.

Let us now describe a parallel processing technique for coherent optics. Place a *multiple* diffraction grating behind an input-F signal transparency $s(x, y)$, at the input plane P_1 of a coherent optical processor, as shown in fig. 8.27. For simplicity of analysis, we assume that the multiple diffraction grating is an additive type and its subgrating spatial frequencies are oriented in N different angular directions. Thus the amplitude transmittance of the multiple diffraction grating can be written as

$$T(x, y) = K\left[1 + \frac{1}{n}\sum_{n=1}^{N} \cos \omega x_n\right], \tag{8.83}$$

where ω is the spatial frequency of the subgratings, (x_n, y_n) are the spatial coordinate systems of the subgratings, and K is a proportionality constant. The resultant complex amplitude transmittance function of the input plane is

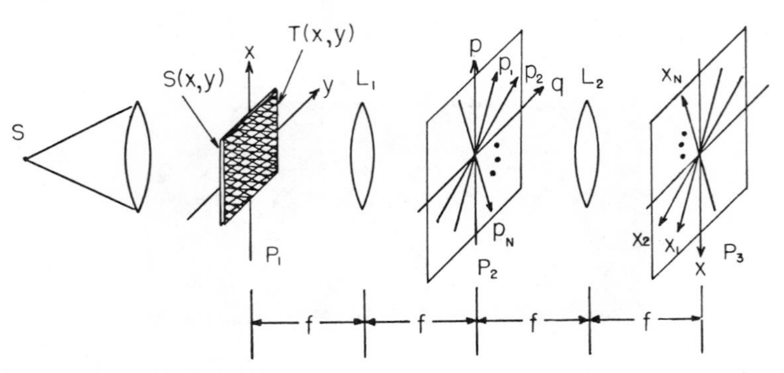

Fig. 8.27 A coherent optical parallel processing system. $s(x, y)$, input signal transparency; $T(x, y)$, high diffraction efficient multiplex grating; S, monochromatic point source.

$$s(x, y)T(x, y) = Ks(x, y)\left[1 + \frac{1}{n}\sum_{n=1}^{N} \cos \omega x_n\right]. \tag{8.84}$$

Since the input plane is illuminated by a collimated monochromatic plane wave, the complex light distribution at the back focal plane of the transform lens is

$$E(p, q) = KS(p, q) + \frac{K}{2n}\sum_{n=1}^{N} S(p_n \pm \omega, q_n), \tag{8.85}$$

where $S(p_n, q_n)$ are the corresponding Fourier transformations of $s(x_n, y_n)$, and (p_n, q_n) are the spatial frequency coordinate systems. From eq. (8.85) we see that multi-Fourier spectra were generated by the multiple diffraction grating at the input plane. In other words, besides the zero-order signal spectrum, there are $2N$ first-order signal spectra diffracted in the neighborhood $(\pm\omega, 0)$ of the (p_n, q_n) coordinate systems. Thus the system is capable of performing parallel complex processing in the spatial frequency plane. In principle, this parallel processing system has $(2N + 1)$-fold processing channels as compared to a single-channel conventional coherent optical processor.

Parallel Complex Spatial Filterings

We now illustrate that multisignal detection can be realized with this parallel processing technique. For simplicity, we assume that $(N + 1)$ complex spatial filters were constructed. The corresponding amplitude transmittances are

$$H_0(p, q) = K_1 + K|S_0(p, q)| \cos[x_0 p + \phi_0(p, q)], \tag{8.86}$$

and

$$H_n(p_n, q_n) = K_1 + K|S_n(p_n - \omega, q_n)| \cos[x_0 p_n + \phi_n(p_n - \omega, q)], \quad n = 1, 2, \ldots, N, \tag{8.87}$$

where x_0 is an arbitrary carrier spatial frequency, K_1 and K are proportionality constants, and

$$S_n(p_n, q_n) = |S_n(p_n, q_n)| \exp[i\phi_n(p_n, q_n)], \qquad n = 0, 1, 2, \ldots, N \tag{8.88}$$

are the detecting signal spectra. We assume that the detecting signal spectra are designed for the detection of different objects.

If we place these complex spatial filters behind each of the signal spectra in the spatial frequency coordinate system, as shown in fig. 8.28, then the complex light fields immediately behind the filters are

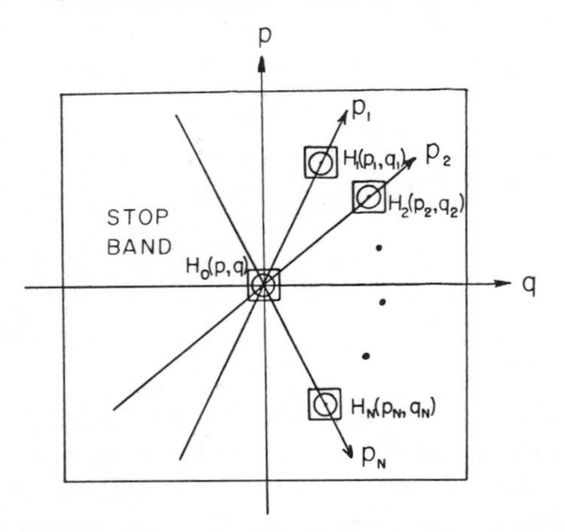

Fig. 8.28 Parallel complex spatial filtering. $H_n(p_n, q_n)$s are complex spatial filters.

$$E_0(p, q) = CS(p, q)H_0(p, q), \tag{8.89}$$

and

$$E_n(p_n, q_n) = CS(p_n - \omega, q_n)H_n(p_n, q_n), \qquad n = 1, 2, \ldots, N. \tag{8.90}$$

The corresponding complex light distribution at the output plane P_3 of the optical parallel processor is

$$g_o(x, y) = s(x, y) + s(x, y) * s_0(x - x_0, y) + s(x, y) * s_0(-x + x_0, y), \tag{8.91}$$

and

$$g_n(x_n, y_n) = s(x_n, y_n)\exp(i\omega x_n) + s(x_n, y_n)\exp(i\omega x_n) * s_n(x_n - x_0, y_n) \times \exp(i\omega x_n) + s(x_n, y_n)\exp(i\omega x_n) * s_n(-x_n + x_0, y_n)\exp(i\omega x_n), \quad n = 1, 2, \ldots, N, \tag{8.92}$$

where (x, y) and (x_n, y_n) are the output plane spatial coordinate system with respect to (p, q) and to (p_n, q_n) and $*$ is the convolution operation. We have ignored the proportionality constants for convenience,.

From the above equations, we see that the first terms are the zero-order terms, which are diffracted in the neighborhood of the optical axis, and the second and third terms are the convolution and correlation terms, which are diffracted in the neighborhood of $(-x_0, 0)$ and $(x_0, 0)$ in the (x_n, y_n) coordinate systems, as shown in fig. 8.29.

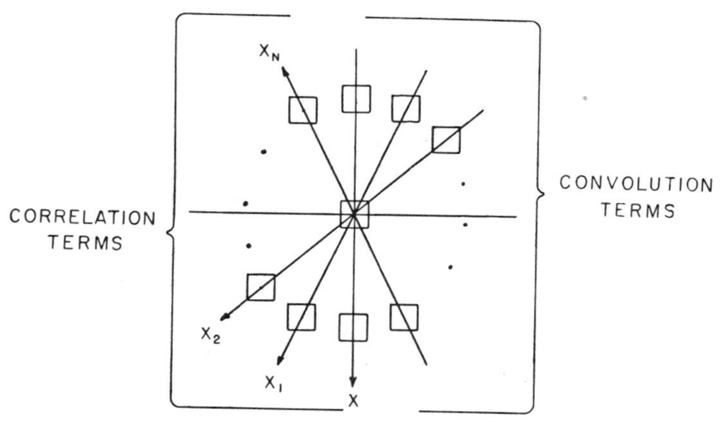

Fig. 8.29 A sketch of output diffraction.

Polychromatic Parallel Processing

We now extend this technique to polychromatic coherent parallel processing. We replace the monochromatic point source of fig. 8.27 by a polychromatic coherent point source, which emits red, green, and blue wavelengths of light, represented by λ_r, λ_g, and λ_b. Then at the back focal plane of the first achromatic transform lens, the complex light distribution is

$$E(p, q) = S_r(p, q) + S_g(p, q) + S_b(p, q) + \frac{1}{2n} \sum_{n=1}^{N} [S_r(p_n \pm \omega, q_n) + S_g(p_n \pm \omega, q_n) + S_b(p_n \pm \omega, q_n)], \qquad (8.93)$$

where $S_r(p_n, q_n)$, $S_g(p_n, q_n)$, and $S_b(p_n, q_n)$ are the corresponding red, green, and blue wavelength signal spectra. For simplicity we have again dropped the proportionality constants. Since $p_n = (2\pi/\lambda f)\alpha_n$, and $q_n = (2\pi/\lambda f)\beta_n$, where (α_n, β_n) are the linear spatial coordinate systems of (p_n, q_n), eq. (8.93) can be written

$$E(\alpha, \beta) = S_r(\lambda_r\alpha, \lambda_r\beta) + S_g(\lambda_g\alpha, \lambda_g\beta) + S_b(\lambda_b\alpha, \lambda_b\beta) + \frac{1}{n} \sum_{n=1}^{N} \left[S_r\left(\alpha_n \pm \frac{\lambda_r r}{2\pi}\omega, \beta_n\right) + S_g\left(\alpha_n \pm \frac{\lambda_g f}{2\pi}\omega, \beta_n\right) + S_b\left(\alpha_n \pm \frac{\lambda_b f}{2\pi}\omega, \beta_n\right)\right], \qquad (8.94)$$

where f is the focal length of the achromatic transform lens. From eq. (8.94), we

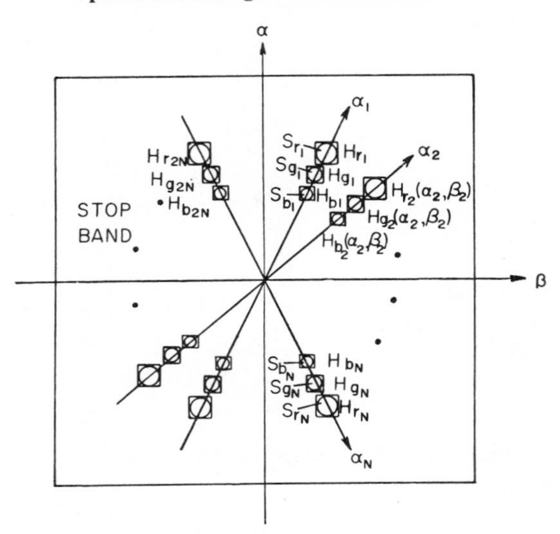

Fig. 8.30 A polychromatic parallel complex spatial filtering. S_{r_n}, S_{g_n}, and S_{b_n}, red, green, and blue signal spectra; H_{r_n}, H_{g_n}, and H_{b_n}, complex spatial filters.

may see that the red, green, and blue color signal spectra are separately diffracted along the α_n axes. If we assume that the spatial frequency ω of the multiplex grating is more than twice as high as the highest spatial frequency content of the input signal $s(x, y)$, the color signal spectra are spatially separated. In parallel signal processing, we place $6N$ complex spatial filters over the first-order color signal spectra, as sketched in fig. 8.30. Note that there are about three times as many processing channels in the polychromatic processor as in the monochromatic parallel processing system, which is about $6N$-fold more than in a conventional processor. In practice, the number of parallel processing channels is generally limited by the practical number of the multiple gratings at the input plane, and by the available power of the light source.

Multisignal Parallel Processing

Let us first describe a spatial encoding technique for multisignal parallel processing, as described by Mueller (ref. 8.30). The encoding is accomplished by sequential spatial modulation of different input signals in a fresh photographic transparency, as shown in fig. 8.31. That is, the number N of input signals can be sequentially encoded in a photographic transparency with a sinusoidal grating oriented at N different angular directions, as illustrated in fig. 8.32. Thus the encoded photographic transparency is

$$T_n(x, y) = K_1\left\{\sum_{n=1}^{N} s_n(x, y)(1 + \cos \omega x_n)\right\}^{-\gamma_1}, \tag{8.95}$$

where $T_n(x, y)$ represents a *negative* signal intensity transmittance, $s_n(x, y)$ are the input signals, ω is the spatial frequency of the grating, (x_n, y_n) are the spatial

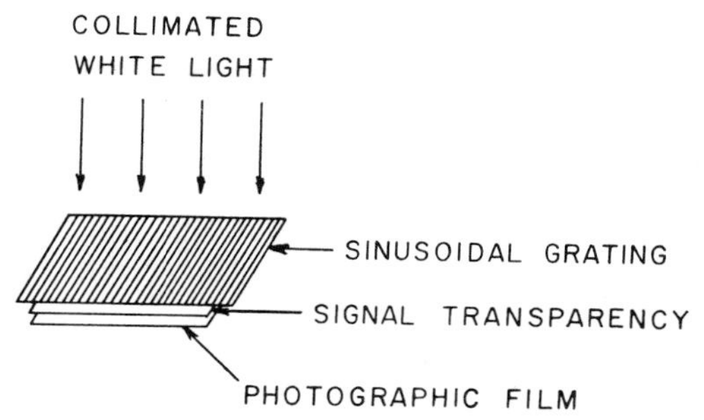

Fig. 8.31 A multisignal encoding technique.

coordinate systems of the grating positions, γ_1 is the *film gamma* of the recorded transparency, and K_1 is a proportionality constant. In order to obtain a positive image transparency, a *contact printing process* can be used. Thus the spatially modulated positive transparency is described by

$$T_p(x, y) = K_2\left\{\sum_{n=1}^{N} s_n(x, y)(1 + \cos \omega x_n)\right\}^{\gamma_1 \gamma_2}, \tag{8.96}$$

where $T_p(x, y)$ represents a *positive* intensity transmittance, γ_2 is the *film gamma* of the second transparency, and K_2 is a proportionality constant. If the product of the film gamma is made equal to two ($\gamma_1 \gamma_2 = 2$), then the corresponding amplitude transmittance is

$$T(x, y) = K\left\{\sum_{n=1}^{N} s_n(x, y)(1 + \cos \omega x_n)\right\}, \tag{8.97}$$

which is a linear function of the input signals $s_n(x, y)$.

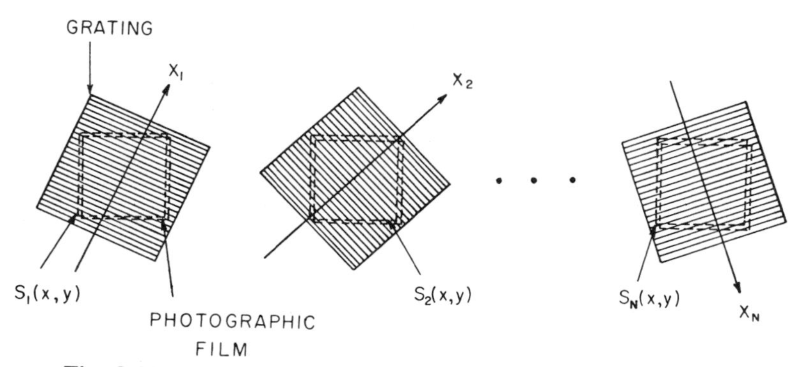

Fig. 8.32 A sketch of multisignal spatial modulation encoding.

For multisignal parallel processing, we place the encoded transparency $T(x, y)$ at the input plane P_1 of a monochromatic coherent processor as shown in fig. 8.27. Then the complex light distribution at the spatial frequency plane P_2 is

$$E(p, q) = \sum_{n=1}^{N} [S_n(p, q) + \tfrac{1}{2}S_n(p_n \pm \omega, q_n)], \tag{8.98}$$

where the proportionality constant is dropped from consideration. Since $p_n = (2\pi/\lambda f)\alpha_n$, and $q_n = (2\pi/\lambda f)\beta_n$, eq. (8.98) can be written in terms of the spatial coordinate systems (α_n, β_n) as

$$E(\alpha, \beta) = \sum_{n=1}^{N} \left[S_n(\alpha, \beta) + \tfrac{1}{2}S_n\left(\alpha_n \pm \frac{\lambda f}{2\pi}\omega, \beta_n\right)\right]. \tag{8.99}$$

Except for the zero-orders, we see that the input signal spectra are diffracted in the neighborhood of $\left(\pm\frac{\lambda f}{2\pi}\omega, \beta_n\right)$ of the (α_n, β_n) coordinate systems. In multisignal parallel processing, we place $2N$ complex spatial filters over the first-order signal spectra, as shown in fig. 8.33.

Furthermore, if we replace the coherent light source by a polychromatic coherent source that emits λ_r, λ_g, and λ_b, representing the red, green, and blue wavelengths, the complex light distribution at the spatial frequency plane P_2 is

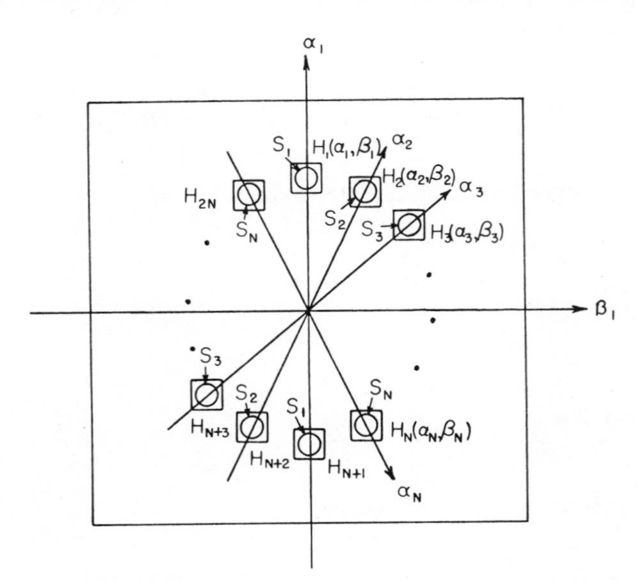

Fig. 8.33 Multisignal parallel processing. S_n, input signal spectra; H_n, complex spatial filters.

$$E(\alpha, \beta) = \sum_{n=1}^{N} \left\{ S_n(\lambda_r \alpha, \lambda_r \beta) + S_n(\lambda_g \alpha, \lambda_g \beta) + S_n(\lambda_b \alpha, \lambda_b \beta) \right.$$
$$+ \tfrac{1}{2}\left[S_n\left(\alpha_n \pm \frac{\lambda_r f}{2} \omega, \beta_n \right) + S_n\left(\alpha_n \pm \frac{\lambda_g f}{2\pi} \omega, \beta_n \right) \right.$$
$$\left.\left. + S_n\left(\alpha_n \pm \frac{\lambda_b f}{2\pi} \omega, \beta_n \right) \right] \right\}. \tag{8.100}$$

From the above equation, again we see that the $6N$-different color signal spectra are separately diffracted over the spatial frequency plane P_2. Thus multisignal parallel processing can be accomplished through complex spatial filtering of these signal spectra. We note again that the number of processing channels of the polychromatic technique is about three times the number of the monochromatic case. There is, however, a disadvantage to this multisignal processing technique; it is not a real-time processing technique, because it requires an initial encoding step. Nevertheless, this disadvantage may be alleviated with ingenuous design of the encoding technique.

In concluding this section, we would like to point out that this technique of optical parallel processing may also be applied to the incoherent optical processing that we have described in previous sections, for example, real-time white-light pseudocolor encoding (ref. 8.31).

PROBLEMS

8.1 Let us consider the white-light point source optical information processing system of fig. 8.1. If we ignore the input signal transparency:

(a) Determine the smeared length of the Fourier spectra as a function of the focal length f of the achromatic transform lens and the spatial frequency p_0 of the sinusoidal phase grating $T(x)$.

(b) If the spatial frequency of the diffraction grating is $p_0 = 80\pi$ rad/mm and $f = 30$ cm, then what is the precise length of the smeared rainbow color spectra?

8.2 With reference to prob. 8.1, if the white-light source is a uniformly circular extended source of diameter D, then:

(a) Determine the size of the smeared Fourier spectra as a function of f and p_0.

(b) If $D = 2$ mm, $p_0 = 80\pi$ rad/mm and $f = 30$ cm, then determine the precise size of the smeared Fourier spectra.

8.3 We now consider the effect due to the input signal transparency of prob. 8.1. If the spatial bandwidth of the input object transparency is p_1, which is smaller as compared with the spatial frequency of the diffraction grating (i.e., $p_1 \ll 2p_0$), then:

(a) What is the smearing length of the Fourier spectra, if the white-light is a point source?

(b) If the white-light source is a uniformly circular extended source of diameter D, then what is the size of the smeared Fourier spectra?

(c) If the spatial bandwidth of the input object is $p_1 = 20\pi$ rad/mm, the spatial frequency of the phase grating is $p_0 = 80\pi$ rad/mm, and the focal length of the achromatic transform lens is $f = 30$ cm, then calculate the size of the smeared Fourier spectra.

8.4 For a coherent optical information operation, the spatial filter function is a binary wedge-shaped filter, as shown in fig. 8.34. Show that the optical processing operation can be used with a white-light source instead of a coherent source.

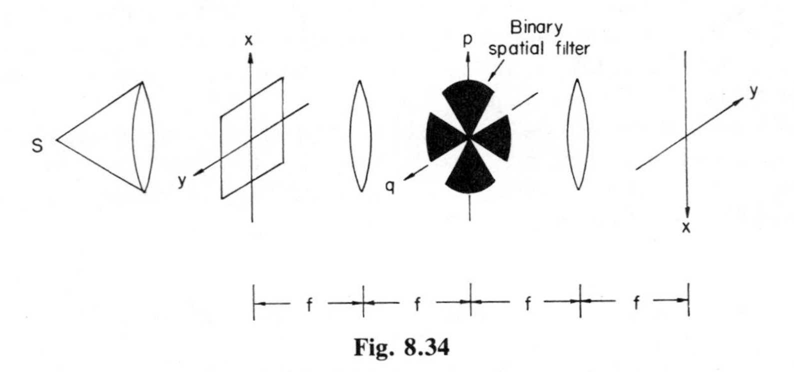

Fig. 8.34

8.5 Let us suppose that the amplitude transmittance of an input object transparency is

$$T(x, y) = s(x, y)[1 + \text{sgn}(\cos p_0 x)],$$

where

$$\text{sgn}(\cos p_0 x) \triangleq \begin{cases} 1, & \cos(p_0 x) \geq 0, \\ -1, & \cos(p_0 x) < 0, \end{cases}$$

p_0 is the spatial sampling frequency, $s(x, y)$ is the input signal, and (x, y) is the spatial coordinate system. We assume that the sampling frequency is more than twice the spatial bandwidth of $s(x, y)$, then show that the input sigal $s(x, y)$ can be recovered with a white-light optical information processing technique. Why is it that this optical processing operation does not require a coherent source?

8.6 Let us consider the white-light multi-image regeneration technique of sec. 8.3.

(a) Design a set of diffraction grating filters so that an $m \times n$ array of images can be generated.

(b) If we assume that a white-light point source is used and that the spatial bandwidth of the input object is about 20 lines/mm, what is the minimum spatial frequency requirement of the diffraction grating $T(x)$ and what is the maximum spectral bandwidth requirement of the grating filters so that minimum color blurring of the regenerated images can be obtained.

8.7 Given the white-light photographic image deblurring of sec. 8.4; if the linear smeared length is about 1 mm, then:

(a) What are the spatial and temporal coherence requirements for the deblurring process with incoherent source?

(b) In order to achieve the above spatial coherence requirement, what is the requirement of the light source?

(c) In order to obtain the temporal coherence requirement for the deblurring, what is the minimum spatial frequency requirement of the diffraction grating $T(x)$?

8.8 With reference to prob. 8.7:

(a) Design a fan-shaped deblurring filter such that the deblurring can take place with the whole spectral band of the white-light source.

(b) For the fan-shaped deblurring filter described in part (a), evaluate the deblurred image intensity at the output image plane.

8.9 If the spatial frequency limit of this blurred image of prob. 8.8 is about 25 lines/mm, then estimate the degree of the deblurring error.

8.10 A distant object moves at a variable speed, and it is recorded on a photographic film with a camera. If the amplitude transmittance of this recorded smeared image is

$$T(x, y) = s(x - x', y) \exp(-x^2), \qquad |x| \leq 1 \text{ mm},$$

where $s(x, y)$ is the object image, then design a white-light optical processing system and an appropriate fan-shaped spatial filter to deblur the image. (Note: the dimension of the source size, the spatial frequency of the grating, and the dimension of the deblurred filter should be specified.)

8.11 Assume that the object image is linearly smeared in the x direction and then in the y direction so that the transmittance of the recorded photographic image is

$$T(x, y) = s(x - x', y - y'), \qquad |x| \leq \Delta x/2, \ |y| \leq \Delta y/2,$$

where $s(x, y)$ is the unsmeared object image, and Δx and Δy are the smeared length in the x and y axes, respectively. Design a white-light optical processor with appropriate deblurring filters to unsmear the image.

8.12 Let us consider the image subtraction with encoded extended source of sec. 8.5. We assume that the spatial frequencies of the input object transparencies O_1 and O_2 are about 10 lines/mm.

(a) Calculate the temporal coherence requirement for the subtraction processing operation.

(b) If the main separation of the two input object transparencies is 20 mm and the focal length of the achromatic transform lenses is 300 mm, then design a diffraction grating G and a source encoding mask to provide a point-pair spatial coherence function, of about 0.75 percent coherence, for the subtraction processing operation.

(c) Calculate the corresponding visibility of the subtracted image.

8.13 We wish to extend the image subtraction of prob. 8.12 to an extended white-light source. If a phase grating $T(x) = \exp[ip_0 x]$, where $p_0 = 100\pi$ rad/mm, is inserted in the input plane with the input object transparencies of fig. 8.19, design a source encoding mask and a sinusoidal grating G so that the subtraction processing operation can take place with the whole spectral band of the light source.

8.14 Given a mutual coherence function, at the input plane of a partially coherent optical information processing system such as shown in fig. 8.23, in the following:

$$\Gamma(x) = \tfrac{1}{2}[\delta(x - h_0) + \delta(x + h_0)] + \delta(x),$$

where $x = x_2 - x_2'$. Evaluate a source encoding mask $s(x_1)$ so that the specified mutual coherence function can be obtained with encoded extended source.

8.15 With reference to Young's experiment of fig. 8.22, if we wish to extend the source encoding concept over a three-dimensional space near the extended source:

(a) Describe the appropriate encoding process so that a high degree of the mutual coherence between Q_1 and Q_2 at plane P_2 may be obtained.

(b) As compared with the encoding over source plant P_1, estimate the increased utilization of available power of the source.

8.16 For linear image deblurring, it is assumed that the smearing length is about 0.5 mm and the input object is spatial frequency limited to about 10 lines/mm.

(a) Design a multisource white-light processor for the image deblurring.

(b) Determine the source size requirement, the spatial frequency of the diffraction grating, and the specification of the fan-shaped deblurring filters used.

REFERENCES

8.1 A. Vander Lugt, "Coherent Optical Processing," *Proc. IEEE*, **62,** 1300 (1974).

8.2 G. L. Rogers, "Non-Coherent Optical Processing," *Opt. Laser Technol.*, **7,** 153 (1975).

8.3 K. Bromley, "An Optical Incoherent Correlation," *Opt. Acta,* **21,** 35 (1974).

8.4 M. A. Monahan, K. Bromley, and R. P. Bocker, "Incoherent Optical Correlations," *Proc. IEEE,* **65,** 121 (1977).

8.5 G. L. Rogers, *Noncoherent Optical Processing,* John Wiley, New York, 1977.

8.6 E. N. Leith and J. Upatnieks, "Holography with Achromatic-Fringe Systems," *J. Opt. Soc. Am.*, **57,** 975 (1967).

8.7 R. E. Brooks, L. O. Weflinger, and R. F. Wuesker, "Pulsed Laser Holograms," *IEEE, J. Quantum Electron., QE-2,* 275 (1966).

8.8 S. Lowenthal and P. Chavel, in R. Wiener and J. Shamer, Eds., *Proc. of ICO Jerusalem 1976, Conference on Holography and Optical Processing,* Plenum Press, New York, 1977.

8.9 A. Lohmann, "Incoherent Optical Processing of Complex Data," *Appl. Opt.*, **16,** 261 (1977).

8.10 E. N. Leith and J. Roth, "White-Light Optical Processing and Holography," *Appl. Opt.*, **16,** 2565 (1977).

8.11 G. M. Morris and N. George, "Space and Wavelength Dependence of a Dispersion-Compensated Matched Filter," *Appl. Opt.,* **19,** 3843 (1980).

8.12 D. Gabor, "Laser Speckle and its Elimination," *IBM J. Res. Develop.*, **14,** 509 (1970).

8.13 F. T. S. Yu, "A New Technique of Incoherent Complex Signal Detection," *Opt. Commun.*, **27,** 23 (1978).

8.14 F. T. S. Yu, "A Technique of White-Light Optical Processing with Diffraction Grating Method," *Proc. SPIE,* **232,** 9 (1980).

8.15 F. T. S. Yu and T. H. Chao, "Experiments of White-Light Processing Utilizing a Diffraction Grating Method," *Optik,* **56,** 423 (1980).

8.16 F. T. S. Yu, S. L. Zhuang, and T. H. Chao, "Multi-Image Regeneration by White-Light Processing," *Opt. Commun.*, **14,** 11 (1980).

8.17 F. T. S. Yu, "Image Restoration, Uncertainty, and Information," *Appl. Opt.*, **8,** 53 (1969).

8.18 F. T. S. Yu, *Optics and Information Theory,* Wiley-Interscience, New York, 1976, chap. 7.

8.19 S. L. Zhuang, T. H. Chao, and F. T. S. Yu, "Smeared Photographic Image Deblurring Utilizing White-Light Processing Technique," *Opt. Lett.*, **6,** 102 (2081).

8.20 W. Swindell, "A Noncoherent Optical Analog Image Processor," *Appl. Opt.*, **9,** 2459 (1970).

8.21 D. Gabor, G. W. Stroke, R. Restrick, A. Funkhouser, and D. Brumm, "Optical Image Synthesis (Complex Amplitude Addition and Subtraction) by Holographic Fourier Transformation," *Phys. Lett.*, **18,** 116 (1965).

8.22 K. Bromley, M. A. Monahan, J. F. Bryant, and B. J. Thompson, "Holographic Subtraction," *Appl. Opt.*, **10,** 174 (1971).

8.23 S. H. Lee, S. K. Yao, and A. G. Milnes, "Optical Image Synthesis (Complex Amplitude Addition and Subtraction) in Real Time by a Diffraction-Grating Interferometric Method," *J. Opt. Soc. Am.*, **60,** 1037 (1970).

8.24 J. F. Ebersole, "Optical Image Subtraction," *Opt. Eng.*, **15,** 436 (1975).

8.25 F. T. S. Yu and A. Tai, "Incoherent Images Addition and Subtraction: A Technique," *Appl. Opt.*, **18,** 2705 (1979).

8.26 S. T. Wu and F. T. S. Yu, "Image Subtraction with Encoded Extended Incoherent Source," *Appl. Opt.*, **20,** 4082 (1981).

8.27 M. Born and E. Wolf, *Principles of Optics,* 2nd rev. ed., Pergamon Press, New York, 1964.

8.28 F. T. S. Yu, S. L. Zhuang, and S. T. Wu, "Source Encoding for Partially Coherent Optical Processing," *Appl. Phys.,* **B27,** 99 (1982).

8.29 F. T. S. Yu and M. S. Dymek, "Optical Information Parallel Processing: A Technique," *Appl. Opt.,* **20,** 1450 (1981).

8.30 P. F. Mueller, "Linear Multiple Image Storage," *Appl. Opt.,* **8,** 267 (1969).

8.31 F. T. S. Yu, S. L. Zhuang, T. H. Chao, and M. S. Dymek, "Real-Time White-Light Spatial Frequency and Density Pseudocolor Encoder," *Appl. Opt.,* **17,** 2986 (1980).

8.32 F. T. S. Yu and J. L. Horner, "Optical Processing of Photographic Images," *Opt. Eng.,* **20,** 666 (1981).

9

Polychromatic Processing with Noncoherent Light

In the preceding chapters we have illustrated several basic principles of coherent and incoherent optical processing techniques. We have shown that the optical processing operations are similar to linear system theory, and that both the coherent and incoherent optical processing systems can be treated as linear processing systems. Although the coherent optical processing technique has been used for most of the optical information processing operations, it is plagued with coherent artifact noise that frequently limits its processing capability. We have, in previous chapters, shown that there are several optical processing operations that can be easily carried out with an incoherent, or white-light, source. Since the white-light source contains the entire spectral band of visible light, it is particularly suitable for polychromatic image processing. In this chapter we illustrate a few applications of incoherent optical procesing as applied to color images. We show that there are several well-known optical information processing operations that can be easily extended to polychromaic processing. For example, color image deblurring, color image subtraction, pseudocolor encoding, and possibly many others. In short, we expect the new trend of white-light optical processing to open a new dimension for color image processing.

9.1 ARCHIVAL STORAGE OF COLOR FILMS

Archival storage of color films has long been an unresolved problem for the film industry. The major reason is that the organic dyes used in color films are usually unstable under prolonged storage, often causing gradual color fading. Although there are several available techniques for preserving the color images, all of them possess certain definite drawbacks. One of the most commonly used techniques involves repetitive application of primary color filters, so that the color images can be preserved in three separate rolls of black-and-white film. To reproduce the color image, a system with three primary color projectors must be used. These films should be projected in perfect unison that the primary color images will be precisely recorded on a fresh roll of color film. However, this technique has two major disadvantages: First, the storage volume for each film is tripled. Second, the reproduction system is rather elaborate and expensive.

In this section, we describe a white-light processing technique for archival storage of color films. This technique may be the most efficient technique existing to date. This technique also allows for direct viewing capability, and may be suitable for library applications.

The use of monochrome transparencies to retrieve color images was first reported by Ives (ref. 9.1) in 1906. He introduced a slide viewer that produced color images by a diffraction phenomenon. Grating either of different spatial frequencies or of azimuthal orientation was used. More recently, Mueller (ref. 9.2) described a similar technique, employing a tricolor grid screen for image encoding. In decoding, he used three quasi-monochromatic sources for color image retrieval. Since then, similar work on color image retrieval has been reported by Macovski (ref. 9.3), Grousson and Kinany (ref. 9.4), and Yu (ref. 9.5).

Let us describe a technique in which color film is spatially encoded onto a black-and-white transparency. The encoding will take place by sequential recordings on a black-and-white photographic film through a color image transparency that has been superimposed by a Ronchi-type grating, as shown in fig. 9.1. The sequential recordings take place for three primary color filters (i.e. red, blue, and green) with respect to three angular grating positions, as illustrated in fig. 9.2. The first exposure is made through a red filter when the grating is at a 0° azimuth position, the second is made through a blue filter when the grating is oriented to a position at 60°, and the last exposure takes place through a green filter when the grating is at a position of 120°. If the three exposures are properly recorded on the photographic film, a multiplex spatially (encoded) modulated black-and-white transparency is produced. The corresponding intensity transmittance can be described as

$$T_n(x, y) = K_1\{T_r(x, y)[1 + \operatorname{sgn}(\cos p_0 x)] + T_b(x', y')[1 + \operatorname{sgn}(\cos p_0 x')] + T_g(x'', y'')[1 + \operatorname{sgn}(\cos p_0 x'')]\}^{-\gamma_{n1}}, \qquad (9.1)$$

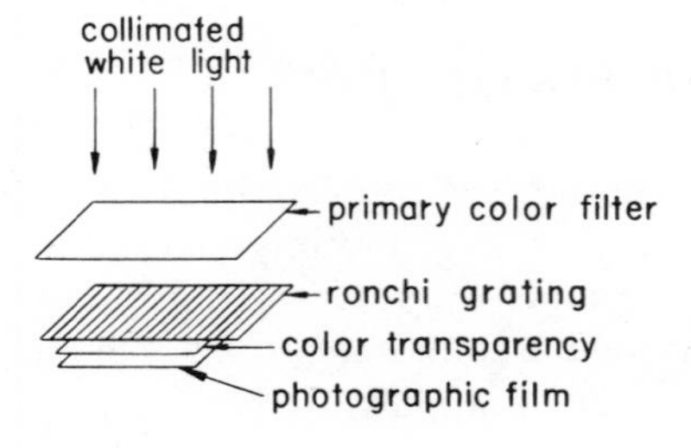

Fig. 9.1 A sequentially recorded, spatial color encoding technique.

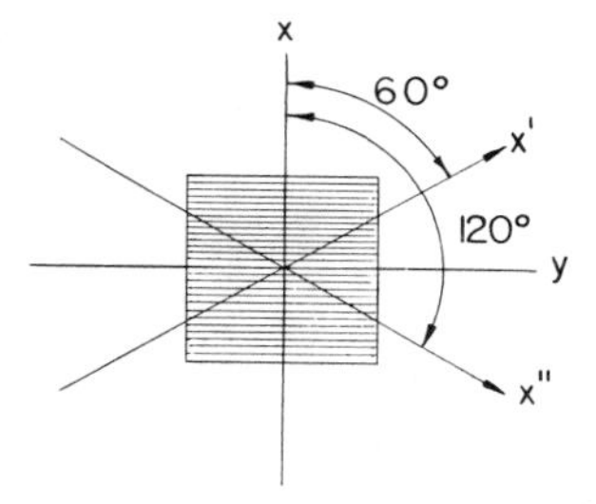

Fig. 9.2 Three angular positions of the Ronchi grating.

where

$$\operatorname{sgn}(\cos p_0 x) \triangleq \begin{cases} 1, & \cos(p_0 x) \geq 0, \\ -1, & \cos(p_0 x) < 0, \end{cases}$$

p_0 is the spatial frequency of the grating, T_n represents a negative image intensity transmittance, T_r, T_b, and T_g are the corresponding red, blue, and green image exposures recorded on the film, (x, y), (x', y'), and (x'', y'') are the coordinate systems with respect to the 0°, 60°, and 120° grating positions, γ_{n1} is the film gamma of the film, and K_1 is a proportionality constant.

To obtain a positive image transparency suitable for optical processing, a contact printing process is used. In this manner a multiplex spatially modulated positive image transparency is obtained and the corresponding intensity transmittance is

$$T_p(x, y) = K_2\{T_r(x, y)[1 + \operatorname{sgn}(\cos p_0 x] + T_b(x', y')$$
$$[1 + \operatorname{sgn}(\cos p_0 x') + T_g(x'', y'')[1 + \operatorname{sgn}(\cos p_0 x'')\}^{\gamma_{n1}\gamma_{n2}}, \qquad (9.2)$$

where T_p represents a positive image intensity transmittance, K_2 is a proportionality constant, and γ_{n_2} is the film gamma of the second transparency. To obtain a linear amplitude image transmittance, we let the product of the film gammas equal two ($\gamma_{n_1}\gamma_{n_2} = 2$), that is

$$t_p(x, y) = KT_r(x, y)[1 + \operatorname{sgn}(\cos p_0 x)] + T_b(x', y')[1 + \operatorname{sgn}(\cos p_0 x')]$$
$$+ T_g(x'', y'')[1 + \operatorname{sgn}(\cos p_0 x'')]. \qquad (9.3)$$

Now let us describe a white-light processing technique that allows faithful color images to be reproduced from a black-and-white encoded transparency. If we place the spatially encoded transparency $T_p(x, y)$ at the input plane P_1 of a white-light processor, as shown in fig. 9.3, the complex light distribution for every wavelength λ at the spatial frequency plane P_2 is

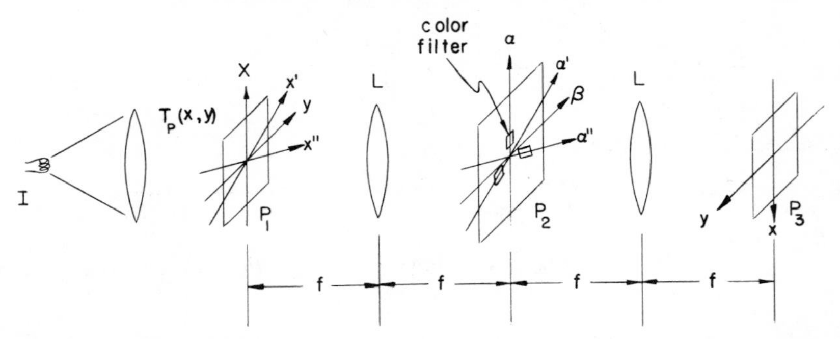

Fig. 9.3 A white-light processor for spatial color decoding. I, extended white-light source; L, transform lens; $T_p(x, y)$, spatially encoded transparency.

$$E(p, q; \lambda) = c \iint t_p(x, y) \exp[-i(px + qy)] \, dx \, dy, \tag{9.4}$$

where the integration is over the spatial domain of the encoded transparency t_p, and C is a complex constant.

For simplicity of illustration, we drop the proportionality constant and evaluate eq. (9.4); we have

$$\begin{aligned} E(p, q; \lambda) = {} & T_r(p, q) + \frac{1}{2} \sum_{n=1}^{\infty} a_n T_r(p \pm np_0, q) + T_b(p', q') \\ & + \frac{1}{2} \sum_{n=1}^{\infty} a_n T_b(p' \pm np_0, q') + T_g(p'', q'') + \frac{1}{2} \sum_{n=1}^{\infty} a_n T_g(p'' \pm np_0, q''), \end{aligned} \tag{9.5}$$

where $T_r(p, q)$, $T_b(p', q')$, and $T_g(p'', q'')$ are the corresponding Fourier transformations of $T_r(x, y)$, $T_b(x', y')$ and $T_g(x'', y'')$, $p = (2\pi/\lambda f)\alpha$, $q = (2\pi/\lambda f)\beta$, $p' = (2\pi/\lambda f)\alpha'$, $q' = (2\pi/\lambda f)\beta'$, $p'' = (2\pi/\lambda f)\alpha''$, and $q'' = (2\pi/\lambda f)\beta''$ are the respective angular spatial frequency coordinates, (α, β), (α', β'), and (α'', β'') are the linear spatial coordinate systems of (p, q), (p', q'), and (p'', q''), f is the focal length of the transform lens, and the a_ns are the Fourier coefficients.

Equation (9.5) can be written in terms of spatial coordinates of α, β, α', β', α'', and β''. Thus we have

$$\begin{aligned} E(\alpha, \beta; \lambda) = {} & T_r(\alpha, \beta) + \frac{1}{2} \sum_{n=1}^{\infty} a_n T_r\left(\alpha \pm \frac{n\lambda f}{2\pi} p_0, \beta\right) \\ & + T_b(\alpha', \beta') + \frac{1}{2} \sum_{n=1}^{\infty} a_n T_b\left(\alpha' \pm \frac{n\lambda f}{2\pi} p_0, \beta'\right) \\ & + T_g(\alpha'', \beta'') + \frac{1}{2} \sum_{n=1}^{\infty} a_n T_g(\alpha'' \pm \frac{n\lambda f}{2\pi} p_0, \beta''). \end{aligned} \tag{9.6}$$

From this equation we see that different orders of the image spectra are linearly dispersed in rainbow color with respect to the α, α', and α'' axes. If the spatial frequency of the grating p_0 is more than twice as high as the highest spatial frequency content of the color image transparency, then the orders of the smeared color image spectrums are physically separated.

In spatial color decoding (or filtering) we would allow a set of three first-order smeared spectra to pass through a red, a blue, and a green filter, respectively, as shown in fig. 9.4. Note that the color filterings take place with broad spatial bands of color image spectra; the marginal resolution loss can be avoided. The corresponding complex light distribution immediately behind the spatial frequency plane P_2 is

$$E(\alpha, \beta) = T_r\left(\alpha - \frac{\lambda_r f}{2\pi} p_0, \beta\right) + T_b\left(\alpha' + \frac{\lambda_b f}{2\pi} p_0, \beta'\right) + T_g\left(\alpha'' - \frac{\lambda_g f}{2\pi} p_0, \beta''\right), \tag{9.7}$$

where λ_r, and λ_b, and λ_g represent red, blue, and green color wavelengths. At the output image plane, the complex light distribution is

$$E(x, y) = T_r(x, y)e^{ixp_0} + T_b(x, y)e^{-ix'p_0} + T_g(x, y)e^{ix''p_0}. \tag{9.8}$$

Since the primary color images of eq. (9.8) are mutually *incoherent*, the corresponding image irradiance would be

$$I(x, y) = T_r^2(x, y) + T_b^2(x, y) + T_g^2(x, y), \tag{9.9}$$

which is essentially the combination of the encoded primary color images. Thus

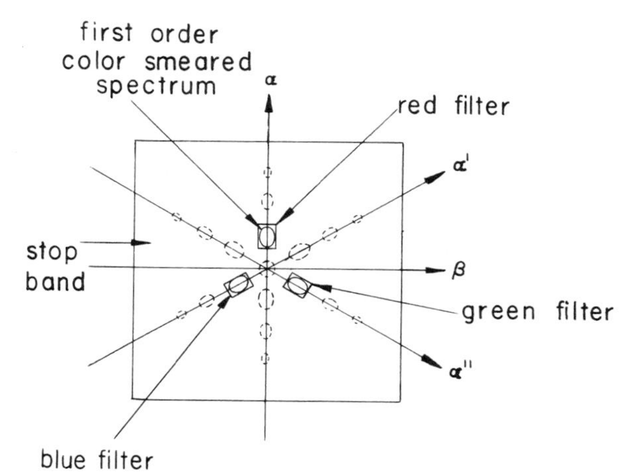

Fig. 9.4 A spatial color decoding filter.

a decoded multicolor image can be reproduced by this simple white-light technique. Note that the system also offers a direct viewing capability if a fine diffuser is introduced at the output image plane P_3.

In experimental demonstrations we provide a color picture obtained by this archival storage technique, as shown in Color Plate I*a*. The picture is a field of tulips with a wide range of colors and the image is focused on the two tulips at the foreground. For simplicity, this experiment used only red and green color filters for the spatial encodings on the x and y axes, respectively. The first negative (multiplex spatial encoded) transparency was made by Kodak 4147 Plus X film, because of its relatively flat spectra response and a lower film gamma. Kodak 4154 Contrast Process Ortho film with a higher film gamma was used to produce the positive image transparency. In the photographic process, the film gamma of the first negative transparency was controlled to be about 0.8 and the second film gamma was about 2.5. Thus a linear positive image transparency was obtained. For comparison, we also provide a color picture of the original color image transparency, as shown in Color Plate I*b*. Although the spatial encoding of this experiment was not optimally controlled to meet the best standard and the white-light processing system was not optimally adjusted, the color reproduction achieved by this archival storage technique is impressively faithful and the resolution of the color image is exceptionally good. Note that the image resolution can be further improved if a higher spatial frequency grating is employed, and to obtain a more faithful color reproduction all three primary color spatial encodings should be utilized.

9.2 COLOR ENHANCEMENT FOR FADED COLOR FILM

In the previous section we have described archival storage of color film with a white-light processing technique. We have shown that white-light processing offers a simple and economical technique for color image retrieval. Since the white-light provides all the visible wavelengths, it is particularly suitable for color image processing.

In this section we describe a white-light processing technique for color enhancement of faded color films (ref. 9.6). We first show that color enhancement can be accomplished by a spatial encoding and color filtering technique. Second, we demonstrate that color enhancement can also be obtained with real-time white-light processing. Although the technique with spatial encoding has the advantage of color encoding, it is not a real-time technique.

Let us assume that the degree of a faded color is known *a priori*. That color can be spatially encoded in a black and white transparency with a higher degree of spatial modulations. With the encoding procedure, as described in previous sections, the resultant amplitude transmittance of the encoded film can be shown to be

$$t_p(x, y) = A_r(x, y)[1 + \operatorname{sgn}(\cos p_0 x)] + A_g(x', y')[1 + \operatorname{sgn}(\cos p_0 x')] + A_b(x'', y'')[1 + \operatorname{sgn}(\cos p_0 x'')], \qquad (9.10)$$

where

$$\operatorname{sgn}(\cos p_0 x) \triangleq \begin{cases} 1, & \cos(p_0 x) \geq 0, \\ -1, & \cos(p_0 x) < 0, \end{cases}$$

p_0 is the spatial sampling frequency, A_r, A_g, and A_b are the red, green, and blue amplitude modulations, and (x, y), (x', y'), and (x'', y'') are the coordinate systems with respect to the spatial encodings. In our illustration we assume that the spatial encodings are 60° apart. Note that, if green is more faded than the other colors then the amplitude modulation A_g should be encoded in a higher degree. In other words, a proper ratio of A_r, A_g, and A_b should be predetermined.

If the encoded transparency of eq. (9.10) is inserted in the input plane of P_1 of a white-light optical processor, as shown in fig.(9.5), various orders of encoded image spectra in rainbow colors come into view at the spatial frequency plane P_2. In color restoration we would place a set of adjustable density color filters (i.e., red, green, and blue filters) over the encoded spectra, as shown in the figure. By further adjusting density color filters, an enhanced color image can be observed at the output plane P_3. We see that this color restoration technique is primarily accomplished with a spatial encoding and color filtering processing.

As an example we provide a black-and-white picture of an enhanced color image obtained by this color restoration technique, fig. 9.6*a*. For simplicity, this experiment uses only red and green color encodings. For comparison, we also

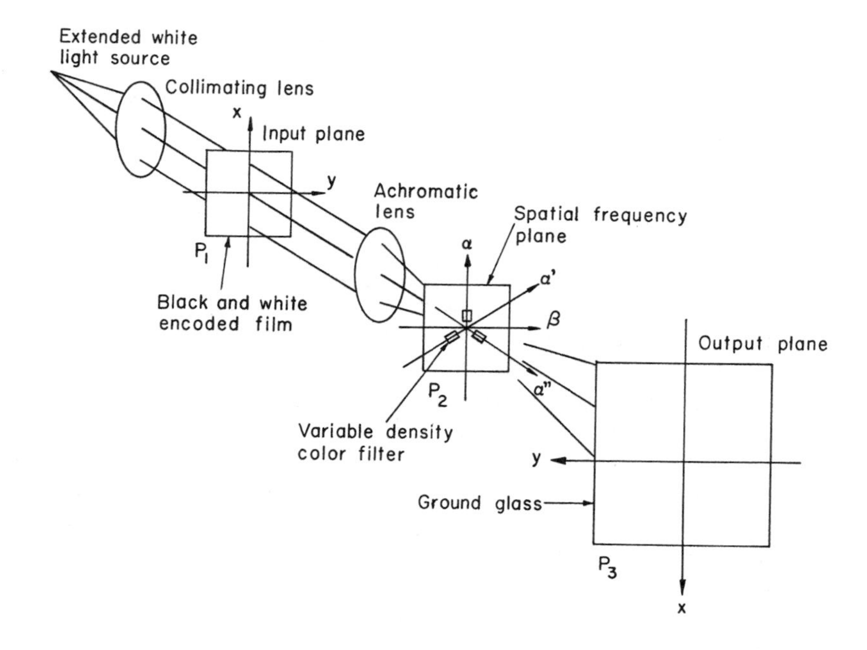

Fig. 9.5 A white-light processing technique for color enhancement.

Fig. 9.6 (*a*) A black-and-white picture of an enhanced color image obtained with the spatial encoding technique. (*b*) A black-and-white version of the original faded color picture.

provide a black-and-white picture of the original *faded* color image, fig. 9.6*b*. Although the experimental demonstration is not optimally manipulated, the restored color is vivid. Note that the noiselike appearance in the image is primarily due to the coarseness of the encoding process and the uncleanness of the optical processing system. In principle, this noiselike appearance can be avoided with a more careful experimental procedure. We stress that, if a color film is completely faded, it may not be possible to enhance the color. However, if the fading is not complete, some enhancement may be possible. In other words, the more the color is faded, the more difficult it becomes to enhance.

Let us now illustrate a real-time color enhancement technique for faded color film. We show that this technique is rather simple and versatile. Place a faded color image transparency in contact with a sinusoidal diffraction grating at the input plane P_1 of a white-light optical procesor, as shown in fig. 9.7. At the spatial frequency plane P_2, three orders of smeared color image spectra can be observed. In color enhancement we place a set of variable density color filters

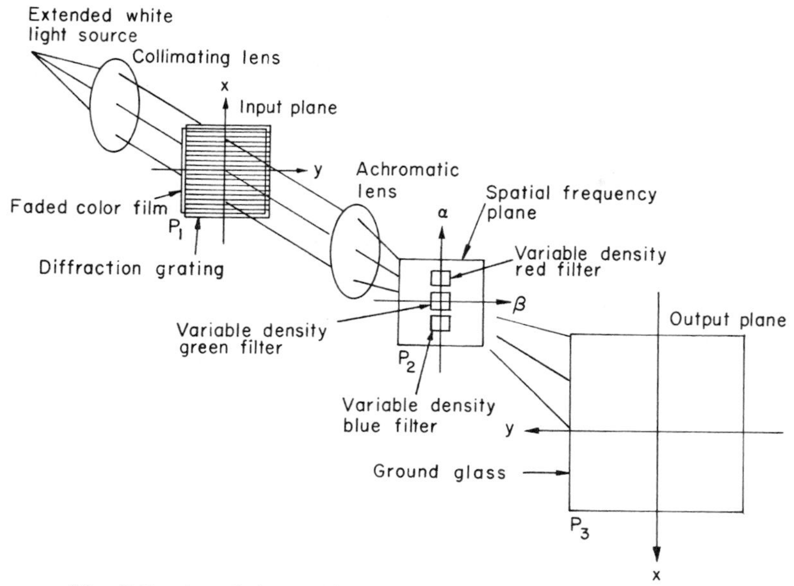

Fig. 9.7 A real-time white-light color enhancement technique.

over the image spectra, as illustrated in the figure. By adjusting the color density filters, a restored color image can be obtained at the output plane P_3. Thus it is possible to enhance a faded color image to some degree with a real-time processing technique.

In an experimental demonstration, we provide a black-and-white copy of an enhanced color picture produced by the real-time processing technique, fig. 9.8.

Fig. 9.8 A black-and-white picture of an enhanced color image obtained with the real-time white-light processing technique.

In comparison with the original faded color picture of fig. 9.6*b*, we see that the enhanced color image is faithfully reproduced. The noiselike appearance in the restored color image is primarily due to uncleanness of the optical processing system, which, in principle, can be eliminated.

We have described a white-light processing technique for color enhancement of faded color films. We have shown that the color enhancement can be achieved by a spatial encoding and color filtering technique. Although the spatial encoding technique offers the advantage of color encoding, it is not a real-time process. We have also demonstrated a real-time white-light color enhancement technique. This technique is very simple, versatile, and economical to use. There is another advantage of the white-light processing technique: it provides a direct viewing capability, which is very suitable for general color image enhancement. Note that, for more sophisticated enhancement, space variant color filters should be used in the spatial frequency plane. Finally, if a color film is totally faded, then it may not be possible to enhance the faded color. On the other hand, if the fading is not complete, it is possible to enhance the color to some degree.

9.3 PSEUDOCOLOR DENSITY ENCODING THROUGH HALFTONE SCREEN

Most of the optical images obtained in various scientific applications are density modulated black-and-white images. Human observation, however, can perceive variations in colors better than gray levels. Thus a color encoded image can often provide better visual discrimination. Pseudocolor encoding by computer technique has been widely used in applications where the images are initially digitized (ref. 9.7). While the computer technique is the logical choice for digital images, optical processing techniques may be more advantageous for applications where the initial images are analog photographic pictures, such as aerial photography and X-ray diagnosis.

Historically, the use of pseudocolor encoding filters in an imaging system was first introduced by Rheinberg (ref. 9.8) in 1896. He reported the application of color filtering in microscopy. More recently, Bescos and Strand (ref. 9.9) proposed a white-light color encoding technique using an extended polychromatic light source. The proposed technique encodes the image by spatial frequency instead of density. We have introduced an alternative technique utilizing the one-step rainbow holographic recording arrangement (refs. 9.10, 9.11). The encoded images can be reconstructed with a simple white-light source but the encoding is also by spatial frequency. In this section we introduce a simple white-light processing technique that can be used to encode black-and-white pictures by density into a large variety of color codes.

In a recent paper, Liu and Goodman (ref. 9.12) describe a technique utilizing a specially fabricated halftone screen (sec. 7.6) to perform pseudocolor encoding by density. The halftone image is processed by two superimposed laser beams of different wavelengths. By selectively combining different diffraction orders of the two wavelengths, it is shown that pseudocolor encoded images can be obtained. We demonstrate that this technique can be implemented with a white-light processor (ref. 9.13). In addition, white-light processing offers some important advantages.

A halftone transparency of the original image is first obtained using a specially constructed halftone screen as described in sec. 7.6. The amplitude transmittance of the halftone image in one dimension can be described as

$$h(x) = \sum_{n=-\infty}^{\infty} \delta(x - nx_0) * \operatorname{rect}\left(\frac{x}{w}\right), \tag{9.11}$$

where $\sigma(x)$ is the Dirac delta function, x_0 is the pulse period, w is the pulse width, and

$$\operatorname{rect}\left(\frac{x}{w}\right) = \begin{cases} 1, & \left|\dfrac{x}{w}\right| \leq \dfrac{1}{2} \\ 0, & \left|\dfrac{x}{w}\right| > \dfrac{1}{2}. \end{cases}$$

Note that the pulse period x_0 is determined by the period of the halftone screen and the pulse width w is a function of the density of the original image at x. The halftone image transparency is then inserted into the white-light processor shown in fig. 9.9. The complex amplitude distribution at the back focal plane of the

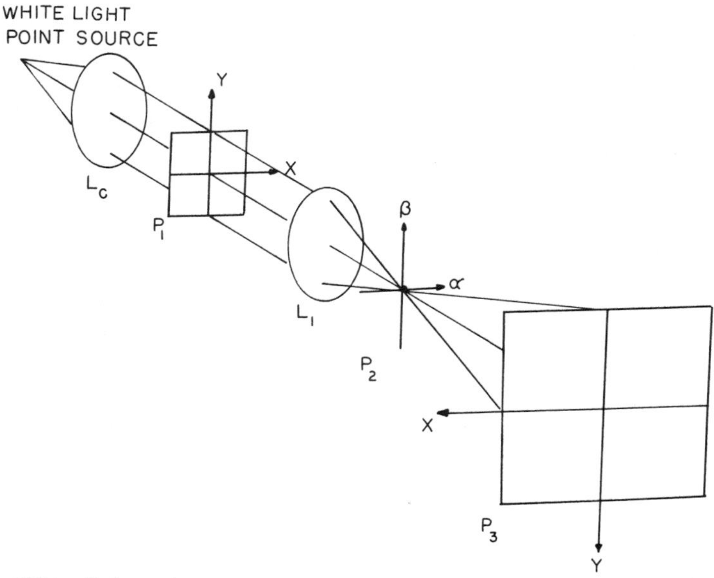

Fig. 9.9 White-light optical processor. L, achromatic collimating lens; L_1, achromatic transform lens; P_1, halftone input transparency; P_2, spatial frequency plane; P_3, magnified output image.

achromatic lens L_1 can be written as

$$E(\alpha, \lambda) = \exp\left[-iK_1\left(\frac{\alpha}{\lambda}\right)^2\right] \int h(x) \exp\left(-i\frac{2\pi}{\lambda f}\alpha x\right) dx, \qquad (9.12)$$

where α is spatial frequency coordinate, K_1 is an appropriate constant, and f is the focal length of the lens. The integration is performed over the spatial limits of the input transparency and the spectral width of the polychromatic light source. Thus for a given wavelength λ, the various orders of diffracted light at the spatial frequency plane are

$$E(\alpha, \lambda) = \exp\left[-iK_1\left(\frac{\alpha}{\lambda}\right)^2\right] \sum_{n=-\infty}^{\infty} \frac{\lambda^2 f^2}{\pi\alpha x_0} \sin\left(\frac{\pi\alpha w}{\lambda f}\right) \delta\left(\alpha - \frac{n\lambda f}{x_0}\right). \qquad (9.13)$$

Thus we can see that, except for the zero-order term, the higher order diffractions are dispersed into rainbow colors along the α direction at the spatial frequency plane p_2. Within the visible region of the diffracted light, there is no overlapping between diffraction orders up to the third order. For a given wavelength the irradiance of the nth diffraction order at $\alpha = n\lambda f/x_0$ is

$$I_n = \left(\frac{\lambda f}{n\pi}\right)^2 \sin^2\left(\frac{n\pi w}{x_o}\right). \qquad (9.14)$$

To perform pseudocolor encoding, we may select two different colors (e.g., red, green) from two different orders of diffraction by placing narrow slits at the appropriate positions to bandpass the desired colors. At the output plane P_3, we have a pseudocolor image formed by the addition of the intensities of the two filtered color images. Note that the result obtained with this method would be essentially the same as that obtained with the coherent processing technique used by Liu and Goodman (ref. 9.12); however, there are several important advantages in using white-light processing. First, the system is significantly cheaper, requiring only a single white-light point source instead of two different lasers. Second, the coherent technique requires a precise alignment of the two laser beams. With the white-light technique, all the colors are produced by the same light source and the precise alignment of the different color lights is, therefore, not necessary. Since all visible colors are contained in the white-light source, we can choose a large variety of color combinations for the different diffraction orders to create many different color codes. The most important advantage of the white-light processing technique is the freedom from coherent noises that plague coherent optical processing systems.

In our experiments a tungsten arc lamp is used as the light source and a specially fabricated one-dimensional multilevel halftone screen is used to produce the halftone input, as shown in fig. 9.10. For simplicity, we use a single slit with adjustable width as the spatial filter. The slit is placed between the

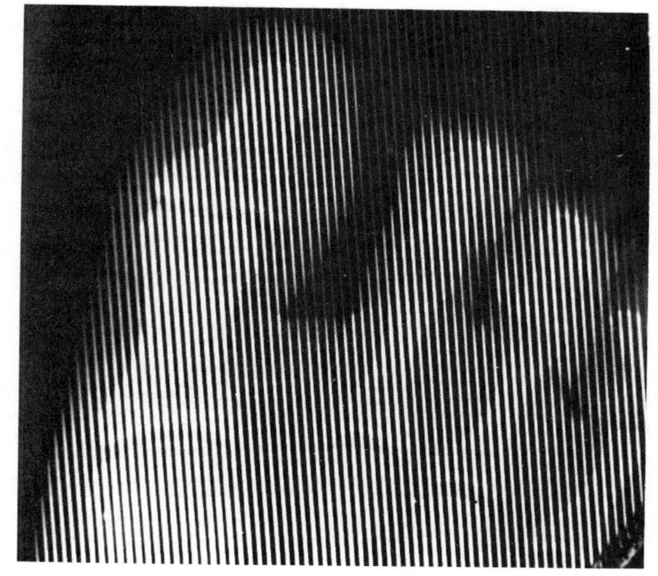

Fig. 9.10 Halftone input transparency of an X-ray picture.

second and third orders as shown in fig. 9.11, such that only the red portion of the second-order diffraction and the blue-green portion of the third-order diffraction can pass through. In Color Plate II*a*, we show a pseudocolor encoded image obtained with the white-light technique. For comparison, we show in Color Plate II*b* the encoded image obtained using a coherent processing system. The image obtained with the white-light technique is much cleaner, due to the lack of coherent noises.

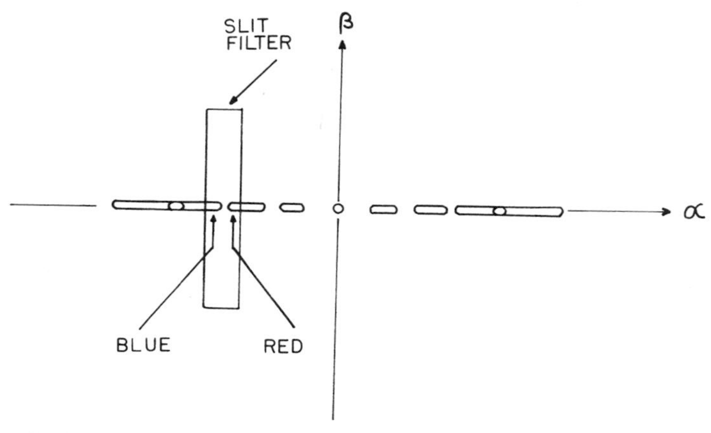

Fig. 9.11 Spatial filtering of dispersed spectrum by adjustable slit filter.

9.4 PSEUDOCOLOR DENSITY ENCODING THROUGH CONTRAST REVERSAL

In the previous section we have shown that pseudocolor density encoding can be achieved by using the halftone technique. However, with this method there is a spatial resolution loss as well as a reduction in the number of discrete lines that are due to image sampling and are generally present in the color coded image. Thus small details and low contrast features of the image can be lost in the halftone technique. A new density pseudocoloring technique through contrast reversal has been reported recently by Santamaria et al. (ref. 9.14). Although this technique offers several advantages over the previous technique, the optical system is quite elaborate and it requires an incoherent and a coherent source for the pseudocoloring. Because it utilizes a coherent source, coherent artifacts cannot be avoided.

In this section we present a simple white-light processing technique for pseudocolor density encoding (ref. 9.15). No marginal resolution is lost by the use of this technique. In addition, the system is simple, versatile, and economical to operate. The system also offers a direct viewing capability, but it is not a real-time pseudocolor encoding system.

Let us now describe a technique of contrast reversal encoding for gray level black-and-white photographic images for the white-light pseudocolor coding. Assume that negative and positive image transparencies of the same object are available. Now let the encodings take place on a fresh photographic film by sequentially recording the negative and the positive image transparencies with a Ronchi grating in two angular positions. For example, the first recording could be exposed with the negative transparency at 0° Ronchi grating position, and the second recording could be made with the positive transparency at a 90° grating position, as illustrated in fig. 9.12. If this spatial encoding is properly recorded in the linear region of the T-E curve of fresh photographic film, then we have

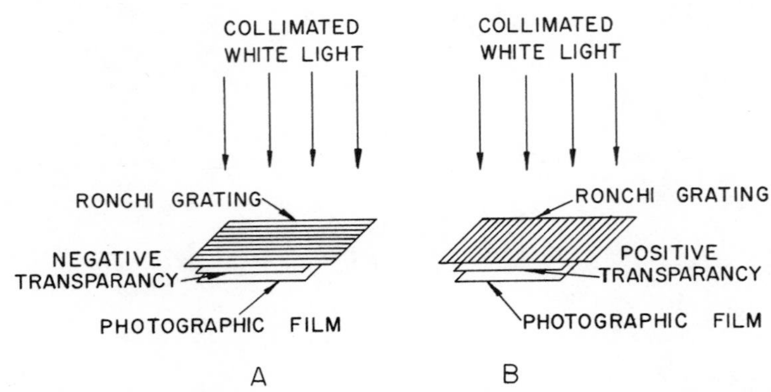

Fig. 9.12 Contrast reversal spatial encoding. (*a*) 0° position, (*b*) 90° position.

a multiplex spatially encoded contrast reversal transparency. The corresponding amplitude transmittance is

$$T(x, y) = K_1\{I_n(x, y)[1 + \operatorname{sgn}(\cos p_0x)] + I_p(x, y)[1 + \operatorname{sgn}(\cos q_0y)]\}, \tag{9.15}$$

where

$$\operatorname{sgn}(\cos p_0x) \triangleq \begin{cases} 1, & \cos(p_0x) \geq 0 \\ -1, & \cos(p_0x) < 0 \end{cases},$$

$p_0 = q_0$ is the spatial frequency of the grating, $I_n(x, y)$ and $I_p(x, y)$ are the corresponding negative and positive image irradiances, (x, y) is the coordinate system of the film, and K_1 is a proportionality constant. From eq. (9.15) we see that the amplitude transmittance of the recorded transparency is linearly related to the spatially encoded negative and positive image irradiances.

Let us describe a white-light processing technique for pseudocolor density encoding. If we place the spatially encoded transparency $T(x, y)$ at the input plane P_1 of a white-light processor, as shown in fig. 9.13, the complex amplitude light distribution for a given wavelength λ at the spatial frequency plane P_2 will be

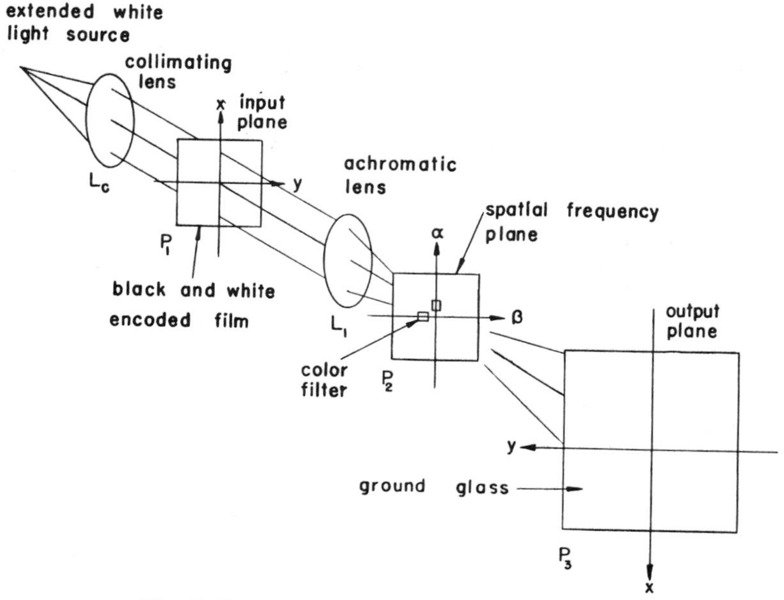

Fig. 9.13 A white-light pseudocolor encoder.

$$E(p, q; \lambda) = I_n(p, q) + \frac{1}{2}\sum_{m=1}^{\infty} a_m\tilde{I}_n(p \pm mp_0, q) + \tilde{I}_p(p, q) + \frac{1}{2}\sum_{m=1}^{\infty} a_m\tilde{I}_p(p, q \pm mq_0), \qquad (9.16)$$

where $\tilde{I}_n(p, q)$ and $\tilde{I}_p(p, q)$ are the corresponding Fourier transformations of $I_n(x, y)$ and $I_p(x, y)$, respectively, and the a_m's are the Fourier coefficients. Equation (9.16) can be written in terms of spatial coordinates (α, β), i.e.,

$$E(\alpha, \beta; \lambda) = \tilde{I}_n(\alpha, \beta) + \frac{1}{2}\sum_{m=1}^{\infty} a_m\tilde{I}_n\left(\alpha \pm \frac{m\lambda f}{2\pi}p_0, \beta\right) + \tilde{I}_p(\alpha, \beta) + \frac{1}{2}\sum_{m=1}^{\infty} a_m\tilde{I}_p\left(\alpha, \beta \pm \frac{m\lambda f}{2\pi}q_0\right). \qquad (9.17)$$

From this equation we see that orders of negative and positive image spectra are dispersed in rainbow colors on the α and β axes, respectively. If the spatial frequency of the grating is more than twice as high as the spatial frequency content of the image transparencies, the smeared image spectra will be spatially separated.

In pseudocolor encoding we would filter *two* first-order smeared spectra through green and red color filters, respectively, as illustrated in fig. 9.14. Then the complex amplitude light distribution immediately behind the filters becomes

$$E(\alpha, \beta) = \tilde{I}_n\left(\alpha - \frac{\lambda_g f}{2\pi}p_0, \beta\right) + \tilde{I}_p\left(\alpha, \beta + \frac{\lambda_r f}{2\pi}q_0\right), \qquad (9.18)$$

where λ_g and λ_r represent the green and red color wavelengths. Thus at the

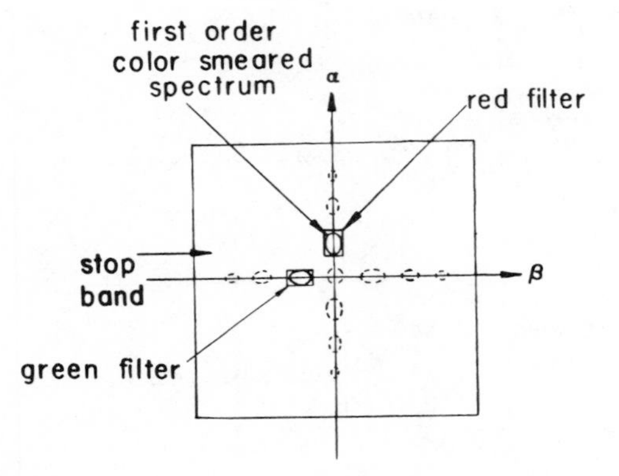

Fig. 9.14 A spatial color encoding filter.

output plane of the processor, we have the image irradiance

$$I(x, y) = I_{ng}^2(x, y) + I_{pr}^2(x, y), \tag{9.19}$$

where I_{ng} and I_{pr} are the *green color negative* and *red color positive* image irradiances, respectively. The output image is the superposition of a green negative image with a red positive image. Thus a broad range of pseudocolor density images can be obtained by this simple white-light processing technique. Note that the selection of color filters is arbitrary, that is, for different color filters we would obtain different shades of pseudocolor coded images. Note also that this white-light processing system offers the advantage of direct viewing capability, which provides an easy image accommodation in the electronic system. For example,a color television camera can easily pick up the pseudocolor coded image and encode it for transmission or storage.

In our experiments a xenon arc lamp is used as the extended light source, and spatially modulated contrast reversal transparencies (as described earlier) are used for the pseudocolor density encoding. In our first experimental demonstration a gray level variation pattern as shown in fig. 9.15*a* is used as a test object. The black-and-white photograph of the color result obtained by this white-light pseudocolor encoding technique is shown in fig. 9.15*b*. The dark region in the original test object appears red in color, the light region appears green, and the intermediate gray level regions appear pink, yellow, and light green. Thus we see that this white-light pseudocoloring technique indeed encodes the gray level images. Since the pseudocolor encoding is primarily obtained from a white-light source, the coherent artifact noise is avoided.

In our second experimental demonstration, we provide two pseudocolor encoded X-ray images, as shown in Color Plate III. From these pseudocolor images we see that different color combinations can be obtained with this technique and that the color coded images appear free from coherent artifact noise.We further note that multicolor filters can also be implemented in the Fourier spectral bands. For example, to obtain a different shade of pseudocolor, we may insert different color filters in different orders of spectral bands. Because different color filtered spectral bands are mutually incoherent, a broad range of color combinations can be obtained by this white-light pseudocoloring technique.

9.5 REAL-TIME PSEUDOCOLOR ENCODING

In previous sections we have described a simple white-light processing technique by which pseudocolor density encoding through contrast reversal can be easily obtained. We have shown that the white-light pseudocoloring technique offers no marginal resolution loss and the system is very simple to operate.

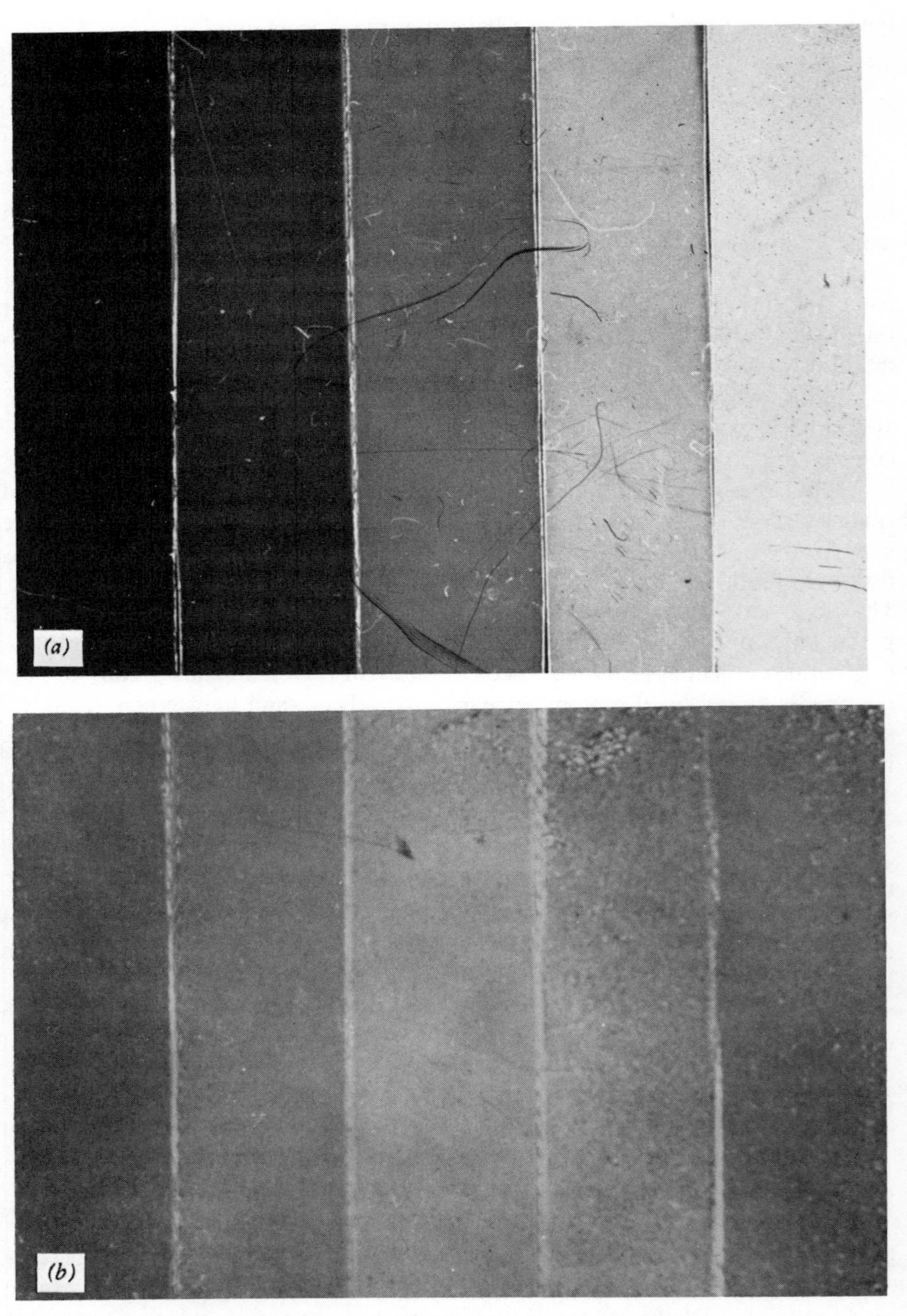

Fig. 9.15 (*a*) Gray level test pattern. (*b*) Black-and-white photograph of a pseudocolor test pattern. In color, the left bar is red, the intermediate bars are pink, yellow, and light green, and the right bar is green.

However, the technique still suffers one major drawback; it is not a real-time pseudocolor encoding technique.

We now describe a real-time white-light pseudocoloring technique for *spatial frequency* and *density* encodings (ref. 9.16). We show that this technique also offers no marginal resolution loss and, in addition, coherent artifact noise can be avoided. The system is very simple to operate, and it provides the advantage of direct-viewing capability. Furthermore, this real-time encoding technique does not involve color filtering nor does it introduce color dye; the pseudocoloring is primarily derived from the colors of the white-light source. Thus natural pseudocolor images can be obtained by this technique. We emphasize that this real-time white-light pseudocoloring technique may offer all the major advantages that previous techniques have offered.

Spatial Frequency Pseudocoloring

In spatial frequency pseudocolor encoding, we place a gray level black-and-white transparency $s(x, y)$ in contact with a two-dimensional diffraction grating $T(x, y)$ at the input plane P_1 of a white-light optical processor, as shown in fig. 9.16. For simplicity, we assume that the amplitude transmittance of the two-dimensional diffraction grating is

$$T(x, y) = 1 + \tfrac{1}{2} \cos p_0 x + \tfrac{1}{2} \cos q_0 y, \tag{9.20}$$

where p_0 and q_0 are the carrier spatial frequencies of the diffraction grating. The

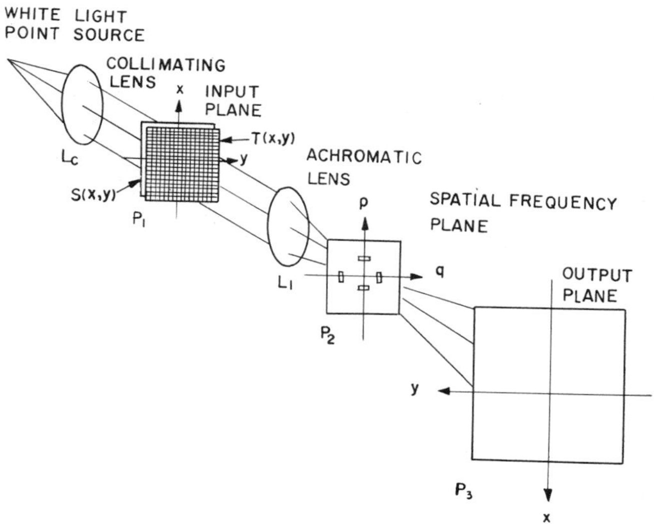

Fig. 9.16 A real-time white-light pseudocolor encoder.

corresponding complex light distribution for a given wavelength λ at the spatial frequency plane P_2 is

$$E(p, q; \lambda) = S(p, q) + \tfrac{1}{4}[S(p - p_0, q) + S(p + p_0, q) + S(p, q - q_0) + S(p, q + q_0)], \qquad (9.21)$$

where $S(p, q)$ is the Fourier spectrum of $s(x, y)$ and, for simplicity, we disregard the proportionality constants.

Since $p = (2\pi/\lambda f)\alpha$, $q = (2\pi/\lambda f)\beta$, and (α, β) is the linear spatial coordinate system of the spatial frequency plane, eq. (9.21) can be written as

$$E(p, q; \lambda) = S(\alpha, \beta) + \tfrac{1}{4}\left[S\left(\alpha - \frac{\lambda f}{2\pi}p_0, \beta\right) + S\left(\alpha + \frac{\lambda f}{2\pi}p_0, \beta\right) + S\left(\alpha, \beta - \frac{\lambda f}{2\pi}q_0\right) + S\left(\alpha, \beta + \frac{\lambda f}{2\pi}q_0\right)\right]. \qquad (9.22)$$

From the above equation, we see that four first-order signal spectra are dispersed in rainbow color proportional to wavelength λ along with the α and β axes. Since the spatial filtering is effective in the direction perpendicular to the color smeared spectrum (sec. 8.1), we adopt one-dimensional spatial filters for pseudocolor encoding as shown in fig. 9.17. The complex light amplitude distribution immediately behind the spatial frequency plane is then

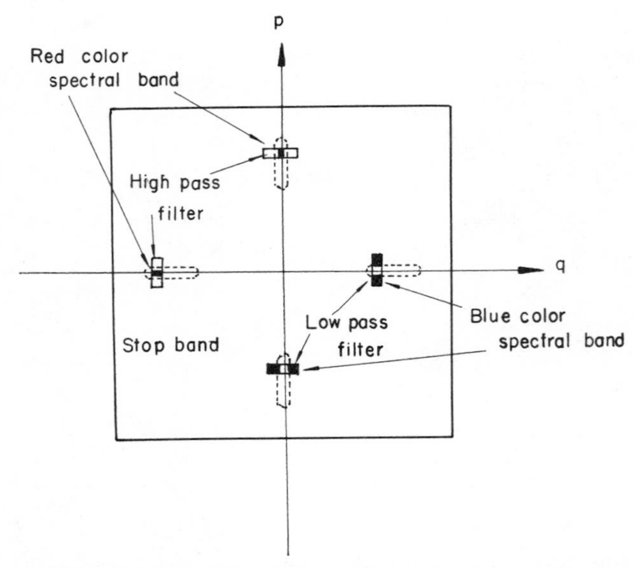

Fig. 9.17 Spatial frequency pseudocolor encoding.

$$E(p, q; \lambda) = S_r(p - p_0, q)H_1(q) + S_r(p, q - q_0)H_1(p)$$
$$+ S_b(p + p_0, q)H_2(q) + S_b(p, q + q_0)H_2(p), \qquad (9.23)$$

where S_r and S_b are the selected color-band signal spectra (e.g., red and blue), and H_1 and H_2 are the one-dimensional spatial filters. The corresponding complex light distribution over the output plane is

$$g(x, y; \lambda) = \iint [S_r(p - p_0, q)H_1(q) + S_r(p, q - q_0)H_1(p)]$$
$$\exp[i(px + qy)]\, dp\, dq + \iint [S_b(p + p_0, q)H_2(q)$$
$$+ S_b(p, q + q_0)H_2(p)] \exp[i(px + qy)]\, dp\, dq, \qquad (9.24)$$

where the integrals are over the spatial frequency plane. We note that, if the spatial carrier frequencies p_0 and q_0 of the grating are sufficiently high, then eq. (9.24) may be approximated by

$$I(x, y) \simeq \Delta\lambda_r |\exp(ip_0x)s_r(x, y) * h_1(y) + \exp(iq_0y)s_r(x, y) * h_1(x)|^2$$
$$+ \Delta\lambda_b |\exp(-ip_0x)s_b(x, y) * h_2(y) + \exp(-iq_0y)s_b(x, y) * h_2(x)|^2, \qquad (9.25)$$

where $\Delta\lambda_r$ and $\Delta\lambda_b$ are the color (e.g., red and blue) spectral bands of the signal spectra and h_1 and h_2 are the corresponding impulse responses of H_1 and H_2. Thus from the above equation, we may see that two spatial filtered images are *incoherently* added together to form a color encoded image at the output plane P_3 of the white-light processor.

In our experiments a xenon arc lamp with a 300-μm pinhole is used as a white-light point source. Figure 9.18*a* shows a black-and-white picture of a spatial frequency pseudocolor encoded radar image obtained by this white-light processing technique. Note that the high spatial frequency components are red and the low spatial frequency components are blue. Figure 9.18*b* shows a black-and-white copy of a spatial frequency color coded image of a *negative* radar image transparency obtained with this technique. Again the high spatial frequency components are red and the low spatial frequency components are blue. It is interesting to note that in fig. 9.18*b* the "lake" of the color coded radar image appears to be blue, while in fig. 9.18*a* the lake is black. Note also that, since the pseudocolor encoding is obtained using a white-light source, the coherent artifact noise is substantially suppressed.

Density Pseudocolor Encoding

We would now like to present a real-time white-light *density* pseudocolor encoding technique. In density pseudocoloring we insert two narrow strips of half-wave phase objects in the centers of the selected color-band signal spectra

Fig. 9.18 Real-time spatial frequency pseudocoloring radar image. In color the high spatial frequency terrains are red and the low spatial frequency terrains are blue. (*a*) A black-and-white picture of spatial frequency pseudocolor encoding radar image. (*b*) A black-and-white picture of a pseudocolor negative radar image.

to provide the image contrast reversal, as shown in fig. 9.19. The complex amplitude light distribution immediately behind the spatial frequency plane is

$$E(p, q, \lambda) = S_r(p - p_0, q) + S_r(p, q - q_0) + S_g(p - p_0, q)\, H(q) + S_g(p, q - q_0)H(p), \qquad (9.26)$$

where S_r and S_g are the selected color-band signal spectra (e.g., red and green) and

$$H(q) = \begin{cases} -1, & q \simeq 0 \\ 1, & \text{otherwise} \end{cases}, \qquad H(p) = \begin{cases} -1, & p \simeq 0 \\ 1, & \text{otherwise} \end{cases} \qquad (9.27)$$

are the narrow strips of half-wave phase objects. At the output imaging plane P_3 of the white-light processor, the complex light amplitude distribution is

$$g(x, y; \lambda) = \iint [S_r(p - p_0, q) + S_r(p, q - q_0)] \exp[i(px + qy)]\, dx\, dy$$
$$+ \iint [S_g(p - p_0,q)H(q) + S_g(p, q - q_0)H(p)] \exp[i(px + qy)]\, dx\, dy, \qquad (9.28)$$

where the integrals are over the spatial frequency plane. Again, if we assume that the spatial carrier frequencies of the grating are sufficiently high, then eq.

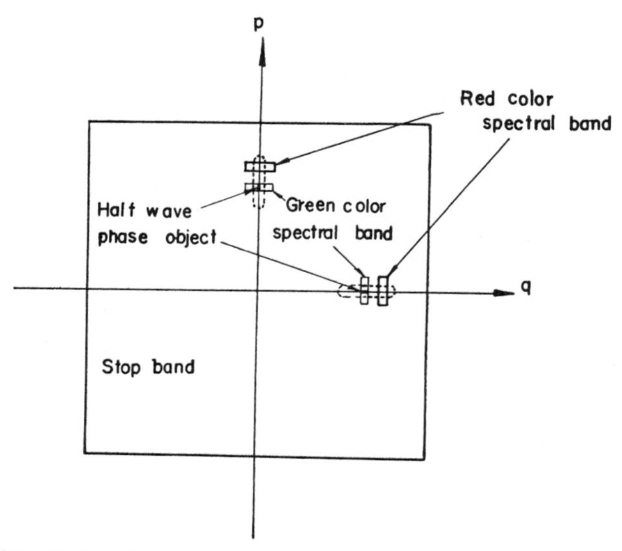

Fig. 9.19 Contract reversal density pseudocolor encoding.

(9.28) can be approximated by

$$g(x, y; \lambda) = [\exp(ip_0x) + \exp(iq_0y)]s_r(x, y) + [\exp(ip_0x) + \exp(iq_0y)]s_{gn}(x, y) \tag{9.29}$$

where $s_{gn}(x, y)$ is the (approximate) color contrast reversed image, i.e.,

$$s_{gn}(x, y) = s_g(x, y) - 2\langle s_g(x, y)\rangle, \tag{9.30}$$

where $\langle s_g(x, y)\rangle$ denotes the spatial ensemble average (i.e., the dc level) of $s_g(x, y)$. Since the two images s_r and s_{gn} are diffracted from two different color spectral bands of the light source, they are mutually incoherent. The output image irradiance is, therefore;

$$I(x, y) = \int |g(x, y; \lambda)|^2 \, d\lambda = \Delta\lambda_r I_r(x, y) + \Delta\lambda_g I_{gn}(x, y), \tag{9.31}$$

where $I_r(x, y)$ is a positive color (e.g., red) image irradiance, $I_{gn}(x, y)$ is an (approximate) contrast reversed or negative color (e.g., green) image irradiance, $\Delta\lambda_r$ and $\Delta\lambda_g$ are the narrow color spectral bands of the signal spectra (e.g., red and green). Thus we see that a density pseudocolor coded image is formed with incoherent addition of a positive image in one color and a negative image in another.

In our experiments a xenon arc lamp with a 300-μm pinhole is again used as a white-light point source,and the phase reversal object is about 200 μm wide. Figure 9.20 shows a black-and-white picture of a real-time pseudocolor density encoded X-ray image of a hand. From this figure we see that a broad range of density pseudocolor encoded images can be obtained with this technique. Since the pseudocolor encoding is accomplished with a white-light source,the color encoded images appear free from the coherent artifact noise. We also stress that, in this white-light pseudocolor encoding technique, there is the freedom to select different color spectral bands. Thus in practice a wide range of different pseudocolor encoded images can easily be obtained.

9.6 VISUALIZATION OF PHASE OBJECTS THROUGH COLOR CODING

The study of phase objects generally depends upon interferometric techniques (refs. 9.17, 9.18). For slow object phase variation, the measurements can be very accurate with such techniques (ref. 9.19). However, the study of phase objects that contain fine structure, such as living cells, is exceedingly difficult, if not impossible, when we try to interpret with the interferometric techniques.

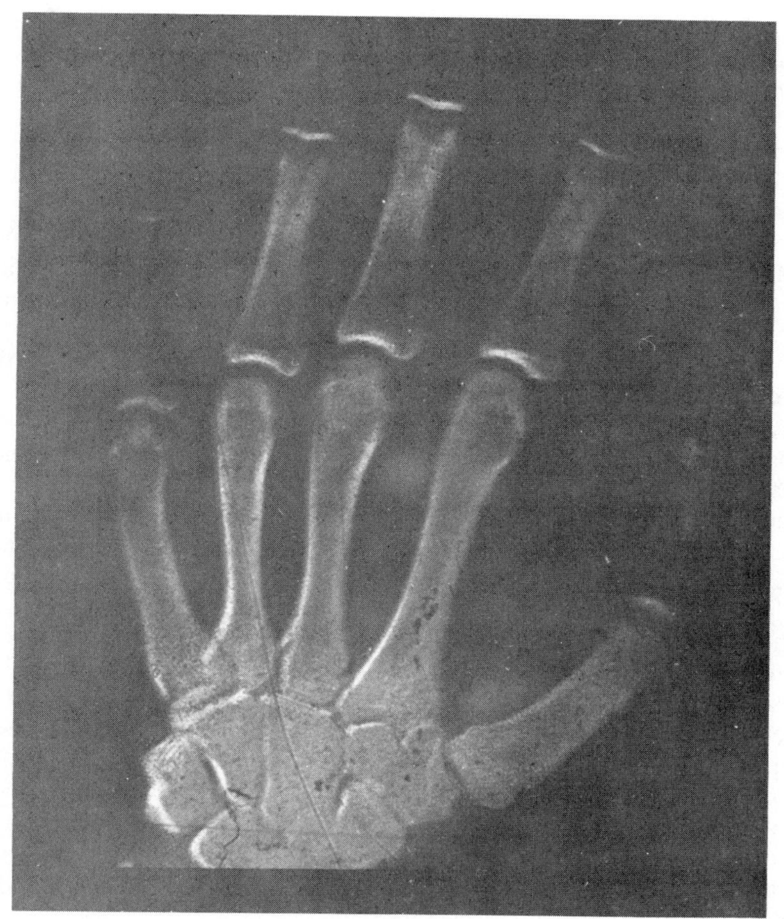

Fig. 9.20 A black-and-white picture of a density pseudocolor encoding image of an X-ray transparency. In color the thicker bones are displayed in red and the fingers are green.

Frequently, the study of this type of phase object requires a special procedure to produce an image irradiance that is related to the phase of the object.

The Schlieren technique (ref. 9.20) and differential shearing interference microscopy (refs. 9.21, 9.22) both produce image irradiances that are proportional to the derivative of the object phase. Although the differential techniques provide visualization of shapes and sizes of the phase objects, they are not able to detect the detail phase variation between fringes. The total shearing interference microscope is, however, able to produce fringe patterns superimposed on a phase object, but the application is limited to evaluation of simple isolated phase structure, and it is difficult to apply to the more general case of phase variation.

There is a technique available for detecting a phase object whose image irradiance is proportional to the phase variation (ref. 9.23). However, this technique also suffers one major drawback: the application is limited to very small phase variation in the order of a fraction of a wavelength.

The concept of detecting a large phase variation with differentiation and integration to obtain an image irradiance proportional to phase variation was proposed by DeVelis and Reynolds (ref. 9.20). This concept has been subsequently tested by Spraque and Thompson (ref. 9.24) with a coherent optical processing technique. They achieved good results, obtaining an image irradiance that was proportional to the large object phase variation. However, the technique is rather elaborate and the phase detection is not a real-time operation. Moreover, the system utilizes a coherent source, and the annoying coherent artifact noise cannot be avoided. Although a multibeam interferometric technique (ref. 9.18) is able to produce sharper fringe patterns, the technique also suffers drawbacks. The technique still cannot provide the phase variation between the fringes with a single monotone fringe pattern. It is capable of detecting the phase variation between the fringes, by simultaneously changing the viewing angle, reference beam, or wavelength of the light source. In a recent paper Roblin and El Sherif (ref. 9.25) describe an electronic scanning technique, based on phase modulation interferometry, for detecting phase variation between the fringes. But this scanning technique requires complicated electronic circuitry and the operation is rather cumbersome.

In sec. 8.6 we have shown a technique of encoding an extended incoherent source for image subtraction. We extend the same basic optical processing concept to the detection of object phase variation with a pseudocolor encoding technique (ref. 9.26). Since this technique utilizes incoherent sources, high quality color coded phase variation patterns can be visualized. We stress that this incoherent processing technique may alleviate certain drawbacks inherent to previous techniques, and the processing system is rather simple and economical to operate.

Let us now describe an incoherent optical processing technique for object phase detection, as depicted in fig. 9.21. From this figure we see that two encoded extended incoherent sources for different colors of light (i.e., red and green) are used for the processing. The purpose of using a source encoding mask to achieve the point-pair spatial coherence requirement for an image subtraction operation has been discussed in sec. 8.6.

In pseudocolor encoding we insert a phase object $\exp[i\phi(x, y)]$ in one of the open apertures in the input plane P_2. Two sinusoidal gratings G_1 and G_2 are properly placed in the spatial frequency plane P_3:

$$G_1 = 1 + \sin(h_0 p_r + \theta) \tag{9.32}$$

and

$$G_2 = 1 + \sin(h_0 p_g), \tag{9.33}$$

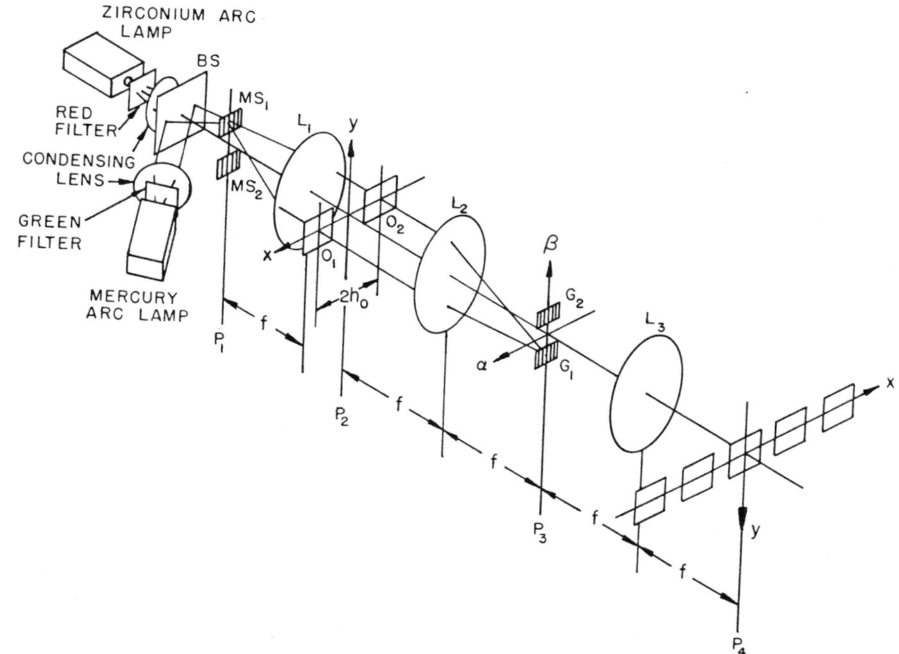

Fig. 9.21 A multicolor incoherent processing system for detecting object phase variation. *BS*, beam spliter; MS_1 and MS_2 source encoding masks; L_1, L_2 and L_3, achromatic transform lenses; G_1 and G_2, diffraction gratings.

where $2h_O$ is the separation of the O_1 and O_2, $p_r = (2\pi\alpha/\lambda_r f)$ and $p_g = (2\pi\alpha)/\lambda_g f)$ are the spatial frequencies of the gratings, λ_r and λ_g are the red and green color wavelengths, α denotes the spatial coordinate in the same direction as p, f is the focal length of the achromatic transform lens, and θ is a phase factor that we have introduced. θ will play an important role in the color encoding process.

By a straightforward computation, shown in sec. 8.6, the irradiance around the origin of the output image plane P_3 can be shown to be

$$\begin{aligned} I(x, y) &= I_r(x, y) + I_g(x, y) \\ &= K\{1 - \cos[\phi_r(x, y) + \theta]\} + K\{1 - \cos[\phi_g(x, y)]\}, \quad (9.34) \end{aligned}$$

where I_r and I_g denote the red and green color image irradiances, $\phi(x, y)$ is the phase distribution of the object, θ is a constant phase factor that we have introduced (by shifting one of the gratings), and K is a proportional constant. The phase distributions $\phi_r(x, y)$ and $\phi_g(x, y)$ are slightly different because of different wavelength illuminations. From eq. (9.34) we see that a broad range color coded phase fringe pattern can be visualized.

In color mixing analysis, we would use I_r and I_g to form a two-dimensional orthogonal coordinate system, as shown in fig. 9.22. The variations of color

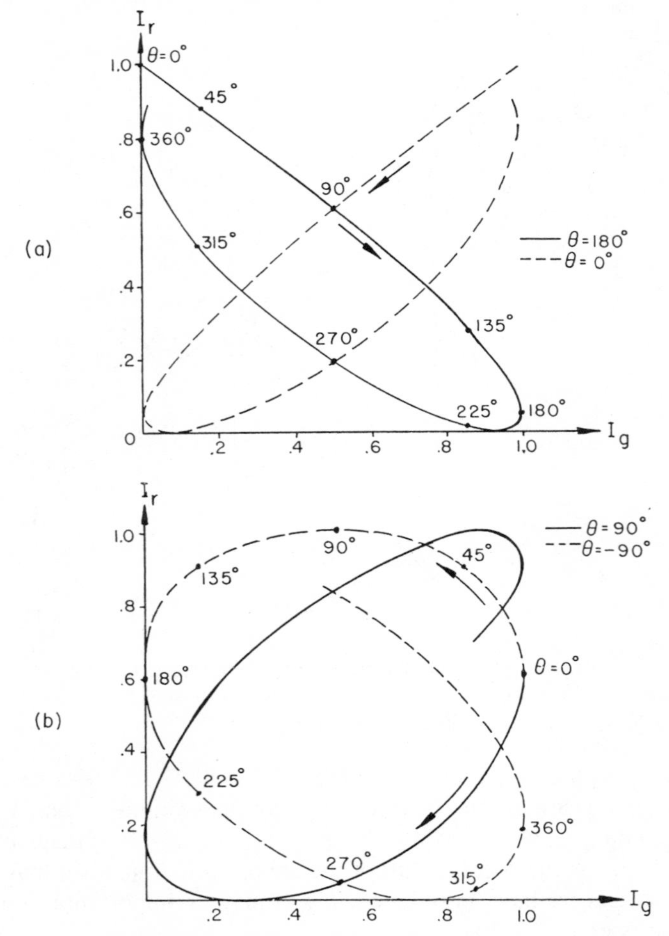

Fig. 9.22 Color encoding mixing curves.

mixing irradiance as a function of object phase variation θ for various values of θ are plotted. Since different wavelengths (i.e., red and green) are used in the color encoding, the color mixing curve is not generally enclosed for every 2π cycle as a function θ, as shown in the figure. In other words, the phase detection is more accurate within $-\pi \leq \phi \leq \pi$.

Consider a few cases of the color mixing procedure. For the example of $\theta = 0$, the locus of the color mixing curve lies near the 45° degree angle of (I_r, I_g) coordinates as shown in fig. 9.22*a*. There is no significant color change in this region to call for distinction of object phase variation. On the other hand, if $\theta = 180°$, the locus of the color mixing curve lies near the $-45°$ degree region (i.e., $I_r = 1 - I_g$) of the (I_r, I_g) coordinate system, and a broad range color variation can be perceived. However, in this region it is difficult to distinguish a positive or a negative phase variation (i.e., $\pm\phi$). If we let $\theta = 90°$, the locus of the color mixing curve extends outward around the edges of the (I_r, I_g) coordinate system, as shown in fig. 9.22*b*. Thus it provides not only a broad range color coded phase variation, but also establishes two different sequences of color coded bands so that the positive and negative phase variations can be detected. In other words, this color encoding technique is capable of detecting finer detail for phase variations, including positive and negative phases.

In experiments a mercury arc lamp with a green filter (5461 Å) and a zirconium arc lamp with a red filter (6328 Å) are used for the color light sources. The intensity ratio of these two color lights is adjusted to unity with a variable beam splitter.

Color Plate IV*a* shows a monotone fringe pattern of a phase object obtained with this incoherent processing technique. From this figure we see that detailed phase variation between the fringes is not perceptible. Color Plate IV*b* is obtained with a multicolor technique described earlier. The phase factor θ between the diffraction gratings G_1 and G_2 is set to about 90°. From this figure we see that a multicolor phase fringe pattern is formed so that a more detailed phase variation between the fringes can be observed.

Between the two yellow color bands there is a color variation from yellow-to- red-dim yellow-green and back to yellow again. The phase angle between these two color bands is 2π. These color variations correspond to the color mixing curve of fig. 9.22*b*; the corresponding phase variations with respect to the color bands are $0° \rightarrow 180° \rightarrow 225° \rightarrow 315° \rightarrow 360°$, as indicated in Color Plate IV*b*. Thus we see that there is a positive increasing phase variation from the top yellow band to the low yellow band. In another experiment, Color Plate V*a* shows the monotone phase fringe pattern obtained with this technique from a phase object; again the detailed phase variation between the fringes cannot be determined. If we take the cross section along with line A-A of the phase object between the two arrows, there are four possible phase variations that may be interpreted, as shown in fig. 9.23. It may not be possible to retrieve the actual object phase variation with a monotone fringe pattern. However, with the multicolor phase fringe pattern shown in Color Plate V*b*, we are able to identify the detailed phase variation between the fringes. If we take the same cross section A-A, a sequence of color bands from left to right, i.e., green-dim yellow-red-yellow-red, can be observed. Corresponding to the color mixing curve of fig. 9.22*b*, this sequence of color bands represents a sequence of phase angles of $315° \rightarrow 250° \rightarrow 180° \rightarrow 90° \rightarrow 180°$. Thus this region represents a phase depression object that corresponds to the case of fig. 9.23.

Object phase variation can be visually studied by a color coding technique with incoherent processing. This technique is accomplished with a complex image subtraction scheme using an incoherent optical processing method. The object phase distribution within $-\pi \leq \phi \leq \pi$ can be detected with a color coded phase variation with lesser ambiguity. This technique offers more finely detailed object phase detection than other interferometric techniques. Since it uses inexpensive incoherent sources, the multicolor technique may offer a wider range of applications. In the case of microscopic phase object observation, the

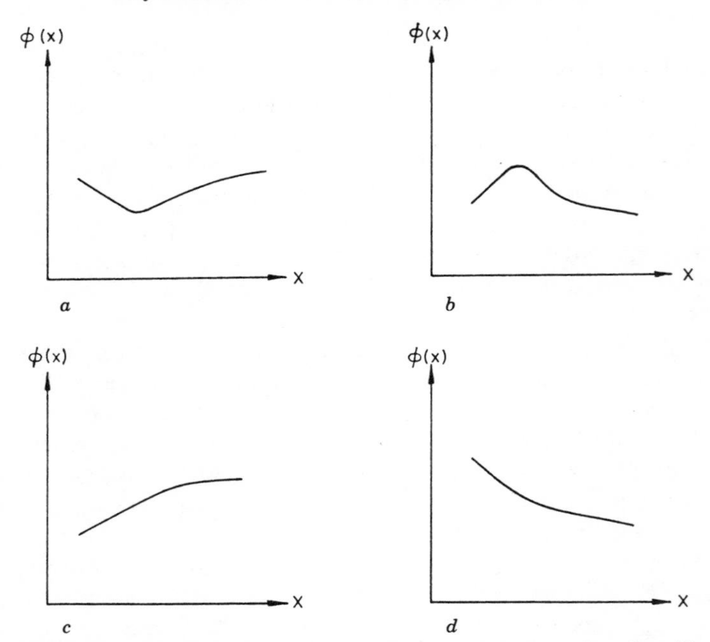

Fig. 9.23 Four possible object phase variations for color plate V*a*.

system can be designed to fit within a microscopic system. This technique may be used in other fields as well: study of birefringence and interference figures in polarizing microscopy and crystallography; distinction between compressed and stretched areas in photoelasticity; analysis of the aerodynamic pressure variations examined inside a wind tunnel by the stereoscopic technique; analysis of wave fronts to test the aberrations of an optical system; and others.

9.7 COLOR PHOTOGRAPHIC IMAGE DEBLURRING

We have pointed out earlier that the white-light processing technique is also suitable for color image processing, since the light source contains all the color spectral lines. Virtually all images to be processed, including those that are black-and-white, are multicolor. The aim of this section is to illustrate a white-light processing technique that will deblur a smeared color photographic image (ref. 9.27). We first describe a deblurring technique for a smeared color image transparency; then an experimental result is provided.

Let us describe a white-light processing technique to deblur a smeared color photographic image. With reference to the white-light optical processor of fig. 9.24, we place a smeared color photographic image transparency in contact with a sinuosidal phase grating at the input plane P_1. The complex light distribution

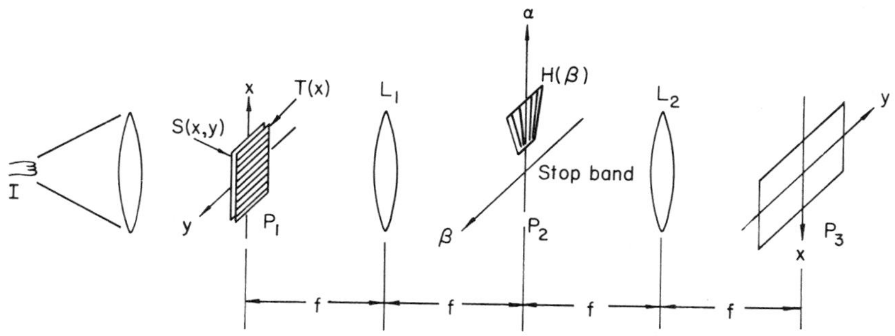

Fig. 9.24 A white-light optical information process. I, white-light point source; $s(x, y)$, smeared image transparency; $T(x)$, diffraction grating; L_1 and L_2, achromatic transform lenses; $H(\beta)$, deblurring spatial filter.

for every wavelength λ in the spatial frequency plane is

$$E(\alpha, \beta; \lambda) = C_1 S(\alpha, \beta; \lambda) + C_2 S\left(\alpha - \frac{\lambda f}{2\pi} p_0, \beta; \lambda\right) + C_3 S\left(\alpha + \frac{\lambda f}{2\pi} p_0, \beta; \lambda\right), \qquad (9.35)$$

where $S(\alpha, \beta; \lambda)$ is the Fourier spectrum of a monochrome smeared color image $s(x, y; \lambda)$, and C_1, C_2, and C_3 are the appropriate complex constants. For simplicity, assume that the smeared color image is spatial frequency limited, that the smearing is in the y direction of the input spatial plane, and that a fan-shaped deblurring filter is available to compensate for the wavelength variation. The complex amplitude transmittance of the deblurring filter may be described as

$$H(\beta; \lambda) = A(\beta; \lambda) \exp[i\phi(\beta, \lambda)], \qquad \alpha_1 \leq \alpha \leq \alpha_2, \qquad (9.36)$$

where $A(\beta; \lambda)$ and $\phi(\beta; \lambda)$ are the corresponding amplitude and phase filter functions (see sec. 8.4), and $\alpha_1 = (\lambda_1 P/2\pi)p_0$ and $\alpha_2 = (\lambda_2 P/2\pi)p_0$ are the lower and upper wavelengths of the filters.

If we insert the deblurring filter in the spatial frequency plane P_2 to cover the entire smeared color spectra $S[\alpha - (\lambda f/2\pi)p_0, \beta; \lambda]$, the complex light field at the output plane P_3, for a given wavelength λ, becomes

$$g(x, y; \lambda) = C \int_{\infty}^{\infty} \int_{\alpha_1}^{\alpha_2} S\left(\alpha - \frac{\lambda f}{2\pi} p_0, \beta; \lambda\right) H(\beta; \lambda) \exp\left[\frac{2\pi}{\lambda f}(\alpha x + \beta y)\right] d\alpha\, d\beta, \qquad (9.37)$$

where C is a complex constant. The corresponding output light intensity distribution is then

$$I(x, y) \simeq \int |g(x, y; \lambda)|^2 \, d\lambda, \tag{9.38}$$

where the integration is over the spectral band of the light source. From the above equation we see that the deblurred color image takes place over the broad spectral band of the light source. In other words, the deblurred images for every λ will be incoherently added to form a multicolor image at the output plane P_3.

Note that the technique of synthesizing a fan-shaped spatial filter is currently under investigation. However, a description of an equivalent multifilter technique for this smeared color image deblurring will be given. Assume that three primary color narrow spectral band deblurring filters are available. The corresponding complex amplitude transmittances are

$$H_r(\beta; \lambda_r) = A(\beta; \lambda_r) \exp[i\phi(\beta; \lambda_r)], \qquad \alpha_1 \le \alpha \le \alpha_2,$$

$$H_g(\beta; \lambda_g) = A(\beta; \lambda_g) \exp[i\phi(\beta; \lambda_g)], \qquad \alpha_1 \le \alpha \le \alpha_2,$$

and

$$H_b(\beta; \lambda_b) = A(\beta; \lambda_b) \exp[i\phi(\beta; \lambda_b)], \qquad \alpha_1 \le \alpha \le \alpha_2, \tag{9.39}$$

where $A(\beta; \lambda)$ and $\phi(\beta; \lambda)$ are the amplitude and phase functions, λ_r, λ_g, and λ_b are the red, green, and blue color wavelengths, $\alpha_1 = (\lambda f/2\pi)(p_0 + \Delta_p)$ and $\alpha_2 = (\lambda f/2\pi)(p_0 - \Delta_p)$ are the upper and lower spatial limits of the filters, and Δp is the spatial frequency limit of this smeared color image.

If we center these deblurring filters at the primary color wavelengths over the smeared Fourier spectra, respectively, as shown in fig. 9.25, the complex light

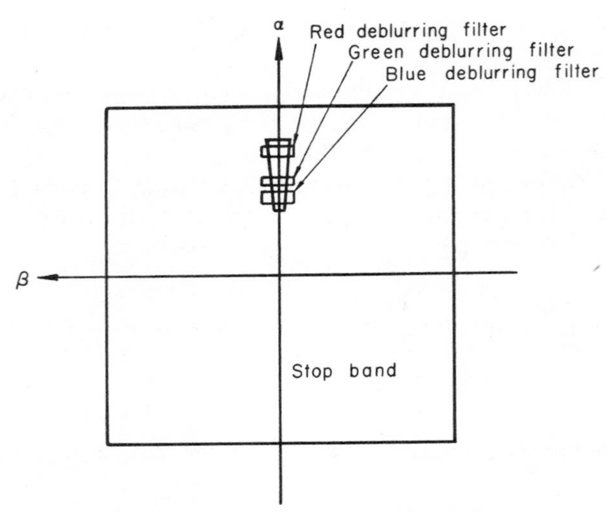

Fig. 9.25 Spatial frequency plane deblurring filtering.

distribution at the output plane p_3, for each primary wavelength, becomes

$$g(x, y; \lambda) = C \iint S\left(\alpha - \frac{\lambda f}{2\pi} p_0, \beta; \lambda\right) H(\beta; \lambda) \exp\left[\frac{2\pi}{\lambda f}(\alpha x + \beta y)\right] d\alpha\, d\beta, \tag{9.40}$$

The corresponding output light intensity distribution is then

$$I(x, y) \simeq \Delta\lambda[|g(x, y; \lambda_r)|^2 + |g(x, y; \lambda_g)|^2 + |g(x, y; \lambda_b)|^2], \tag{9.41}$$

where $\Delta\lambda \simeq \lambda(4\Delta p/p_0)$ is the narrow spectral band of the deblurring filters. From the above equation we see that a multicolor deblurred image is produced with the incoherent addition of the red, green, and blue deblurred images.

In our experimental demonstration a color object in motion is photographed. The assumption is made that the smeared color image is recorded in the linear amplitude region of a color photographic film. A xenon arc lamp with a 100μm pinhole is used as a white-light point source. For simplicity, only red and green primary color deblurring filters are used for our experimental illustration. The center wavelengths of this set of deblurring filters are about 6328 and 5461 Å, respectively. The filter bandwidth contains five main lobes and the spectral widths, which are about 100 Å.

An experimental result obtained with this white-light deblurring technique is depicted in fig. 9.26. From this result we see that the deblurring effect on the green letters "P" and "C" seems to be more effective than on the other colored letters. This is primarily due to the linear amplitude transmittance characteristic of these smeared letters, which can be seen in Fig. 9.26*a*. In other words, for the red and yellow letters, the colors are more saturated. If the smeared color photographic image is linearly recorded, a better color deblurred image can be obtained.

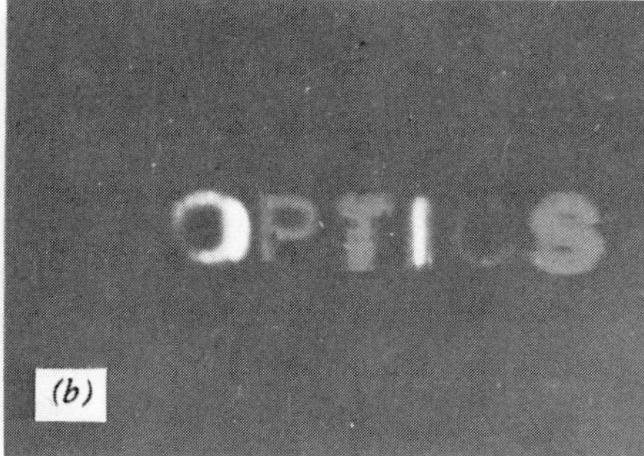

Fig. 9.26 Smeared color image restoration of the word "OPTICS." (*a*) A black-and-white picture of original smeared color image. In color the "O" and "I" are in yellow, "P" and "C" are in green, and "T" and "S" are in red. (*b*) A black-and-white picture of the deblurred color image.

9.8 COLOR IMAGE SUBTRACTION

In sec. 8.5 a technique of image subtraction has been demonstrated that utilizes an extended incoherent source. Because image subtraction is a one-dimensional processing operation, and because the spatial coherence requirement for the subtraction operation is a point-pair coherence requirement, it is possible to encode an extended source to obtain the required spatial coherence. In sec. 8.6 we have shown a Fourier transform relationship between the spatial coherence and source intensity distribution. In principle, it is possible to encode an extended source to obtain an appropriate spatial coherence for specific information processing operations.

Strictly speaking, all images in the visible wavelengths, which include the black-and-white images, are color images. Therefore, it is of interest to us, in this section, to extend this incoherent processing technique for color image subtraction (ref. 9.28).

A color image subtraction operation with encoded extended incoherent sources has been outlined in fig. 9.21 and sec. 9.6. For simplicity, two incoherent light sources, each for a different color of light (i.e., red and green), are used for the subtraction operation. For a detailed analysis of image subtraction with an encoded source, we refer to the reader to sec. 8.5.

In color image subtraction, the object transparencies $O_1(x, y)$ and $O_2(x, y)$ are inserted in the open apertures of the input plane P_2, which can be described as

$$f(x, y) = O_1(x - h_O, y) + O_2(x + h_O, y), \tag{9.42}$$

where $2h_O$ is the separation between the two input transparencies O_1 and O_2. Two sinusoidal gratings G_1 and G_2 designed for the red and the green colored wavelengths, respectively, are inserted in the spatial frequency plane P_3, and can be written as

$$G_1 = \tfrac{1}{2}[1 + \sin(h_O p_r)], \tag{9.43}$$

and

$$G_2 = \tfrac{1}{2}[1 + \sin(h_O p_g)], \tag{9.44}$$

where $p_r = (2\pi\alpha)/(\lambda_r f)$ and $p_g = (2\pi\alpha)/(\lambda_g f)$ are the spatial frequencies of the gratings, λ_r and λ_g are the red and green colored wavelengths, α denotes the spatial coordinate in the same direction as p, and f is the focal length of the achromatic transform lens L_2. By a straightforward but rather cumbersome evaluation, the irradiance around the origin of the output image plane P_4 can be shown to be (see sec. 8.5)

$$\begin{aligned} I(x, y) &= I_r(x, y) + I_g(x, y) \\ &= K|O_{1r}(x, y) - O_{2r}(x, y)|^2 + K|O_{1g}(x, y) - O_{2g}(x, y)|^2, \end{aligned} \tag{9.45}$$

where $I_r(x, y)$ and $I_g(x, y)$ denote the red and green subtracted color image irradiances, and O_{1r}, O_{2r}, O_{1g}, and O_{2g} are the corresponding red and green colored input objects. From eq. (9.45) we see that the subtracted color image can be obtained at the output image plane. Since the color image subtraction is obtained with extended incoherent sources, the coherent artifact noise is suppressed.

Strictly speaking, light sources emitting all primary colors (i.e., red, green, and blue) should be used for the color image subtraction. For simplicity of experimental demonstration, a mercury arc lamp with a green filter (5461 Å) and a zirconium arc lamp with a red filter (6328 Å) are used for the color light sources. The intensity ratio of the two resulting light sources is adjusted to approximate unity with a variable beam splitter.

The slit widths for the source encoding masks are about 2.5 μ and the spacing of the slits is 25 μ for the green wavelength and 29 μ for the red wavelength. The overall size of the source encoding masks is about 3×3 mm^2; thus the mask for the green wavelength contains about 120 slits and that for the red has about 100 slits. The focal length of the transform lenses is 300 mm. A liquid gate is placed behind the collimator. It contains the color image transparencies, each about 6×8 mm^2; they are separated by about 13.2 mm. Two sinusoidal gratings with spatial frequencies of $1/(25\ \mu)$ and $1/(29\ \mu)$ are used for the green and red color image subtraction operation, as shown in fig. 9.21.

In the first experimental demonstration, two sets of differently colored words are provided as input objects, as shown in black-and-white in figs. 9.27*a* and 9.27*b*. Figure 9.27*b* shows the subtracted color image obtained with the color subtraction technique. The input words "STATE UNIV." are red and green, respectively, in one set, but they are green and red in the other set, while in the subtracted image, as shown in fig. 9.27*c*, both words are yellow in color, which is consistent with the result we would expect. In other words, the subtracted image of red and green produces yellow, since the red and green wavelengths are incoherent and will add incoherently to produce yellow.

For a second demonstration, black-and-white pictures of two continuous tone color images of two sets of fruit are provided in figs. 9.28*a* and 9.28*b*. By comparing these two figures, we see that a dark green cucumber and a red tomato are missing in fig. 9.28*b*. Figure 9.28*c* shows the subtracted color image obtained with this incoherent color image subtraction technique. In this result the profiles of the cucumber and tomato can be seen at the output image plane.

For the final demonstration, two continuous tone color images of a parking lot are used as input color object transparencies, as shown in Color Plate VI*a* and *b*. From these input transparencies, we see that a red passenger car, shown in the parking lot in Color Plate VI*a*, is missing in Color Plate VI*b*. Color Plate VI*c* is the color subtracted image obtained from the incoherent color image subtraction technique previously described. In this figure a red passenger car can clearly be seen at the output image plane. It is also interesting to point out that the parking line (in yellow) on the right side of the red car can readily be seen with the subtracted image.

In concluding this chapter we stress that virtually all images are color images.

Fig. 9.27 Color image subtraction: binary object. (*a*) and (*b*) Black-and-white pictures of input color English words. (*a*) The word "STATE" is in red and "UNIV." is in green. (*b*) The word "PENN" is in yellow, "STATE" in green, and "UNIV." in red. (*c*) A black-and-white picture of the subtracted color image. In color all the words are in yellow.

The simplicity of the polychromatic processing technique with incoherent light will make broad range of applications of color image processing possible.

In spite of the flexibility of digital image processing, optical methods offer the advantages of capacity, simplicity,and cost. Instead of conflicting with each other, we expect a gradual merging of the optical and digital techniques. The continued development of optical-digital interfaces and input devices will lead to a fruitful result of hybrid optical-digital processing techniques, utilizing the strengths of both processing operations.

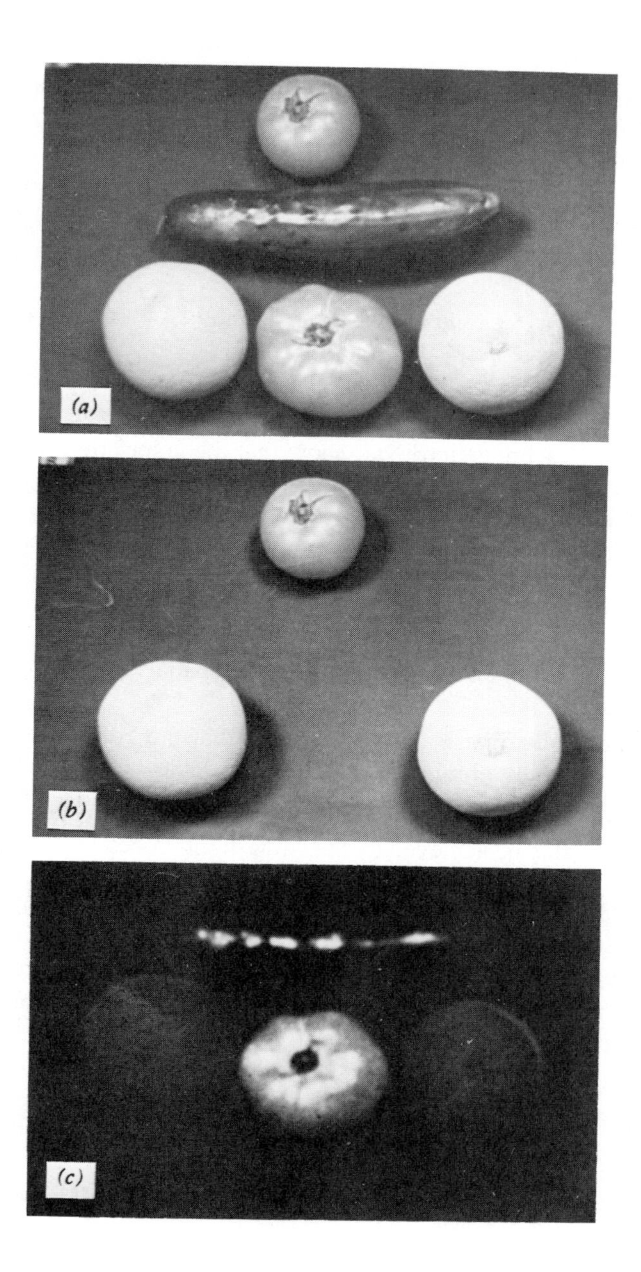

Fig. 9.28 Color image subtraction: continuous tone object. (*a*) and (*b*) Black-and-white pictures of input color object transparencies. In color, tomatoes are red, cucumber is green, and oranges are yellow. (*c*) A black-and-white subtracted color image. In color the images are reddish.

PROBLEMS

9.1 For simplicity, we assume that the image encoding for archival storage of color films takes place with sinusoidal grating. Let the amplitude transmittance of this encoded film be

$$t_p(x, y) = [T_r(x, y)(1 + \cos p_0x) + T_b(x', y')(1 + \cos p_0x') + T_g(x'', y'')(1 + \cos p_0x'')]^{\gamma/2},$$

where $\gamma = \gamma_{n1}\gamma_{n2}$, and (x, y), (x', y'), and (x'', y'') are 0°, 60°, and 120° coordinate systems.

(a) With reference to fig. 9.3, evaluate the corresponding complex Fourier spectra at the spatial frequency plane for $\gamma = 4$.

(b) In color image retrieval, show that the color cross talk may not be completely eliminated.

9.2 Given an image transparency containing only two primary color images (e.g., red and green). If the encoding takes place sequentially with a sinusoidal grating rotated by 90° so that the amplitude transmittance of the encoded film is $t_p(x, y) = [T_r(x, y)(1 + \cos p_0x) + T_g(x, y)(1 + \cos p_0y)]^{\gamma/2}$, where $\gamma = \gamma_{n1}\gamma_{n2}$, then:

(a) Evaluate the Fourier spectra at the spatial frequency plane for the case $\gamma \neq 2$.

(b) For color image decoding, show that the color cross talk can be completely eliminated.

9.3 With reference to the color image encoding process of sec. 9.1, the same objective may be achieved by the white-light processing technique shown in fig. 9.29, where W is the extended white-light source, $s(x, y)$ is the color transparency, $T(x, y)$ is a two-dimensional sinusoidal grating, and P_3 is the output image plane. For simplicity, we assume that the input color object transparency $s(x, y)$ contains only red and green colors.

(a) Calculate the corresponding Fourier spectra as a function of the spectral wavelength of the light source at plane p_2.

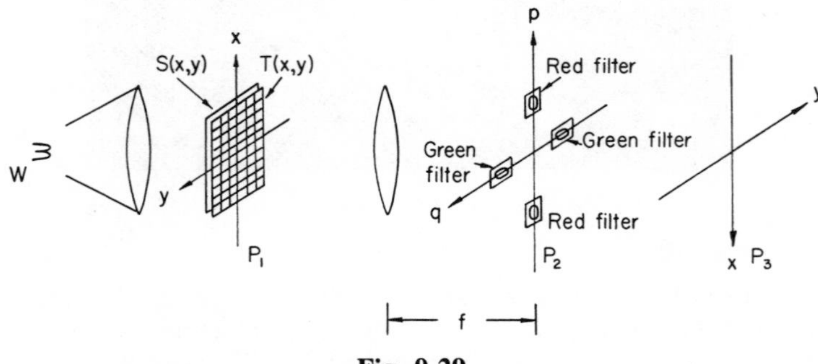

Fig. 9.29

(b) Evaluate the irradiance of the encoded image at the output plane P_3.

9.4 With reference to the real-time white-light color enhancement of fig. 9.7, design[a] three-color spatial filter with appropriate grating frequencies so that three primary color images can be simultaneously displayed at the output image plane.

9.5 Let us consider a spatial frequency pseudocoloring of a black- and-white object transparency with a conventional white-light processor. We assume that the low spatial frequency content of the input object is known *a priori*.

(a) Design a multicolor spatial filter so that the low and high spatial frequency signals can be encoded in two different color transparencies.

(b) Evaluate the intensity distribution of the pseudocolor image at the output plane.

(c) If precise spatial frequency content of the input object is known, is it possible to color code the object with a multicolor spatial filter? If yes, provide an example.

9.6 **(a)** Design a system for real-time white-light density pseudocolor encoding through phase contrast reversal, utilizing a one-dimensional diffraction grating technique and color spatial filtering.

(b) Describe a technique of obtaining a phase contrast reversal filter, and design a density pseudocolor spatial filter.

(c) Through a step-by-step evaluation, show that real-time density pseudocolor encoding may be obtained with a white-light optical processing technique.

9.7 With reference to the nonlinear processing with halftone screen of sec. 7.6:

(a) Evaluate the effect of using a white-light point source instead of a coherent source.

(b) Show that the equidensity images can be obtained with the white-light processing technique. (To avoid the cross talk of the smeared Fourier spectra, primary color filters can be used in the spatial filtering.)

(c) Show that the density pseudocolor encoding can be obtained through color spatial filtering of the smeared Fourier spectra.

(d) State the major advantages of using a white-light source as compared with a coherent source.

9.8 Let us extend the visualization of phase object technique of sec. 9.6 to a broad-band white-light source. For simplicity, we assume that the input object is a slow phase variation type.

(a) Sketch a phase color coding system with a white-light source.

(b) Design an appropriate source encoding mask and a diffraction grating such that a color coded phase object can be observed at the output plane.

(c) Analyze the white-light color encoder system and show that color coding can take place with the broad-band light source.

9.9 It is known that the integrated circuit mask is generally a two-dimensional cross-grating-type pattern; the corresponding Fourier spectrum may be represented in a finite region of the spatial frequency plane as depicted in fig. 9.30.

(a) Design a white-light optical processing technique for the integrated ciruit mask inspection.

(b) Carry out a detailed analysis to show that the defects can be easily detected and identified through color coding.

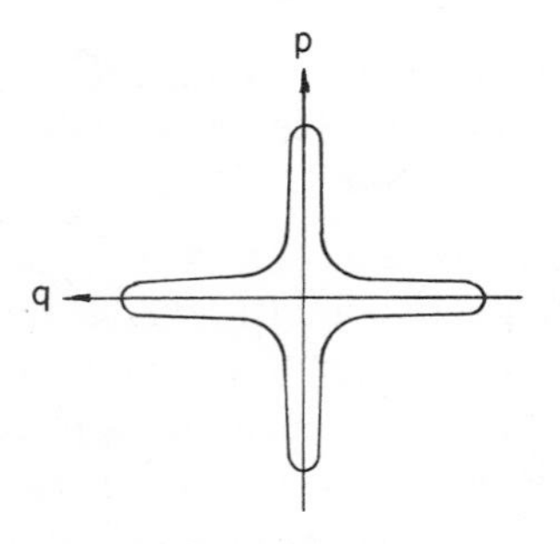

Fig. 9.30

9.10 With reference to the linear smeared color image deblurring of fig. 9.24, we assume that the smeared length is 1 mm, the spatial frequency of the phase grating $T(x)$ is 100π rad/mm, and the focal length of the transformer lens is 300 mm.

(a) Design a color deblurring filter that takes the effects of the whole spectral band of the white-light source.

(b) Evaluate the spatial coherence and temporal requirement for the color image deblurring.

(c) Evaluate the requirement of source size to achieve an optimum utilization of the light source.

9.11 If the color image deblurring of sec. 9.7 is performed by a fan-shaped deblurring filter, then:

(a) Evaluate the deblurred color image that takes place with the entire color spectral band of the white-light source.

(b) Show that the degree of deblurred errors depends upon the spatial frequency of the diffraction grating $T(x)$ and the size of the light source.

9.12 Let us consider color image subtraction with encoded incoherent sources, as discussed in sect. 9.8.

(a) Show that the color image subtraction technique may be extended to an encoded white-light source.

(b) Let us assume the input objects are spatial frequency limited to ω rad/mm; what is the minimum spatial frequency requirement of the diffraction grating at the input plane, the requirement of this source encoding mask, and the sopecification of the diffraction grating G in the spatial frequency plane?

(c) Evaluate the processing system to show that complex color image subtraction can be observed at the output plant of this white-light processor.

REFERENCES

9.1 H. E. Ives, "Improvement in the Diffraction Process of Color Photography," *Br. J. Photog.*, 609 (1906).

9.2 P. F. Mueller, "Color Image Retrieval from Monochrome Transparencies," *Appl. Opt.*, **8,** 2051 (1969).

9.3 A. Macovski, "Encoding and Decoding of Color Information," *Appl. Opt.*, **11,** 416 (1972).

9.4 R. Grousson and R. S. Kinany, "Multi-Color Image Storage on Black and White Film Using a Crossed Grating," *J. Opt.*, **9,** 333 (1978).

9.5 F. T. S. Yu, "White-Light Processing Technique for Archival Storage of Color Films," *Appl. Opt.*, **19,** 2457 (1980).

9.6 F. T. S. Yu, G. G. Mu, and S. L. Zhuang, "Color Restoration of Faded Color Films," *Optik*, **58,** 389 (1981).

9.7 H. C. Andrews, A. B. Tescher, and R. P. Kruger, "Image Processing by Digital Computer," *IEEE Spectrum*, **9,** 20 (1972).

9.8 J.Rheinberg, "On an Addition to the Method of Microscopical Research by a New Way of Optically Producing Color-Contrast Between an Object and its Background or Between Definite Parts of the Object Itself," *J. Roy. Micros. Soc.*, 333 (1896).

9.9 J. Bescos and T. C. Strand, "Optical Pseudocolor Encoding of Spatial Frequency Information," *Appl. Opt.*, **17,** 2524 (1978).

9.10 F. T. S. Yu, "New Techniques of Pseudocolor Encoding of Holographic Imaging,"*Opt. Lett.*, **3,** 57 (1978).

9.11 F. T. S. Yu, A. Tai, and H. Chen, "Spatial Filtered Pseudocolor Holographic Imaging," *J. Opt.*, **9,** 269 (1978).

9.12 H. K. Liu and J. W. Goodman, "A New Coherent Optical Pseudocolor Encoder," *Nouv. Rev. Opt.*, **7,** 285 (1976).

9.13 A Tai, F. T. S. Yu, and H. Chen, "White-Light Pseudocolor Density Encoder," *Opt. Lett.*, **3,** 190 (1978).

9.14 J. Santamaria, M. Gea, and J. Bescos, "Optical Pseudocoloring Through Contrast Reversal Filtering," *J. Opt.*, **10,** 151 (1979).

9.15 T. H. Chao, S. L. Zhuang, and F. T. S. Yu, "White-Light Pseudocolor Density Encoding Through Contrast Reversal," *Opt. Lett.*, **5,** 230 (1980).

9.16 F. T. S. Yu, S. L. Zhuang, T. H. Chao, and M. S. Dymek, "Read-Time White-Light Spatial Frequency and Density Pseudocolor Encoder," *Appl. Opt.*, **19,** 2986 (1980).

9.17 M. Born and E. Wolf, *Principles of Optics,* 2nd rev. ed., Pergamon Press, New York, 1964.

9.18 D. Malacara, *Optical Shop Testing,* John Wiley, New York, 1978.

9.19 W. H. Steel, *Interferometry,* Cambridge University Press, London, 1967.

9.20 J. B. DeVelis and G. O. Reynolds, *Theory and Application of Holography,* Addison-Wesley, Reading, Mass., 1967.

9.21 M. Françon, *Progress in Microscopy,* Pergamon Press, New York, 1961.

9.22 L. C. Martin, *The Theory of the Microscope,* American Elsevier, New York, 1966.

9.23 A. Bennett, H. Jupink, H. Osterberg, and O. Richards, *Phase Microscopy,* John Wiley, New York, 1951.

9.24 R. A. Spraque and B. J. Thompson, "Quantitative Visualization of Large Variation Phase Objects," *Appl. Opt.,* **11,** 1969 (1972).

9.25 G. Roblin and M. El Sherif, "Restoration of the Complex Amplitude of a Phase Object in Microscopy by Phase Modulation Interferometry," *Appl. Opt.,* **19,** 4247 (1980).

9.26 S. T. Wu and F. T. S. Yu, "Visualization of Color Coded Phase Variation with Incoherent Optical Processing Technique," *J. Opt.,* **13** (1982).

9.27 F. T. S. Yu, S. L. Zhuang, and T. H. Chao, "Color-Photographic-Image Deblurring Utilization White-Light Processing Technique," *J. Opt.,* **13,** 52 (1982).

9.28 F. T. S. Yu and S. T. Wu, "Color Image Subtraction with Extended Incoherent Sources," *J. Opt.,* **13** (1982).

9.29 F. T. S. Yu and J. L.Horner, "Optical Processing of Photographic Images," *Opt. Eng.,* **20,** 666 (1981).

10

Introduction to Holography

The theory of wave front reconstruction was introduced by D. Gabor in 1948 (ref. 10.1), who further developed it in a series of classic articles (refs. 10.2, 10.3). At the time he encountered two difficulties. The main difficulty was that a high-intensity coherent source suitable for wave front recording was not available; the second difficulty was that the virtual and real images could not be separated. Nevertheless, he set down the basic foundation for modern three-dimensional photography or holography. As a matter of fact, the word "holography" was first applied to this process by Gabor. The word is derived from the combination of two Greek words: "holos," meaning "whole," and "graphein," meaning "to write."* Thus, holography means a complete writing (i.e., recording).

This new imaging technique was at first received with only mild interest. In the 1950s, a number of investigators, including G. L. Rogers (ref. 10.4), H. M. A. El-Sum (ref. 10.5), and A. Lohmann (ref. 10.6), significantly extended the theory and understanding of this new imaging technique. With the invention of the laser, coherent light sources of adequate intensity were at last available. The difficulty of separating the overlapping real and virtual images was overcome by Leith and Upatnieks, who added a high spatial frequency carrier to the recording wave front (refs. 10.7–10.9). Since then numerous articles on holography and its engineering applications have been published. The purpose of this chapter is to approach holography from an elementary engineering point of view. That is to say, we employ impulse-exitation and other concepts from system theory. It is well known that optical instruments exhibit similarities with some electrical systems. It follows that holography may be approached as an input-output system analog.

In the following sections, wave front construction and reconstruction are demonstrated, and holographic magnifications, resolution limits, and bandwidth

*The same term had long been applied to documents written wholly by hand.

requirements are calculated. Third-order holographic aberrations, spatially incoherent holography, and color holography are also discussed.

10.1 WAVE FRONT CONSTRUCTION AND RECONSTRUCTION

In this seciton, on- and off-axis wave front constructions and reconstructions of a simple point object are demonstrated, extension toward a more complicated object is given, and a system analog of holographic recording and reconstruction is illustrated.

To show the wave front construction of a point object, let a monochromatic point radiator be located at a distance R away from a photographic plate, as shown in fig. 10.1. The complex light amplitude at a distance r from the radiator is

$$u = \frac{A}{r} \exp[i(kr - \omega t)], \tag{10.1}$$

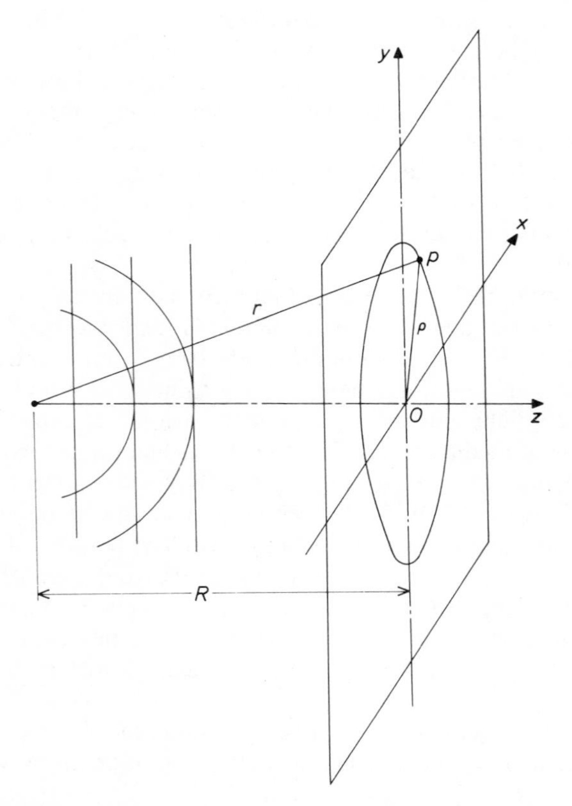

Fig. 10.1 Wave front recording of a simple point object.

where A is a complex constant, k is the wave number, and ω is the radian frequency of the light source. Let the *reference wave,* a monochromatic plane wave front* of the same frequency, travel perpendicular to the recording medium. The complex light amplitude of the reference wave is given by

$$v = B \exp\{i[k(R + z) - \omega t]\}, \tag{10.2}$$

where B is again a complex constant. At this point, and without loss of generality, the time-varying factor $\exp(i\omega t)$ will be dropped from the calculation.

At the surface of the recording medium ($z = 0$), eqs. (10.1) and (10.2) become

$$u(x, y) = \frac{A}{(R^2 + \rho^2)^{1/2}} \exp[ik(R^2 + \rho^2)^{1/2}] \tag{10.3}$$

and

$$v = B \exp(ikR), \tag{10.4}$$

where $\rho^2 = x^2 + y^2$. The resultant complex light distribution on the plate due to these two wave fronts is thus

$$u(x, y) + v = \frac{A}{(R^2 + \rho^2)^{1/2}} \exp[ik(R^2 + \rho^2)^{1/2}] + B \exp(ikR). \tag{10.5}$$

However, if the separation R is large compared to the aperture of the recording medium, then r can be replaced by the paraxial approximation

$$r \simeq R + \frac{\rho^2}{2R} \tag{10.6}$$

in the exponent, and by R in the denominator of eq. (10.5). Thus eq. (10.5) becomes

$$u(x, y) + v = \exp(ikR)\left[\frac{A}{R} \exp\left(ik \frac{\rho^2}{2R}\right) + B\right]. \tag{10.7}$$

The correxponding irradiance is therefore

*As will be seen, a *plane* reference wave is not required; a spherical wave would serve as well. We use a plane wave here for simplicity.

$$I(x, y) = [u(x, y) + v][u(x, y) + v]^* = \left(\frac{|A|}{R}\right)^2 + |B|^2 + \frac{2|A|\,|B|}{R}\cos\left(\frac{k\rho^2}{2R} + \phi\right), \tag{10.8}$$

where ϕ is the phase angle between the complex amplitudes A and B.

The exposure during recording (i.e., encoding) can be assumed to be proportional to the I of eq. (10.8). Indeed, eq. (10.8) can be recognized as describing a *Fresnel zone lens* construction (ref. 10.10). Moreover, if the wave front construction is properly recorded in the linear region of the T-E characteristic of the emulsion (sec. 8.1), then the transmittance of the recorded hologram will be

$$T(\boldsymbol{\rho}; k) = K_1 + K_2 \cos\left(\frac{k\rho^2}{2R} + \phi\right), \tag{10.9}$$

where K_1 and K_2 are proportionality constants.

If the point-object hologram of eq. (10.9) is illuminated (i.e., decoded) by a *normally incident* monochromatic plane wave of the same wavelength λ (fig. 10.2), then by the Fresnel–Kirchhoff theory (sec. 3.7) the complex light distribution behind the hologram can be determined from the convolution theorem,

$$E(\boldsymbol{\sigma}; k) = B \iint_S T(\boldsymbol{\rho}; k) E_l^+(\boldsymbol{\sigma} - \boldsymbol{\rho}; k)\, dx\, dy, \tag{10.10}$$

where

$$E_l^+(\boldsymbol{\rho}; k) = -\frac{i}{\lambda l} \exp\left[ik\left(l + z + \frac{\rho^2}{2l}\right)\right]$$

is the free-space impulse response, S denotes integration over the entire hologram surface, $\boldsymbol{\rho}(x, y, z)$ is the coordinate system at the hologram, and $\boldsymbol{\sigma}(\alpha, \beta, \gamma)$ is a separate system at a distance l from the first.

If we evaluate eq. (10.10) at a distance $l = R$ behind the hologram, then the solution is

$$E(\boldsymbol{\sigma}; k) = C_1 + C_2 \exp\left(i\frac{k}{4R}\sigma^2\right) + C_3\delta(\alpha, \beta), \tag{10.11}$$

where C_1, C_2, and C_3 are the appropriate complex constants, $\sigma^2 = \alpha^2 + \beta^2$ and $\delta(\alpha, \beta)$ is the two-dimensional Direac delta function. The three terms of eq. (10.11) may be interpreted as follows: C_1 represents the zero-order (i.e., dc) diffraction, the second term is the first-order virtual image diffraction, and the third term is the first-order real image.

As shown in fig. 10.2, all these diffractions are overlapping; thus spurious distortions are introduced into the reconstructed image. Overlapping will occur even if the hologram is illuminated by an oblique plane wave. For such oblique

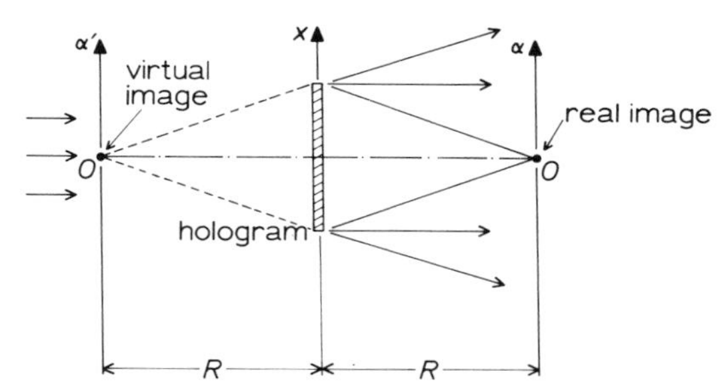

Fig. 10.2 Wave front reconstruction of a point object. Plane monochromatic illumination.

illumination, the complex light distribution is given by

$$E(\boldsymbol{\sigma}; k) = B \iint_S T(\boldsymbol{\rho}; k) \exp(ikx \sin \theta) E_l^+(\boldsymbol{\sigma} - \boldsymbol{\rho}; k)\, dx\, dy. \quad (10.12)$$

By substituting eq. (10.9) in this, the solution for $l = R$ is seen to be

$$E(\boldsymbol{\sigma}; k) = C_1' + C_2' \exp\left[i \frac{k}{4R} (\alpha + R \sin \theta)^2 + \beta^2 \right] + C_3' \delta(\alpha - R \sin \theta, \beta), \quad (10.13)$$

where C_1', C_2', and C_3' are the appropriate complex constants. The last two terms can be recognized again as virtual and real images, as shown in fig. 10.3. Apparently, oblique illumination is not able to separate the diffractions.

In order to make the diffractions separable, an oblique reference wave may be used during recording, as shown in fig. 10.4. The complex light amplitude distribution on the surface of the recording medium due to the reference wave

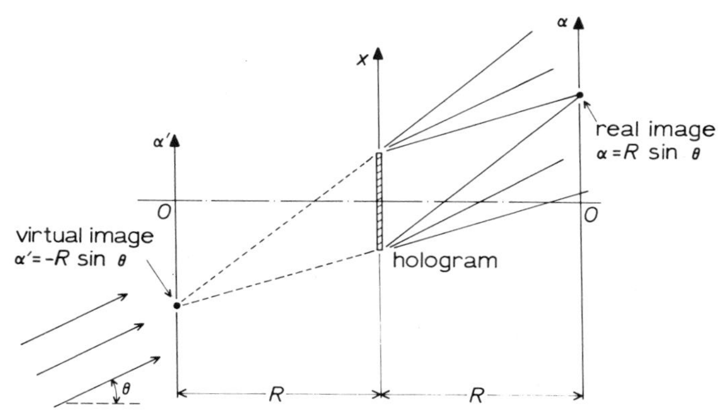

Fig. 10.3 Oblique wave front reconstruction of a point-object hologram. Plane monochromatic illumination.

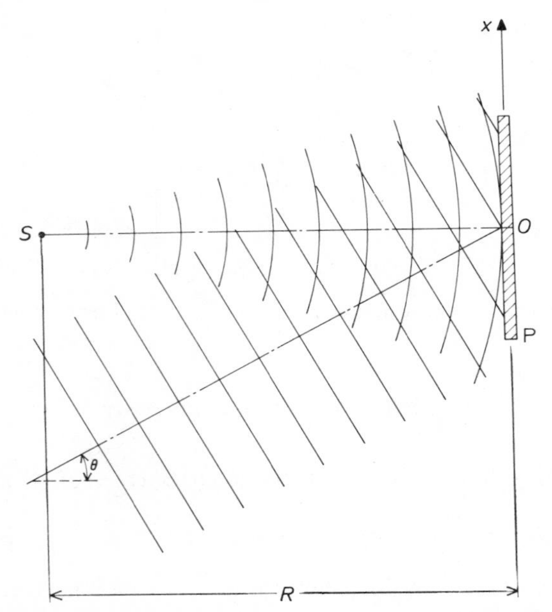

Fig. 10.4 Spatial carrier wave front recording of a point object. Oblique plane reference wave; S, monochromatic point source (object); P, photographic plate.

is

$$v(x) = B \exp[ik(R + x \sin\theta)]. \tag{10.14}$$

By the usual paraxial approximation, the resultant complex light distribution on the recording medium is

$$u(x, y) + v(x) = \frac{A}{R} \exp\left[ik\left(R + \frac{\rho^2}{2R}\right)\right] + B \exp[ik(R + x \sin\theta)]. \tag{10.15}$$

The corresponding irradiance is

$$\begin{aligned} I(x, y) &= [u(x, y) + v(x)][u(x, y) + v(x)]^* \\ &= |A_1|^2 + |B|^2 + 2|A_1|\,|B| \cos\left[k\left(\frac{\rho^2}{2R} - x \sin\theta\right) + \phi\right], \end{aligned} \tag{10.16}$$

where $A_1 = A/R$, $\rho^2 = x^2 + y^2$, and ϕ is the phase angle between the complex amplitudes A and B. Again, if we assume the wave front recording is linear, the transmittance of the off-axis hologram is

$$T(\boldsymbol{\rho}; k) = K_1 + K_2 \cos\left[k\left(\frac{\rho^2}{2R} - x \sin\theta\right) + \phi\right], \tag{10.17}$$

where K_1 and K_2 are proportionality constants.

If this off-axis hologram is illuminated by a plane wave of wavelength λ and obliquity $-\theta$ (fig. 10.5), then the complex light amplitude distribution behind the hologram is

$$E(\boldsymbol{\sigma}; k) = B \iint_S T(\boldsymbol{\rho}; k) \exp(-ikx \sin \theta) E_l^{\dagger}(\boldsymbol{\sigma} - \boldsymbol{\rho}; k)\, dx\, dy. \quad (10.18)$$

(Note: The readout angle could be any other than $-\theta$, which we use here for simplicity.)

Again the wave front reconstruction at distance $l = R$ behind the hologram is

$$E(\boldsymbol{\sigma}; k) = C_1 \exp(-ik\alpha \sin \theta) + C_2 \exp\left\{ i\frac{k}{4R}[(\alpha - 2R \sin \theta)^2 + \beta^2] \right\} + C_3\delta(\alpha, \beta), \quad (10.19)$$

where C_1, C_2, and C_3 are the appropriate complex constants. Figure 10.5 shows the real and virtual image reconstructions. It is clear that for a properly chosen angle of incidence of the reference wave, the real image may be separated from the zero-order and the virtual image diffractions.

We may also let the angle of incidence of the reference wave be positive, as shown in fig. 10.6. Again, the complex light distribution at $l = R$ can be written as

$$E(\boldsymbol{\sigma}; k) = C_1' \exp(ik\alpha \sin \theta) + C_2' \exp\left(i\frac{k}{4R}\sigma^2 \right) + C_3'\delta(\alpha - 2R \sin \theta, \beta), \quad (10.20)$$

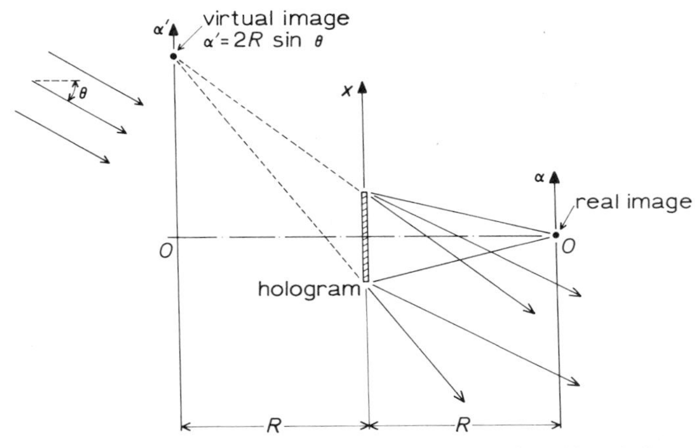

Fig. 10.5 Wave front reconstruction of a spatial carrier point-object hologram. Monochromatic plane wave, negative θ.

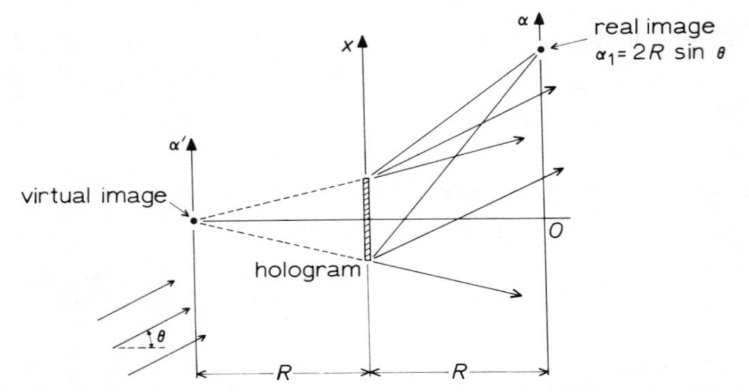

Fig. 10.6 Wave front reconstruction of a spatial carrier point-object hologram. Monochromatic plane wave, positive θ.

where C_1', C_2' and C_3' are the appropriate complex constants. From fig. 10.6. we see that the virtual image diffraction is again separated from the zero-order and the real image diffractions.

The complex light distribution from an extended object (fig. 10.7) may be obtained from the convolution theorem, and is

$$u(x, y) = \iint_{S_o} O(\xi, \eta, \zeta) E_l^{\dagger}(\boldsymbol{\rho} - \boldsymbol{\xi}; k)\, d\xi\, d\eta, \qquad (10.21)$$

where $O(\xi, \eta, \zeta)$ is the object function, and S_o denotes the surface integral over the object, viewed from the hologram aperture.

Again, under the assumption of the spatial linearity of the recording medium, the transmittance of the recorded hologram can be shown to be

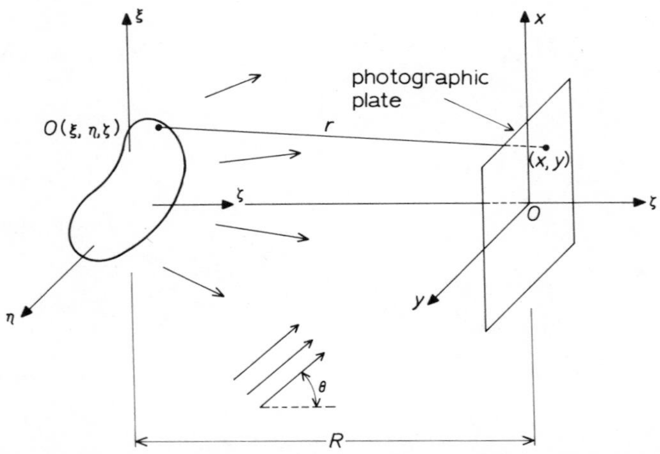

Fig. 10.7 Wave front construction of an extended object. Monochromatic plane reference wave.

$$T(\boldsymbol{\rho}; k) = K[u(x, y) + v(x)][u(x, y) + v(x)]^*$$
$$= K\{|u(x, y)|^2 + B^2 + 2\,|u(x, y)|\, B \cos[\phi(x, y) - kx \sin \theta]\}, \quad (10.22)$$

where K is a proportionality constant, $v(x)$ is the oblique reference wave given in eq. (10.14), and $u(x, y) = |u(x, y)| \exp[i\phi(x, y)]$. Alternatively, eq. (10.22) could be written as

$$T(\boldsymbol{\rho}; k) = K[|u(x, y)|^2 + B^2 + Bu(x, y) \exp(-ikx \sin \theta) + Bu^*(x, y) \exp(ikx \sin \theta)]. \quad (10.23)$$

If the hologram is obliquely illuminated by the reference wave, as shown in fig. 10.8, then the complex light distribution behind the hologram can be written as eq. (10.18). If we consider only the term giving the real image diffraction [i.e., the last term of eq. (10.23)], then we have

$$E_r(\boldsymbol{\sigma}; k) = C \exp\left(\frac{k}{2l'}\sigma^2\right) \iint\limits_{S} \left\{ \iint\limits_{S_o} O^*(\xi, \eta, \zeta) \exp\left[-i\frac{k}{2l}(\xi^2 + \eta^2)\right] \times \exp\left[i\frac{k}{l}(\xi x + \eta y)\right] d\xi\, d\eta \right\}$$
$$\times \exp\left[i\frac{k}{2}\rho^2\left(\frac{1}{l'} - \frac{1}{l}\right)\right] \exp\left[-i\frac{k}{l'}(\alpha x + \beta y)\right] dx\, dy, \quad (10.24)$$

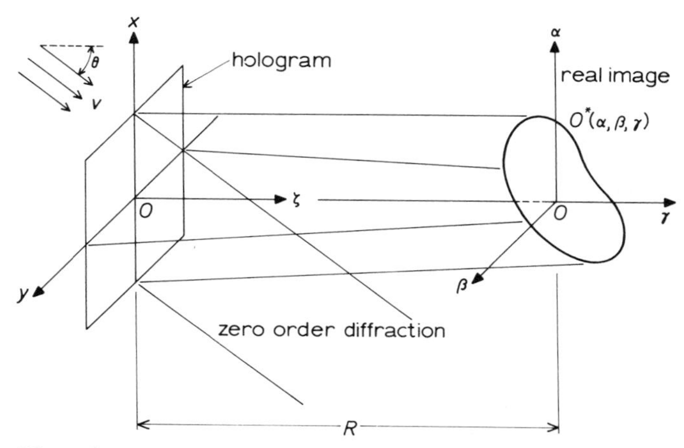

Fig. 10.8 Wave front reconstruction of an extended object. Oblique plane monochromatic illumination, negative θ.

where C is an appropriate complex constant, the subscript r denotes the real image diffraction, and $l' = R + \gamma$ is the separation between the holographic plate and the image coordinate $\boldsymbol{\sigma}(\alpha, \beta, \gamma)$. From eq. (10.24) the real hologram image can be shown reconstructed uniquely at $l = l'$ (i.e., $R = R'$ and $\zeta = -\gamma$), as pictured in fig. 10.8. Thus

$$E_r(\boldsymbol{\sigma}; k) = CO^*(\alpha, \beta, \gamma), \qquad \text{for } l = l'. \tag{10.25}$$

Similarly, if the hologram is illuminated at a positive oblique angle, the virtual hologram image can be shown constructed uniquely at $l = l'$ (i.e., $R = -R'$, and $\zeta = y'$), as shown in fig. 10.9. That is,

$$E_v(\boldsymbol{\sigma}; k) = CO(\alpha', \beta' \gamma'), \qquad \text{for } l = l'. \tag{10.26}$$

Photographs of virtual and real hologram images are pictured in figs. 10.10 and 10.11, respectively. The transmittance, eq. (10.22), may be written as

$$T(\boldsymbol{\rho}; k) = K[|u(x, y)|^2 + |v(x)|^2 + u(x, y)v^*(x) + u^*(x, y)v(x)], \tag{10.27}$$

where

$$u^*(x, y) = \iint_{S_o} O^*(\xi, \eta, \zeta)E_l^{+*}(\boldsymbol{\rho} - \boldsymbol{\xi}; k)\, d\xi\, d\eta$$

and

$$v^*(x) = \exp(-ikx \sin\theta).$$

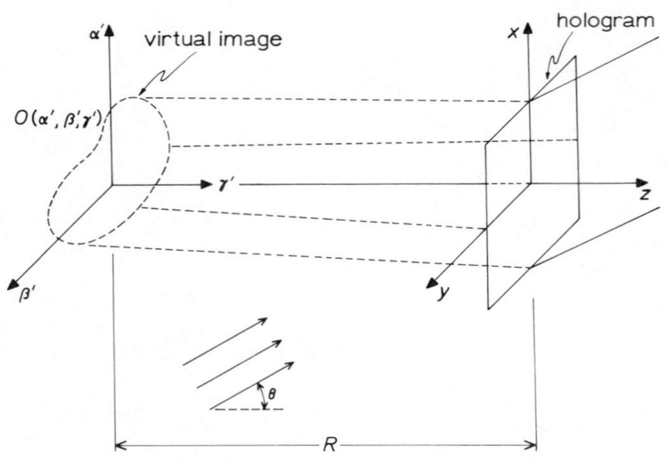

Fig. 10.9 Virtual image reconstruction. Oblique plane monochromatic illumination.

Fig. 10.10 A virtual three-dimensional hologram image.

Fig. 10.11 A real three-dimensional hologram image.

The Fourier transform of $u(x, y)$ is, therefore,

$$U(p, q) = O(p, q)E_1(p, q), \tag{10.28}$$

where p and q are the spatial frequencies, and $U(p, q)$, $O(p, q)$, and $E_1(p, q)$ are the respective Fourier transforms of $u(x, y)$, $O(x, y)$, and $E_l^+(x, y)$.

Similarly, we can write the Fourier transform of $u^*(x, y)$ as

$$U^*(p, q) = O^*(p, q)E_l^*(p, q). \tag{10.29}$$

If the hologram of eq. (10.27) is illuminated by an oblique plane wave of negative θ, then the complex light amplitude at the immediate opposite side of the hologram can be written as

$$\begin{aligned} v^*(x)T(\boldsymbol{\rho}; k) = K\{v^*(x)[|u(x, y)|^2] \\ + |v(x)|^2 + u(x, y)[v^*(x)]^2 + u^*(x, y)\,|v(x)|^2\}, \end{aligned} \tag{10.30}$$

where $|v(x)| = B$ is a constant.

Thus the Fourier transform of eq. (10.30) may be written as

$$\begin{aligned} \mathscr{F}[v^*(x)T(\boldsymbol{\rho}; k)] = K\mathscr{F}\{v^*(x)[|u(x, y)|^2 + |v(x)|^2] + u(x, y)[v^*(x)]^2\} \\ + KB^2O^*(p, q)E_l^*(p, q). \end{aligned} \tag{10.31}$$

where $\mathscr{F}$ denotes the Fourier transform.

Since the complex light diffraction behind the hologram can be determined by the Fresnel–Kirchhoff theory, as shown in eq. (10.18), the Fourier transform of the diffraction can be written as

$$E(p, q) = \mathscr{F}[v^*(x)T(\boldsymbol{\rho}; k)]E_l(p, q), \tag{10.32}$$

where $E(p, q)$ is the Fourier transform of $E(\boldsymbol{\sigma}; k)$.

If we substitute eq. (10.31) in eq. (10.32), we have

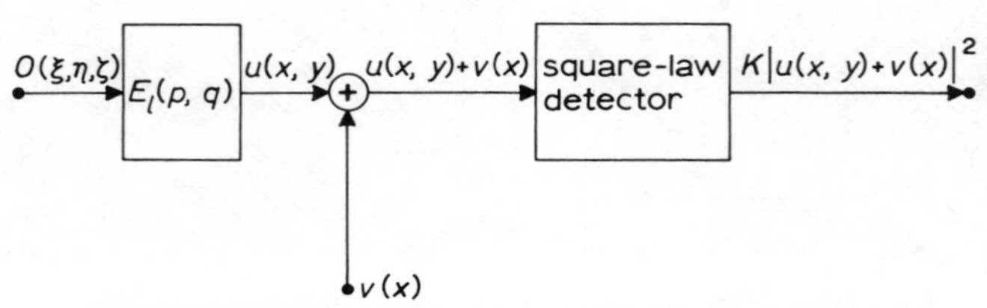

Fig. 10.12 System analog of wave front recording.

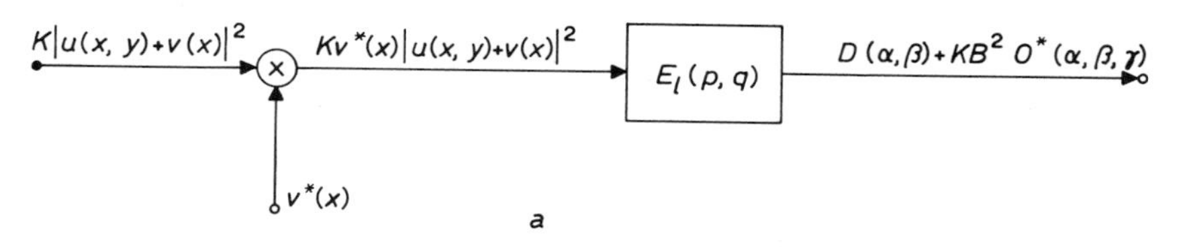

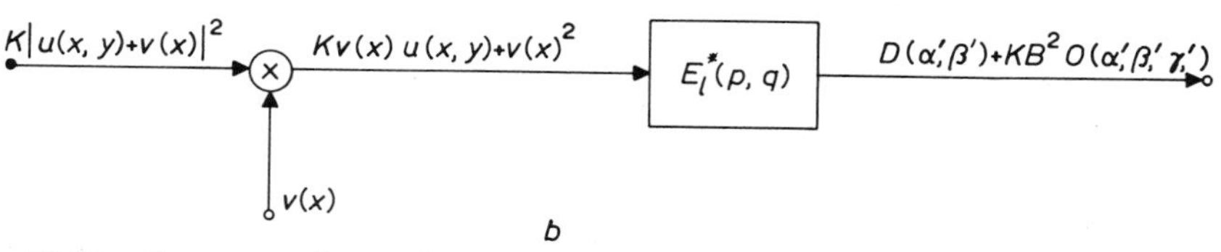

Fig. 10.13 System analogs of wave front reconstruction. (*a*) For real image with $D(\alpha, \beta)$ a complex function; (*b*) For virtual image with $D(\alpha', \beta')$ a complex function.

$$E(p, q) = K\mathscr{F}\{v^*(x)[|u(x, y)|^2 + |v(x)|^2] + u(x, y)[v^*(x)]^2\}E_l(p, q) + KB^2O^*(p, q). \tag{10.33}$$

Note that the last term of this equation is proportional to $O^*(p, q)$, the Fourier transform of the conjugate of the object function.

Similarly, if the illumination is done by a wave of positive θ, then the Fourier transform of the complex light diffraction behind the hologram can be shown to be

$$E(p, q) = K\mathscr{F}\{v(x)[|u(x, y)|^2 + |v(x)|^2] + u^*(x, y)v^2(x)\}E_l^*(p, q) + KB^2O(p, q). \tag{10.34}$$

The last term of this equation represents the virtual hologram image diffraction, which is proportional to the Fourier transform of the object function.

System-analog diagrams representing wave front recording and reconstruction are given in figs. 10.12 and 10.13, respectively.

10.2 HOLOGRAPHIC MAGNIFICATIONS

Wave front reconstructions are generally three-dimensional in nature. The lateral and longitudinal magnifications in wave front reconstruction may be discussed separately.

Lateral Magnifications

To obtain the lateral holographic magnifications, we may start from a wave front recording of the three monochromatic point radiators of wave length λ_1

(fig. 10.14). If the paraxial approximation (eq. 10.6) holds, then the complex light distributions from the three radiators are

$$
\begin{aligned}
u_1(\boldsymbol{\rho}; k_1) &\simeq A_1 \exp\left(ik_1\left\{R_1 + \frac{1}{2R_1}\left[\left(x - \frac{h}{2}\right)^2 + y^2\right]\right\}\right), \\
u_2(\boldsymbol{\rho}; k_1) &\simeq A_2 \exp\left(ik_1\left\{R_1 + \frac{1}{2R_1}\left[\left(x + \frac{h}{2}\right)^2 + y^2\right]\right\}\right), \\
u_3(\boldsymbol{\rho}; k_1) &\simeq A_3 \exp\left(ik_1\left\{L_1 + \frac{l}{2L_1}[(x + a)^2 + y^2]\right\}\right),
\end{aligned}
\tag{10.35}
$$

where $k_1 = 2\pi/\lambda_1$, and A_1, A_2 and A_3 are real constants. The corresponding irradiance is

$$
\begin{aligned}
I(\boldsymbol{\rho}; k_1) = {} & (u_1 + u_2 + u_3)(u_1 + u_2 + u_3)^* = A_1^2 + A_2^2 + A_3^2 + 2A_1A_2 \cos\frac{k_1}{R_1} hx \\
& + 2A_1A_3 \cos\left\{k_1(R_1 - L_1) + \frac{k_1}{2R_1}\left[\left(x - \frac{h}{2}\right)^2 + y^2\right]\right. \\
& \qquad \left. - \frac{k_1}{2L_1}[(x + a)^2 + y^2]\right\} \\
& + 2A_2A_3 \cos\left\{k_1(R_1 - L_1) + \frac{L_1}{2R_1}\left[\left(x + \frac{h}{2}\right)^2 + y^2\right]\right. \\
& \qquad \left. - \frac{k_1}{2L_1}[(x + a)^2 + y^2]\right\}.
\end{aligned}
\tag{10.36}
$$

The above equation represents two overlapping Fresnel zone lens constructions. Again, if the wave front recording is linear, then the transmittance of the hologram is

$$
T(\boldsymbol{\rho}; k_1) = K_0 + K_1 \cos\frac{k_1}{R_1} hx + K_2[e^{i\{*\}} + e^{-i\{*\}}] + K_3[e^{i\{**\}} + e^{-i\{**\}}],
\tag{10.37}
$$

where the K's are real proportionality constants, and

$$
\begin{aligned}
\{*\} &= \left\{k_1(R_1 - L_1) + \frac{k_1}{2R_1}\left[\left(x - \frac{h}{2}\right)^2 + y^2\right] - \frac{k_1}{2L_1}[(x + a)^2 + y^2]\right\}, \\
\{**\} &= \left\{k_1(R_1 - L_1) + \frac{k_1}{2R_1}\left[\left(x + \frac{h}{2}\right)^2 + y^2\right] - \frac{k_1}{2L_1}[(x + a)^2 + y^2]\right\}.
\end{aligned}
$$

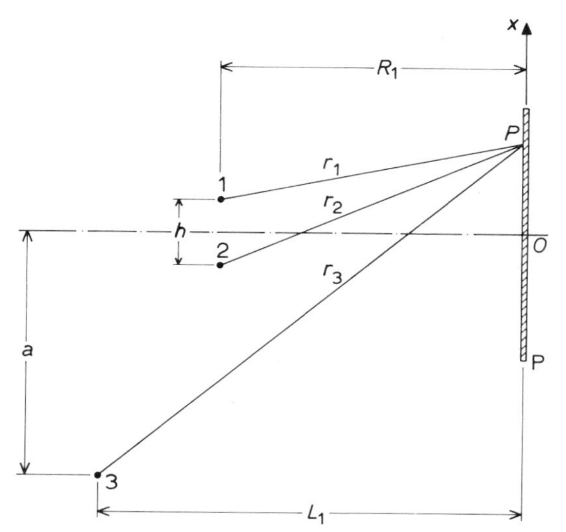

Fig. 10.14 Recording geometry for determining the lateral magnifications. 1,2, monochromatic point sources; 3, divergent monochromatic reference source; P, photographic plate.

If this hologram is illuminated by divergent light of wavelength λ_2, as shown in fig. 10.15,

$$u_4(\boldsymbol{\rho}; k_2) = A_4 \exp\left\{ik\left[L_2 + \frac{1}{2L_2}[(x - b)^2 + y^2]\right]\right\}, \qquad (10.38)$$

then the complex light distribution behind the hologram is

$$E(\boldsymbol{\sigma}; k_2) = \iint_S T(\boldsymbol{\rho}; k_1)u_4(\boldsymbol{\rho}; k_2)E_l^+(\boldsymbol{\sigma} - \boldsymbol{\rho}; k_2)\, dx\, dy. \qquad (10.39)$$

Since the third and fifth terms of eq. (10.37) contribute to the virtual image diffractions, and the fourth and sixth terms correspond to the real image reconstructions, the evaluation of eq. (10.39) can be performed termwise with respect to the real and virtual image reconstructions. Thus for the reconstruction of the real images we have,

$$E_r(\boldsymbol{\sigma}; k_2) = \iint_S [K_2 e^{-i\{*\}} + K_3 e^{-i\{**\}}]u_4(\boldsymbol{\rho}; k_2)E_l^+(\boldsymbol{\rho} - \boldsymbol{\sigma}; k_2)\, dx\, dy, \qquad (10.40)$$

where the subscript r denotes the real image.

By substitution, the above equation can be written as

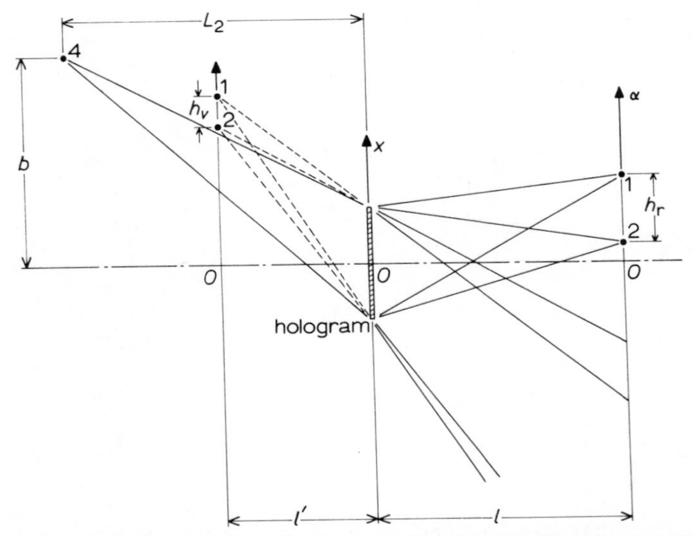

Fig. 10.15 Reconstruction geometry for determining the lateral magnifications. A monochromatic point source is at 4.

$$E_r(\boldsymbol{\sigma}; k_2) = C_1 \iint_S \exp\left\{-i\frac{k_2}{2}\left[\left(\frac{\lambda_2}{\lambda_1 R_1} - \frac{\lambda_2}{\lambda_1 L_1} - \frac{1}{L_2} - \frac{1}{l}\right)\rho^2 + \frac{2}{l}\left(\alpha + l\left(\frac{b}{L_2} - \frac{\lambda_2 h}{2\lambda_1 R_1} - \frac{\lambda_2 a}{\lambda_1 L_1}\right)\right)x + \frac{2\beta}{l}y\right]\right\} dx\, dy$$
$$+ C_2 \iint_S \exp\left\{-i\frac{k_2}{2}\left[\left(\frac{\lambda_2}{\lambda_1 R_1} - \frac{\lambda_2}{\lambda_1 L_1} - \frac{1}{L_2} - \frac{1}{l}\right)\rho^2 + \frac{2}{l}\left(\alpha + l\left(\frac{b}{L_2} + \frac{\lambda_2 h}{2\lambda_1 R_1} - \frac{\lambda_2 a}{\lambda_1 L_1}\right)\right)x + \frac{2\beta}{l}y\right]\right\} dx\, dy, \quad (10.41)$$

where C_1 and C_2 are the appropriate complex constants. From this equation, it is clear that the real images will be uniquely reconstructed at

$$l = \frac{\lambda_1 R_1 L_1 L_2}{\lambda_2 L_1 L_2 - \lambda_2 R_1 L_2 - \lambda_1 R_1 L_1}. \quad (10.42)$$

The solution of eq. (10.41), evaluated at a distance behind the hologram given by (10.42), is

$$E_r(\boldsymbol{\sigma}; k_2) = C_1'\delta\left[\alpha + l\left(\frac{b}{L_2} - \frac{\lambda_2 h}{2\lambda_1 R_1} - \frac{\lambda_2 a}{\lambda_1 L_1}\right), \beta\right] + C_2'\delta\left[\alpha + l\left(\frac{b}{L_2} + \frac{\lambda_2 h}{2\lambda_1 R_1} - \frac{\lambda_2 a}{\lambda_1 L_1}\right), \beta\right]. \quad (10.43)$$

where C_1' and C_2' are the appropriate complex constants, and δ is the Dirac delta function. From the construction of figs. 10.14 and 10.15, the lateral magnification of the real image can be shown to be (refs. 10.11–10.14)

$$M_{\text{lat}}^{r} = \frac{h_r}{h} = \left(1 - \frac{\lambda_1 R_1}{\lambda_2 L_2} - \frac{R_1}{L_1}\right)^{-1}. \tag{10.44}$$

For the virtual image reconstructions, we can similarly show that:

$$\begin{aligned} E_v(\boldsymbol{\sigma}; k_2) = {} & C_1'\delta\left[\alpha' + l'\left(\frac{b}{L_2} + \frac{\lambda_2 h}{2\lambda_1 R_1} + \frac{\lambda_2 a}{\lambda_1 L_1}\right), \beta'\right] \\ & + C_2'\delta\left[\alpha' + l'\left(\frac{b}{L_2} - \frac{\lambda_2 h}{2\lambda_1 R_1} + \frac{\lambda_2 a}{\lambda_1 L_1}\right), \beta'\right]. \end{aligned} \tag{10.45}$$

where the subscript v denotes the virtual image diffraction, and

$$l' = \frac{\lambda_1 R_1 L_1 L_2}{\lambda_2 R_1 L_2 - \lambda_2 L_1 L_2 - \lambda_1 L_1 R_1}.$$

The corresponding lateral magnification is therefore (refs. 10.11–10.14)

$$M_{\text{lat}}^{v} = \frac{h_v}{h} = \left(1 + \frac{\lambda_1 R_1}{\lambda_2 L_2} - \frac{R_1}{L_1}\right)^{-1}. \tag{10.46}$$

From eqs. (10.44) and (10.46) we can conclude that

$$M_{\text{lat}}^{r} \geq M_{\text{lat}}^{v}. \tag{10.47}$$

The equality holds where the reference and illuminating beams are both plane waves.

Longitudinal Magnifications

The longitudinal magnifications can also be obtained for the holographic process, as shown in fig. 10.16. Again by paraxial approximation, the complex light distributions of the three monochromatic point sources of wavelength λ_1 are

$$\begin{aligned} u_1(\boldsymbol{\rho}; k_1) &= A_1 \exp\left\{ik_1\left[R_1 + \frac{1}{2R_1}(x^2 + y^2)\right]\right\}, \\ u_2(\boldsymbol{\rho}; k_1) &= A_2 \exp\left\{ik_1\left[(R_1 + d) + \frac{1}{2(R_1 + d)}(x^2 + y^2)\right]\right\}, \\ u_3(\boldsymbol{\rho}; k_1) &= A_3 \exp\left\{ik_1\left[L_1 + \frac{1}{2L_1}[(x + a)^2 + y^2]\right]\right\}. \end{aligned} \tag{10.48}$$

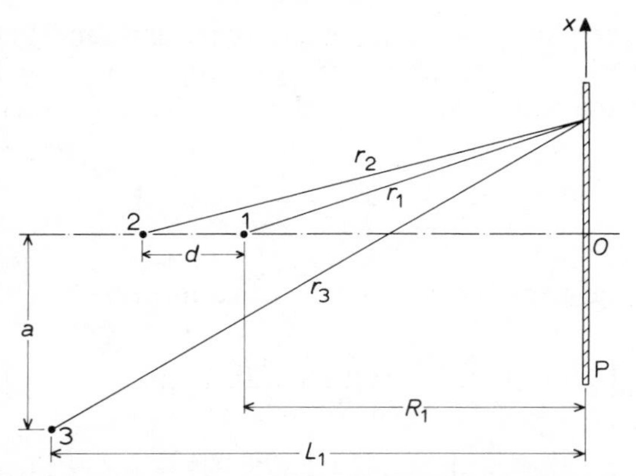

Fig. 10.16 Construction geometry for determining the longitudinal magnifications. 1,2, monochromatic point sources; 3, divergent reference source; P, photographic plate.

The corresponding irradiance is

$$\begin{aligned} I(\boldsymbol{\rho}; k_1) &= (u_1 + u_2 + u_3)(u_1 + u_2 + u_3)^* \\ &= A_1^2 + A_2^2 + A_3^2 + 2A_1A_2 \\ &\quad \times \cos\left\{k_1\left[-d + \frac{1}{2}\left(\frac{1}{R_1} - \frac{1}{R_1 + d}\right)(x^2 + y^2)\right]\right\} \\ &\quad + 2A_1A_3 \cos\left\{k_1\left[R_1 - L_1 + \frac{1}{2R_1}(x^2 + y^2) \right.\right. \\ &\quad \left.\left. - \frac{1}{2L_1}[(x + a)^2 + y^2]\right]\right\} \\ &\quad + 2A_2A_3 \cos\left\{k_1\left[R_1 + d - L_1 + \frac{1}{2(R_1 + d)}(x^2 + y^2) \right.\right. \\ &\quad \left.\left. - \frac{1}{2L_1}[(x + a)^2 + y^2]\right]\right\}. \end{aligned} \tag{10.49}$$

Once again we assume the recording to be properly biased; then the transmittance is

$$\begin{aligned} T(\boldsymbol{\rho}; k_1) = K_0 + K_1 \cos\left\{k_1\left[-d + \frac{1}{2}\left(\frac{1}{R_1} - \frac{1}{R_1 + d}\right)(x^2 + y^2)\right]\right\} \\ + K_2[e^{i\{\Delta\}} + e^{-i\{\Delta\}}] + 2K_3[e^{i\{\Delta\Delta\}} + e^{-i\{\Delta\Delta\}}]. \end{aligned} \tag{10.50}$$

where the K's are real proportionality constants, and

$$\{\Delta\} = \left\{k_1\left[R_1 - L_1 + \frac{1}{2R_1}(x^2 + y^2) - \frac{1}{2L_1}[(x + a)^2 + y^2]\right]\right\},$$

$$\{\Delta\Delta\} = \left\{k_1\left[R_1 + d - L_1 + \frac{1}{2(R_1 + d)}(x^2 + y^2) - \frac{1}{2L_1}[(x + a)^2 + y^2]\right]\right\}.$$

If this hologram is illuminated by a divergent beam of wavelength λ_2 (fig. 10.17), then

$$u_4(\boldsymbol{\rho}; k_2) = A_4 \exp\left\{ik_2\left[L_2 + \frac{1}{2L_2}[(x - b)^2 + y^2]\right]\right\}, \tag{10.51}$$

and the complex light distribution due to the real image diffractions [i.e., the fourth and sixth terms of eq. (10.50)] gives

$$E_r(\boldsymbol{\sigma}; k_2) = \iint_S [K_2 e^{-i\{\Delta\}} + K_3 e^{-i\{\Delta\Delta\}}] u_4(\boldsymbol{\rho}; k_2) E_l^{+}(\boldsymbol{\rho} - \boldsymbol{\sigma}; k_2)\, dx\, dy. \tag{10.52}$$

By substitution, we can put this equation into the form

$$\begin{aligned}
E_r(\boldsymbol{\sigma}; k_2) = C_1 \iint_S \exp&\left\{-i\frac{k_2}{2}\left[\left(\frac{\lambda_2}{\lambda_1 R_1} - \frac{\lambda_2}{\lambda_1 L_1} - \frac{1}{L_2} - \frac{1}{l}\right)\rho^2\right.\right. \\
&\left.\left. + \frac{2}{l}\left(\alpha + l\left(\frac{b}{L_1} - \frac{\lambda_2 a}{\lambda_1 L_1}\right)\right)x + \frac{2\beta}{l}y\right]\right\} dx\, dy \\
+ C_2 \iint_S \exp&\left\{-i\frac{k_2}{2}\left[\left(\frac{\lambda_2}{\lambda_1 (R_1 + d)} - \frac{\lambda_2}{\lambda_1 L_1} - \frac{1}{L_2} - \frac{1}{l}\right)\rho^2\right.\right. \\
&\left.\left. + \frac{2}{l}\left(\alpha + l\left(\frac{b}{L_2} - \frac{\lambda_2 a}{\lambda_1 L_1}\right)\right)x + \frac{2\beta}{l}y\right]\right\} dx\, dy.
\end{aligned} \tag{10.53}$$

where C_1 and C_2 are the appropriate complex constants. From the above equation, it is clear that the real images will be uniquely reconstructed at $l = l_1$,

$$l_1 = \frac{\lambda_1 R_1 L_1 L_2}{\lambda_2 L_1 L_2 - \lambda_2 R_1 L_2 - \lambda_1 R_1 L_1}, \tag{10.54}$$

and at $l = l_2$,

$$l_2 = \frac{\lambda_1 L_1 L_2 (R_1 + d)}{\lambda_2 L_1 L_2 - \lambda_2 L_2 (R_1 + d) - \lambda_1 L_1 (R_1 + d)}. \tag{10.55}$$

Thus the solution of eq. (10.53) may be expressed termwise,

$$E_r(\boldsymbol{\sigma}; k_2) = C_1'\delta\left[\alpha + l_1\left(\frac{b}{L_1} - \frac{\lambda_2 a}{\lambda_1 L_1}\right), \beta\right]_{l=l_1} + C_2'\delta\left[\alpha + l_2\left(\frac{b}{L_2} - \frac{\lambda_2 a}{\lambda_1 L_1}\right), \beta\right]_{l=l_2}, \quad (10.56)$$

where C_1' and C_2' are the appropriate complex constants.

From eqs. (10.54) and (10.55), the longitudinal separation of the real images can be shown to be

$$d_r = l_2 - l_1$$

$$d_r = \frac{\lambda_1\lambda_2(L_1L_2)^2 d}{[\lambda_2L_1L_2 - \lambda_2(R_1 + d)L_2 - \lambda_1(R_1 + d)L_1][\lambda_2L_1L_2 - \lambda_2R_1L_2 - \lambda_1R_1L_1]}. \quad (10.57)$$

If the separation d is small compared with R_1, then the longitudinal magnification can be written (refs. 10.11–10.14) as

$$M^r_{\text{long}} = \frac{d_r}{d} \simeq \frac{\lambda_1\lambda_2(L_1L_2)^2}{[\lambda_2L_1L_2 - \lambda_2R_1L_2 - \lambda_1R_1L_1]^2}, \qquad \text{for } d \ll R_1. \quad (10.58)$$

Furthermore, if we recall the lateral magnification for the real image reconstructions of eq. (10.44), then we can write down the following relation (refs. 10.11–10.14):

$$M^r_{\text{long}} \simeq \frac{\lambda_1}{\lambda_2}(M^r_{\text{lat}})^2, \qquad \text{for } d \ll R_1. \quad (10.59)$$

Similarly, for the virtual image reconstructions, we can show that

$$E_v(\boldsymbol{\sigma}; k_2) = C_1'\delta\left[\alpha' + l_1'\left(\frac{b}{L_1} + \frac{\lambda_2 a}{\lambda_1 L_1}\right), \beta'\right]_{l=l_1'} + C_2'\delta\left[\alpha' + l_2'\left(\frac{b}{L_2} + \frac{\lambda_2 a}{\lambda_1 L_1}\right), \beta'\right]_{l=l_2'}, \quad (10.60)$$

where

$$l_1' = \frac{\lambda_1L_1L_2R_1}{\lambda_2R_1L_2 - \lambda_2L_1L_2 - \lambda_1L_1R_1},$$

$$l_2' = \frac{\lambda_1L_1L_2(R_1 + d)}{\lambda_2L_2(R_1 + d) - \lambda_2L_1L_2 - \lambda_1L_1(R_1 + d)}.$$

The corresponding longitudinal magnification is

$$M_{\text{long}}^{\text{v}} = \frac{d_{\text{v}}}{d} = \frac{\lambda_1\lambda_2(L_1L_2)^2}{[\lambda_2L_2R_1 - \lambda_2L_1L_2 - \lambda_1L_1R_1]^2}, \qquad \text{for } d \ll R_1. \quad (10.61)$$

From eq. (10.46), once again we can write (refs. 10.11–10.14)

$$M_{\text{long}}^{\text{v}} \simeq \frac{\lambda_1}{\lambda_2}(M_{\text{lat}}^{\text{v}})^2, \qquad \text{for } d \ll R_1. \quad (10.62)$$

Furthermore, from eqs. (10.58) and (10.61), we have

$$M_{\text{long}}^{\text{v}} \leq M_{\text{long}}^{\text{r}}, \quad (10.63)$$

where the equality holds for plane reference and illuminating beams.

It is interesting to note that, as can be seen from eqs. (10.56) and (10.60) or from fig. 10.17, there is a translational distortion (i.e., the image is twisted for a three-dimensional object). In practice it is possible to remove these translational distortions of the real and virtual images by setting the illuminating beam at

$$b = \pm\frac{\lambda_2L_2}{\lambda_1L_1}a, \quad (10.64)$$

where the $+$ and $-$ signs are for the removal of the real and virtual image

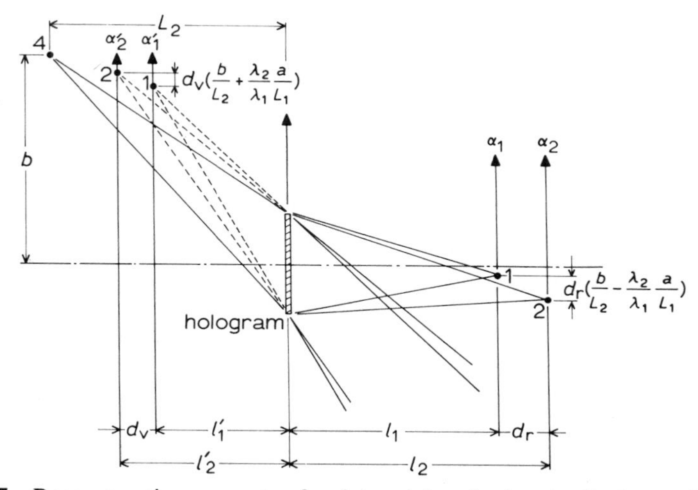

Fig. 10.17 Reconstruction geometry for determining the longitudinal magnifications. A divergent monochromatic source is at 4.

translational distortions, respectively. The translational distortion for both real and virtual images may be removed by setting $a = b = 0$. However, by doing this, we lose the separation of the real, virtual, and zero-order diffractions.

From eqs. (10.59) and (10.62), a distortion due to different lateral and longitudinal magnifications may be expected for a three-dimensional object. However, this distortion will be minimized if we set $M_{\text{lat}} = \lambda_2/\lambda_1$. In this case, eqs. (10.59) and (10.62) give

$$M_{\text{lat}} = M_{\text{long}}, \qquad \text{for } d \ll R_1. \tag{10.65}$$

10.3 RESOLUTION LIMITS

In general, the lateral holographic resolutions are limited by the size of the hologram aperture, the spatial frequency limit of the recording medium, and aberrations in the wave front reconstruction. Only the first two effects are discussed here. The longitudinal resolution is limited by the frequency bandwidth of the illuminating beam, which is also demonstrated.

Lateral Resolution Limits

Let us consider first the resolution limit imposed by the size of the hologram aperture. Recall eq. (10.41), where the surface integral over the hologram aperture S is not assumed to be of infinite extent, but finite within the hologram aperture. Thus

$$T(\boldsymbol{\rho}; k) = 0 \qquad \text{for} \qquad |x| > \frac{Lx}{2},\ |y| > \frac{Ly}{2}. \tag{10.66}$$

Then at the distance where the real images are reconstructed, the solution of eq. (10.41) is

$$\begin{aligned} E(\boldsymbol{\sigma}: k_2) = {} & C_1 L_x L_y \frac{\sin\left[\dfrac{\pi L_x}{l\lambda_2}(\alpha + \alpha_1)\right]}{\dfrac{\pi L_x}{l\lambda_2}(\alpha + \alpha_1)} \frac{\sin\left(\dfrac{\pi L_y}{l\lambda_2}\beta\right)}{\dfrac{\pi L_y}{l\lambda_2}\beta} \\ & + C_2 L_x L_y \frac{\sin\left[\dfrac{\pi L_x}{l\lambda_2}(\alpha + \alpha_2)\right]}{\dfrac{\pi L_x}{l\lambda_2}(\alpha + \alpha_2)} \frac{\sin\left(\dfrac{\pi L_y}{l\lambda_2}\beta\right)}{\dfrac{\pi L_y}{l\lambda_2}\beta}, \end{aligned} \tag{10.67}$$

where

$$\alpha_1 = l\left(\frac{b}{L_2} - \frac{\lambda_2 h}{2\lambda_1 R_1} - \frac{\lambda_2 a}{\lambda_1 L_1}\right),$$

$$\alpha_2 = l\left(\frac{b}{L_2} + \frac{\lambda_2 h}{2\lambda_1 R_1} - \frac{\lambda_2 a}{\lambda_1 L_1}\right),$$

C_1 and C_2 are the appropriate complex constants and l is given by eq. (10.42).

If we use the Rayleigh criterion (sec. 4.6) for the lateral resolution limit, then the minimum resolvable distance for the real images (fig. 10.18) can be shown to be

$$h_{\mathrm{r,\,min}} = \frac{l\lambda_2}{L_x}. \tag{10.68}$$

Accordingly, from the real-image lateral magnification, we may conclude that

$$h \geq \frac{l\lambda_2}{L_x}(M_{\mathrm{lat}}^{\mathrm{r}})^{-1} = \frac{\lambda_1 R_1}{L_x}. \tag{10.69}$$

Here equality holds for the minimum resolvable distance of h (refs. 10.13, 10.14), i.e.,

$$h_{\mathrm{r,\,min}} = \frac{\lambda_1 R_1}{L_x}. \tag{10.70}$$

Similarly, for the lateral resolution limit of the virtual images, we can show that:

$$h \geq -\frac{l'\lambda_2}{L_x}(M_{\mathrm{lat}}^{\mathrm{v}})^{-1} = \frac{\lambda_1 R_1}{L_x}, \tag{10.71}$$

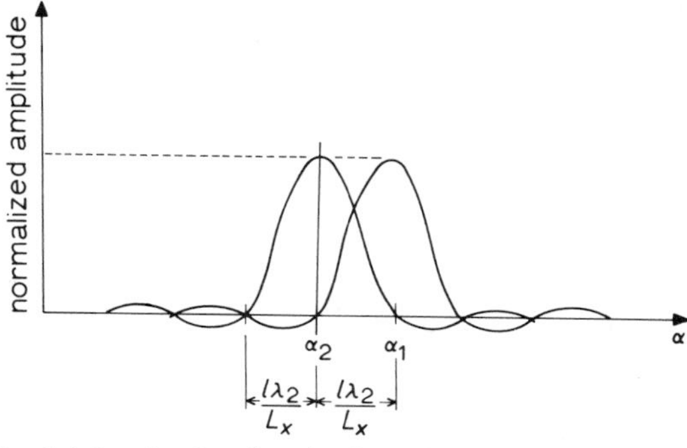

Fig. 10.18 Rayleigh criterion for the lateral resolution limit of a hologram.

where l' is defined by eq. (10.45). The corresponding minimum resolvable distance is therefore

$$h_{v,\,min} = \frac{\lambda_1 R_1}{L_x}, \tag{10.72}$$

which is identical to eq. (10.70), as expected. Accordingly, the minimum resolvable distance in holography is proportional to the wavelength of the coherent source used for the wave front construction and to the distance between the object and the recording medium, and it is inversely proportional to the size of the hologram aperture. In other words, the lateral resolution limit is set during the recording (encoding), not during the reconstruction (decoding).

Now let us consider the resolution limit set by the spatial frequency limit of the film. In this case, the limits of the surface integral of eq. (10.41) do not extend over the entire hologram aperture, but only from $x = x_2$ to $x = x_1$ and $y = -(x_1 - x_2)/2$ to $y = (x_1 - x_2)/2$ for the first integral, and $x = x_2'$ to $x = x_1'$ and $y = -(x_1' - x_2')/2$ to $y = (x_1' - x_2')/2$ for the second integral. The limiting parameters are defined by

$$\begin{aligned} x_1 &= \frac{\lambda_1 \nu_2 R_1 L_1 + \frac{1}{2} h L_1 + R_1 a}{L_1 - R_1}, \\ x_2 &= \frac{-\lambda_1 \nu_2 R_1 L_1 + \frac{1}{2} h L_1 + R_1 a}{L_1 - R_1}, \\ x_1' &= \frac{\lambda_1 \nu_2 R_1 L_1 - \frac{1}{2} h L_1 + R_1 a}{L_1 - R_1}, \\ x_2' &= \frac{-\lambda_1 \nu_2 R_1 L_1 - \frac{1}{2} h L_1 + R_1 a}{L_1 - R_1}, \end{aligned} \tag{10.73}$$

where ν_2 is the high spatial frequency limit of the recording medium. Then at the distance where the real image is reconstructed, the solution of eq. (10.41) is

$$\begin{aligned} E(\boldsymbol{\sigma}; k_2) = {} & C_1 (\Delta x)^2 \frac{J_1\left\{\frac{\pi \Delta x}{l \lambda_2}[(\alpha + \alpha_1)^2 + \beta^2]^{1/2}\right\}}{\frac{\pi \Delta x}{l \lambda_2}[(\alpha + \alpha_1)^2 + \beta^2]^{1/2}} \\ & + C_2 (\Delta x)^2 \frac{J_1\left\{\frac{\pi \Delta x}{l \lambda_2}[(\alpha + \alpha_1)^2 + \beta^2]^{1/2}\right\}}{\frac{\pi \Delta x}{l \lambda_2}[(\alpha + \alpha_2)^2 + \beta^2]^{1/2}}, \end{aligned} \tag{10.74}$$

where C_1 and C_2 are the appropriate complex constants, α_1 and α_2 are defined in eq. (10.67), J_1 is the first-order Bessel function, and

$$\Delta x = x_1 - x_2 = x_1' - x_2' = \frac{2\lambda_1 \nu_2 R_1 L_1}{L_1 - R_1}.$$

Accordingly, we can show that the minimum resolvable distance is

$$h_{\mathrm{r,\,min}} = 1.22 \frac{\lambda_1 R_1}{\Delta x}, \tag{10.75}$$

and similarly,

$$h_{\mathrm{v,\,min}} = 1.22 \frac{\lambda_1 R_1}{\Delta x}. \tag{10.76}$$

Again, eqs. (10.75) and (10.76) are identical, and the minimum resolvable distance is inversely proportional to the spatial frequency limit ν_2 of the recording medium.

Longitudinal Resolution Limit

Let us now consider the holographic longitudinal resolution limit, which is set by the finite frequency bandwidth (i.e., the quasi-monochromaticity) of the illuminating beam. We recall the two-point-object hologram of eq. (10.50). If it is assumed that the hologram is illuminated by a quasi-monochromatic divergent source with a finite bandwidth $\Delta\nu$, then the minimum resolvable longitudinal distance of the real image reconstruction may be shown to be

$$d_{\mathrm{r,min}} \simeq \Delta l_{\mathrm{r}}, \quad \text{for } d \ll R_1, \tag{10.77}$$

where

$$\Delta l_{\mathrm{r}} = l_{\mathrm{r}}' - l_{\mathrm{r}}'',$$

$$l_{\mathrm{r}}' = \frac{\lambda_1 R_1 L_1 L_2}{\lambda' L_1 L_2 - \lambda' R_1 L_2 - \lambda_1 R_1 L_1},$$

$$l_{\mathrm{r}}'' = \frac{\lambda_1 R_1 L_1 L_2}{\lambda'' L_1 L_2 - \lambda'' R_1 L_2 - \lambda_1 R_1 L_1},$$

and λ' and λ'' are the respective low and high cutoff wavelengths of the source. From eq. (10.58) we can conclude that

$$d \geq \Delta l_{\mathrm{r}} (M_{\mathrm{long}}^{\mathrm{r}})^{-1}, \tag{10.78}$$

where

$$M_{\text{long}}^{\text{r}} = \frac{\lambda_1 \lambda_2 (L_1 L_2)^2}{[\lambda_2 L_1 L_2 - \lambda_2 R_1 L_2 - \lambda_1 R_1 L_1]^2},$$

and $\lambda_2 = (\lambda' \lambda'')^{1/2}$ is the mean wavelength of the source. Therefore the minimum resolvable longitudinal distance may be shown to be

$$d_{\text{r, min}} = \Delta l_{\text{r}} (M_{\text{long}}^{\text{r}})^{-1}. \tag{10.79}$$

Similarly, for virtual image reconstruction, we can show that the minimum resolvable longitudinal distance is

$$d_{\text{r, min}} = \Delta l_{\text{v}} (M_{\text{long}}^{\text{v}})^{-1}, \tag{10.80}$$

where

$$\Delta l_{\text{v}} = l'_{\text{v}} - l''_{\text{v}},$$

$$l'_{\text{v}} = \frac{\lambda_1 L_1 R_1 L_2}{\lambda' R_1 L_2 - \lambda' L_1 L_2 - \lambda_1 L_1 R_1},$$

$$l''_{\text{v}} = \frac{\lambda_1 L_1 R_1 L_2}{\lambda'' R_1 L_2 - \lambda'' L_1 L_2 - \lambda_1 L_1 R_1},$$

$$M_{\text{long}}^{\text{v}} = \frac{\lambda_1 \lambda_2 (L_1 L_2)^2}{[\lambda_2 L_2 R_1 - \lambda_2 L_1 L_2 - \lambda_1 L_1 R_1]^2},$$

$$\lambda_2 = (\lambda' \lambda'')^{1/2}.$$

10.4 BANDWIDTH REQUIREMENTS

From the Fresnel–Kirchhoff theory (sec. 3.7), the complex light distribution on the recording medium from a diffuse object is

$$u(x, y) = \iint_{s_0} O(\xi, \eta) E_l^{\dagger}(\boldsymbol{\rho} - \boldsymbol{\xi}; k)\, d\xi\, d\eta, \tag{10.81}$$

where $O(\xi, \eta,)$ is the object function projected onto the (ξ, η) plane of the $\boldsymbol{\xi}(\xi, \eta, \zeta)$ coordinates, and s_0 denotes integration over the object surface that gives rise to O.

If we take the Fourier transform of eq. (10.81), then it becomes

$$U(p, q) = O(p, q) E_l(p, q), \tag{10.82}$$

where

$$E_l(p, q) = -\frac{i}{\lambda l}\exp\left\{-i\frac{l}{2k}(p^2 + q^2)\right\},$$

and p and q are the corresponding spatial frequencies.

Since $|E_l(p, q)| = 1/\lambda l$, we have

$$|U(p, q)|^2 = \left(\frac{1}{\lambda l}\right)^2 |O(p, q)|^2. \quad (10.83)$$

Equation (10.83) implies that the spatial power spectrum of the light distribution is proportional to the spatial power spectrum of the object function $O(\xi, \eta)$. Therefore the preservation of the spatial frequency spectrum of the object depends on the frequency spectrum response of the recording medium. The spatial frequency bandwidth is most often limited by the size of the hologram aperture but not by the film spatial frequency limit. However, as we have seen in the last section, the resolution limits in holography are set both by the size of the hologram aperture and by the spatial frequency limit of the film. It is clear that whichever of these resolution limits comes first will set the resolution limit of the hologram.

As shown in fig. 10.19, the resolution requirements may be determined from the field angles as follows. We let the tips of the object act like secondary point radiators; then the complex light distributions on the recording medium due to

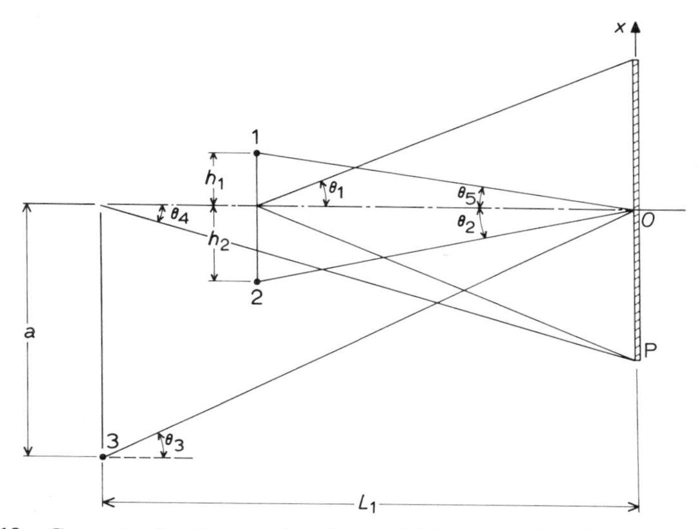

Fig. 10.19 Geometry for determining the spatial frequency bandwidth of a hologram. 3, divergent source; P, photographic plate.

these radiators are:

$$
\begin{aligned}
u_1(\boldsymbol{\rho}; k) &= A_1 \exp\left\{ ik\left[R_1 + \frac{1}{2R_1}[(x - h_1)^2 + y^2] \right] \right\}, \\
u_2(\boldsymbol{\rho}; k) &= A_2 \exp\left\{ ik\left[R_1 + \frac{1}{2R_1}[(x + h_2)^2 + y^2] \right] \right\}, \\
u_3(\boldsymbol{\rho}; k) &= A_3 \exp\left\{ ik\left[L_1 + \frac{1}{2L_1}[(x + a)^2 + y^2] \right] \right\}.
\end{aligned}
\tag{10.84}
$$

For simplicity, the A's are assumed to be positive real constants. Then the light intensity distribution on the recording medium is

$$
\begin{aligned}
I(\boldsymbol{\rho}; k) &= (u_1 + u_2 + u_3)(u_1 + u_2 + u_3)^* \\
&= A_1^2 + A_2^2 + A_3^2 + 2A_1A_2 \cos\frac{k}{R_1}(h_1 + h_2)x \\
&\quad + 2A_1A_3 \cos\left\{ k(R_1 - L_1) + \frac{k}{2R_1}[(x - h_1)^2 + y^2] \right. \\
&\quad \left. - \frac{k}{2L_1}[(x + a)^2 + y^2] \right\} \\
&\quad + 2A_2A_3 \cos\left\{ k(R_1 - L_1) + \frac{k}{2R_1}[(x + h_2)^2 + y^2] \right. \\
&\quad \left. - \frac{k}{2L_1}[(x + a)^2 + y^2] \right\}.
\end{aligned}
\tag{10.85}
$$

From eq. (10.85), the spatial phase shift due to u_1 and u_3 is

$$
\phi_{13}(x, y) = k\left\{ R_1 - L_1 + \frac{l}{2R_1}[(x - h_1)^2 + y^2] - \frac{1}{2L_1}[(x + a)^2 + y^2] \right\},
\tag{10.86}
$$

and that due to u_2 and u_3 is

$$
\phi_{23}(x, y) = k\left\{ R_1 - L_1 + \frac{1}{2R_1}[(x + h_2)^2 + y^2] - \frac{1}{2L_1}[(x + a)^2 + y^2] \right\}.
\tag{10.87}
$$

The corresponding spatial frequencies in the x coordinate can be determined to be

$$p_{13}(x) = \frac{\partial \phi_{13}(x, y)}{\partial x} = k\left[\left(\frac{1}{R_1} - \frac{1}{L_1}\right)x - \frac{h_1}{R_1} - \frac{a_1}{L_1}\right] \tag{10.88}$$

and

$$p_{23}(x) = \frac{\partial \phi_{23}(x, y)}{\partial x} = k\left[\left(\frac{1}{R_1} - \frac{1}{L_1}\right)x + \frac{h_2}{R_1} - \frac{a_1}{L_1}\right]. \tag{10.89}$$

Thus the highest positive spatial frequency allowed by the size of the hologram aperture is

$$\nu_{\text{high}} = \frac{1}{2\pi} p_{23}(x)\bigg|_{x=L_x/2} = \frac{1}{\lambda}\left[\left(\frac{1}{R_1} - \frac{1}{L_1}\right)\frac{L_x}{2} + \frac{h_2}{R_1} - \frac{a}{L_1}\right]. \tag{10.90}$$

This can be written in terms of the field angles,

$$\nu_{\text{high}} = \frac{1}{\lambda}[\tan\theta_1 - \tan\theta_4 + \tan\theta_2 - \tan\theta_3]. \tag{10.91}$$

Similarly, the negative spatial frequency limit is

$$\nu_{\text{low}} = \frac{1}{2\pi} p_{13}(x)\bigg|_{x=-L_x/2} = \frac{1}{\lambda}\left[\left(-\frac{1}{R_1} + \frac{1}{L_1}\right)\frac{L_x}{2} - \frac{h_1}{R_1} - \frac{a}{L_1}\right], \tag{10.92}$$

or equivalently,

$$\nu_{\text{low}} = \frac{1}{\lambda}[-\tan\theta_1 + \tan\theta_4 - \tan\theta_5 - \tan\theta_3]. \tag{10.93}$$

Therefore, the spatial frequency bandwidth limited by the size of the hologram aperture is

$$\Delta\nu = \nu_{\text{high}} - \nu_{\text{low}} = \frac{1}{\lambda}\left[\left(\frac{1}{R_1} - \frac{1}{L_1}\right)L_x + \frac{1}{R_1}(h_1 + h_2)\right], \tag{10.94}$$

or equivalently,

$$\Delta\nu = \frac{1}{\lambda}[2\tan\theta_1 - 2\tan\theta_4 + \tan\theta_2 + \tan\theta_5]. \tag{10.95}$$

It is interesting to note that, if the size of the hologram aperture is sufficiently small (i.e., as it approaches a point), the spatial frequency bandwidth reduces to

$$\Delta\nu_1 \simeq \frac{1}{\lambda R_1}(h_1 + h_2) = \frac{1}{\lambda}[\tan\theta_2 + \tan\theta_5]. \tag{10.96}$$

On the other hand, if the object is sufficiently small, the spatial frequency bandwidth becomes

$$\Delta\nu_2 \simeq \frac{L_x}{\lambda}\left(\frac{1}{R_1} - \frac{1}{L_1}\right) = \frac{2}{\lambda}[\tan\theta_1 - \tan\theta_4]. \tag{10.97}$$

Thus we conclude that

$$\Delta\nu \leq \Delta\nu_1 + \Delta\nu_2. \tag{10.98}$$

It may be emphasized that the spatial frequency bandwidth in wave front recording depends upon two factors: the field angles from the object to the recording medium, and the angles of the beams received from the tips of the object (i.e., s_1 and s_2).

Obviously, if the object is large compared to the hologram aperture, then eq. (10.98) reduces to

$$\Delta\nu \simeq \Delta\nu_1 = \frac{1}{\lambda}[\tan\theta_2 + \tan\theta_5]. \tag{10.99}$$

On the other hand, if the size of the hologram aperture is large compared with the object, then

$$\Delta\nu \simeq \Delta\nu_2 = \frac{2}{\lambda}[\tan\theta_1 - \tan\theta_4]. \tag{10.100}$$

Furthermore, from eqs. (10.94) or (10.95) it can be seen that a reduction of spatial frequency bandwidth is possible, if we place the divergent reference beam (i.e., s_3), on the same plane as the object (ref. 10.15). The spatial frequency bandwidth is then

$$\Delta\nu = \frac{1}{\lambda}(\tan\theta_2 + \tan\theta_5) = \Delta\nu_1, \tag{10.101}$$

which is independent of the size of the object relative to the hologram aperture. If the size of the object is large compared with the hologram aperture, then the reduction in spatial frequency is not significant. However if the size of the hologram aperture is large compared with the size of the object, then the spatial frequency bandwidth reduction would be considerable. This second condition may be of some importance in the application of wave front reconstruction to microscopy.

Similarly, if the divergent reference beam in fig. 10.19 is replaced by an oblique plane wave, then it can be shown that the positive spatial frequency limit is

$$\nu'_{\text{high}} = \frac{1}{\lambda}[\tan \theta_1 + \tan \theta_2 - \sin \theta], \tag{10.102}$$

and the negative spatial frequency limit is

$$\nu'_{\text{low}} = \frac{1}{\lambda}[-\tan \theta_1 - \tan \theta_5 - \sin \theta], \tag{10.103}$$

where θ is the oblique angle of incidence of the reference wave. Therefore, the spatial frequency bandwidth allowed by the size of the hologram aperture is

$$\Delta\nu' = \Delta\nu'_{\text{high}} - \Delta\nu'_{\text{low}} = \frac{1}{\lambda}[2 \tan \theta_1 + \tan \theta_2 + \tan \theta_5]. \tag{10.104}$$

Again, if the size of the hologram aperture is sufficiently small, then the spatial frequency bandwidth is

$$\Delta\nu'_1 = \Delta\nu_1 \simeq \frac{1}{\lambda}[\tan \theta_2 + \tan \theta_5]. \tag{10.105}$$

On the contrary, if the size of object is very small, then the spatial frequency bandwidth is

$$\Delta\nu'_2 \simeq \frac{2}{\lambda} \tan \theta_2. \tag{10.106}$$

Thus we can conclude that

$$\Delta\nu' \leq \Delta\nu'_1 + \Delta\nu'_2. \tag{10.107}$$

Obviously, the reduction of spatial frequency bandwidth for a plane reference wave is impossible.

Applying the preceding considerations to a general three-dimensional object (fig. 10.20), we obtain the spatial frequency bandwidths

$$\Delta\nu_x = \frac{1}{\lambda}[2 \tan \theta_1 - 2 \tan \theta_4 + \tan \theta_2 + \tan \theta_5] \tag{10.108}$$

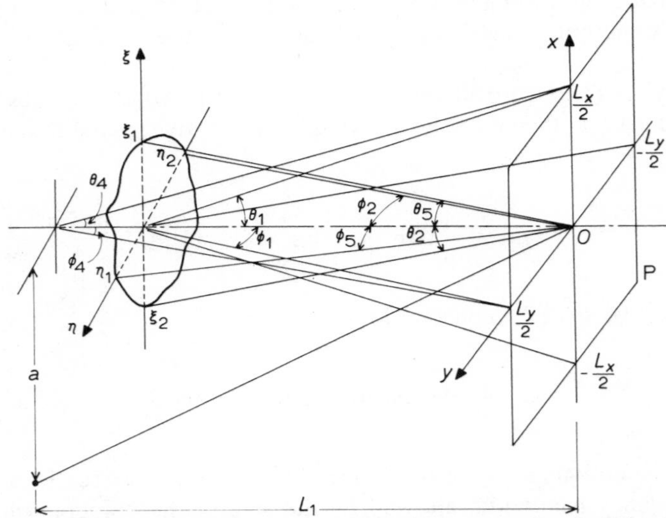

Fig. 10.20 Geometry for determining the spatial frequency bandwidth of a hologram of an extended object. 3, divergent reference source; P, photographic plate.

and

$$\Delta\nu_y = \frac{1}{\lambda}[2 \tan \phi_1 - 2 \tan \phi_4 + \tan \phi_2 + \tan \phi_5], \qquad (10.109)$$

where $\Delta\nu_x$ and $\Delta\nu_y$ are the respective bandwidths in the x and y coordinate axes. If the divergent reference beam is replaced by a plane reference wave, then the corresponding x and y spatial frequency bandwidths are

$$\Delta\nu_x' = \frac{1}{\lambda}[2 \tan \theta_1 + \tan \theta_2 + \tan \theta_5] \qquad (10.110)$$

and

$$\Delta\nu_y' = \frac{1}{\lambda}[2 \tan \phi_1 + \tan \phi_2 + \tan \phi_5]. \qquad (10.111)$$

10.5 HOLOGRAPHIC ABERRATIONS

In sec. 10.2 we developed a general and simple procedure for calculating the holographic image magnification or demagnification. Those calculations, however, were based on the paraxial approximation, for which the reconstructed image exhibits no aberrations. In practice, however, magnified (or demagnified) hologram images frequently suffer from aberrations. It is the purpose of this section to evaluate the five primary aberrations (ref. 10.16) of holographic images, namely: spherical aberration, coma, astigmatism, curvature of field, and distortion. We also discuss the conditions under which these aberrations may be minimized or eliminated.

Let us now identify the essential geometrical parameters of wave front construction and reconstruction, as shown in fig. 10.21. In this figure a plane wave front is used for recording, and a spherical wave front is used for reconstruction. The explicit form of the complex light field of the hologram image on the $\boldsymbol{\sigma}$ coordinate system, for a two-dimensional object, is

$$E(\boldsymbol{\sigma}; k_2) = C \iint_{S_2} \left\{ \iint_{S_1} O^*(\xi, \eta) \exp[ik_1(x \sin \theta_1 - r_1)] \, d\xi \, d\eta \right\} \times \exp\left[ik_2\left(\frac{\rho^2}{2R} - x \sin \theta_2\right)\right] \exp(ik_2 r_2) \, dx \, dy. \tag{10.112}$$

where C is a complex constant, $O^*(\xi, \eta)$ is the two-dimensional conjugate object function, $k_1 = 2\pi/\lambda_1$, with λ_1 the recording wavelength, $k_2 = 2\pi/\lambda_2$, with λ_2 the reconstructing wavelength, $\rho^2 = x^2 + y^2$, and S_1 and S_2 denote the surface integrals of the object function and the transmission function of the hologram, respectively. It is also clear that the first surface integral represents the wave front construction, the second exponential represents the hologram illumination, and the last exponential represents the diffraction from the hologram.

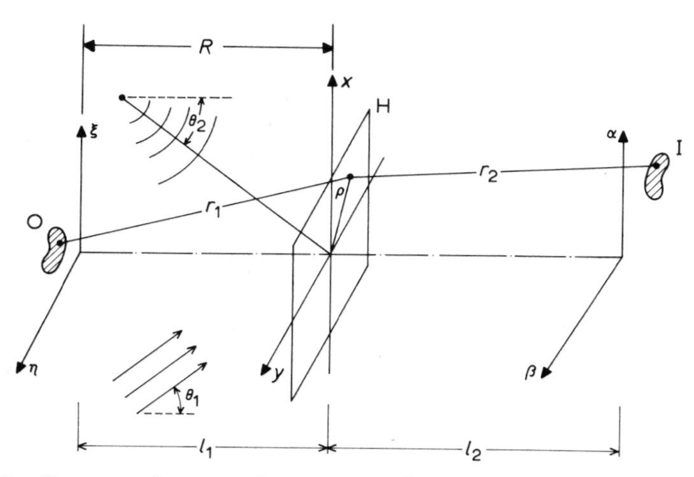

Fig. 10.21 Geometry for wave front construction and reconstruction. O, object; H, hologram; I, image. Divergent reconstruction beam incident at angle θ, and plate reference wave incident at θ_1.

Referring to fig. 10.21, we can write the distances r_1 and r_2 as

$$r_1 = l_1\left[1 + \frac{(x-\xi)^2 + (y-\eta)^2}{l_1^2}\right]^{1/2} \tag{10.113}$$

and

$$r_2 = l_2\left[1 + \frac{(\alpha-x)^2 + (\beta-y)^2}{l_2^2}\right]^{1/2}. \tag{10.114}$$

Then by binomial expansion (ref. 10.17), eqs. (10.113) and (10.114) can be written as

$$r_1 = l_1 + \frac{1}{2l_1}[(x-\xi)^2 + (y-\eta)^2] - \frac{1}{8l_1^3}[(x-\xi)^2 + (y-\eta)^2]^2 + \ldots, \tag{10.115}$$

$$r_2 = l_2 + \frac{1}{2l_2}[(\alpha-x)^2 + (\beta-y)^2] - \frac{1}{8l_2^3}[(\alpha-x)^2 + (\beta-y)^2]^2 + \ldots. \tag{10.116}$$

If the first two terms of eqs. (10.115) and (10.116) are retained, we have the paraxial approximations, and the explicit form of eq. (10.112) will be

$$\begin{aligned} E(\boldsymbol{\sigma}; k_2) = C' \iint_{S_2} \Bigg\{ \iint_{S_1} & O(\xi,\eta) \\ & \times \exp\left[ik_1\left(x \sin\theta_1 - \frac{(x-\xi)^2 + (y-\eta)^2}{2l_1}\right)\right] d\xi\, d\eta\Bigg\} \\ & \times \exp\left[ik_2\left(\frac{\rho^2}{2R} - x\sin\theta_2\right)\right] \\ & \times \exp\left[ik\frac{(\alpha-x)^2 + (\beta-y)^2}{2l_2}\right] dx\, dy, \end{aligned} \tag{10.117}$$

where C is an appropriate complex constant. Equation (10.117) is the form we have used in the previous sections. However, if we retain the first three terms of eqs. (10.115) and (10.116), then the third-order aberrations may be calculated.

It may be emphasized that aberrations of the wave front construction and reconstruction process depend on the exponential argument containing r_1 and r_2. In the usual paraxial approximation, the quadratic exponential factor ρ^2 [for example, in eq. (10.53)] is eliminated by imposing the lens condition

$$\frac{1}{R} + \frac{1}{l_2} = \frac{\lambda_2}{\lambda_1} \frac{1}{l_1}. \tag{10.118}$$

However, in the nonparaxial case it is no longer sufficient to impose the condition expressed by eq. (10.118) in order to eliminate the higher order exponential term. These nonvanishing terms in the exponent constitute the aberrations in hologram images. To investigate the aberrations, we can begin with the evaluation of the phase factor $\Delta\phi = k_2 r_2 - k_1 r_1$ of the construction-reconstruction process. After a long but straightforward calculation, $\Delta\phi$ is seen to be

$$\Delta\phi = -\frac{1}{8}\left(\frac{k_2}{l_2^3} - \frac{k_1}{l_1^3}\right)\rho^4 + \frac{1}{2}\left(\frac{Mk_2}{l_2^3} - \frac{k_1}{l_1^3}\right)\rho^2 K^2$$
$$- \frac{1}{2}\left(\frac{M^2 k_2}{l_2^3} - \frac{k_1}{l_1^3}\right)K^4 - \frac{1}{4}\left(\frac{M^2 k_2}{l_2^3} - \frac{k_1}{l_1^3}\right)\rho^2\tau^2 + \frac{1}{2}\left(\frac{M^3 k_2}{l_2^3} - \frac{k_1}{l_1^3}\right)\tau^2 K^2. \tag{10.119}$$

where $\rho^2 = x^2 + y^2$, $\tau^2 = \xi^2 + \eta^2$, $K^2 = \xi x + \eta y$, and $M = \lambda_2 l_2/\lambda_1 l_1$. The lateral magnification M can be derived from eq. (10.44). By comparing eq. (10.119) with the general treatment of lens aberrations given in sec. 5.3 of ref. 10.16, we can see that the first term of eq. (10.119) is the spherical aberration; the second term is the coma; the third term is the astigmatism; the fourth term is the curvature of field; and the last term is the distortion.

We see that the five primary aberrations that occur for physical lenses also occur in holography. It may also be noted that the higher order (beyond the third-order) aberrations in holography may be calculated; however, we do not attempt that here.

Now let us consider the conditions under which these primary holographic aberrations may be corrected. In order to do so, we set each term of eq. (10.119) equal to zero. The results are tabulated in table 10.1 (taken from ref. 10.13).

From table 10.1 it is clear that, if we correct one of the aberrations, the others generally cannot be corrected. However, there are two exceptions, namely: for unity lateral magnification ($M = 1$), all the aberrations will vanish; and in any case, the astigmatism and curvature of field can be corrected together.

We should note in conclusion that for a more general holographic process the reference and the reconstructing processes will both employ spherical wave fronts. The third-order aberrations in this general case can be determined by a procedure similar to the foregoing. For such an investigation the reader can refer to the article by Meier (ref. 10.11). According to this paper, all five primary aberrations can be made to disappear simultaneously if, and only if, the condition of unity magnification is met. It is also clear that to achieve unity magnification the construction reference beam and the reconstruction beam are both required to be plane waves and to have the same wavelength.

Table 10.1
Aberrations and Conditions for Their Correction

Aberration	Condition
Spherical aberration	$\frac{\lambda_2}{\lambda_1}\left(\frac{l_2}{l_1}\right)^3 = 1$
Coma	$\frac{\lambda_2}{\lambda_1}\left(\frac{l_2}{l_1}\right)^3 = M;\ l_1 = l_2$
Astigmatism and curvature of field	$\frac{\lambda_2}{\lambda_1}\left(\frac{l_2}{l_1}\right)^3 = M^2;\ \frac{\lambda_2}{\lambda_1} = \frac{l_2}{l_1}$
Distortion	$\frac{\lambda_2}{\lambda_1}\left(\frac{l_2}{l_1}\right)^3 = M^3;\ \lambda_1 = \lambda_2$

10.6 SPATIALLY INCOHERENT HOLOGRAPHY

The holographic process was originally conceived to be a coherent imaging process. However, several techniques are available in which a spatially incoherent source may be used. Such a technique was first suggested by Mertz and Young (ref. 10.18), and later the theory and experiment were extended by Lohmann (ref. 10.19), Stroke and Restrick (ref. 10.20), and Cochran (ref. 10.21). In this section we adopt the technique suggested by Cochran.

It is well known that the light scattered by a point on an incoherently illuminated object will not interfere with the light scattered by any other point of the object. However, with special arrangement of the optical setup, it is possible to split the light field from each point of the illuminated object and then rejoin the parts in such a way that an interference pattern is formed. Thus each object point will be recorded in a suitable interference pattern. If the resulting hologram is illuminated by a coherent source, then each of the fringe patterns will be reconstructed into a unique image point. In other words, in making an incoherent hologram, each object point is made to form its own reference beam.

Figure 10.22 is a diagram of the triangular interferometer devised by Cochran. This device consists of two lenses (L_1 and L_2) of different focal lengths (f_1 and f_2). The lenses are separated by a path length $f_1 + f_2$ and their focal points coincide at P, as shown in the figure. The object plane O and the recording plane H are each located at a path length f_1 from lens L_1, and a path length f_2 from lens L_2. Light may travel from plane O to plane H by two different paths, namely the clockwise and counterclockwise paths around the interferometer. For example, for the clockwise path, light travels a distance f_1 from plane O to lens L_1 by means of the reflection from the beam splitter BS. From lens L_1 to lens L_2, the light ray travels a distance of $f_1 + f_2$. From lens L_2 to plane H the light

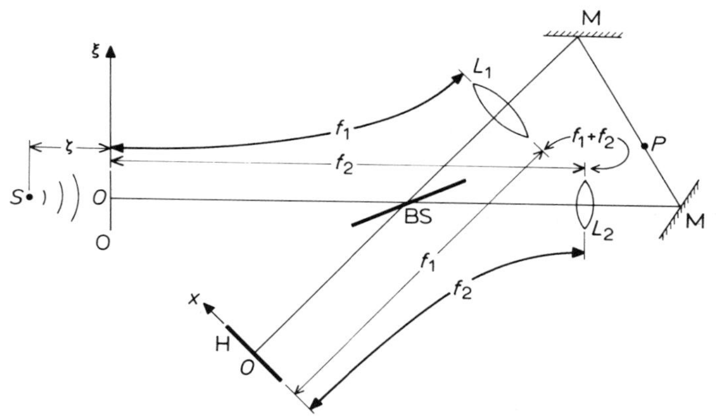

Fig. 10.22 Triangular interferometer for producing incoherent holograms. H, hologram; Ms, mirrors; BS, beam splitter.

travels a distance of f_2 by means of BS. Because of this arrangement of the optical system, it is clear that an illuminated object located at plane O will be imaged onto plane H. It is also clear that, with this particular arrangement of the lenses, the image produced on plane H is magnified by an amount

$$M_1 = -\frac{f_2}{f_1}, \qquad \text{for the clockwise path.} \tag{10.120}$$

In a similar manner, it can be seen that the illuminated object at plane O will also be imaged onto plane H on a counterclockwise path. However, the image magnification is

$$M_2 = -\frac{f_1}{f_2} = \frac{1}{M_1}, \qquad \text{for the counterclockwise path.} \tag{10.121}$$

For simplicity in illustration, let us consider a single point object S, located at distance ζ behind O. The complex light distribution on O due to S is

$$O(\xi, \eta) = C \exp\left[i\frac{k}{2\zeta}(\xi^2 + \eta^2) \right], \tag{10.122}$$

where C is a complex constant and $k = 2\pi/\lambda$.

From the discussion given previously it is clear that, due to the light distribution given by eq. (10.122), two spherical wave fronts will be formed at plane H, with magnifications of M_1 and M_2. The resultant complex light field at plane H is the combination of these two wave fronts.

$$u(x, y) = C_1 \exp\left\{i\frac{k}{2\zeta}\left[\left(\frac{x}{M_1}\right)^2 + \left(\frac{y}{M_1}\right)^2\right]\right\} + C_2 \exp\left\{i\frac{k}{2\zeta}\left[\left(\frac{x}{M_2}\right)^2 + \left(\frac{y}{M_2}\right)^2\right]\right\}, \quad (10.123)$$

where C_1 and C_2 are the appropriate complex constants. The corresponding irradiance is

$$I(x, y) = |C_1|^2 + |C_2|^2 + 2|C_1|\,|C_2| \cos\left[\frac{k}{2\zeta}(M_2^2 - M_1^2)\rho^2 + \phi\right], \quad (10.124)$$

where $\rho^2 = x^2 + y^2$, $\phi =$ phase angle between C_1 and C_2, and $M_1 = M_2^{-1}$.

If we insert a photographic plate at plane H to record the interference pattern given by eq. (10.124), the resultant transmittance function of the recorded plate is

$$T(x, y) = K_1 + K_2 \cos\left[\frac{k}{2\zeta}(M_2^2 - M_1^2)\rho^2 + \phi\right], \quad (10.125)$$

where K_1 and K_2 are the proportionality constants.

It can be seen that eq. (10.125) describes a Fresnel zone lens construction with a focal length of

$$f = \zeta\left(\frac{1}{M_2^2 - M_1^2}\right). \quad (10.126)$$

By the substitution of eqs. (10.120) and (10.121), eq. (10.126) can be written as

$$f = \zeta\left(\frac{f_1^2 f_2^2}{f_1^4 - f_2^4}\right). \quad (10.127)$$

If the transparency of eq. (10.125) is illuminated by a coherent source, it is clear that virtual and real hologram images will be reconstructed.

Let us now extend this concept to a more complicated case; that of a multitude of mutually incoherent point sources. Each of these sources will generate an interference pattern of its own; however, the patterns are incoherent with respect to each other. The total irradiance is the sum of the individual irrandiances due to each of the point sources. The resulting transmittance function of the recorded transparency will therefore be the summation of these interference patterns. Each point source determines the center and focal length of a Fresnel zone lens, and thus a three-dimensional holograph image will be formed. A photograph of an incoherent hologram image reconstruction is given in fig. 10.23.

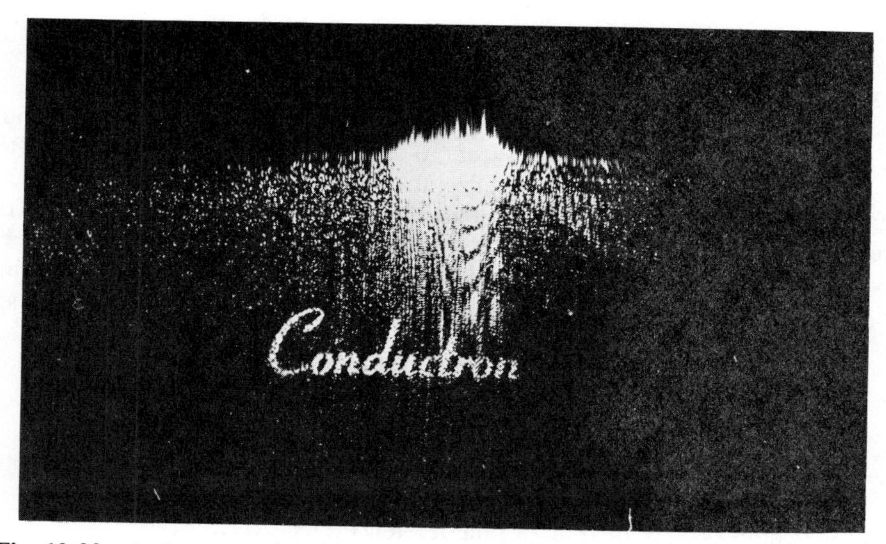

Fig. 10.23 An image produced by a spatially incoherent hologram. (By permission of P. J. Peters).

Although incoherent holography is an attractive technique, there are several drawbacks in practice at present. One of the main disadvantages is that each of the elementary Fresnel zone lenses is formed by the interference of two extremely small portions of the total light incident on the recording medium, whereas in coherent holography the light field from each object point interferes with the whole reference beam incident on the film. Another disadvantage is the accumulation of the bias level of the wave front recording; thus in practice incoherent holography has only been successfully applied to a relatively small number of resolvable point objects. These major drawbacks may become insignificant, depending on the future research and newly developed techniques in incoherent holography.

10.7 REFLECTION HOLOGRAPHY

By a simple rearrangement of the optical setup for the coherent wave front construction process, it is possible to obtain holographic image reconstruction by means of incoherent white-light illumination. This image reconstruction process is entirely dependent upon reflection from the recorded hologram, rather than on transmission through the hologram. Since this new technique mainly utilizes the thick emulsion of the photographic plate, reflection (or white-light) holography is also known as thick emulsion holography.

The wave front construction of a reflection hologram, which we illustrate in a moment, is very similar to the basic concept underlying Lippmann's color photography (ref. 10.23). Thus reflection holography is also known as color holography. The basic theory of reflection holography was first described by Denisyuk (ref. 10.24) in 1962. However, the concept was not fully appreciated here until 1966, when a sequence of papers was published by Stroke and Labeyrie (ref. 10.25), by Lin et al. (ref. 10.26), and by Leith et al. (ref. 10.27).

In this section we first study the reflection holography of a simple point object and then extend the technique to holography of a three-dimensional object. To record a reflection hologram, the coherent object light field and the reference waves are introduced from opposite sides of the recording medium, as shown in fig. 10.24. From this figure, we may write the complex light distribution in the photographic emulsion as

$$u(\boldsymbol{\rho}; k) = A \exp\left\{ik\left[(R + z) + \frac{\rho^2}{2(R + z)}\right]\right\}, \qquad \text{for } -\Delta z \leq z \leq 0, \tag{10.128}$$

and the complex light field due to the reference plane wave as

$$v(\boldsymbol{\rho}; k) = B \exp[-ik(R + z - x \sin \theta)], \qquad \text{for } -\Delta z \leq z \leq 0, \tag{10.129}$$

where A and B are complex constants, $\rho^2 = x^2 + y^2$, and $k = 2\pi/\lambda$. The corresponding irradiance is therefore,

$$\begin{aligned} I(\boldsymbol{\rho}; k) &= (u + v)(u + v)^* \\ &= |A|^2 + |B|^2 + 2|A|\,|B| \\ &\quad \times \cos\left\{k\left[2(R + z) + \frac{\rho^2}{2(R + z)} - x \sin \theta\right] + \phi\right\}, \end{aligned}$$
$$\text{for } -\Delta z \leq z \leq 0, \tag{10.130}$$

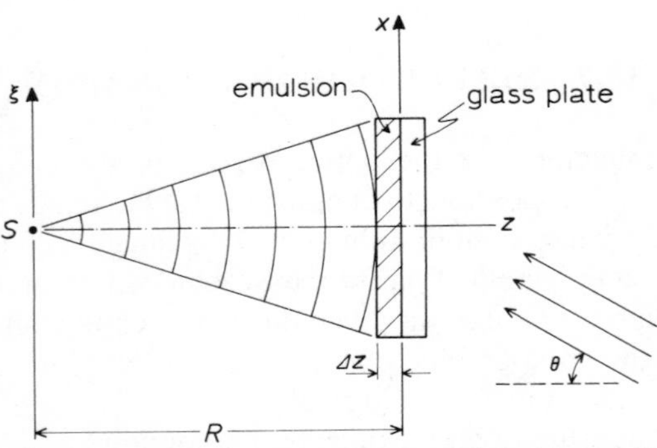

Fig. 10.24 Coherent wave front construction of a point-object reflection hologram. S, monochromatic point source. An oblique monochromatic reference wave is incident from behind the photographic plate.

where ϕ is the phase angle between A and B. For $\Delta z \gg \lambda$, it may be seen from eq. (10.130) that the irradiance is sinusoidally varying along the z direction within the emulsion. If this wave front recording is linear in the developed photographic grain density, then the density function of the recorded reflection hologram may be written as

$$D(\boldsymbol{\rho}; k) = K_1 + K_2 \cos\left\{k\left[2(R + z) + \frac{\rho^2}{2(R + z)} - x \sin \theta\right] + \phi\right\},$$
$$\text{for } -\Delta z \leq z \leq 0, \quad (10.131)$$

where K_1 and K_2 are the appropriate positive constants.

Indeed it can be seen that there is a sequence of very thin holograms arranged in parallel in the photographic emulsion, which act as reflecting planes. Since in practice the reference angle θ is very small, the spacing of these reflecting planes is about $\lambda/2$. If we assume that the reflectance of these thin holograms is proportional to the density of the developed photographic grains, then the reflectance function of the overall hologram may be written as

$$r(\boldsymbol{\rho}; k) = K_1' + K_2' \cos\left\{k\left[2(R + z) + \frac{\rho^2}{2(R + z)} - x \sin \theta\right] + \phi\right\},$$
$$\text{for } -\Delta z \leq z \leq 0, \quad (10.132)$$

where K_1' and K_2' are the appropriate positive constants.

Now, if the hologram described by eq. (10.132) is illuminated by incoherent white light (fig. 10.25), then only a single reconstructing wavelength that satisfies Bragg's law (ref. 10.28) will be strongly reflected. Thus by means of the Fresnel–Kirchhoff theory, the reflected complex light field of this selected wavelength can be calculated to be

$$\begin{aligned} E(\boldsymbol{\sigma}; k) = {} & C_1(z) \exp(-ik\alpha \sin \theta) \\ & + C_2(z) \exp\left\{\frac{k}{4(R + z)}[(\alpha - 2(R + z) \sin \theta)^2 + \beta^2]\right\} \\ & + C_3(z)\delta(\alpha, \beta), \qquad \text{for } -\Delta z \leq z \leq 0, \end{aligned} \quad (10.133)$$

where $C_1(z)$, $C_2(z)$, and $C_3(z)$ are complex functions of z, and $\delta(\alpha, \beta)$ is the Dirac delta fuanction. Needless to say, the first term of eq. (10.133) is the zero-order diffraction, the second term is the first-order divergent term, and the last term obviously is the real hologram image term.

To reconstruct the virtual image, we illuminate the reflection hologram from behind with an oblique white light (fig. 10.26). The geometry shown in this figure for virtual image reconstruction is similar to that shown previously for the real image.

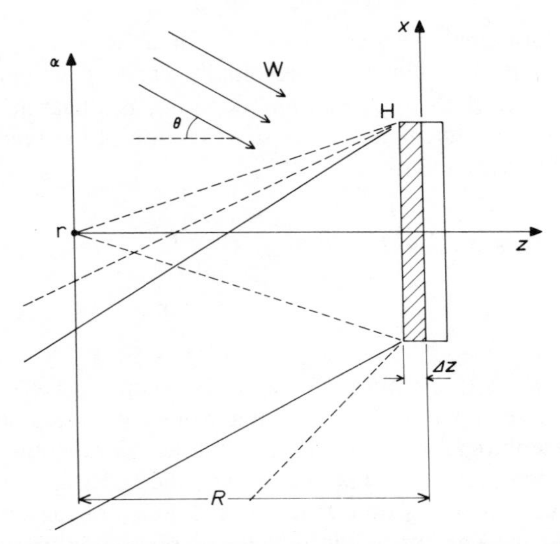

Fig. 10.25 Real image reconstruction, reflection hologram. H, hologram; W, white-light illumination; r, real image.

The above considerations can be extended to multiwavelength recording. To do so, let the object field and the reference wave of fig. 10.24 be derived from two coherent light sources of wavelengths λ_1 and λ_2. Then the complex light field in the emulsion, due to the two-wavelength point source, is

$$\begin{aligned} u(\boldsymbol{\rho}; k_1, k_2) &= u_1(\boldsymbol{\rho}; k_1) + u_2(\boldsymbol{\rho}; k_2) \\ &= A_1 \exp\left\{ik_1\left[z + \frac{\rho^2}{2(R+z)}\right]\right\} + A_2 \exp\left\{ik_2\left[z + \frac{\rho^2}{2(R+z)}\right]\right\}, \end{aligned} \tag{10.134}$$

and that due to the two-wavelength reference field is

$$\begin{aligned} v(\boldsymbol{\rho}; k_1, k_2) &= v_1(\boldsymbol{\rho}; k_1) + v_2(\boldsymbol{\rho}; k_2) \\ &= B_1 \exp[-ik_1(z - x \sin\theta)] + B_2 \exp[-ik_2(z - x \sin\theta)], \end{aligned} \tag{10.135}$$

where the A's and B's are complex constants and $\rho^2 = x^2 + y^2$. Since u and v are derived from two independent coherent sources, the time average over the exposure of $u_1 u_2^*$ and $u_1^* u_2$ may be assumed negligible. Thus the reflectance of the hologram may be written as

$$
\begin{aligned}
r(\boldsymbol{\rho}; k_1, k_2) &= K\langle I\rangle \\
&= K\Delta t\Big\{|A_1|^2 + |A_2|^2 + |B_1|^2 + |B_2|^2 \\
&\quad + 2|A_1|\,|B_1| \cos\left[k_1\left(2z + \frac{\rho^2}{2(R+z)} - x\sin\theta\right) + \phi_1\right] \\
&\quad + 2|A_2|\,|B_2| \cos\left[k_2\left(2z + \frac{\rho^2}{2(R+z)} - x\sin\theta\right) + \phi_2\right]\Big\},
\end{aligned}
\tag{10.136}
$$

where K is a proportionality constant, Δt is the exposure time, $\langle I\rangle$ denotes the time average of the irradiance over Δt, and ϕ_1 and ϕ_2 are the constant phase angles.

It may be seen from eq. (10.136) that this two-wavelength reflection hologram essentially consists of two sets of thin holograms within the emulsion, such that the spacing in one set is $\lambda_1/2$, and in the other set, $\lambda_2/2$. For real image reconstruction, a multiwavelength reflection hologram can be illuminated by an oblique white light, as shown in fig. 10.25. Then by Bragg's law and the Fresnel–Kirchhoff theory, the complex reflected field can be evaluated. The solution is

$$
\begin{aligned}
E(\boldsymbol{\sigma}; k_1, k_2) &= C_{11}(z)\exp(-ik_1\alpha\sin\theta) + C_{12}(z)\exp(-ik_2\alpha\sin\theta) \\
&\quad + C_{21}(z)\exp\left\{\frac{k_1}{4(R+z)}[(\alpha - 2(R+z)\sin\theta)^2 + \beta^2]\right\} \\
&\quad + C_{22}(z)\exp\left\{\frac{k_2}{4(R+z)}[(\alpha - 2(R+z)\sin\theta)^2 + \beta^2]\right\} \\
&\quad + C_{31}(z)\delta(k_1; \alpha, \beta) + C_{32}(z)\delta(k_2; \alpha, \beta),
\end{aligned}
\tag{10.137}
$$

where the $C(z)$'s are complex functions of z.

It can be seen from this equation that the real image is constructed by the two wavelengths λ_1 and λ_2. The image irradiance due to Δz is therefore,

$$
I_r(\boldsymbol{\sigma}; k_1, k_2) = \begin{cases} \left|\int_{-\Delta z}^{0} C_{31}(z)\,dz\right|^2 + \left|\int_{-\Delta z}^{0} C_{32}(z)\,dz\right|^2, & \alpha = \beta = 0, \\ 0, & \text{otherwise}. \end{cases}
\tag{10.138}
$$

which is the sum of the image irradiances contributed by λ_1 and λ_2.

$$
I_r(\boldsymbol{\sigma}; k_1, k_2) = I_{r,1}(\boldsymbol{\sigma}; k_1) + I_{r,2}(\boldsymbol{\sigma}; k_2),
\tag{10.139}
$$

where the subscript r refers to the real image reconstruction. For virtual image reconstruction, the hologram can be illuminated by an oblique white light, as shown in fig. 10.26.

From this simple point-object reflection hologram, it is easy to extend the concept to the case of a three-dimensional object. Let us replace the point object S of fig. 10.24 by a three-dimensional object function $O(\xi, \eta, \zeta)$, and let the object and reference fields be of wavelengths λ_1 and λ_2. Then the complex light field within the photographic emulsion due to the object field is

$$u(\boldsymbol{\rho}; k_1, k_2) = u_1(\boldsymbol{\rho}; k_1) + u_2(\boldsymbol{\rho}; k_2)$$
$$= \iint_{S_o} O(\xi, \eta, \zeta)[E_l^{+}(\boldsymbol{\rho} - \boldsymbol{\xi}; k_1) + E_l^{+}(\boldsymbol{\rho} - \boldsymbol{\xi}; k_2)] \, d\xi \, d\eta, \tag{10.140}$$

where S_o denotes the surface integral over the object function, and E_l^{+} denotes the spatial impulse response.

Retaining the assumptions we have made previously, we may write the reflectance of a full-color (i.e., one constructed from light of wavelengths $\lambda_1, \lambda_2, \ldots, \lambda_N$) hologram as

$$r(\boldsymbol{\rho}; k_n) = K\left\{\sum_{n=1}^{N} |u_n(\boldsymbol{\rho}; k_n)|^2 + |B_n|^2 + 2|u_n(\boldsymbol{\rho}; k_n)| \, |B_n| \cos[\phi_n(\boldsymbol{\rho}; k_n) - k_n x \sin\theta]\right\}, \tag{10.141}$$

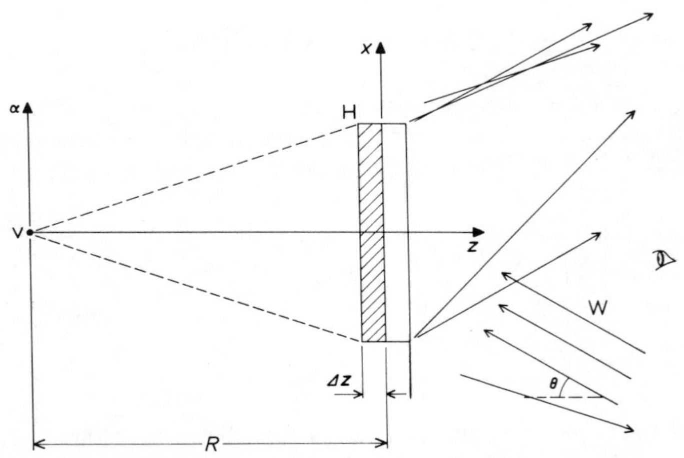

Fig. 10.26 Virtual image reconstruction, reflection hologram. H, hologram; W, white-light illumination; v, virtual image.

where

$$u_n(\boldsymbol{\rho}; k_n) = |u_n(\boldsymbol{\rho}; k_n)| \exp[i\phi_n(\boldsymbol{\rho}; k_n)], \qquad \text{for } n = 1, 2, \ldots, N.$$

If this full-color hologram is illuminated by an incoherent white light, in the manner of fig. 10.25, then by means of the Fresnel–Kirchhoff theory the real image light field may be obtained:

$$E_{\mathrm{r}}(\boldsymbol{\sigma}; k_n) = \sum_{n=1}^{N} C_n(z) O(k_n; \alpha, \beta, \gamma), \tag{10.142}$$

where the subscript r denotes the real image term, and the $C_n(z)$'s are complex functions of z.

The corresponding irradiance is

$$I_{\mathrm{r}}(\boldsymbol{\rho}; k_n) = \sum_{n=1}^{N} K_n |O(k_n; \alpha, \beta, \gamma)|^2, \tag{10.143}$$

which is the sum of the image irradiance due to $\lambda_1, \lambda_2, \ldots, \lambda_N$. By a similar procedure, we can obtain the virtual hologram image.

We might expect the colors of the hologram image to be the same as those of the object. However, in practice, the wavelengths reflected are shorter than those of the recordings. This is due to emulsion shrinkage after the development and fixing process. In order to maintain the original wavelengths, we must prevent this shrinkage. There are certain techniques available for reswelling the emulsion after the fixing process. By careful application of these techniques, it is possible to prevent the emulsion shrinkage enough to preserve the original wavelengths over a broad spectral range.

Figure 10.27 is a photograph of a virtual image reconstruction produced from a single-wavelength reflection hologram under white-light illumination.

The analysis of linear holography given in this chapter has been based upon the point-source concept and linear system theory. Although some of the expressions seem rather long, the calculations were made in a straightforward fashion. The main use of this approach, of course, is for those problems that are simple enough to be directly evaluated. In a manner similar to that of geometrical optics, the holographic magnifications, resolutions, bandwidth requirements, and so on, can in such cases be obtained in this simple way.

10.8 GENERATION OF COLOR HOLOGRAM IMAGES WITH WHITE-LIGHT PROCESSING

True color holographic images that can be reconstructed using white-light illumination have, in the past, been basically of two types—the reflection hologram (sec. 10.7) and the rainbow hologram (sec. 12.3), both of which suffer certain

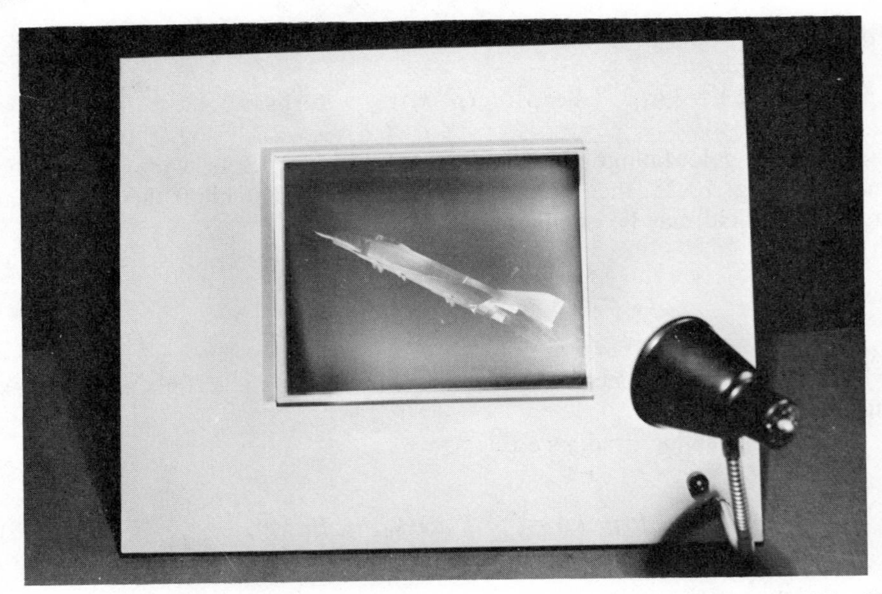

Fig. 10.27 A virtual image produced by white-light illumination of a single-wavelength reflection hologram. (Courtesy of C. Charnetski, Conduction Corp.)

shortcomings. For example the construction of a color reflection hologram requires the multiplexing of three color images onto the recording film, and in order to form a faithful color reproduction of the original image, an elaborate film processing technique must be utilized to prevent emulsion shrinkage (ref. 10.24). Besides this major disadvantage, a reflection hologram generally exhibits a lower diffraction efficiency than transmission-type holograms. Although the rainbow hologram does not possess these disadvantages, the required use of a slit during construction results in a marginal resolution loss along the axis perpendicular to the slit. In addition, the smeared slit images produced during reconstruction cause some degree of color blur in the rainbow hologram image (sec. 12.5). In this section we introduce a method for generating color holographic images that utilizes a white-light processing technique. We show that multiple color images can be encoded onto a photographic plate and reproduced with a white-light optical processor technique (sec. 9.2). Since there is no slit involved in the construction process, in principle, this technique will eliminate those disadvantages posed by the rainbow holographic technique. However, as compared with the rainbow color hologram image reconstruction, the optical setup is more elaborate.

Let us now consider that the color hologram is constructed by sequentially illuminating the object using three primary color wavelengths (i.e., red, green, and blue) and three corresponding reference beams, each oriented at equiangular positions about the object beam axis, as shown in fig. 10.28. By making three primary color holographic exposures, a multiplexed image plane hologram is

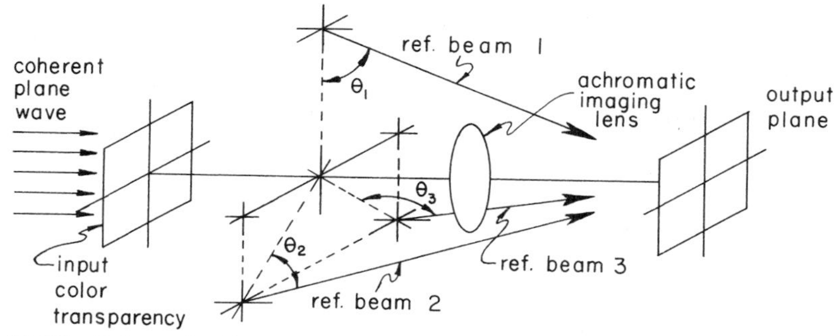

Fig. 10.28 Schematic diagram of a multiplex color holographic construction with various angular reference beams.

recorded (ref. 10.33). If we assume that the multiplexed hologram is linearly recorded in the *T-E* curve of the photographic plate, then the corresponding amplitude transmittance can be written

$$\begin{aligned} T(x, y) = K_1 + K_2\{&|s_r(x, y)| \cos[k_r x \sin\theta + \phi_r(x, y)] \\ &+ |s_g(x', y')| \cos[k_g x' \sin\theta + \phi_g(x', y')] \\ &+ |s_b(x'', y'')| \cos[k_b x'' \sin\theta + \phi_b(x'', y'')], \end{aligned} \qquad (10.144)$$

where K_1 and K_2 are arbitrary constants; the subscripts r, g, and b represent the red, green, and blue color lights, respectively; θ is the oblique angle of the reference beam; $K_r = 2\pi/\lambda_r$, $k_g = 2\pi/\lambda_g$, $k_b = 2\pi/\lambda_b$; $s_r(x, y) = |s_r(x, y)| \exp[i\phi_r(x, y)]$, $s_g(x', y') = |s_g(x', y')| \exp[i\phi_g(x', y')]$, and $s_b(x'', y'') = |s_b(x'', y'')| \exp[i\phi_b(x'', y'')]$ are the corresponding red, green, and blue complex image distributions; and (x, y), (x', y') and (x'', y'') are the coordinate systems with respect to the equiangular directions of the reference beams. If this multiplexed color hologram is inserted in the input plane of a white-light optical processor, as shown in fig. 10.29, the complex light distribution in the back focal plane P_2 of the achromatic lens can be written

$$\begin{aligned} E(\alpha, \beta; \lambda) = C_1\delta(\alpha, \beta; \lambda) &+ C_2 S_r\left[\alpha \pm \frac{\lambda f}{2\pi} P_r, \beta; \lambda\right] \\ &+ C_2 S_g[\alpha' \pm \frac{\lambda f}{2\pi} P_g, \beta'; \lambda] + C_2 S_b[\alpha'' \pm \frac{\lambda f}{2\pi} P_b, \beta''; \lambda], \end{aligned}$$

$$(10.145)$$

where the C's are the appropriate complex constants; $P_r = k_r \sin\theta$, $P_g = k_g \sin\theta$, and $P_b = k_b \sin\theta$ are the spatial carrier frequencies with respect to the equiangular reference beams; f is the transform lens focal length; (α, β), (α', β'), and (α'', β'') are the coordinates systems of the spatial frequency plane

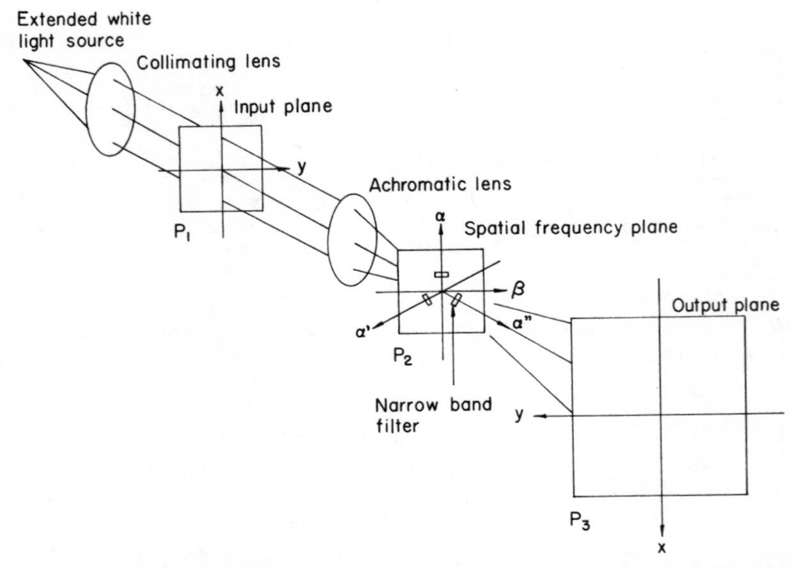

Fig. 10.29 White-light optical processor for reconstruction of the encoded holographic image.

P_2 with respect to the input plane (x, y), (x', y') and (x'', y'') coordinate systems; $S_r(\alpha, \beta; \lambda)$, $S_g(\alpha', \beta'; \lambda)$, and $S_b(\alpha'', \beta''; \lambda)$ are the smeared Fourier spectra of the red, green, and blue color images; and λ is the wavelength of the light source. Since the light source is a broad spectral band white-light source, from this equation we see that the Fourier spectra of the color images are smeared into rainbow color bands along the α, α', and α'' axes, respectively, in the spatial frequency plane P_2.

In color hologram image retrieval, we spatially filter the smeared Fourier spectra by using narrow-band spatial filters placed in the appropriate locations corresponding to the three primary color Fourier spectra, as illustrated in fig. 10.30. These spatially filtered color Fourier spectra are incoherently added at the output image plane P_3 to form a multicolor holographic image. The corresponding irradiance of the color image can be approximated by the following expression:

$$I(x, y) \simeq \Delta\lambda\,[|s_r(x, y; \lambda_r)|^2 + |s_g(x, y; \lambda_g)|^2 + |s_b(x, y; \lambda_b)|^2], \tag{10.146}$$

where $\Delta\lambda$ is the narrow spatial band of the spatial filter. From this equation we see that a true color hologram image can indeed be reconstructed with a white-light processing technique.

For experimental demonstration fig. 10.31 shows a black-and-white photo-

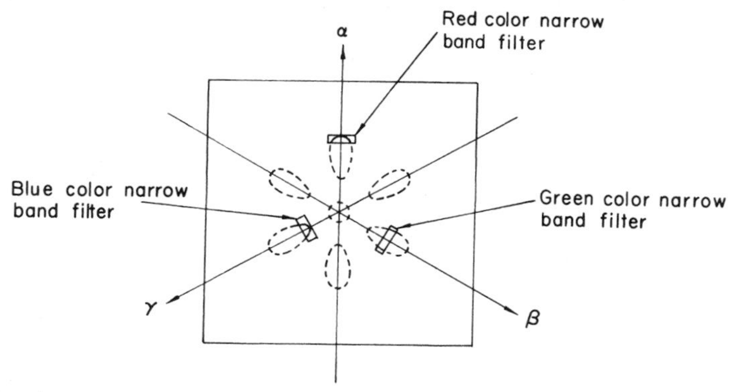

Fig. 10.30 Narrow-band filtering of the smeared Fourier spectra in the frequency plane.

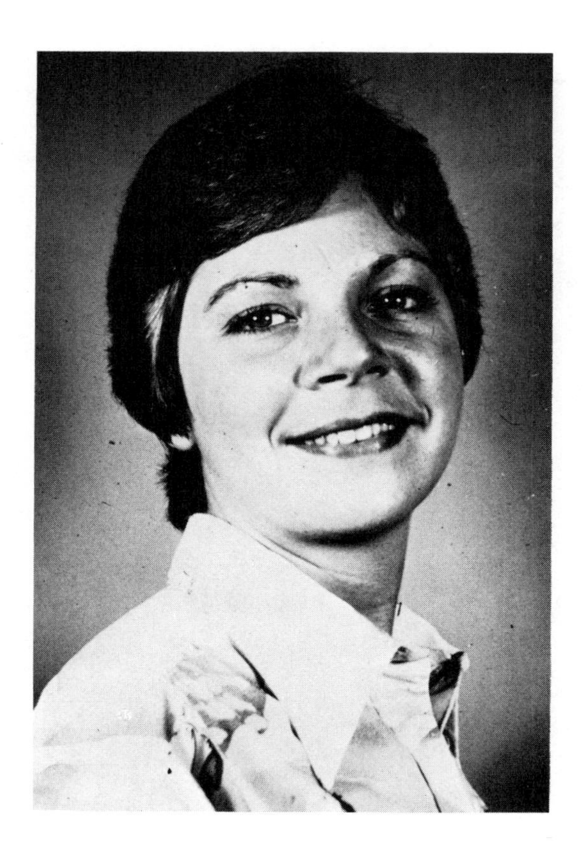

Fig. 10.31 Black-and-white photograph of the original color transparency.

graph of the multicolor input transparency used for the holographic construction. After processing the encoded hologram by the described technique, the reconstructed color image was recorded at the output plane. Figure 10.32 shows a black-and-white photograph of this color image. In concluding this section, we note that this new technique allows a color holographic image to be reconstructed using a white-light source, while, in principle, eliminating the marginal resolution loss and color blur found in the rainbow holographic system (sec. 12.5). However, some of the problems still remain. These include an elaborate recording scheme and the fact that the undesirable speckle effect inherent to coherent construction cannot be totally eliminated, although it is somewhat reduced during white-light reconstruction. This new technique also has a limited range of object-to-reference beam angles available for holographic construction. During the recording process, the three angular carrier spatial frequencies also introduce moire fringe patterns. But, by an ingenious design of the holographic encoding process, it may be possible to eliminate these fringe patterns.

Fig. 10.32 Black-and-white photograph of the reconstructed color holographic image.

10.9 COLOR IMAGE RETRIEVAL FROM COHERENT SPECKLES WITH WHITE-LIGHT PROCESSING

The use of monochrome transparencies to retrieve color images using a white-light source was first reported by Ives (ref. 10.34) as early as 1906. Later, techniques for generating true color images by reflection holographic (refs. 10.24, 10.25) and rainbow holographic processes (refs. 10.35, 10.36) were developed. Nevertheless, each of these techniques has its own advantages and drawbacks. Over a decade ago a technique, similar to that of Ives, was described by Mueller (ref. 10.37); this technique employed a tricolor grid screen for image encoding. Decoding was performed using three quasi-monochromatic sources to retrieve the color image. Since then similar work has been reported by Macovski (ref. 10.38), Grousson and Kinany (ref. 10.39), and Yu (ref. 10.40). More recently, we proposed a different method for retrieving color holographic images using a white-light optical processor (sec. 10.8).

In this section we demonstrate a very simple technique that utilizes coherent speckles to encode color images onto black-and-white film for color image retrieval with a white-light processor (ref. 10.41). In principle, this technique could be simpler than the previous techniques for producing color images. For encoding, a diffuse color object illuminated by coherent light is imaged onto a photographic plate through a narrow slit by an aerial imaging lens, as shown in fig. 10.33. Let us assume that the recording was sequentially performed, using only red and green coherent illuminations, with the slit first oriented in one direction and then rotated by 90°. This would give us a red and a green encoded color image, both multiplexed onto one photographic film. Due to the different orientations of the slit aperture, the red encoded image will have speckles elongated in one direction, while the green encoded image has speckles elongated in the other direction (90° apart). Thus an encoded monochrome multiplex specklegram can be recorded. Retrieving the color image, we insert the multiplex specklegram in the input plane P_1 of a white-light optical processor, depicted in fig. 10.34. Since the elongated speckles of each specklegram are orientated about 90° apart, the corresponding Fourier spectra would be distributed in confined directions perpendicular to these elongations in the spatial frequency plane P_2, as shown in fig. 10.35. That is, the broad spectra of the red

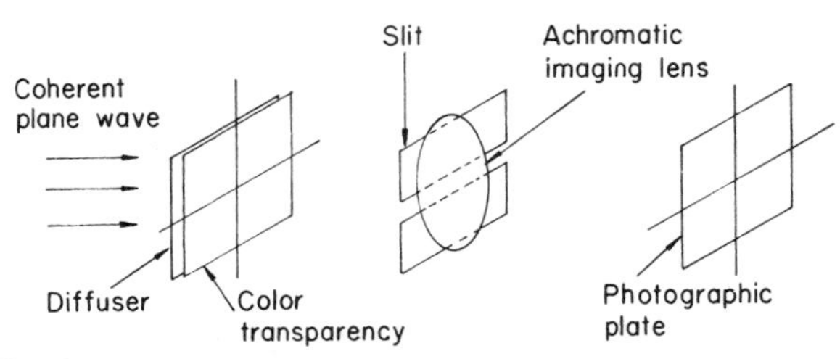

Fig. 10.33 Optical process for construction of a multiplexed specklegram.

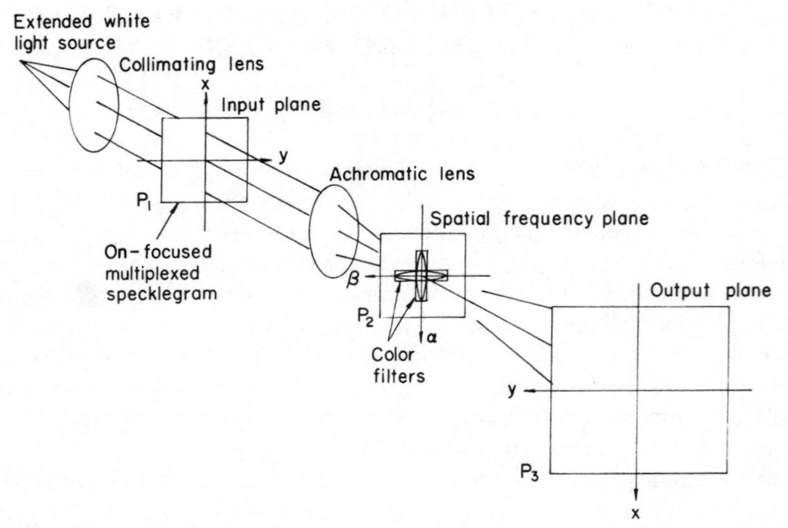

Fig. 10.34 White-light optical processor for reconstructing a color image from an encoded specklegram.

color image is distributed out in one direction, while the spectra of the green color image is spread out in the other direction. By color filtering each set of Fourier spectra with respect to the red and green color filters, as shown in figs. 10.34 and 10.35, a full color image can be reproduced at the output image plane P_3.

For experimental demonstration, we provide in fig. 10.36 a black-and-white photograph of the multicolor image obtained by this specklegraphic technique.

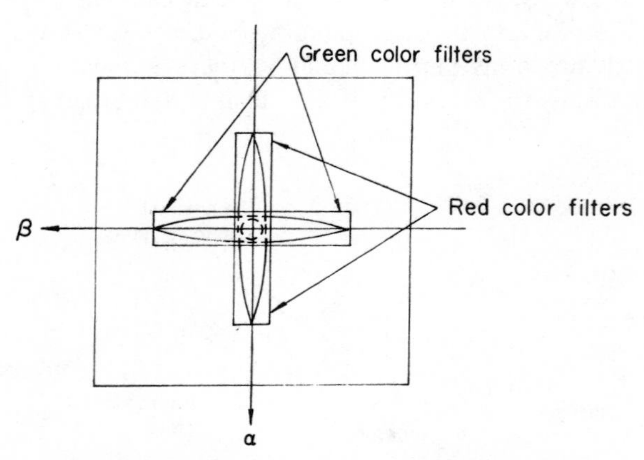

Fig. 10.35 Color spatial filtering of the smeared Fourier spectra.

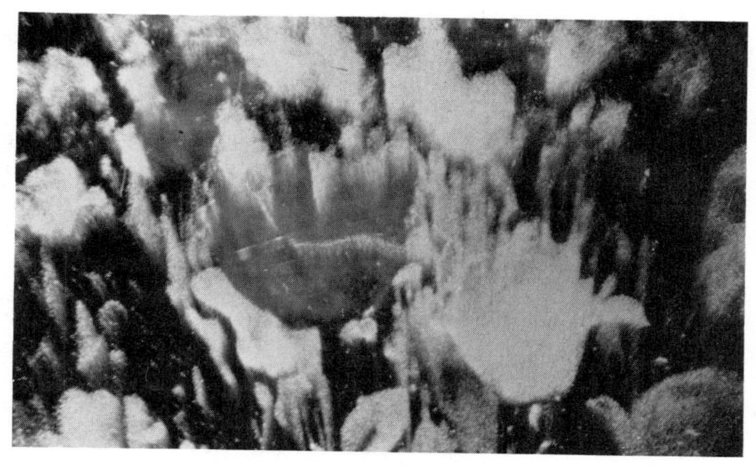

Fig. 10.36 Black-and-white photograph of the reconstructed color image of a field of tulips.

Although the resolution of the reconstructed color image suffers a severe drawback, the color reproduction is relatively faithful. With the use of a finer diffuser and appropriate slit size in the construction process, optimum color image reproduction may be obtained.

PROBLEMS

10.1 Referring to eqs. (10.44), (10.46), (10.58), and (10.61):

(a) Derive the lateral and longitudinal magnifications for the case where a hologram is illuminated by an oblique monochromatic *plane* (rather than divergent) wave of wavelength λ_2.

(b) Derive the lateral and longitudinal magnifications if the hologram is made and reconstructed by *convergent* reference and illumination beams. Show the relationship between the virtual and real image magnifications.

10.2 In a certain wave front recording, the coherent light is derived from an argon laser of wavelength 4880 Å. The reference wave is from a divergent point source, which is located 1 m away from the x axis of the photographic plate and 0.5 m below the optical axis of the recording system, as shown in fig. 10.37, and the object to be recorded is located at about 0.5 m away from the recording aperture. If the recorded transparency (i.e., hologram) is illuminated by a normally incident divergent point source, using light from a helium-neon laser of wavelength 6328 Å, and located 0.5 m away from the hologram, then:

(a) Determine the corresponding lateral and longitudinal magnifications for the real and virtual images.

(b) Compute the locations of the images.

(c) Calculate the position of the divergent point source so that the translational distortions described in sec. 10.2 may be eliminated.

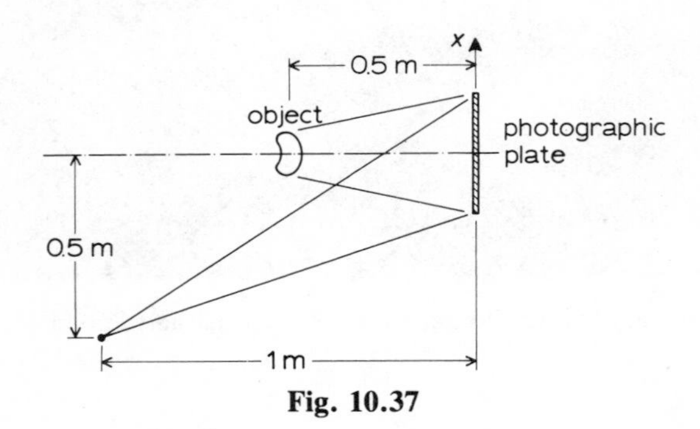

Fig. 10.37

10.3 In a certain holographic process, the hologram is made by a divergent reference beam and the hologram images are reconstructed by a convergent source of the same wavelength. Calculate the locations and relationship between the divergent and convergent beams in order to have unity lateral magnification.

10.4 Suppose a hologram transparency is recorded in the manner shown in fig. 10.14. If this hologram is uniformly enlarged by a factor m, and this enlarged hologram is illuminated by a divergent monochromatic source of different wavelength, as shown in fig. 10.38, then determine the real and virtual hologram image positions and their image magnifications.

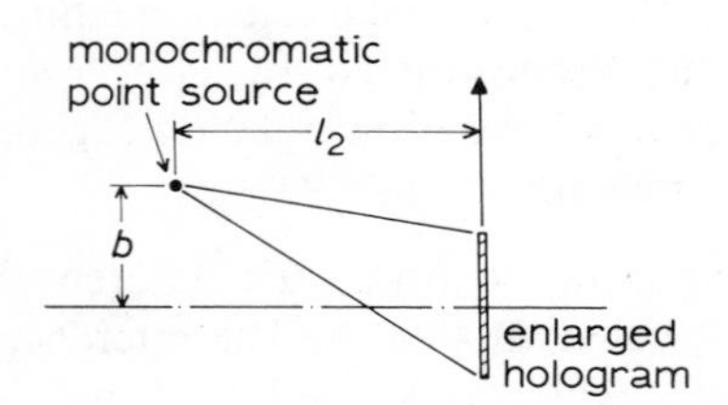

Fig. 10.38

10.5 Given a 10 × 12-cm high-resolution photographic plate. If holographic recording takes place with the illuminated object located on the optical axis of the recording system, at a distance of 24 cm in front of the plate, determine the lateral resolution limits of the hologram images.

10.6 In a certain holographic process, the recording aperture (i.e., the photographic plate) is assumed to be an 8×10-cm rectangle. The center of a circular object transparency of 6-cm diameter is placed on the optical axis, parallel to the plate, at a distance of 20 cm. If a plane reference wave oriented at 45° with respect to the major axis of the recording plate is used, determine the spatial frequency bandwidths of the major and minor axes of the hologram aperture.

10.7 With reference to fig. 10.21, by interchanging the reference plane wave and the reconstruction divergent beam for the holographic process, determine the third-order primary aberrations, as well as the conditions for their corrections. Compare the results with eq. (10.119) and table 10.1.

10.8 If the separation between the object plane and the recording aperture of a holographic process is large enough to satisfy the Fraunhofer diffraction condition, show that the recorded wave front is essentially a Fourier transform hologram (i.e., a hologram of the corresponding Fourier transform).

10.9 With reference to prob. 10.8, if the reference wave is not used during the recording, show that the resulting transparency is the power spectrum of the object function. If this transparency is illuminated by a normally incident monochromatic plane wave, and the transmitted light field is imaged at the back focal length of a positive lens (fig. 10.39), show that the image irradiance is proportional to the square of the autocorrelation function of the object.

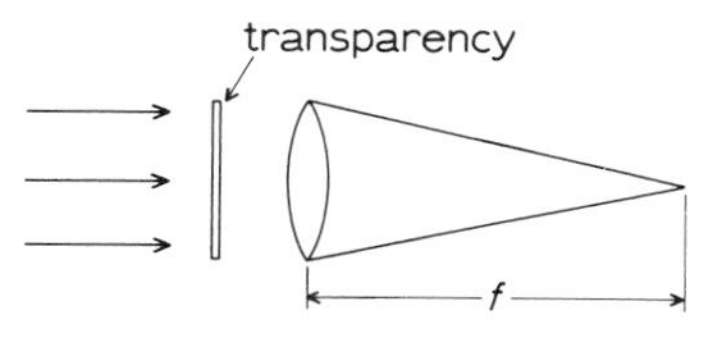

Fig. 10.39

10.10 Let a monochromatic wave front traveling from left to right, as shown in fig. 10.40, distributed over the coordinate plane P_3 be

$$g(\alpha, \beta) = C \exp[i\frac{k}{2l}(\alpha^2 + \beta^2)].$$

(a) Draw a system analog diagram to evaluate this problem.

(b) Calculate the complex light distributions over planes P_1 and P_2.

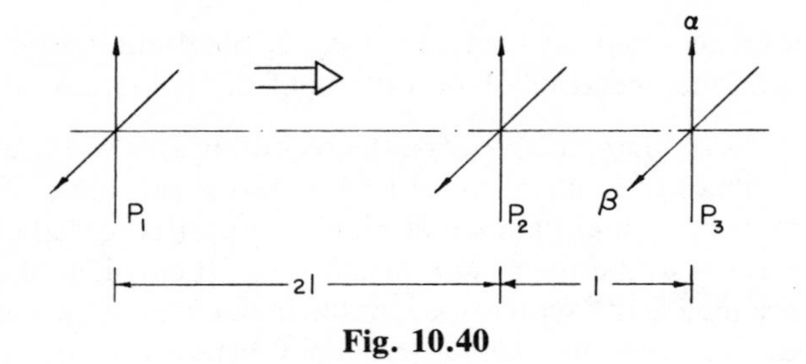

Fig. 10.40

10.11 With reference to the point-object holographic construction of fig. 10.41:

(a) Sketch a system analog diagram to represent the holographic construction.

(b) Calculate the complex light distribution of the reference and object beams on the holographic plate.

(c) If the holographic recording is linear, calculate the corresponding amplitude transmittance.

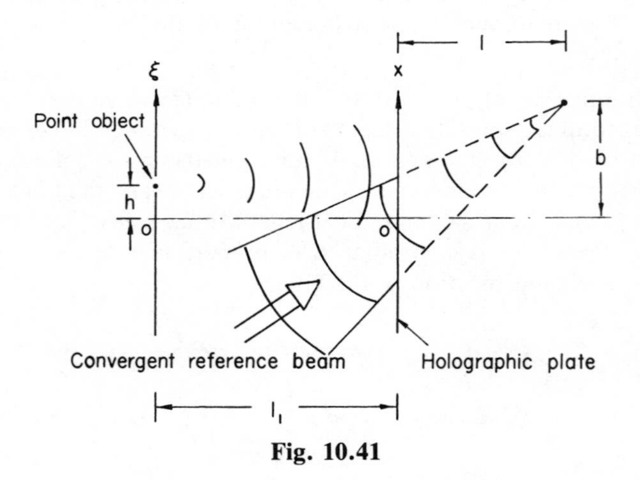

Fig. 10.41

10.12 With reference to prob. 10.11, we assume that the hologram aperture is limited to a finite extent, for example, from $-D_x/2$ to $D_x/2$ and $-D_y/2$ to $D_y/2$.

(a) Evaluate the minimum resolution limit of Rayleigh, if the hologram image is reconstructed by monochromatic conjugate plane wave of the same wavelength.

(b) Are the resolution limits affected by the reconstructed wavelength? Explain your answer.

(c) If the hologram is illuminated by a divergent wave front, what are the resolution limits of the hologram images?

10.13 Let us consider an off-axis point-object hologram constructed with wavelength λ_1; the corresponding amplitude transmittance of the hologram is

$$T(x, y; k_1) = K_1 + K_2 \cos\left[k_1\left(\frac{\rho^2}{2R} - x \sin \theta\right)\right],$$

where $k_1 = 2\pi/\lambda_1$, $\rho^2 = x^2 + y^2$, and (x, y) is the spatial coordinate system of the hologram. If the hologram is illuminated by a conjugate monochromatic plane wave of wavelength λ, then:

(a) Calculate the position of the real hologram image as function of λ.

(b) If the hologram is reconstructed by a conjugate plane wave of all visible wavelengths, then sketch a schematic diagram to show that the effect of the real hologram image takes place due to the spectral wavelengths of the light source.

10.14 Let us consider a transmission hologram constructed with wavelength λ_1 by an oblique reference plane wave of obliquity θ. If the hologram image is reconstructed by a monochromatic conjugate plane wave of λ_2, then:

(a) Determine the corresponding lateral magnifications for the real and the virtual holographic images.

(b) If we assume that the object is very small as compared with the distance between the object and the holographic plate, what are the corresponding longitudinal magnifications for the real and the virtual hologram images?

10.15 Given the point-object reflection holographic construction of fig. 10.42:

(a) Draw a system analog diagram to represent the holographic construction process.

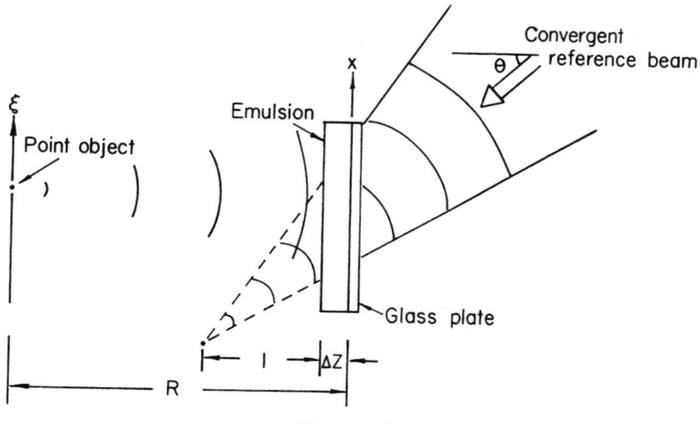

Fig. 10.42

(b) Evaluate the corresponding intensity distribution of the holographic recording inside the film emulsion.

10.16 If the point-object reflection hologram of prob. 10.15 is illuminated with a white-light point source, as shown in fig. 10.43, then:

(a) Draw a system analog diagram to represent the hologram image reconstruction process.

(b) Evaluate where the real image would be reconstructed.

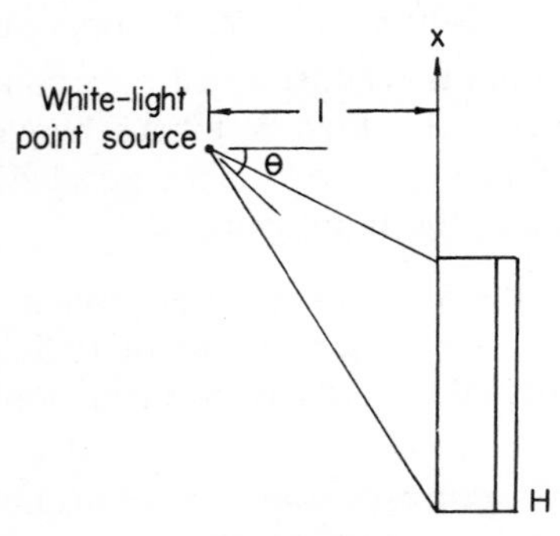

Fig. 10.43

10.17 We consider a holographic recording by an imaging lens, as shown in fig. 10.44. We assume that the recording plate is inserted within the object image during the holographic constructions.

(a) If the hologram is illuminated by the same reference beam, show the location of the hologram image reconstruction.

(b) If the hologram is illuminated by a divergent coherent light of the same wavelength, evaluate the location of the divergent point source to produce a real hologram image.

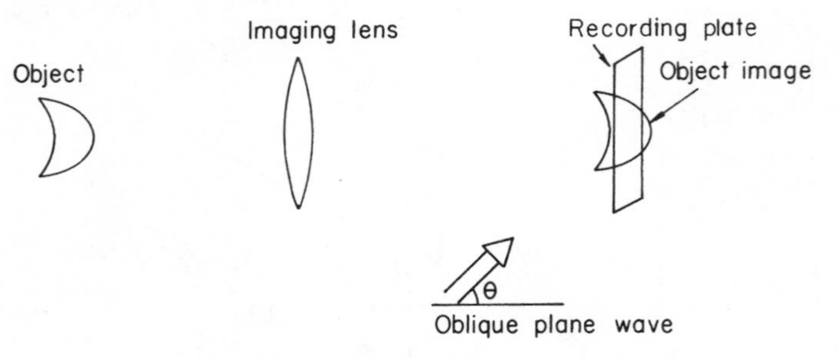

Fig. 10.44

10.18 With reference to the on-focused hologram of prob. 10.17, if the hologram is illuminated by a white-light source of the same oblique angle θ, then:

(a) Evaluate the effect of hologram image reconstruction observed by naked eye.

(b) Show the effect of image blur due to wavelength spread.

10.19 With reference to the color hologram image reconstruction with coherent speckles of sec. 10.9:

(a) Show that the encoded specklegram may be obtained with a white-light source instead of a coherent source.

(b) What are the advantages and disadvantages of using white-light illumination as compared with coherent source?

10.20 **(a)** What are the essential differences between the color hologram imaging of sec. 10.8 and the white-light processing technique for archival storage of color films of sec. 9.1?

(b) Sketch the advantages and disadvantages of both techniques of part (a).

REFERENCES

10.1 D. Gabor, "A New Microscope Principle," *Nature,* **161,** 777 (1948).

10.2 D. Gabor, "Microscopy by Reconstructed Wavefronts," *Proc. Roy. Soc.*, ser. A, **197,** 454 (1949).

10.3 D. Gabor, "Microscopy by Reconstructed Wavefronts, II," *Proc. Phys. Soc.*, ser. B, **64,** 449 (1951).

10.4 G. L. Rogers, "Gabor Diffraction Microscopy: The Hologram as a Generalized Zone Plate," *Nature,* **166,** 237 (1950).

10.5 H. M. A. El-Sum, "Reconstructed Wavefront Microscopy," doctoral dissertation, Stanford University, 1952 (available from University Mirofilms, Ann Arbor, Mich.).

10.6 A. Lohmann, "Optical Single-Sideband Transmission Applied to the Gabor Microscope," *Opt. Acta,* **3,** 97 (1956).

10.7 E. N. Leith and J. Upatnieks, "Reconstructed Wavefront and Communication Theory," *J. Opt. Soc. Am.*, **52,** 1123 (1962).

10.8 E. N. Leith and J. Upatnieks, "Wavefront Reconstruction with Continuous-Tone Objects," *J. Opt. Soc. Am.*, **53,** 1377 (1963).

10.9 E. N. Leith and J. Upatnieks, "Wavefront Reconstruction with Diffused Illumination and Three-Dimensional Objects," *J. Opt. Soc. Am.*, **54,** 1295 (1964).

10.10 J. B. DeVelis and G. O. Reynolds, *Theory and Applications of Holography,* Addison-Wesley, Reading, Mass., 1967.

10.11 R. W. Meier, "Magnification and Third-Order Aberrations in Holography," *J. Opt. Soc. Am.*, **55,** 987 (1965).

10.12 E. N. Leith, J. Upatnieks, and K. A. Haines, "Microscopy by Wavefront Reconstruction," *J. Opt. Soc. Am.*, **55,** 981 (1965).

10.13 J. A. Armstrong, "Fresnel Holograms: Their Imaging Properties and Abberations," *IBM J. Develop.*, **9,** 171 (1965).

10.14 F. I. Diamond, "Magnification and Resolution in Wavefront Reconstruction," *J. Opt. Soc. Am.*, **57,** 503 (1967).

10.15 J. T. Winthrop and C. R. Worthington, "X-Ray Microscopy by Successive Fourier Transformation," *Phys. Lett.*, **15**, 124 (1965).

10.16 M. Born and E. Wolf, *Principles of Optics,* 2nd rev. ed., Pergamon Press, New York, 1964, pp. 211 ff.

10.17 H. B. Dwight, *Table of Integrals and Other Mathematical Data,* 3rd ed., Macmillan, New York, 1957.

10.18 L. Mertz and N. O. Young, "Fresnel Transformations of Optics," in K. J. Habell, Ed., *Proceedings of the Conference on Optical Instruments and Techniques,* Wiley, New York, 1963, pp. 305 ff.

10.19 A. W. Lohmann, "Wavefront Reconstruction for Incoherent Objects," *J. Opt. Soc. Am.*, **55,** 1555 (1965).

10.20 G. W. Stroke and R. C. Restrick III, "Holography with Spatially Noncoherent Light," *Appl. Phys. Lett.*, **7,** 229 (1965).

10.21 G. Cochran, "New Method of Making Fresnel Transforms with Incoherent Light," *J. Opt. Soc. Am.*, **56,** 1513 (1966).

10.22 P. J. Peters, "Incoherent Holograms with Mercury Light Source," *Appl. Phys. Lett.*, **8,** 209 (1966).

10.23 M. G. Lippmann, "La Photographie des Couleurs," *Compt. Rend.*, **112,** 274 (1891).

10.24 Y. N. Denisyuk, "Photographic Reconstruction of the Optical Properties of an Object in its Own Scattered Radiation Field," *Sov. Phys.-Doklady,* **7,** 543 (1962).

10.25 G. W. Stroke and A. E. Labeyrie, "White-Light Reconstruction of Holographic Images Using the Lippmann-Bragg Diffraction Effect," *Phys. Lett.*, **20,** 368 (1966).

10.26 L. H. Lin et al., "Multicolor Holographic Image Reconstruction with White-Light Illumination," *Bell Syst. Tech. J.*, **45,** 659 (1966).

10.27 E. N. Leith et al., "Holographic Data Storage in Three-Dimensional Media," *Appl. Opt.*, **5,** 1303 (1966).

10.28 F. W. Sears, *Optics,* Addison-Wesley, Cambridge, Mass., 1949, p. 243.

10.29 R. J. Collier, C. B. Burckhardt, and L. H. Lin, *Optical Holography,* Academic, New York, 1971.

10.30 J. C. Wyant, "Image Blur for Rainbow Holograms," *Opt. Lett.*, **1,** 130 (1977).

10.31 H. Chen, "Color Blur of the Rainbow Hologram," *Appl. Opt.*, **17,** 3290 (1978).

10.32 S. L. Zhuang, P. H. Ruterbusch, Y. W. Zhang, and F. T. S. Yu, "Resolution and Color Blur of the One-Step Rainbow Hologram," *Appl. Opt.*, **20,** 872 (1981).

10.33 F. T. S. Yu and P. H. Ruterbusch, "Color Holographic Images Reconstructed by White-Light Processing," *Opt. Laser Tech.*, (in press, 1982).

10.34 H. E. Ives, "Improvement in the Diffraction Process of Color Photography," *Br. J. Photog.*, **1906,** 609.

10.35 P. Hariharan, W. H. Steel, and Z. S. Hegedus, "Multi-Color Holographic Imaging with a White-Light Source," *Opt. Lett.*, **1,** 8 (1977).

10.36 H. Chen, A. Tai, and F. T. S. Yu, "Generation of Color Images with One-Step Rainbow Holograms," *Appl. Opt.*, **17,** 1490 (1978).

10.37 P. F. Mueller, "Color Image Retrieval from Monochrome Transparencies," *Appl. Opt.*, **8,** 2051 (1969).

10.38 A. Macovski, "Encoding and Decoding of Color Information," *Appl. Opt.*, **11,** 416 (1972).

10.39 R. Grousson and R. S. Kinany, "Multi-Color Image Storage on Black and White Film Using a Crossed Grating," *J. Opt.*, **9,** 333 (1978).

10.40 F. T. S. Yu, "White-Light Processing Technique for Archival Storage of Color Films," *Appl. Opt.*, **19,** 2457 (1980).

10.41 F. T. S. Yu and P. H. Ruterbusch, "Color-Image Retrieval from Coherent Speckles by White-Light Processing," *Appl. Opt.*, **21,** 2300 (1982).

11

Analysis of Nonlinear Holograms

The detailed mechanisms of the transfer modulation functions and nonlinear effects in wave front reconstruction processes have been discussed by Kozma (ref. 11.1), Friesem and Zelenka (ref. 11.2), Goodman and Knight (ref. 11.3), and Bryngdahl and Lohmann (ref. 11.4). However, the purpose of this chapter (which is based upon refs. 11.5–11.7) is to study the nonlinear effects in wave front reconstructions from an elementary system theory point of view. As we have seen in the previous chapters, a hologram may be conveniently represented by a black-box input-output system analog. Therefore, the study of nonlinear effects in holography may be approached as the study of a nonlinear system. Since the wave front reconstruction of an extended three-dimensional object may be regarded as being composed of a large number of resolvable point-object constructions, we may begin with an elementary point concept, and then extend to the more general case.

11.1 FINITE-POINT ANALYSIS

In the on-axis wave front construction depicted in fig. 10.1, a monochromatic point source is located a distance R away from the recording medium, and a plane reference wave of the same wavelength is traveling from left to right toward the photographic plate. The exposure due to the combination of these two wave fronts on the recording medium may be approximated by [eq. (10.8)]

$$E(\boldsymbol{\rho}; k) = I(\boldsymbol{\rho}; k)t \simeq \left[\left(\frac{|A|^2}{R}\right) + |B|^2 + \frac{2|A|\,|B|}{R}\cos\left(\frac{k\rho^2}{2R} + \phi\right)\right]t, \quad (11.1)$$

where A and B are the complex amplitudes of the point source and reference waves, respectively, ϕ is the phase angle between A and B, $\rho^2 = x^2 + y^2$, $k = 2\pi/\lambda$, with λ the wavelength, and t is the exposure time.

From the amplitude transmittance characteristic of a physical photographic plate, as shown in fig. 11.1, we note that the linear region is very narrow. From eq. (11.1), the quiescent point (the bias) depends on the amplitudes of the two waves arriving at the recording plate, i.e.,

$$E_Q = \left[\left(\frac{|A|}{R}\right)^2 + |B|^2\right]t. \tag{11.2}$$

If the amplitude of the reference wave is much greater than that of the spherical wave distributed on the photographic film, i.e., if $|B| \gg |A|/R$, then eq. (11.1) becomes

$$E(\boldsymbol{\rho}; k) \simeq \left[|B|^2 + \frac{2|A|\,|B|}{R}\cos\left(\frac{k\rho^2}{2R} + \phi\right)\right]t. \tag{11.3}$$

The quiescent value of the exposure for this case is

$$E_Q \simeq |B|^2 t, \qquad \text{for } |B| \gg \frac{|A|}{R}, \tag{11.4}$$

which is independent of $|A|$. Thus the linear operating region of the transmission

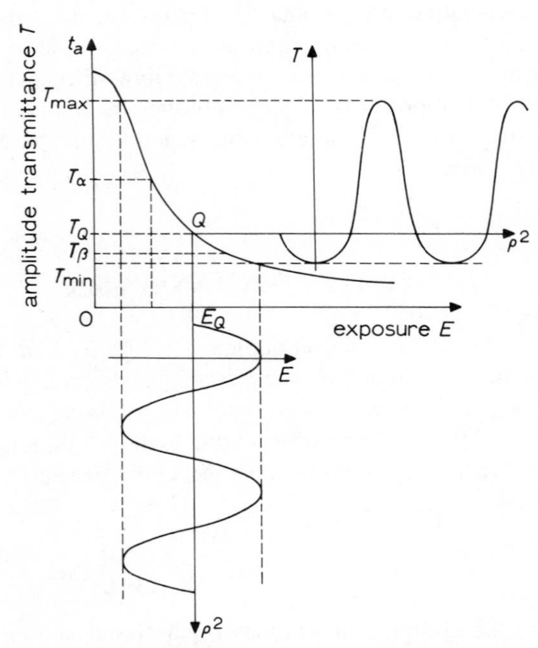

Fig. 11.1 Five-point analysis of a nonlinear hologram.

amplitude versus exposure intensity can be achieved by properly adjusting the magnitude of B, and hence E_Q, of the reference beam.

On the other hand, if $|A|/R$ is either comparable to or greater than $|B|$, then it is difficult to maintain the holographic process within the bounds of the linear region of the amplitude transmittance characteristic. Thus nonlinear distortion of the amplitude transmission will result, as shown in fig. 11.1.

Either from eq. (11.1) or from fig. 11.1, it is apparent that amplitude transmittance of the point-object hologram is periodic in the ρ^2 axis. From the elementary Fourier theorem, the amplitude transmittance of this hologram may be expanded in a Fourier series on the ρ^2 axis, or

$$T(\boldsymbol{\rho}; k) = \frac{a_0}{2} + \sum_{n=1}^{\infty} a_n \cos \frac{nk}{2R}\rho^2, \tag{11.5}$$

where $\boldsymbol{\rho}(x, y, z)$ denotes the given coordinate system, and a_n is the corresponding Fourier coefficient:

$$a_n = \frac{2}{\lambda R}\int_0^{\lambda R} T(\boldsymbol{\rho}; k) \cos \frac{nk}{2R}\rho^2\, d\rho^2, \qquad n = 0, 1, 2, 3, \ldots. \tag{11.6}$$

Note: Without loss of generality, we can drop the phase angle ϕ from the calculation. From eq. (11.5), the amplitude transmittance $T(\boldsymbol{\rho}; k)$ is that of the sum of an infinite number of Fresnel zone lenses (sec. 4.5):

$$T(\boldsymbol{\rho}; k) = T_0 + \sum_{n=1}^{\infty} T_n(\boldsymbol{\rho}; k), \tag{11.7}$$

where

$$T_0 = \frac{a_0}{2},$$

$$T_n(\boldsymbol{\rho}; k) = a_n \cos \frac{nk}{2R}\rho^2, \qquad n = 1, 2, 3, \ldots.$$

The focal lengths of the zone lenses are, therefore,

$$f_n = \frac{R}{n}, \qquad n = 1, 2, 3, \ldots. \tag{11.8}$$

If a recorded photographic plate (hologram) having an amplitude transmittance given by eq. (11.7) is illuminated by a monochromatic plane wave of wavelength λ (fig. 11.2), the complex light field behind the hologram may be

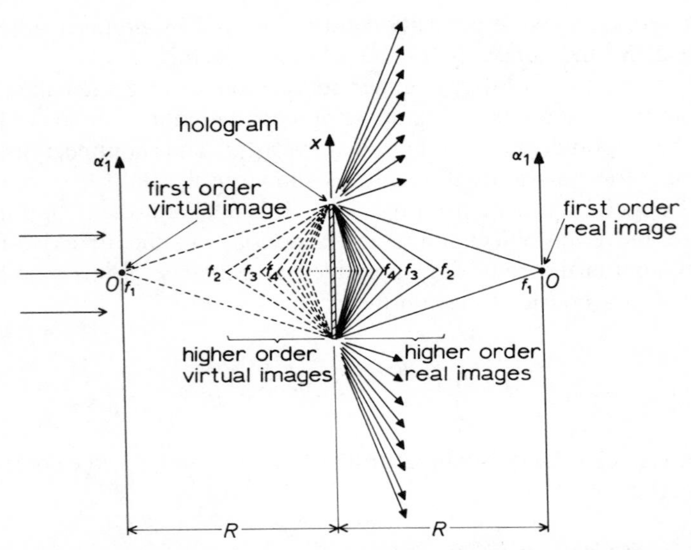

Fig. 11.2 Wave front reconstruction of an on-axis point-object nonlinear hologram. Monochromatic plane wave illumination.

determined from the convolution theorem,

$$E(\boldsymbol{\sigma}; k) = \iint_S T(\boldsymbol{\rho}; k) E_l^+(\boldsymbol{\sigma} - \boldsymbol{\rho}; k)\, dx\, dy, \tag{11.9}$$

where

$$E_l^+(\boldsymbol{\rho}; k) = -\frac{i}{\lambda l} \exp\left[ik\left(l + \gamma - z + \frac{\rho^2}{2l}\right)\right]$$

is the free space impulse response, S denotes the integration over the entire surface of $T(\boldsymbol{\rho}; k)$, and $\boldsymbol{\sigma}(\alpha, \beta, \gamma)$ and $\boldsymbol{\rho}(x, y, z)$ denote the coordinate systems.

By substituting eq. (11.7) in eq. (11.9), we obtain

$$E(\boldsymbol{\sigma}; k) = \iint_S [T_0 + \sum_{n=1}^{\infty} T_n(\boldsymbol{\rho}; k)] E_l^+(\boldsymbol{\sigma} - \boldsymbol{\rho}; k)\, dx\, dy. \tag{11.10}$$

Without loss of generality, we may take the surface S to be of infinite extent. Since we are interested in the hologram image reconstruction due to each of the Fresnel zone lenses, we will evaluate the above integral term by term with respect to the focal length of the corresponding lens. Thus

$$E(\boldsymbol{\sigma}_n; k) = -i\frac{na_n}{2\lambda R}\exp\left(i\frac{kR}{n}\right)\left\{\frac{\lambda R}{2n}\exp\left[i\left(\frac{nk}{4R}\sigma_n^2 + \frac{\pi}{2}\right)\right]\right.$$
$$\left. + \exp\left(i\frac{nk}{2R}\sigma_n^2\right)\delta(\alpha_n, \beta_n)\right\},$$
$$n = 1, 2, 3, \ldots, \qquad (11.11)$$

where $E(\boldsymbol{\sigma}_n; k)$ is the complex light field due to the zone lens $T_n(\boldsymbol{\rho}; k)$ alone at its focal length f_n and where $\sigma_n^2 = \alpha_n^2 + \beta_n^2$.

Obviously, the first term of eq. (11.11) represents the virtual image (i.e., divergent) term, and the second term is the real image (convergent) term. The higher order reconstructed images lie between the first-order real and virtual images, as shown in fig. 11.2. These higher order images are due to the nonlinearity of the hologram. The degree of the nonlinearity of the hologram may be defined as

$$\text{nonlinearity (percent)} = \frac{\left(\sum_{n=2}^{\infty} a_n^2\right)^{1/2}}{a_1} \times 100. \qquad (11.12)$$

The analysis of nonlinear holograms is similar to that used in nonlinear amplifier theory; and to determine the Fourier coefficients a five-point analysis (ref. 11.8), as shown in fig. 11.1, may be adequate. Thus

$$\frac{a_0}{2} = \frac{T_{\max} + T_{\min}}{6} + \frac{T_\alpha + T_\beta}{3},$$
$$a_1 = \frac{T_{\max} - T_{\min}}{3} + \frac{T_\alpha - T_\beta}{3},$$
$$a_2 = \frac{T_{\max} + T_{\min}}{4} - \frac{T_Q}{2},$$
$$a_3 = \frac{T_{\max} - T_{\min}}{6} - \frac{T_\alpha - T_\beta}{3},$$
$$a_4 = \frac{T_{\max} + T_{\min}}{12} - \frac{T_\alpha + T_\beta}{3} + \frac{T_Q}{2}. \qquad (11.13)$$

It is clear that, for a more accurate result, a higher order analysis can be applied. Furthermore, for extreme distortion, the hologram is composed of totally transparent and opaque Fresnel zones. The corresponding amplitude transmittance may be written as

$$T(\boldsymbol{\rho}; k) = \frac{1}{2} + \frac{2}{\pi}\sum_{n=1}^{\infty}\frac{-(-1)^n}{2n-1}\cos\frac{k}{2R}\rho^2. \qquad (11.14)$$

Equation (11.14) differs from eq. (11.7) only in the Fourier coefficients. The image reconstructions can be shown to be located at f_n, which is identical to the case of eq. (11.8). However, the degree of nonlinearity may be shown to be somewhat higher than all the unsaturated cases. On the other hand, for extreme linearity in the holographic recording, it can be seen that the irradiances of the higher order reconstructions vanish.

Let us suppose that the recorded photographic plate of eq. (11.5) is illuminated by an oblique monochromatic plane wave of wavelength λ, as show in fig. 11.3. Again by the Fresnel–Kirchhoff theory, the complex light field behind the hologram is

$$E(\boldsymbol{\sigma}; k) = \iint_S T(\boldsymbol{\rho}; k)\exp(ikx\sin\theta)E_l^{+}(\boldsymbol{\sigma} - \boldsymbol{\rho}; k)\,dx\,dy. \qquad (11.15)$$

By substituting eq. (11.5) in eq. (11.15), we see that the complex light field $E(\boldsymbol{\sigma}_n; k)$ due to $T_n(\boldsymbol{\rho}; k)$ at f_n is

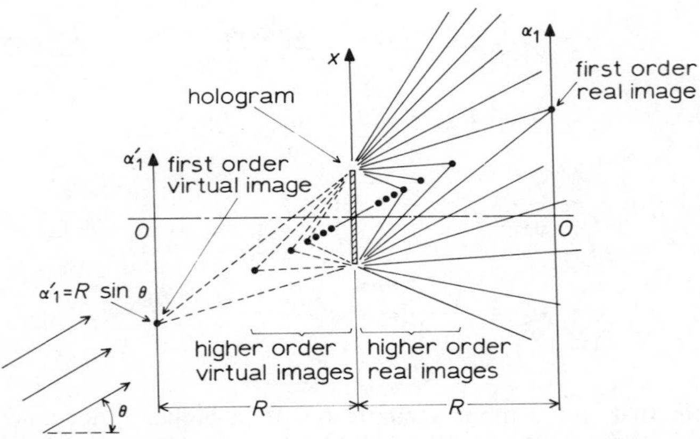

Fig. 11.3 Oblique plane monochromatic illumination of an on-axis nonlinear hologram.

$$E(\boldsymbol{\sigma}_n;\, k) = -i\frac{a_n}{2}\exp\left(i\frac{kR}{n}\right)\cdot\left\{\frac{1}{2}\exp\left[i\frac{\pi}{2} - \frac{kR}{2n}\sin^2\theta\right]\right.$$
$$\times \exp\left[i\frac{nk}{2R}\left(\sigma_n + \frac{R}{n}\sin\theta\right)^2\right]$$
$$\left. + \exp\left(i\frac{nk}{2R}\sigma_n^2\right)\delta\left(\alpha_n - \frac{R}{n}\sin\theta,\, \beta_n\right)\right\},$$
$$n = 1, 2, \ldots . \qquad (11.16)$$

The two terms of eq. (11.16) can be recognized again as virtual and real hologram images, as shown in fig. 11.3. Apparently, oblique illumination is not able to separate the first-order hologram image from the higher order image reconstructions. In order to do so, an oblique reference wave should be used during the wave front construction, as is seen in the next section.

11.2 OFF-AXIS NONLINEAR HOLOGRAM

In order to separate the zero-order, virtual, and real hologram image diffractions, an oblique reference wave may be used for recording, as shown in fig. 11.4. The resulting exposure may be approximated by

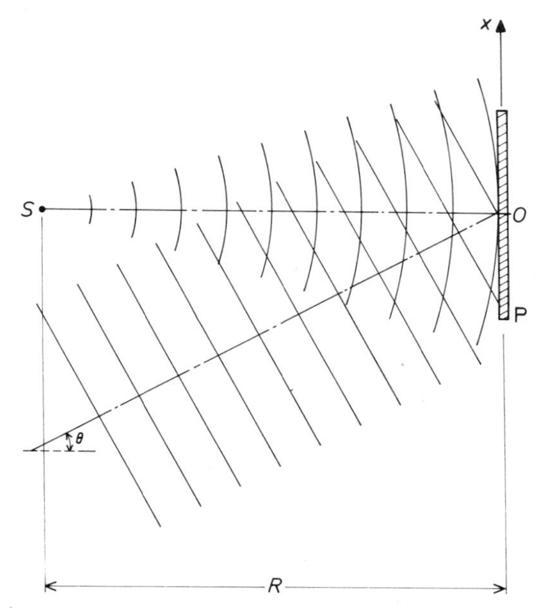

Fig. 11.4 Off-axis wave front construction of a nonlinear hologram. Oblique plane monochromatic reference wave: P, photographic plate.

$$E(\boldsymbol{\rho}; k) = \left\{|A_1|^2 + |B|^2 + 2|A_1|\,|B| \cos\left[k\left(\frac{\rho^2}{2R} - x \sin\theta\right) + \phi\right]\right\}t, \tag{11.17}$$

where A_1 and B are the complex light amplitude distributions from the point source and the reference beam, respectively, ϕ is the phase angle between them, and t is the exposure time.

It is a simple procedure to show that the exposure described by eq. (11.17) consists of concentric circles with a common center at

$$x = R \sin\theta, \qquad y = 0;$$

the corresponding maximum and minimum exposures occur at

$$\rho = \left[R\left(n\lambda - \frac{\lambda}{\pi}\phi\right) + R\sin^2\theta\right]^{1/2}, \qquad n = 0, 1, 2, \ldots.$$

If we let $x = x_1 + R\sin\theta$, then we may write eq. (11.17) as

$$E(\boldsymbol{\rho}; k) = \left\{|A_1|^2 + |B|^2 + 2|A_1|\,|B| \cos\left[\frac{k}{2R}\rho_1^2 - \frac{kR}{2}\sin^2\theta + \phi\right]\right\}t. \tag{11.18}$$

Except for a constant phase factor, eq. (11.18) is identical to eq. (11.1).

Therefore, by the same argument as used in the previous section, the amplitude transmittance of the hologram is

$$T(\boldsymbol{\rho}; k) = \frac{a_0}{2} + \sum_{n=1}^{\infty} a_n \cos\left[nk\left(\frac{\rho^2}{2R} - x\sin\theta\right) + n\phi\right], \tag{11.19}$$

where

$$a_n = \frac{1}{\lambda R}\int_{-\lambda R}^{\lambda R} T(\boldsymbol{\rho}; k) \cos\left[nk\left(\frac{\rho^2}{2R} - x\sin\theta\right) + n\phi\right] d\rho^2,$$
$$n = 0, 1, 2, \ldots.$$

Then by proper translation of the x axis, we can show that the Fourier components in eq. (11.19) are identical to those in eq. (11.5). Hence, the amplitude transmittance $T(\boldsymbol{\rho}; k)$ can be expressed as the sum of an infinite number of Fresnel zone lenses, that is

$$T(\boldsymbol{\rho}; k) = T_0 + \sum_{n=1}^{\infty} T_n(\boldsymbol{\rho}; k), \tag{11.20}$$

where $T_0 = a_0/2$, and

$$T_n(\boldsymbol{\rho}; k) = a_n \cos\left[nk\left(\frac{\rho^2}{2R} - x \sin\theta\right) + n\phi\right], \qquad n = 1, 2, \ldots .$$

If this hologram is illuminated by an oblique monochromatic plane wave of wavelength λ (fig. 11.5), then the complex light amplitude distribution may be evaluated by the convolution equation,

$$E(\boldsymbol{\rho}; k) = \iint_S T(\boldsymbol{\rho}; k) \exp(-ikx\sin\theta) E_l^+(\boldsymbol{\sigma} - \boldsymbol{\rho}; k)\, dx\, dy. \quad (11.21)$$

Note that the readout angle could be any other than θ, but for simplicity we have used the same θ. Again, the wave front reconstructions due to the individual Fresnel zone lenses are exhibited by the termwise evaluation

$$E(\boldsymbol{\sigma}_n; k) = K_1 \exp\left\{i\frac{nk}{4R}\left[\left(\alpha_n - R\left(1 + \frac{1}{n}\right)\sin\theta\right)^2 + \beta_n^2\right]\right\}$$
$$+ K_2\delta\left[\alpha_n - R\left(1 - \frac{1}{n}\right)\sin\theta, \beta_n\right],$$
$$n = 1, 2, 3, \ldots, \quad (11.22)$$

where K_1 and K_2 are the appropriate complex constants.

As shown in fig. 11.5, the sequence of higher order images lies between the primary front and back focal points of the hologram. The same figure shows that, by a proper choice of the angle θ of the reference wave, the first-order real image can be separated from all of the unwanted images.

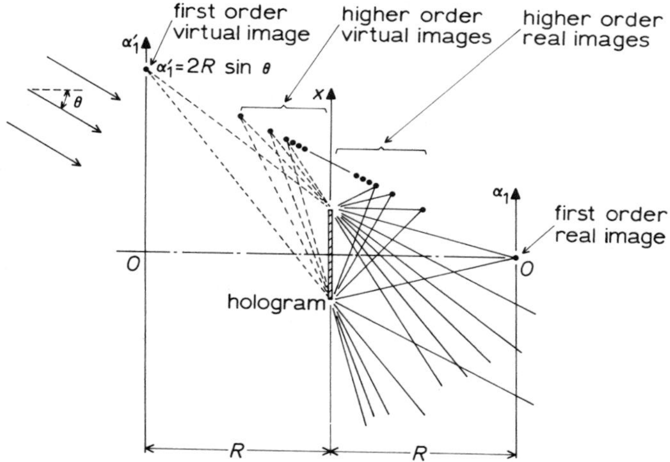

Fig. 11.5 Reconstruction of an off-axis nonlinear hologram. Plane monochromatic illumination, negative θ.

On the other hand, if the point-object hologram is illuminated by a monochromatic plane wave with the same wavelength but positive obliquity, as shown in fig. 11.6, then the complex light amplitude distribution at the focal plane of the associated zone lens is

$$E(\boldsymbol{\sigma}_n; k) = K_1' \exp\left\{ i \frac{nk}{4R}\left[\left(\alpha_n - R\left(1 - \frac{1}{n}\right)\sin\theta\right)^2 + \beta_n^2\right]\right\} + K_2'\delta\left[\alpha_n - R\left(1 + \frac{1}{n}\right)\sin\theta,\ \beta_n\right], \quad n = 1, 2, 3, \ldots, \tag{11.23}$$

where K_1' and K_2' are the appropriate complex constants. From eq. (11.23), once again a sequence of higher order diffracted images is determined. As before, it is possible to separate the first-order virtual image from the higher order diffractions.

Since in most holographic applications the light amplitude distribution on the recording medium from the object is approximately uniform, the transmittance of a nonlinear hologram can be written as

$$T(\boldsymbol{\rho}; k) = \sum_{n=0}^{\infty} T_n(\boldsymbol{\rho}; k), \tag{11.24}$$

where

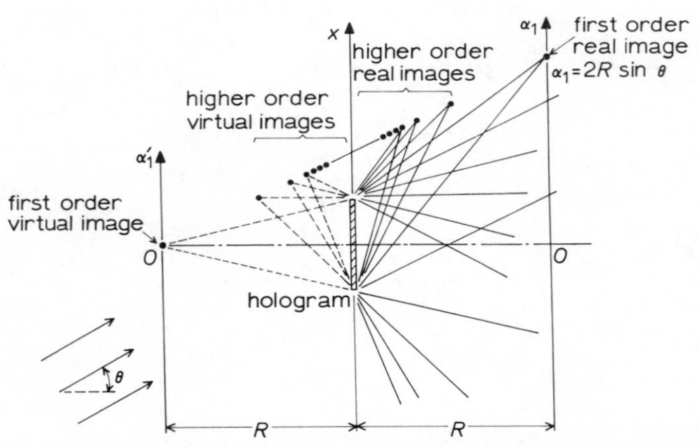

Fig. 11.6 Reconstruction of an off-axis nonlinear hologram. Plane monochromatic illumination, positive θ.

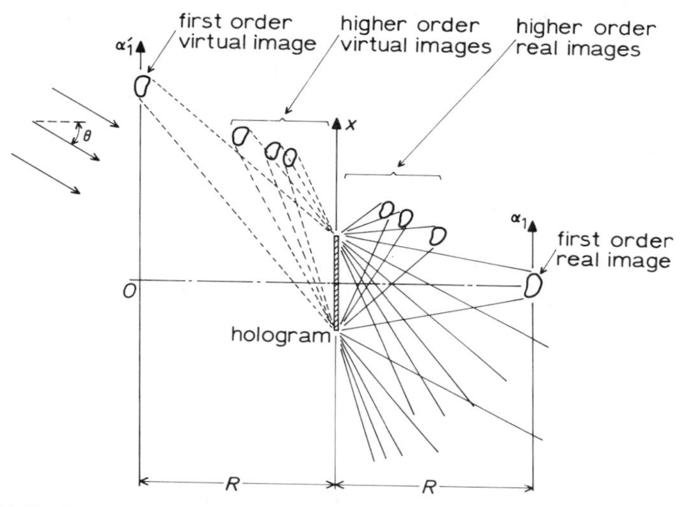

Fig. 11.7 Reconstruction of an off-axis nonlinear hologram of an extended object.

$$T_n(\boldsymbol{\rho}; k) = a_n \cos\{n[kx \sin\theta + \phi(x, y)]\},$$

$$a_n = \frac{1}{d}\int_0^d T(\boldsymbol{\rho}; k) \cos\{n[kx \sin\theta + \phi(x, y)]\}\, d\rho,$$

and d is the corresponding spatial period. Of course, the Fourier components can be determined by graphical analysis, as in the previous section.

First-order image reconstruction can therefore be separated from the unwanted higher order diffractions by off-axis wave front recording. An illustration of the image reconstructions from an off-axis nonlinear hologram is given in fig. 11.7. It can be seen from this figure that the minimum oblique angle of the reference wave can be determined from the size of the object and the hologram aperture. A photograph of a nonlinear hologram image, reconstructed by means of a beam of laser light, is given in fig. 11.8. It can be seen from this

Fig. 11.8 Nonlinear hologram images, reconstructed by means of a laser beam.

figure that, besides the two first-order image reconstructions near the zero-order diffraction in the center, there are two sets of the higher order reconstructions.

11.3 SPURIOUS DISTORTION

At this point it is possible to calculate the distortion due to all the unwanted signals (i.e., the zero and higher order, as well as first-order virtual or real hologram images), which are defined as spurious distortion. To do this, we may recall the case of fig. 11.5. Equation (11.21) is evaluated at $l = R$,

$$E(\boldsymbol{\sigma}; k)\bigg|_{l=R} = \iint_S T(\boldsymbol{\rho}; k)\exp(-ikx\sin\theta)E_l^+(\boldsymbol{\sigma} - \boldsymbol{\rho}; k)\bigg|_{l=R} dx\,dy; \tag{11.25}$$

then the complex light amplitude distribution at $l = R$ is

$$\begin{aligned} E(\boldsymbol{\sigma}; k)\bigg|_{l=R} &= \frac{a_0}{2}\exp\left[ik\left(R - \frac{R}{2}\sin^2\theta - \alpha\sin\theta\right)\right] \\ &+ \frac{1}{2}\exp[ik(R - \alpha\sin\theta)]\sum_{n=1}^{\infty}\frac{a_n}{n+1} \\ &\qquad \times \exp\left[i\left(n\phi - \frac{n+1}{2}kR\sin\theta + \frac{kn}{2R(n+1)}\sigma^2\right)\right] \\ &+ \frac{a_1}{2}\exp\left[i\left(kR + \frac{k}{2R}\sigma^2 - \phi - \frac{\pi}{2}\right)\right]\delta(\alpha, \beta) \\ &- \frac{1}{2}\exp[ik(R - \alpha\sin\theta)]\sum_{n=2}^{\infty}\frac{a_n}{n-1} \\ &\qquad \times \exp\left[i\left(\frac{n-1}{2}kR\sin^2\theta - n\phi + \frac{kn}{2R(n-1)}\sigma^2\right)\right]. \end{aligned} \tag{11.26}$$

The degree of spurious distortion D_s may be defined as the total unwanted irradiance in the neighborhood of a reconstructed image divided by the irradiance of that image, i.e.,

$$D_s = \frac{\sum_{n=2}^{\infty}\left(\frac{a_n}{n-1}\right)^2 + \sum_{n=1}^{\infty}\left(\frac{a_n}{n+1}\right)^2 + a_0^2}{a_1^2}, \tag{11.27}$$

for $\theta = 0$. If there is no nonlinear distortion, then for $\theta = 0$ the distortion is

$$D_s = \frac{\left(\frac{a_1}{2}\right)^2 + a_0^2}{a_1^2}.$$

If the unwanted signals are called noise,* then the signal-to-noise ratio may be defined by

$$\text{signal-to-noise ratio} = \frac{1}{D_s}. \tag{11.28}$$

It is a simple matter to show that the degree of spurious distortion for the case shown in fig. 11.6 is given by eq. (11.27).

Furthermore, as implied in figs. 11.5, 11.6, and 11.7, for a given finite hologram aperture, it is possible to reduce the spurious distortion to a minimum by an appropriate choice of the reference-beam angle.

We should note that Fourier-decomposition analysis of nonlinear holograms may be difficult to apply when the irradiance of the recording field is non-uniform. However, the method not only provides a direct picture of what causes the distortion, but also provides a simple graphical procedure for obtaining the behavior of the hologram. In any event, the location of the reconstructed higher order images can be determined by means of this method; qualitatively, the degree of nonlinear and spurious distortion can be easily determined. Of course, an analysis of nonlinear holograms can be obtained by utilizing the exact mathematical relation between the exposure and the amplitude transmittance of the recording emulsion. This may be accomplished by expanding the transfer characteristic of the amplitude transmittance in a power series (refs. 11.1–11.4), or in a suitable set of orthogonal functions. However, this method is more complicated, and is difficult to apply in practice.

11.4 EFFECT OF EMULSION THICKNESS VARIATION ON WAVE FRONT RECONSTRUCTION

Precise holographic-image reconstruction is required in certain engineering applications, e.g., in contouring, holographic interferometry, microscopic wave front reconstruction, and so on. We now investigate the effects of photographic emulsion thickness variations on the precision of image reconstruction (ref. 11.9). In the following, a general mathematical approach is developed for wave front recording and reconstruction on a holographic plate of nonuniform emul-

* Reluctantly called "noise," since the unwanted signals are deterministic. Strictly speaking, noise should be probabilistic in nature.

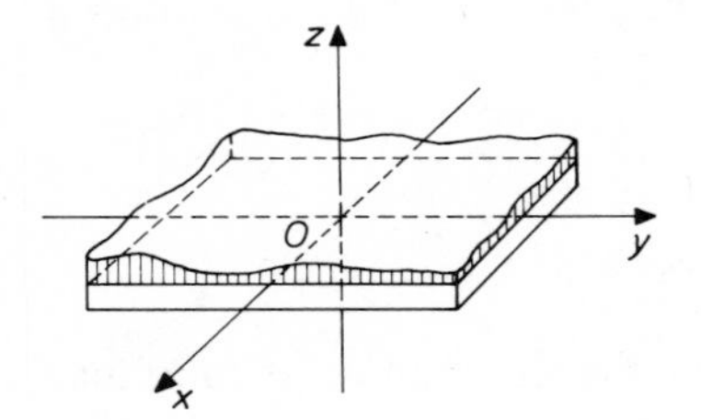

Fig. 11.9 Cross section of a photographic plate, showing emulsion thickness variation.

sion thickness. We see that the emulsion thickness variation does not affect wave front recording in any essential way. However, the emulsion thickness variation does appreciably affect the precision of the hologram image. A simplified example of the effect is illustrated.

Effect on Holographic Recording

Given a photographic plate in which the thickness of the thin emulsion* is nonuniform over the plate (fig. 11.9). Then the phase delay due to the emulsion thickness at points in the coordinate system (x, y) may be written as

$$\phi(x, y) = k[z_0 + (\eta - 1)z(x, y)], \tag{11.29}$$

where $z(x, y)$ is the thickness variation of the emulsion, z_0 is the maximum thickness of the emulsion, η is the index of refraction of the emulsion, and k is the wave number.

If the thickness variation $z(x, y)$ varies smoothly over the surface of the photographic plate, then the corresponding phase transmittance may be approximated by a Taylor expansion.

$$\phi(x, y) = \phi(0, 0) + \frac{\partial\phi(0, 0)}{\partial x}x + \frac{\partial\phi(0, 0)}{\partial y}y + \frac{1}{2}\frac{\partial^2\phi(0, 0)}{\partial x^2}x^2 + \frac{1}{2}\frac{\partial^2\phi(0, 0)}{\partial y^2}y^2 + \frac{\partial^2\phi(0, 0)}{\partial x\,\partial y}xy + \cdots . \tag{11.30}$$

If this photographic plate is used as the recording medium in the geometry shown in fig. 11.10, then the effect of the emulsion thickness variation on the scattered light is given by

$$u'(x, y) = u(x, y)\exp[i\phi(x, y)], \tag{11.31}$$

*A thin emulsion is defined in the same way as a thin lens: a light ray entering a point of the coordinate system (x, y) from one side of the emulsion emerges at approximately the same point on the other side of the coordinate system.

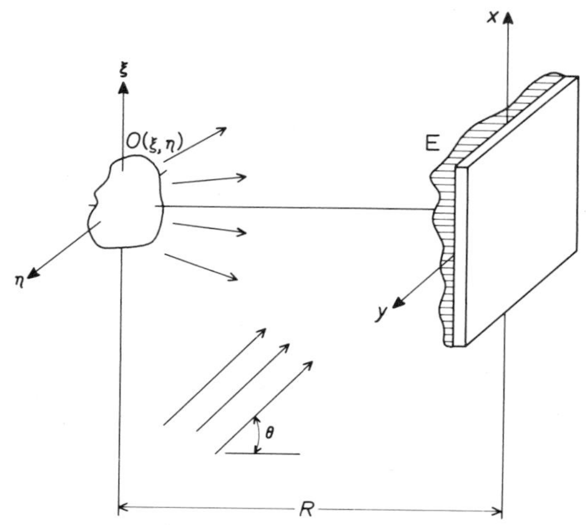

Fig. 11.10 Geometry of wave front recording. Oblique plane monochromatic reference wave.

and

$$v'(x, y) = v \exp\{i[kx \sin \theta + \phi(x, y)]\}, \tag{11.32}$$

where $u(x, y)$ and $v \exp(ikx \sin \theta)$ are the complex light amplitude distributions scattered from the object and from the reference wave, respectively, without the emulsion thickness effect, with v a positive real constant. The resulting irradiance of the photographic plate is

$$\begin{aligned} I(\boldsymbol{\rho}; k) &= [u'(x, y) + v'(x, y)][u'(x, y) + v'(x, y)]^* \\ &= |u(x, y)|^2 + v^2 + vu(x, y) \exp(-ikx \sin \theta) \\ &\quad + vu^*(x, y) \exp(ikx \sin \theta). \end{aligned} \tag{11.33}$$

If the wave front construction is biased so that the transmittance of the plate remains always linear in irradiance, and if the lateral emulsion shrinkage does not occur during the photographic process, then the transmittance of the recorded hologram is

$$\begin{aligned} T(\boldsymbol{\rho}; k) = K[&|u(x, y)|^2 + v^2 + u(x, y)v \exp(-ikx \sin \theta) \\ &+ vu^*(x, y) \exp(ikx \sin \theta)]. \end{aligned} \tag{11.34}$$

From either of eqs. (11.33) or (11.34), it is clear that the emulsion thickness variation has not affected the wave front construction.

Effect on Image Reconstruction

If the recorded hologram of eq. (11.34) is illuminated by a plane wave front of the same wavelength, with the geometry as shown in fig. 11.11, then the light amplitude distribution behind the hologram is

$$E(\boldsymbol{\sigma}; k) = \iint_S T(\boldsymbol{\rho}; k) v \exp(-ikx \sin\theta) \exp[i\phi(x, y)] E_l^+(\boldsymbol{\sigma} - \boldsymbol{\rho}; k)\, dx\, dy, \tag{11.35}$$

where

$$E_l^+(\boldsymbol{\rho}; k) = -\frac{i}{\lambda l} \exp[ik(l - z + \gamma)] \exp\left[i\frac{k}{2l}(x^2 + y^2)\right]$$

is the spatial impulse response, l is the separation between the two coordinate systems $\boldsymbol{\rho}(x, y, z)$ and $\boldsymbol{\sigma}(\alpha, \beta, \gamma)$, and S denotes the integration over the hologram surface.

Equation (11.35) can be evaluated termwise with respect to the order of image diffraction. Thus for real image reconstruction, the complex light amplitude at $l = R$ can be written as

$$E_{\mathrm{r}}(\boldsymbol{\sigma}; k) = v^2 \iint_S u^*(x, y) \exp[i\phi(x, y)] E_l^+(\boldsymbol{\sigma} - \boldsymbol{\rho}; k)\, dx\, dy. \tag{11.36}$$

where the subscript r denotes the real image.

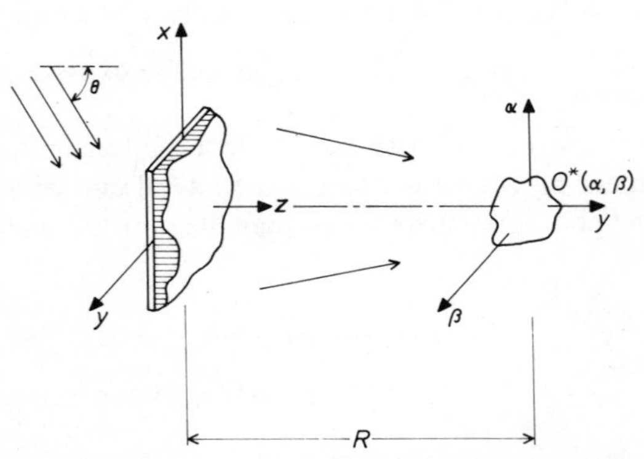

Fig. 11.11 Reconstruction of the real image. Plane monochromatic illumination, negative θ.

Let the conjugate light amplitude distribution be

$$u^*(x, y) = \iint_{S_1} O^*(\xi, \eta) E_l^{+*}(\boldsymbol{\rho} - \boldsymbol{\xi}; k)\, d\xi\, d\eta, \tag{11.37}$$

where $O(\xi, \eta)$ is the object function and S_1 denotes the surface integral of the object, viewed from the hologram aperture. By substitution of eq. (11.37), eq. (11.36) becomes

$$\begin{aligned} E_r(\boldsymbol{\sigma}; k) = K_1 \iint_S \Big\{ \iint_{S_1} & O^*(\xi,\eta) \exp\left[-i\frac{k}{2R}(\xi^2 + \eta^2)\right] \\ & \times \exp\left[i\frac{k}{R}(\xi x + \eta y)\right] d\xi\, d\eta \Big\} \exp[i\phi(x, y)] \\ & \times \exp\left[i\frac{k}{2R}(\alpha^2 + \beta^2)\right] \exp\left[-i\frac{k}{R}(\alpha x + \beta y)\right] dx\, dy, \end{aligned} \tag{11.38}$$

where K_1 is an appropriate complex constant.

Alternatively, eq. (11.38) can be written as

$$\begin{aligned} E_r(\boldsymbol{\sigma}; k) = K_1 \exp\left[i\frac{k}{2R}(\alpha^2 + \beta^2)\right] \iint_S & \left\{ G^*(x_0, y_0) \exp\left[i\frac{R}{2k}(x_0^2 + y_0^2)\right]\right\} \\ & \times \exp[i\phi(x, y)] \exp\left[-i\frac{k}{R}(\alpha x + \beta y)\right] dx\, dy, \end{aligned} \tag{11.39}$$

where $G^*(x_0, y_0)$ is the Fourier transform of the conjugate object function $O^*(\xi, \eta)$.

If there is no emulsion thickness variation, i.e., if $\phi(x, y) = 0$, the image reconstruction will be unique:

$$E_r(\boldsymbol{\sigma}; k) = K_1 O^*(\alpha, \beta). \tag{11.40}$$

Furthermore, if the phase delay is a first-degree equation (i.e., a perfect wedge),

$$\phi(x, y) = b_0 + b_1 x + b_2 y, \tag{11.41}$$

then the precision of the image reconstruction will also be unique, with a lateral translation

$$E_r(\boldsymbol{\sigma}; k) = K_1 O^*\left(\alpha - \frac{l}{k} b_1, \beta - \frac{l}{k} b_2\right). \tag{11.42}$$

On the other hand, if the higher order terms of the phase delay due to the emulsion thickness variation are considered, then exact precision of the image reconstruction would not be expected. In the following, a simplified example is given. We see that a very small amount of emulsion thickness variation can severely affect the precision of image reconstruction, and would have to be restricted in some engineering applications.

Simplified Example

We have seen that impulse-response techniques are applicable to linear holography, since superposition holds. Therefore, in the following example the effect due to the emulsion thickness variation is derived for a discrete-point source. Let us consider the wave front recording of two point objects separated by a distance h (fig. 11.12). If these monochromatic point sources are assumed to have identical wavelength λ, then the light amplitude distribution on the (x, y) coordinate system may be approximated by

$$u_1(\boldsymbol{\rho}; k) = A_1 \exp\left\{i\left[kR + \frac{k}{2R}\left(\left(x - \frac{h}{2}\right)^2 + y^2\right) + \phi(x, y)\right]\right\} \tag{11.43}$$

and

$$u_2(\boldsymbol{\rho}; k) = A_2 \exp\left\{i\left[kR + \frac{k}{2R}\left(\left(x + \frac{h}{2}\right)^2 + y^2\right) + \phi(x, y)\right]\right\}, \tag{11.44}$$

where $\phi(x, y)$ is the phase delay due to the emulsion thickness variation, and A_1 and A_2 are arbitrary positive constants.

If a plane reference wave (of the same wavelength as that of the object wave) is used [i.e., the effect of the emulsion is simply that given by eq. (11.32)], then the transmittance of the recorded plate is

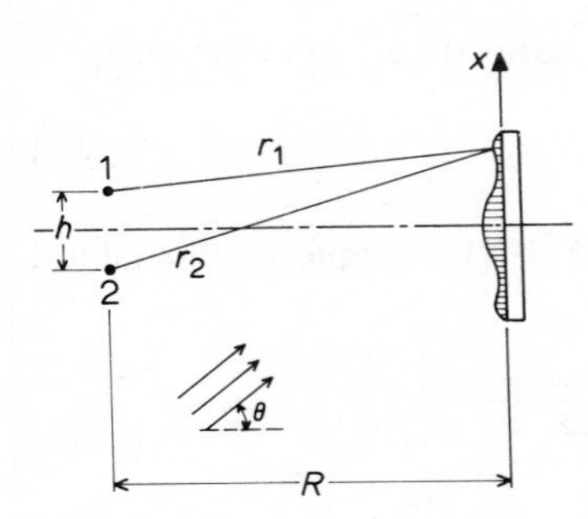

Fig. 11.12 Geometry for recording two point objects. Plane monochromatic reference wave, positive θ.

$$T(\boldsymbol{\rho}; k) = \left\{ a_0 + a_1 \cos \frac{k}{R} hx + \frac{a_2}{2} [\exp(i\wedge) + \exp(-i\wedge)] \right.$$
$$\left. + \frac{a_3}{2} [\exp(i\curlywedge) + \exp(-i\curlywedge)] \right\} \exp[i\phi(x, y)], \qquad (11.45)$$

where

$$\wedge = kR + \frac{k}{2R}\left[\left(x - \frac{h}{2}\right)^2 + y^2\right],$$

and

$$\curlywedge = kR + \frac{k}{2R}\left[\left(x + \frac{h}{2}\right)^2 + y^2\right].$$

If this hologram is illuminated by a plane wave of the same wavelength (fig. 11.13), then the light amplitude distribution behind the hologram may be determined by the Fresnel–Kirchhoff theory, i.e., by eq. (11.35). However, the evaluation of eq. (11.35) is in general very complicated, since the phase delay $\phi(x, y)$ is not generally specified. In order to see some of the effect of the emulsion thickness variation, we carry out the computation for a simplified case. Let the phase delay be approximated by a second-order polynomial,

$$\phi(x, y) \simeq b_0 + b_1 x + b_2 y + b_3 x^2 + b_4 y^2, \qquad (11.46)$$

where the b's are arbitrary real constants.

It can be recognized that the third and fifth terms of eq. (11.45) contribute to the virtual images, and the fourth and sixth terms to the real images. Thus if we

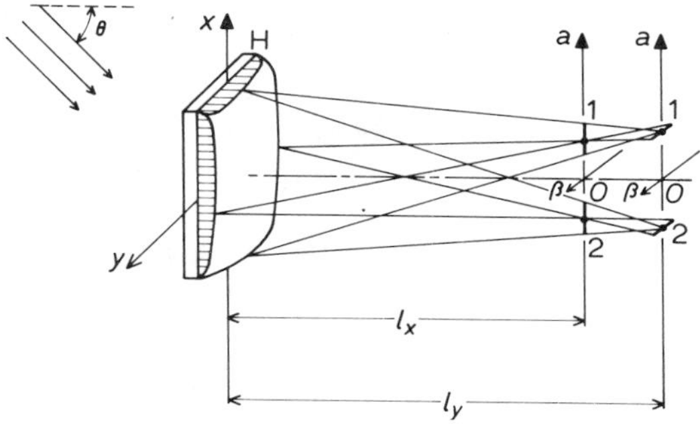

Fig. 11.13 Reconstruction of the real two-point image. Plane monochromatic illumination, negative θ.

separate the Fresnel–Kirchhoff integral for the real image terms, we obtain

$$
\begin{aligned}
E_r(\boldsymbol{\sigma}; k) = K_1 \int_{S_x} &\exp\left[-i\frac{k}{2}\left(\frac{1}{R} - \frac{1}{l} - \frac{2}{k}b_3\right)x^2\right] \\
\times \exp\left[-i\frac{k}{l}\left(\alpha - \frac{lh}{2R} - \frac{l}{k}b_1\right)\right] dx &\int_{S_y} \exp\left[-i\frac{k}{2}\left(\frac{1}{R} - \frac{1}{l} - \frac{2}{k}b_4\right)y^2\right] \\
\times \exp\left[-i\frac{k}{l}\left(\beta - \frac{l}{k}b_2\right)\right] dy &+ K_2 \int_{S_x} \exp\left[-i\frac{k}{2}\left(\frac{1}{R} - \frac{1}{l} - \frac{2}{k}b_3\right)x^2\right] \\
\times \exp\left[-i\frac{k}{l}\left(\alpha + \frac{lh}{2R} - \frac{l}{k}b_1\right)\right] dx &\int_{S_y} \exp\left[-i\frac{k}{2}\left(\frac{1}{R} - \frac{1}{l} - \frac{2}{R}b_4\right)y^2\right] \\
&\times \exp\left[-i\frac{k}{l}\left(\beta - \frac{l}{k}b_2\right)\right] dy. \qquad (11.47)
\end{aligned}
$$

where K_1 and K_2 are the appropriate complex constants, and S_x and S_y are integrations in the x and y directions, respectively. Without loss of generality, in the following the surface integral S is assumed to be of infinitely large extent.

From eq. (11.47) the unique image reconstructions with respect to the x and y coordinates will occur at

$$
l_x = \frac{kR}{k - 2Rb_3} \qquad (11.48)
$$

and

$$
l_y = \frac{kR}{k - 2Rb_4}, \qquad (11.49)
$$

where l_x and l_y are the separations between the hologram coordinate system $\boldsymbol{\rho}(x, y)$ and the image reconstruction coordinate system $\boldsymbol{\sigma}(\alpha, \beta)$. Thus at $l = l_x$ for the x integral, and at $l = l_y$ for the y integral, eq. (11.47) becomes

$$
\begin{aligned}
E_r(\boldsymbol{\sigma}; k) = K_1' \delta\left(\alpha - \frac{lh}{2R} - \frac{l}{k}b_1\right)\Big|_{l=l_x} &\delta\left(\beta - \frac{l}{k}b_2\right)\Big|_{l=l_y} \\
+ K_2' \delta\left(\alpha + \frac{lh}{2R} - \frac{l}{k}b_1\right)\Big|_{l=l_x} &\delta\left(\beta - \frac{l}{k}b_2\right)\Big|_{l=l_y}, \qquad (11.50)
\end{aligned}
$$

where K_1' and K_2' are the appropriate complex constants.

From eq. (11.50) or fig. 11.13, there is an astigmatic effect in the image reconstructions. It is clear that this astigmatism will vanish for the case $l_x = l_y$ (i.e., $b_3 = b_4$). The quadratic nature of the phase delay characteristic will also cause some degree of lateral magnification,

$$M^{r}_{lat,x} = \frac{k}{k - 2Rb_3} \tag{11.51}$$

and

$$M^{r}_{lat,y} = \frac{k}{k - 2Rb_4}, \tag{11.52}$$

where the superscript r denotes the real images, and the subscripts lat,x and lat,y are for the lateral magnifications with respect to the x and y axes.

The longitudinal magnifications can be determined with respect to the lateral magnifications due to the quadratic phase delay. Thus

$$M^{r}_{long,x} \simeq (M^{r}_{lat,x})^2 \tag{11.53}$$

and

$$M^{r}_{long,y} \simeq (M^{r}_{lat,y})^2. \tag{11.54}$$

From eqs. (11.53) and (11.54), the astigmatic effect can be seen. It is also clear that, as we have stated, the astigmatism disappears for $b_3 = b_4$.

Further, the angular magnification of the image reconstruction can be shown to be independent of the quadratic variation of the emulsion thickness. If higher order variations of the phase delay are considered (which is beyond the scope of the present work), more complicated image distortions may be determined.

In a similar manner, the virtual image lateral magnifications can be shown to be

$$M^{v}_{lat,x} = \frac{k}{k + 2Rb_3} \tag{11.55}$$

and

$$M^{v}_{lat,y} = \frac{k}{k + 2Rb_4}. \tag{11.56}$$

The corresponding longitudinal magnifications are

$$M^{v}_{long,x} \simeq (M^{v}_{lat,x})^2 \tag{11.57}$$

and

$$M^{v}_{long,y} \simeq (M^{v}_{lat,y})^2. \tag{11.58}$$

Again, the astigmatic effect disappears for $b_3 = b_4$.

In order to have some practical insight into the effect of emulsion thickness variation on hologram images, we may assume that the emulsion thickness variation of the holographic plate duplicates that of a positive lens. The phase delay is then

$$\phi(x, y) = k\left[\eta z_0 - (\eta - 1)\frac{x^2 + y^2}{2R_1}\right], \tag{11.59}$$

where η is the refractive index of the emulsion, z_0 is the maximum thickness of the emulsion, and R_1 is the radius of curvature of the lens. Then the lateral magnifications for the corresponding real and virtual images are

$$M_{\text{lat}}^{\text{r}} = \frac{R_1}{R_1 + R(\eta - 1)}, \tag{11.60}$$

$$M_{\text{lat}}^{\text{v}} = \frac{R_1}{R_1 - R(\eta - 1)}. \tag{11.61}$$

Suppose we let an object of lateral dimension $h = 30$ cm be holographed at a distance $R = 30$ cm away from the photographic plate. If the reconstructed real image is measured to be 0.2 cm shorter, then (assuming $\eta = 1.5$) the radius of curvature of the emulsion is seen to be $R_1 = 2240$ cm. Thus from this very simple example, it is clear that a very small variation of the emulsion thickness does affect the accuracy of the image reconstruction.

11.5 LINEAR OPTIMIZATION IN HOLOGRAPHY

The application of a conventional optimization technique to holography is discussed in this section (ref. 11.10). It is then demonstrated that this conventional optimization may not, in fact, be generalizable to most holographic processes (refs. 11.11, 11.12). Therefore, following our investigation of this process, a general optimal linearization method for the photographic emulsion is discussed. The application of this linearization technique to a simple point-object hologram is demonstrated, and its extension to more complicated objects is illustrated.

For the conventional photographic process (sec. 5.1), there is only one linear optimum condition of the H and D curve. In this section we apply the conventional photographic optimization to holography. We see that, in fact, there is only one linear optimum condition for wave front recording; however, this conventional optimization process may not be required or even obtainable in most holographic processes.

Note the conventional wave front recording geometry shown in fig. 11.10.

The complex light field scattered from the illuminated object onto the emulsion is

$$u(x, y) = A(x, y) \exp[i\psi(x, y)], \tag{11.62}$$

and the reference wave field on the emulsion is

$$v(x, y) = B \exp(ikx \sin \theta). \tag{11.63}$$

Then the corresponding irradiance of the wave front recording is

$$I(x, y) = A^2(x, y) + B^2 + 2A(x, y)B \cos[kx \sin \theta - \psi(x, y)].$$

If the recording is on the linear portion of the H and D curve (sec. 5.1), the amplitude transmittance of the hologram may be written as

$$T(x, y) = K[I(x, y)]^{\gamma/2}, \tag{11.64}$$

where K is a proportionality constant, and γ is the slope of the linear region of the H and D curve.

If the hologram described by eq. (11.64) is illuminated by a monochromatic plane wave, then the transmitted light field immediately behind the hologram becomes

$$E(x, y) = K_1[A^2(x, y) + B^2 + 2A(x, y)B \cos \phi(x, y)]^{\gamma/2}, \tag{11.65}$$

where K_1 is an appropriate positive constant, and $\phi = kx \sin \theta - \psi(x, y)$. Alternatively, eq. (11.65) may be expressed as

$$E(x, y) = K_2\left\{1 + \left[\frac{A(x, y)}{B}\right]^2 + 2\frac{A(x, y)}{B} \cos \phi(x, y)\right\}^{\gamma/2}, \tag{11.66}$$

where $K_2 = K_1 B^{\gamma}$.

If we restrict A and B by the inequality

$$\left[\frac{A(x, y)}{B}\right]^2 + 2\frac{A(x, y)}{B} \cos \phi(x, y) < 1, \qquad \text{for every } (x, y), \tag{11.67}$$

that is,

$$\left|\frac{A(x, y)}{B}\right| < 0.414,$$

then eq. (11.66) may be written as the binomial expansion

$$E(x, y) = K\Bigg\{1 + \frac{\gamma}{2}\left[\left(\frac{A(x, y)}{B}\right)^2 + 2\left(\frac{A(x, y)}{B}\right) \cos \phi(x, y)\right]$$
$$+ \frac{1}{2!}\frac{\gamma}{2}\left(\frac{\gamma}{2} - 1\right)\left[\left(\frac{A(x, y)}{B}\right)^2 + 2\left(\frac{A(x, y)}{B}\right)^2 \cos \phi(x, y)\right]^2$$
$$+ \frac{1}{3!}\frac{\gamma}{2}\left(\frac{\gamma}{2} - 1\right)\left(\frac{\gamma}{2} - 2\right)\left[\left(\frac{A(x, y)}{B}\right)^2\right.$$
$$\left. + 2\left(\frac{A(x, y)}{B}\right)^2 \cos \phi(x, y)\right]^3 + \cdots\Bigg\}. \tag{11.68}$$

After expanding the bracket [ ] terms of eq. (11.68), the equation can be written in the form

$$E(x, y) = K\Bigg\{\Bigg[1 + \left(\frac{\gamma}{2}\right)^2\left(\frac{A(x, y)}{B}\right)^2 + \frac{1}{2!}\frac{\gamma}{2}\left(\frac{\gamma}{2} - 1\right)\left(\frac{\gamma}{2} - 3\right)\left(\frac{A(x, y)}{B}\right)^4$$
$$+ \frac{1}{3!}\frac{\gamma}{2}\left(\frac{\gamma}{2} - 1\right)\left(\frac{\gamma}{2} - 2\right)\left(\frac{A(x, y)}{B}\right)^6 + \cdots\Bigg]$$
$$+ 2\left[\frac{\gamma}{2}\left(\frac{A(x, y)}{B}\right) + \frac{1}{2!}\left(\frac{\gamma}{2}\right)^2\left(\frac{\gamma}{2} - 1\right)\left(\frac{A(x, y)}{B}\right)^3 + \cdots\right]$$
$$\times \cos \phi(x, y) + 2\left[\frac{1}{2!}\frac{\gamma}{2}\left(\frac{\gamma}{2} - 1\right)\left(\frac{A(x, y)}{B}\right)^2\right.$$
$$\left. + \frac{3}{3!}\frac{\gamma}{2}\left(\frac{\gamma}{2} - 2\right)\left(\frac{A(x, y)}{B}\right)^4 + \cdots\right]$$
$$\times \cos\,[2\phi(x, y)]$$
$$+ 2\left[\frac{1}{3!}\frac{\gamma}{2}\left(\frac{\gamma}{2} - 1\right)\left(\frac{\gamma}{2} - 2\right)\left(\frac{A(x, y)}{B}\right)^3 + \cdots\right]$$
$$\times \cos[3\phi(x, y)] + \cdots\Bigg\}. \tag{11.69}$$

We see that eq. (11.69) may be written as

$$E(x, y) = K \sum_{n=1}^{\infty} a_n(\gamma) \cos\,[n\phi(x, y)], \tag{11.70}$$

where the $a_n(\gamma)$'s are the corresponding coefficients. Then the degree of nonlinearity of the hologram (eq. 11.12) is

$$\text{nonlinearity } \% = \frac{\left[\sum_{n=2}^{\infty} a_n^2(\gamma)\right]^{1/2}}{a_1(\gamma)} \times 100. \tag{11.71}$$

It can be seen from eq. (11.69) that the degree of nonlinearity approaches zero whenever γ approaches two. Thus $\gamma = 2$ is the optimum condition in the sense of minimum nonlinear distortion. to achieve the value $\gamma = 2$ in practice, a two-step contact printing process, as described in sec. 5.1, is applied.

The conventional linear optimization process described above minimizes the nonlinear distortion. However, the linear optimization technique that is described in the following sections maximizes the reconstruction of the first-order linear hologram image. In other words, in the holographic process we allow a certain degree of nonlinear wave front recording, and still have the best first-order hologram image reconstruction.

Linear Optimization

The T-E curve for a photographic emulsion is a monotonic decreasing function (sec. 5.1), which may be expanded into a finite power series:

$$T(E) = \sum_{n=0}^{N} a_n E^n, \qquad \text{for } E \geq 0, \tag{11.72}$$

where T is the amplitude transmittance, E is the exposure, and the a_n's are the real coefficients. Clearly, from the boundary conditions on the amplitude transmittance [$T(0) = 1$ and $T(\infty) = 0$], we can conclude

$$a_0 = 1$$

and

$$\lim_{E \to \infty} \sum_{n=1}^{N} a_n E^n = -1.$$

Let us now replace eq. (11.72) with a linear approximated transmittance:

$$T^{\dagger}(E) = \lambda_0 + \lambda_1 E. \tag{11.73}$$

Then we choose the parameters λ_0 and λ_1 to approximate the nonlinear transmittance of eq. (11.72) as closely as possible. Let us now consider the difference of eqs. (11.72) and (11.73):

$$T - T^{\dagger} = (a_0 - \lambda_0) + (a_1 - \lambda_1)E + a_2 E^2 + a_3 E^3 + \cdots + a_N E^N. \tag{11.74}$$

We seek to minimize the mean-square integral,

$$\iint_S (T - T^\dagger)^2 \, dx \, dy, \tag{11.75}$$

where s denotes integration over the surface of the photographic plate. Let us define the spatial ensemble average as

$$\langle E^n \rangle \triangleq \iint_S E^n(x, y) \, dx \, dy, \qquad n = 1, 2, 3, \ldots, N. \tag{11.76}$$

Then the optimal choice of the parameters λ_0 and λ_1 can be determined by

$$\frac{\partial}{\partial \lambda_n} \iint_S (T - T^\dagger)^2 \, dx \, dy, \qquad n = 0, 1. \tag{11.77}$$

The solution of eq. (11.77) is

$$\lambda_0 = \frac{\sum_{n=0}^{N} a_n(\langle E^n \rangle \langle E^2 \rangle - \langle E^{n+1} \rangle \langle E \rangle)}{\langle E^2 \rangle - \langle E \rangle^2}, \tag{11.78}$$

and

$$\lambda_1 = \frac{\sum_{n=0}^{N} a_n(\langle E^{n+1} \rangle - \langle E^n \rangle \langle E \rangle)}{\langle E^2 \rangle - \langle E \rangle^2}. \tag{11.79}$$

The corresponding matrix representation is

$$\begin{bmatrix} 1 & \langle E \rangle \\ E & \langle E^2 \rangle \end{bmatrix} \begin{bmatrix} \lambda_0 \\ \lambda_1 \end{bmatrix} = \begin{bmatrix} \sum_{n=0}^{N} a_n \langle E^n \rangle \\ \sum_{n=0}^{N} a_n \langle E^{n+1} \rangle \end{bmatrix}. \tag{11.80}$$

Equations (11.78) and ((11.79) or eq. (11.80) are the best linear approximations with respect to eq. (11.73), as determined by the least-mean-square-error criterion of eq. (11.75). Equation (11.79) can be considered to be a generalized first-order transmittance (i.e., first-order transfer function) for the photographic emulsion. The application of this optimal linearization method in holography is discussed in the following passages.

The object recorded in a hologram may be considered to be composed of a large number of infinitesimal point objects. Thus in optimal linearization, we can start from a point object and then extend the results by superposition.

Suppose a monochromatic point source and an oblique plane reference wave are of the same wavelength, and suppose that the recording geometry is as shown in fig. 11.4. Then the irradiance contributed by these two wave fronts on the recording medium can be approximated by

$$I(x, y) = |A|^2 + |B|^2 + 2|A|\,|B| \cos\left[k\left(\frac{\rho^2}{2R} - x \sin\theta\right) + \psi\right], \quad (11.81)$$

where A and B are the complex amplitudes of the spherical and plane wave fronts, respectively, ψ is the phase angle between A and B, $\rho^2 = x^2 + y^2$, and k is the wave number.

For convenience in notation, in the following A and B are used to represent $|A|$ and $|B|$, respectively, unless otherwise specified. Rewrite eq. (11.81) as

$$I(x, y) = (A^2 + B^2)\left[1 + \frac{2AB}{A^2 + B^2}\cos\left(\frac{k}{2R}\rho_1^2 + \phi\right)\right], \quad (11.82)$$

where $\phi = -\frac{1}{2}kR \sin^2\theta + \psi$, $\rho_1^2 = x_1^2 + y^2$, and $x_1 = x - R\sin\theta$. Thus

$$E^n = (It)^n = t^n(A^2 + B^2)^n\left[1 + \frac{2AB}{A^2 + B^2}\cos\left(\frac{k}{2R}\rho_1^2 + \phi\right)\right]^n,$$

$$n = 1, 2, \ldots, N + 1, \quad (11.83)$$

where t is the exposure time. By performing the integration of eq. (11.83) over a period of the ρ_1^2 axis (ref. 11.1), the ensemble average can be obtained,

$$\langle E^n\rangle = \langle I^n\rangle t^n = t^n \sum_{2r=0}^{n} \frac{n!}{(n - 2r)!\,(r!)^2}(A^2 + B^2)^{n-2r}(AB)^{2r}. \quad (11.84)$$

The parameters λ_0 and λ_1 can be obtained by substitution of eq. (11.84) into eqs. (11.78) and (11.79), respectively. The optimal linear amplitude transmittance is, then,

$$T^{\dagger}(x, y) = \sum_{n=0}^{N} a_n\langle E^n\rangle + 2\lambda_1 tAB \cos\left(\frac{k}{2R}\rho_1^2 + \phi\right). \quad (11.85)$$

Equation (11.85) is a linear approximation for the given A and B, and it is also the best approximation with respect to eq. (11.73) and to the mean-square-error criterion of eq. (11.77).

Furthermore, from eqs. (11.78) and (11.84), λ_1 is a function of three variables, i.e., $\lambda_1 = \lambda_1(A, B, t)$. Therefore, for a fixed amplitude of the reference wave front B, the optimum value of λ_1 may be determined by the partial derivatives

$$\frac{\partial \lambda_1}{\partial t} = 0, \qquad \frac{\partial \lambda_1}{\partial A} = 0 \tag{11.86}$$

On the other hand, if the values of A and B are given, the optimum value of λ_1 can be determined from eq. (11.86). The variable λ_1 can be considered to be a generalized first-order transmittance or the first-order diffraction for this point-object hologram. Thus the optimum value of λ_1 is also the best first-order diffraction of this wave front recording, with respect to the optimal linearization of eqs. (11.73) and (11.77).

If the irradiance of the reference wave is much larger than that of the spherical wave (resulting in only weak nonlinear distortion), then eq. (11.85) may be approximated by

$$T^{\dagger}(x, y) \simeq \sum_{n=0}^{N} a_n t^n B^{2n} + 2 \sum_{n=1}^{N} n a_n t^n B^{2n-1} A \cos\left(\frac{k}{2R}\rho_1^2 + \phi\right), \qquad B \gg A. \tag{11.87}$$

It must be emphasized that, in this weak distortion, it is possible to define a linearization that will make the distortion of the hologram negligible.

As an example, for Kodak 649F (D-19, 5 minutes) photographic emulsion, commonly used in holography, the T-E curve may be approximated by the third-order polynomial (refs. 11.3, 11.4)

$$T = \sum_{n=0}^{3} (-1)^n a_n E^n, \qquad \text{for } E \geq 0, \tag{11.88}$$

where the a_n's are the positive real coefficients. If the input signal (the irradiance) is that of eq. (11.81), we have

$$\begin{aligned} \langle E \rangle &= t(A^2 + B^2), \\ \langle E^2 \rangle &= t^2[(A^2 + B^2)^2 + 2(AB)^2], \\ \langle E^3 \rangle &= t^3[(A^2 + B^2)^3 + 6(AB)^2(A^2 + B^2)], \\ \langle E^4 \rangle &= t^4[(A^2 + B^2)^4 + 12(AB)^2(A^2 + B^2)^2 + 6(AB)^4]. \end{aligned} \tag{11.89}$$

Hence,

$$\lambda_0 = a_0 + a_2 t^2[2(AB)^2 - (A^2 + B^2)^2] + a_3 t^3(A^2 + B^2)[2(A^2 + B^2)^2 - 3(AB)^2] \tag{11.90}$$

and

$$\lambda_1 = -a_1 + 2a_2t(A^2 + B^2) - 3a_3t^2[(A^2 + B^2)^2 + (AB)^2], \quad (11.91)$$

which in these cases depend on amplitudes of A and B on the input irradiance, but not on the spatial frequency. (In some cases they may depend on both the amplitude and the spatial frequency.)

For a given amplitude B, the optimum value of λ_1 can be obtained by

$$\frac{\partial \lambda_1}{\partial A} = a_2(A^2 + B^2) - 3a_3t[(A^2 + B^2)^2 + (AB)^2] = 0 \quad (11.92)$$

and

$$\frac{\partial \lambda_1}{\partial A} = 2a_2 - 3a_3t(2A^2 + 3B^2) = 0. \quad (11.93)$$

The solutions for eqs. (11.92) and (11.93) are

$$A = B, \qquad t = \frac{2a_2}{15a_3B^2}. \quad (11.94)$$

The optimum value of λ_1 is then

$$(\lambda_1)_{\text{op}} = -a_1 + \frac{4a_2^2}{15a_3}, \quad (11.95)$$

and the corresponding optimal linear transmittance (for $A = B$) is

$$T^{\dagger}(x, y) = 1 - \frac{4a_1a_2}{15a_3} + \frac{32a_2^3}{675a_3^2} + 2\left(-\frac{2a_1a_2}{15a_3} + \frac{8a_2^3}{225a_3^2}\right)\cos\left(\frac{k}{2R}\rho_1^2 + \phi\right). \quad (11.96)$$

On the other hand, if A and B are given, then the optimum value of λ_1 occurs at

$$t = \frac{a_2}{3a_3}\frac{A^2 + B^2}{(A^2 + B^2)^2 + (AB)^2}. \quad (11.97)$$

For $B \gg A$, i.e., weak nonlinear distortion, eq. (11.97) may be approximated by

$$t \simeq \frac{a_2}{3a_3B^2}. \quad (11.98)$$

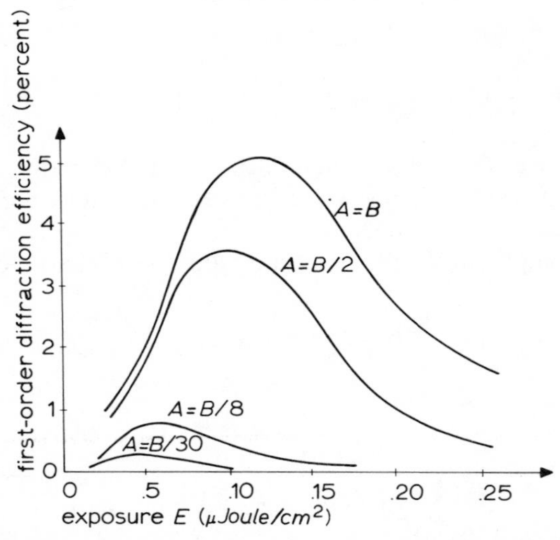

Fig. 11.14 First-order diffraction efficiency as a function of exposure.

The corresponding optimum value of λ_1 is

$$(\lambda_1)_{\mathrm{op}} \simeq -a_1 + \frac{a_2^2}{3a_3}, \tag{11.99}$$

and the optimum linear transmittance is therefore,

$$T^{\dagger}(x, y) = 1 - \frac{a_1 a_2}{3a_3} + \frac{2a_2^3}{27a_3^2} + 2\frac{A}{B}\left[-\frac{a_1 a_2}{3a_3} + \frac{a_2^3}{9a_3^2}\right]\cos\left(\frac{k}{2R}\rho_1^2 + \phi\right), \quad \text{for } B \gg A. \tag{11.100}$$

In this case, the nonlinear distortion in the hologram is negligible.

From the result of eq. (11.94), note that, in most holographic linear optimization processes, a one-to-one object-reference beam ratio and an appropriate optimum exposure are required. Figure 11.14 gives the experimentally determined first-order diffraction efficiency* as a function of exposure for various object-reference beam ratios. The recording medium used is a Kodak 649F(D-19, 5 minutes) photographic plate. It can be seen from this figure that the optimum first-order hologram image occurs for the unity object-reference beam ratio. Figure 11.15 is a series of photographs of three-dimensional first-order optimum hologram images for various object-reference beam ratios.

*The diffraction efficiency is defined as

$$\text{diffraction efficiency} \overset{\Delta}{=} \frac{\text{output (hologram image irradiance)}}{\text{input (incident irradiance)}}.$$

Fig. 11.15 Hologram images for various object-reference beam ratios.

Again, it can be seen that the optimum image reconstruction occurs when the object-reference beam ratio equals one.

To recapitulate: In nonlinear holographic processes the separation of the first-order hologram image from the higher order images is possible (sec. 11.2). Therefore, instead of confining the recording to the linear region of the photographic emulsion, it is possible to obtain optimum hologram image reconstruction from a nonlinear holographic recording. It may also be emphasized that in most holographic processes the phase information is far more important than the amplitude information. Therefore, as long as the phase information is preserved in the wave front construction, an amplitude distortion will not cause any defect in the hologram image reconstruction, but instead generates higher order diffractions, which can be spatially separated from the first-order hologram image diffraction.

11.6 ANALYSIS OF SIGNAL DETECTION BY OPTIMUM NONLINEAR SPATIAL FILTERING

In sec. 6.6 we have learned that a complex matched filter can be synthesized by means of coherent optical processes. It can be seen that the filter recording is indeed a Fourier transform hologram. Since the signal in complex spatial filter synthesis can be considered to be the summation of signals from a large number of resolvable point sources, the Fourier transform of the signal can be treated as the superposition of a large number of monochromatic plane waves. By reasoning along the lines developed in the previous section, it can be shown that an optimum spatial filter can be synthesized (refs. 11.3, 11.4) with the same condition as that for optimization in holography in eq. (11.94), a one-to-one signal-reference beam ratio with an appropriate exposure.

In this section we show that optimum signal detection can be achieved with an optimum nonlinear spatial filter. From sec. 6.6, we note that the intensity distribution of a complex matched filter recording is

$$I(p, q) = S^2(p, q) + R^2 + 2RS(p, q) \cos[x_0 p + \phi(p, q)], \qquad (11.101)$$

where $S(p, q) = |S(p, q)| \exp[i\phi(p, q)]$ is the signal spectrum, and (p, q) is the angular spatial frequency coordinate system. From this equation it is clear that the complex quantity $S(p, q)$ is spatially modulated by a spatial carrier frequency x_0. In order to insure the separation of the crosscorrelation from the zero-order diffraction, the spatial carrier frequency must be sufficiently high, and may be approximated by

$$x_0 > l_f + \tfrac{3}{2} l_s. \qquad (11.102)$$

where 1_f and 1_s are the spatial lengths in the x direction of the input signal $f(x, y)$ and the detecting signal s(x, y), respectively. Moreover, it is clear from eq. (11.101) that the nonlinear distortion due to the film transfer characteristic can occur only in the amplitude modulation and not in the phase modulation. Therefore, the nonlinear spatial filter yielding the signal $s(x, y)$ may be approximated, by means of the Fourier decomposition theorem, as

$$H(p, q) \simeq H_0(p, q) + \sum_{n=1}^{\infty} H_n(p, q), \qquad (11.103)$$

where $H_0(p, q) = \frac{1}{2}B_0(p, q)$ and $H_n(p, q) = B_n(p, q) \cos\{n[x_0 p + \phi(p, q)]\}$. The spatial Fourier components $B_n(p, q)$ may be found by finite-point analysis (sec. 11.1) on a cycle-by-cycle basis with respect to the spatial carrier frequency x_0.

It may be emphasized that, although $B_1(p, q)$ is not quite identical to $S(p, q)$, there is a great deal of similarity between them, particularly between their corresponding inverse Fourier transforms:

$$\mathcal{F}^{-1}\{B_1(p, q) \exp[i\phi(p, q)]\} \sim Ks(x, y), \tag{11.104}$$

where $\mathcal{F}^{-1}$ denotes the inverse Fourier transform, $\sim$ denotes the similarity, and K is an arbitrary constant.

If the complex nonlinear filter is inserted in the frequency plane of the coherent optical processing system of fig. 11.16, the output signal may be approximated by

$$g(\alpha, \beta) \simeq \sum_{n=0}^{\infty} \iint_{-\infty}^{\infty} B_n(p, q) F(p, q) \times \cos\{[x_0 p + \phi(p, q)] \exp[-i(p\alpha + q\beta)]\}\, dp\, dq \tag{11.105}$$

or

$$g(\alpha, \beta) \simeq f(x, y) * b_0(x, y) + \sum_{n=1}^{\infty} f(x, y) * b_n(x + nx_0, y) + \sum_{n=1}^{\infty} f(x, y) * b_n(-x - nx_0, -y), \tag{11.106}$$

where the $b_n(x, y)$ are the corresponding inverse Fourier transforms of $\frac{1}{2}B_n(p, q) \exp[\phi(p, q)]$. From eq. (11.106), it can be seen that the first summation covers the corresponding convolutions, and that the last summation covers the crosscorrelations.

If the input signal is assumed to be

$$f(x, y) = s(x, y) + n(x, y), \tag{11.107}$$

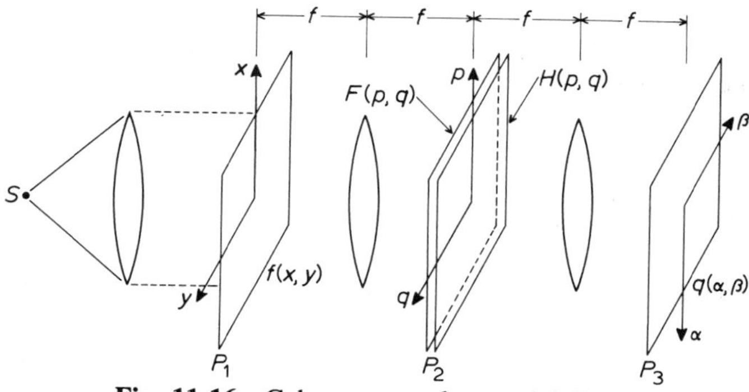

Fig. 11.16 Coherent complex spatial filtering.

where $s(x, y)$ is the detecting signal, and $n(x, y)$ is additive white Gaussian noise, then the crosscorrelation terms of eq. (11.107) may be written as

$$R_{12}(\alpha, \beta) = \sum_{n=1}^{\infty} [s(x, y) + n(x, y)] * b_n(-x - nx_0, y). \quad (11.108)$$

The crosscorrelations with respect to $n(x, y)$ and $b_n(x, y)$ are approximately zero; also, $b_1(x, y)$ has a great deal of similarity to $s(x, y)$. That is,

$$b_1(x, y) \sim Ks(x, y), \quad (11.109)$$

but

$$b_n(x, y) \nsim Ks(x, y), \qquad \text{for } n \neq 1, \quad (11.110)$$

where K is an arbitrary constant. Therefore, the crosscorrelation of eq. (11.108) may be approximated as

$$\begin{aligned} R_{12}(\alpha, \beta) &\simeq s(x, y) * b_1(-x - x_0, -y) \\ &\simeq s(x, y) * Ks(-x - x_0, -y), \end{aligned} \quad (11.111)$$

that is, R_{12} is proportional to the autocorrelation of $s(x, y)$.

It may be emphasized that eq. (11.110) is true in practice but not necessarily required, since the higher order crosscorrelation between $s(x, y)$ and $b_n(x, y)$ can be spatially separated from the first-order crosscorrelation. Moreover, from eq. (11.111) it is clear that an optimum detection requires an optimum value of $b_1(x, y)$, which may be obtained by the linear optimization technique proposed in the previous section.

Figures 11.17 and 11.18 demonstrate the separation of signal from random noise. The input transparency consists of a signal embedded in random noise (fig. 11.17). The results of the signal detection for various values of signal to reference beam ratio are given in fig. 11.18. The recording medium used for the filter synthesis is a Kodak 649F(D-19, 5 minutes) photographic plate. These filters are synthesized at the linear optimization conditions obtained from interpolations of fig. 11.14. From fig. 11.18 it can be seen that the optimum signal detection occurs at a one-to-one signal to reference beam ratio.

Instead of being restricted to the linear region of the transfer characteristic of the photographic emulsion, an optimum nonlinear spatial filter can be achieved form the system theory point of view. Although it may be difficult to apply this theory to the case of complicated signals, we may always obtain a generalized linear optimization. In the analysis of nonlinear spatial filtering, it has been shown that the signal detection (represented by the first-order crosscorrelation) can be spatially separated from the unwanted higher order correlations. Thus the

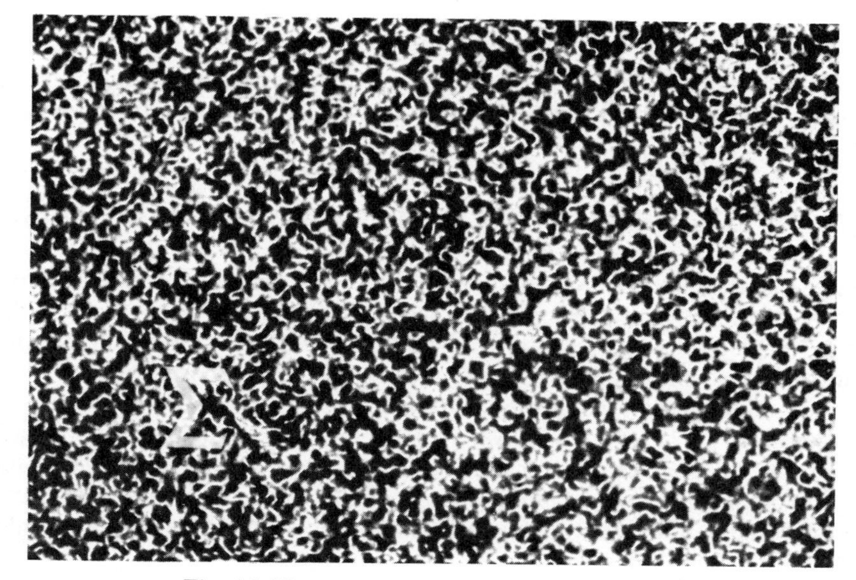

Fig. 11.17 Signal embedded in random noise.

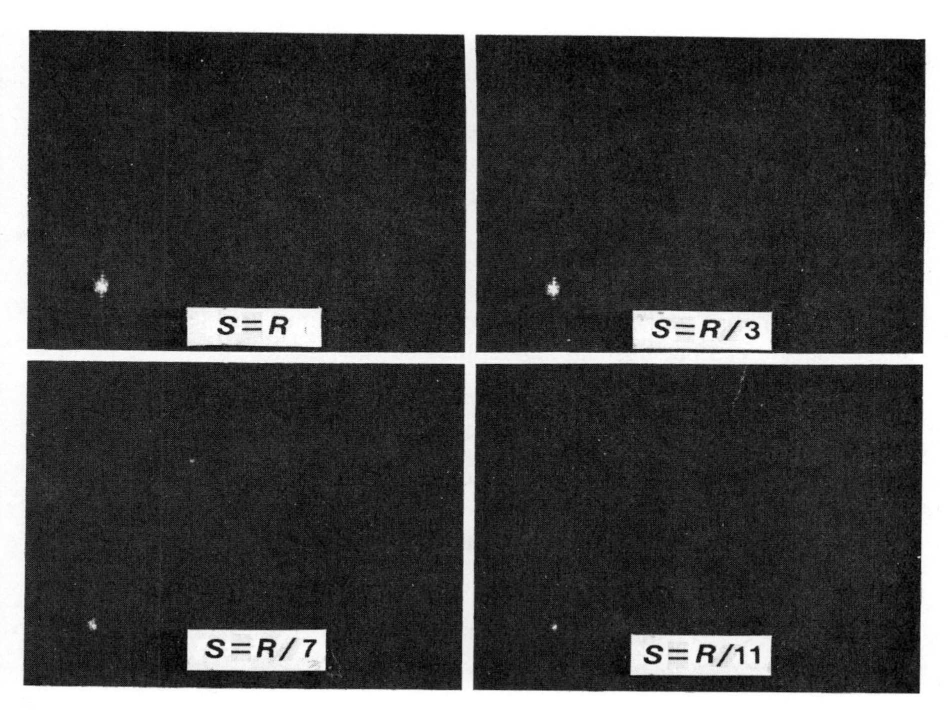

Fig. 11.18 Signal detection by means of complex filtering, for various values of signal-reference beam ratios.

ambiguity of identifying the first-order from the higher order correlations can be resolved. In fact, it has even been shown that the higher order correlations are generally very weak.

11.7 SPECKLE REDUCTION BY RANDOM SPATIAL SAMPLING

Speckle noise in the hologram image has been commonly regarded as the number one enemy of modern holography. The speckle effect, which is discussed subjectively, arises from the finite holographic aperture in which a diffused object is recorded. Historically, almost as soon as lasers became available this image granularity or speckling effect was first described by Rigden and Gordon (ref. 11.15) and by Oliver (ref. 11.16). These authors gave a basic account of the speckle phenomenon, that the image granularity arises by the diffraction of the finite aperture of the optical instrument (e.g., the eye). Later, a more complete mathematical analysis of this speckle effect was given by Enloe (ref. 11.17), and others (ref. 11.18).

Techniques of speckle reduction in holography, from the objective point of view, have been suggested by Gerritsen et al. (ref. 11.19), Gabor (ref. 11.20), Caulfield (ref. 11.21), and Van Ligten (ref. 11.22). These techniques are primarily based on the redundancy multibeam recording principle; the method proposed by Caulfield carries the idea even further, spatially multiplexing the hologram. The Van Ligten method is based on sequential illumination, in which a mask with a circular aperture is placed on the Fourier domain.

A different technique is discussed here, in which the speckle in the hologram image can be reduced by means of random spatial sampling (ref. 11.23). This sampling technique can be implemented during the hologram image reconstruction by means of a movable random aperture (i.e., sampling function). If the spatial sampling function is made uncorrelated in time variable, the hologram image speckle can be substantially suppressed. In practice, however, an ideal sampling function is easy to obtain.

In most holographic processes the objects to be holographed generally do not have a perfectly smooth surface (i.e., they are diffuse). In the conventional off-axis wave front recording, as shown in fig. 11.10, the light scattered from the object can be regarded as composed of a large number of infinitesimal point radiators. The complex light field distributed on the recording aperture can be written as

$$u(x, y) = \frac{1}{R} \sum_{n=1}^{N} a_n \exp[ik(R - \zeta_n)] \times \exp\left\{ i \frac{k}{2R} [(x - \xi_n)^2 + (y - \eta_n)^2] \right\}, \qquad (11.112)$$

where the a_n's are arbitrary constants, R is the separation between the coordinate systems $\boldsymbol{\zeta}(\zeta, \eta, \zeta)$ and $\boldsymbol{\rho}(x, y, z)$, $(\zeta_n, \eta_n, \zeta_n)$ are the position vectors of the point radiators, and k is the wave number. The corresponding irradiance of the wave front recording is

$$I(\boldsymbol{\rho}; k) = (u + v)(u + v)^*$$
$$= b^2 + \frac{1}{R^2} \sum_{i=1}^{N} \sum_{j=1}^{N} a_i a_j \cos\{k[\phi_i(x, y) - \phi_j(x, y)]\}$$
$$+ \frac{2b}{R} \sum_{n=1}^{N} a_n \cos\{k[x \sin\theta + \phi_n(x, y)]\}, \qquad i \neq j, \qquad (11.113)$$

where $v(x, y) = b \exp(ikx \sin\theta)$, the oblique reference beam, and

$$\phi_n(x, y) = R - \zeta_n + \frac{1}{2R}[(x - \zeta_n)^2 + (y - \eta_n)^2].$$

The first summation of eq. (11.113) yields a random holographic intermodulation and the second summation gives rise to a superposition of Fresnel zone lens constructions. If it is assumed that the wave front recording is linear, the transmittance of the recorded hologram is

$$T(\boldsymbol{\rho}; k) = KI(\boldsymbol{\rho}; k), \qquad (11.114)$$

where K is a proportional constant.

If the hologram is illuminated by an oblique plane wave of negative θ, the complex light for the real image reconstruction can be written as

$$E_r(\boldsymbol{\sigma}; k) = K \int_{-Ly/2}^{Ly/2} \int_{-Lx/2}^{Lx/2} \sum_{n=1}^{N} a_n \exp[ik\phi_n(x, y)] E_l^{+}(\boldsymbol{\sigma} - \boldsymbol{\rho}; k)\, dx\, dy, \qquad (11.115)$$

where $\boldsymbol{\sigma}(\alpha, \beta, \gamma)$ represents the image coordinate system, Lx and Ly are the corresponding dimensions of the hologram aperture, and $E_l^{+}(\boldsymbol{\rho}; k)$ is the spatial impulse response.

Because of the finite aperture of the hologram, it can be shown that an image point will be reconstructed as a spread function rather than a point image (i.e., delta function). Since the spread images were derived from a common coherent source, they mutually interfere. Thus they produce a random interference pattern known as speckle noise. This speckle effect is mainly caused by the finite aperture of the recording medium.

This type of speckle effect can be reduced by means of random spatial sampling of the hologram.

With reference to fig. 11.19, let us assume a spatial sampling function (i.e.,

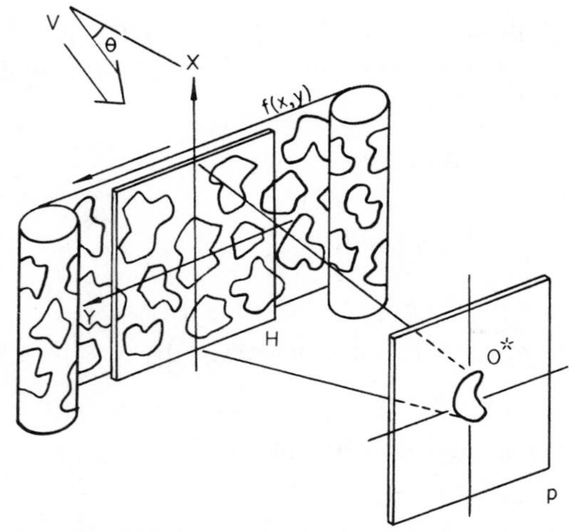

Fig. 11.19 Schematic diagram of speckle reduction by random spatial sampling. $f(x, y)$, the sampling function; H, hologram; v, monochromatic illumination; p, photographic plate.

random masking) is

$$f(x, y; t) = \begin{cases} 1, & \text{over the open aperture} \\ 0, & \text{otherwise} \end{cases}, \tag{11.116}$$

where t is a time variable. It may be noted that the aperture of the sampling function is assumed to be a few times larger than the wavelength of the source. As shown in fig. 11.19, the sampling function of eq. (11.116) is located at the front of the hologram, during the reconstruction process. The complex light field illuminating the hologram is proportional to the spatial sampling, i.e.,

$$u(x, y; t) = Kf(x, y; t), \tag{11.117}$$

where K is a proportionality constant. The complex light field at the back focal distance of the hologram can be written by the following convolution integral:

$$E(\boldsymbol{\sigma}; t; k) = \iint f(x, y; t)T(x, y)E_l^{+}(\boldsymbol{\sigma} - \boldsymbol{\rho}; k)\, dx\, dy, \tag{11.118}$$

where $T(x, y)$ is the transmittance of the hologram.

As we have noted previously, the speckle effect arises from the mutual interference of the spread images. The amplitudes of these spread functions are proportional to the amplitude of the hologram image. Thus this speckle noise due to a finite hologram aperture can be assumed to be multiplicative. The real image construction at a given t is therefore,

$$E_r(\boldsymbol{\sigma}; t; k) = CO^*(\boldsymbol{\sigma}; k)N^*(\boldsymbol{\sigma}; t; k), \tag{11.119}$$

where C is an appropriate complex constant, $O(\boldsymbol{\sigma}; k)$ is the object function, $N(\boldsymbol{\sigma}; t; k)$ is a time-dependent speckle noise, and * denotes the complex conjugate.

The illumination of the spatial sampling hologram corresponds to the open apertures of the sampling function $f(x, y; t)$. The spread of the point images varies in accordance with the random spatial ensemble of the hologram aperture. Therefore, it may be seen from eq. (11.118) that $N(\boldsymbol{\sigma}; t; k)$ varies as the spatial sampling function $f(x, y; t)$.

The corresponding real image irradiance is

$$I(\boldsymbol{\sigma}; t; k) = K|O(\boldsymbol{\sigma}; k)|^2|N(\boldsymbol{\sigma}; t; k)|^2. \tag{11.120}$$

If the hologram image is recorded on a photographic plate, as shown in fig. 11.19, then the recording exposure is

$$\begin{aligned} E(\alpha, \beta) &= \int_0^T I(\boldsymbol{\sigma}; t; k)\, dt \\ &= K|O(\boldsymbol{\sigma}; k)|^2 \int_0^T |N(\boldsymbol{\sigma}; t; k)|^2\, dt, \end{aligned} \tag{11.121}$$

where T is the exposure time.

It can be shown that, if the sampling time correlation is

$$\int_\epsilon^T f(\boldsymbol{\rho}; t)f(\boldsymbol{\rho}; t + \tau)\, dt \simeq 0, \qquad \epsilon > 0, \tag{11.122}$$

then the time ensemble of the speckle irradiance will approach a constant, that is,

$$\int_\epsilon^T |N(\boldsymbol{\sigma}; t, k)|^2\, dt \simeq K_1. \tag{11.123}$$

Thus the recording exposure of eq. (11.121) becomes

$$E(\alpha, \beta) \simeq K_2|O(\boldsymbol{\sigma}; k)|^2, \tag{11.124}$$

which is proportional to the object irradiance. Thus the speckle effect in the hologram image is eliminated.

However, an ideal sampling function may not be obtainable in practice. Since the hologram image resolution is limited by the sampling apertures, a reduction of hologram image resolution is expected with this technique.

For experimental demonstration, a hologram image obtained with the sampling technique is shown in fig. 11.20*a*. In comparison with the hologram image obtained without sampling in fig. 11.20*b*, we see that the one obtained with sampling technique offers less speckle noise.

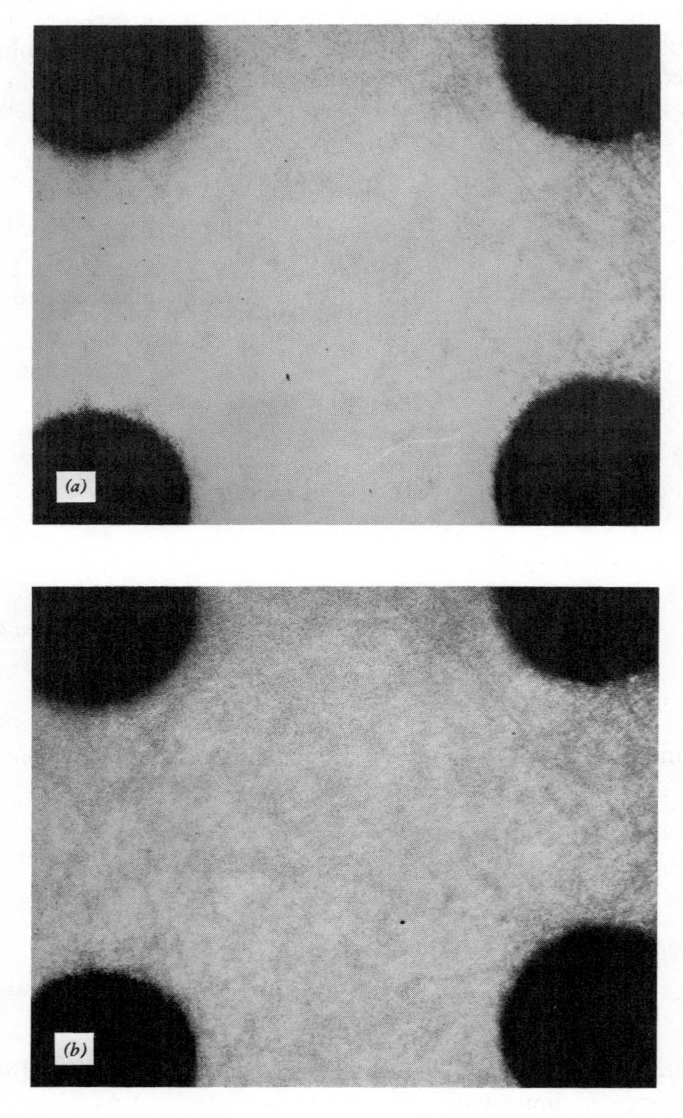

Fig. 11.20 Real hologram images. (*a*) Obtained with spatial sampling. (*b*) Obtained without sampling.

PROBLEMS

11.1 Given the T-E transfer characteristics of fig. 5.35 and a sinusoidal grating input exposure

$$E(x, y) = 1000 + 1000 \cos (px) \text{ erg/cm}^2, \qquad \text{for every } y.$$

Determine, by means of a five-point analysis, the corresponding higher order amplitude transmittances and the degree of nonlinearity.

11.2 Consider a holographic recording process in which the complex light field incident on the recording aperture from a coherently illuminated object is

$$u(x, y) = A(x, y) \exp[i\phi(x, y)],$$

where $A(x, y)$ is uniformly distributed over the hologram aperture, and the complex light field from the reference beam is

$$v(x, y) = B \exp(ikx \sin \theta),$$

where B is an appropriate constant. Using the T-E transfer characteristic of fig. 11.21, with an object-reference beam ratio of 75 percent (i.e., $A/B = 0.75$) and a quiescent bias exposure $E_Q = 37.5\ \mu\text{J/cm}^2$, determine the degree of nonlinearity for this holographic process, and the first-, second-, and third-order diffraction efficiencies.

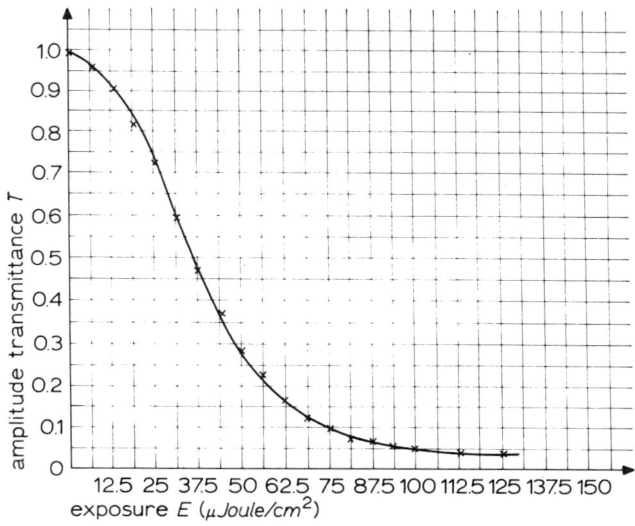

Fig. 11.21

11.3 Repeat prob. 11.2, for unity object-reference beam ratio.

11.4 For the holographic process described in prob. 11.2, by means of a five-point analysis:

(a) Make a rough plot of the degree of nonlinearity as a function of exposure for the following object-reference beam ratios: $A/B = 1$, $A/B = \frac{3}{4}$, $A/B = \frac{1}{2}$, and $A/B = \frac{1}{4}$.

(b) Plot the corresponding first-order diffraction efficiency as a function of exposure.

(c) Discuss the significance of the graphic results obtained in parts (a) and (b).

11.5 In certain holographic processes (e.g., for some Fourier holograms), the light amplitude field $A(x, y)$ may not be uniformly distributed over the recording aperture.

(a) May the finite-point analysis, as described in sec. 11.1, be applied for a qualitative analysis? Show what a qualitative finite-point analysis would or would not be able to uncover.

(b) Illustrate that, for a more accurate analysis, the nonlinear holographic analysis can be performed on a cycle-to-cycle basis.

11.6 Given the following binary object transparency:

$$T(x, y) = \tfrac{1}{2} + \tfrac{1}{2}\,\mathrm{sgn}[\cos p(x^2 + y^2)],$$

where

$$\mathrm{sgn}[\quad] \triangleq \begin{cases} 1, & \cos(\quad) \geq 0 \\ -1, & \cos(\quad) < 0 \end{cases},$$

and p is the spatial frequency.

(a) Show that the transparency consists of an infinite sum of positive and negative lenses.

(b) Determine also the corresponding focal length of these lenses.

11.7 If the object tranparency, as described in prob. 11.6, is illuminated by a monochromatic plane wave of λ, then by the Fresnel–Kirchhoff theory, calculate where the virtual and the real images are to be located.

11.8 Figure 11.22 shows the geometrical parameters in a certain nonlinear

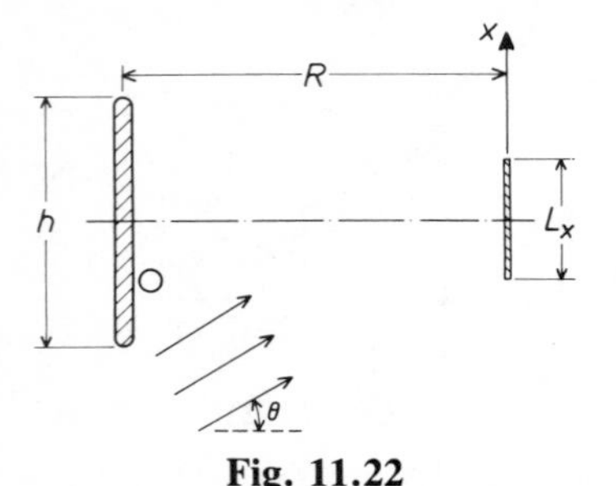

Fig. 11.22

holographic recording. The object O is of height h, the hologram aperture is of length L_x, and the separation between the object and the hologram is R. Determine the minimum angle θ of the plane reference wave for which the first-order real image can be entirely separated from the higher order diffractions as well as the zero-order diffraction.

11.9 Let us assume that the phase delay of a certain photographic emulsion may be approximated by a second-order polynomial,

$$\phi(x, y) \simeq b_0 + b_1 x + b_2 y + b_3 x^2 + b_4 y^2.$$

If the holographic image is constructed and reconstructed by plane wave fronts (i.e., both the reference and illuminating beams are plane), then by means of a technique similar to that described in sec. 10.5, determine the five primary holographic aberrations.

11.10 Using the *T-E* transfer characteristic of fig. 11.21:

(a) By a trial-and-error curve fitting process, approximate a third-order polynomial to the corresponding *T-E* curve.

(b) With the polynomial obtained in part (a), compute the optimum exposures of a holographic process for the object-reference beam ratios $A/B = 1$, $A/B = \frac{1}{2}$, $A/B = \frac{1}{10}$, and $A/B = \frac{1}{30}$.

(c) Determine the corresponding optimum first-order linear transmittances for the reference-object beam ratios specified in part (b).

11.11 Given the same *T-E* transfer characteristic as described in prob. 11.10 for a certain holographic process, consider the following problems:

(a) By means of a five-point analysis (sec. 11.1), make a rough plot of the first-order diffraction efficiency as a function of exposure, for the object-reference beam ratios $A/B = 1$, $A/B = \frac{1}{2}$, $A/B = \frac{1}{10}$, and $A/B = \frac{1}{30}$.

(b) From the plotted curves of part (a), determine the corresponding optimum exposures. Compare these results with the calculated values obtained in prob. 11.10.

11.12 Let us assume that the transmittance of a certain recording medium may be approximated by the nonlinear equation

$$T(E) = 1 + a_1 E + a_3 E^3, \qquad \text{for } E \geq 0,$$

where E is the exposure. If the input exposure is a sinusoidal grating,

$$E(x, y) = E_0 + E_1 \cos px, \qquad \text{for every } y,$$

where E_0 and E_1 are arbitrary constants and p is the spatial frequency. Determine the first-order and the higher order transmittances.

11.13 Repeat prob. 11.12 if the input exposure is a spatial random noise, which follows the one-sided Gaussian probability distribution

$$p(E) = \frac{2}{\sqrt{2\pi}\sigma} \exp\left(-\frac{E^2}{2\sigma^2}\right), \qquad \text{for } E \geq 0 \text{ and every } (x, y),$$

where σ^2 is the variance of the Gaussian statistics. By means of the mean-square error criterion of eq. (11.75), determine the optimum linear transmittance.

11.14 In the synthesis of an optimum complex spatial filter, the Fourier transform of a detecting signal is $S(p, q) \exp[i\phi(p, q)]$. Take $S(p, q)$ as approximately uniform over the spatial bandwidth of interest, and use the *T-E* transfer characteristic of fig. 11.21.

(a) Determine the approximate optimum linear transmittance of the complex filter.

(b) Analyze the complex spatial filtering of fig. 11.16; show that the output signal detection with respect to the filter synthesis in part (a) is indeed optimum.

REFERENCES

11.1 A. Kozma, "Photographic Recording of Spatially Modulated Coherent Light," *J. Opt. Soc. Am.*, **56,** 428 (1966).

11.2 A. A. Friesem and J. S. Zelenka, "Effects of Film Nonlinearities in Holography," *Appl. Opt.*, **6,** 1755 (1967).

11.3 J. W. Goodman and G. R. Knight, "Effects of Film Nonlinearities on Wavefront Reconstruction Images of Diffuse Objects," *J. Opt. Soc. Am.*, **58,** 1276 (1968).

11.4 O. Bryngdahl and A. Lohmann, "Nonlinear Effects in Holography," *J. Opt. Soc. Am.*, **58,** 1325 (1968).

11.5 F. T. S. Yu, "Analysis of Nonlinear Holograms," *J. Opt. Soc. Am.*, **58,** 1550 (1968).

11.6 F. T. S. Yu, "Five-Point Analysis of Nonlinear Holograms," *J. Opt. Soc. Am.*, **59,** 360 (1969).

11.7 F. T. S. Yu, "Off-Axis Nonlinear Hologram and Spurious Distortion," *Opt. Commun.*, **1,** 427 (1970).

11.8 F. E. Terman, *Electronic and Radio Engineering*, 4th ed., McGraw-Hill, New York, 1955, p. 326.

11.9 F. T. S. Yu and A. D. Gara, "Effect of Emulsion Thickness Variations on Wavefront Reconstruction," *J. Opt. Soc. Am.*, **59,** 1530 (1969); and *Appl. Opt.*, **10,** 1324 (1971).

11.10 H. W. Rose, "Effect of Carrier Frequency on Quality of Reconstructed Wavefronts," *J. Opt. Soc. Am.*, **55,** 1565 (1965).

11.11 F. T. S. Yu, "Optimal Linearization in Holography," *J. Opt. Soc. Am.*, **59,** 490 (1969); and *Appl. Opt.*, **8,** 2483 (1969).

11.12 F. T. S. Yu, "Method of Linear Optimization in the Wavefront Recording of Nonlinear Holograms," ICO Symposium on application of Holography, Besancon, France, 1970.

11.13 F. T. S. Yu, "Linear Optimization in the Synthesis of Nonlinear Spatial Filters," *IEEE Trans. Inform. Theory*, **IT-17,** 524 (1971).

11.14 F. T. S. Yu and G. G. Kung, "Sythesis of Optimum Complex Spatial Filters," *J. Opt. Soc. Am.*, **62,** 147 (1972).

11.15 J. D. Rigden and E. I. Gordon, "The Granularity of Scattered Optical Maser Light," *Proc. IRE,* **50,** 2367 (1962).

11.16 B. M. Oliver, "Sparkling Spots and Random Diffraction," *Proc. IEEE,* **51,** 220 (1963).

11.17 L. H. Enloe, "Noise-like Structure in the Image of Diffusely Reflecting Objects in Coherent Illumination," *Bell Syst. Tech. J.,* **46,** 1479 (1967).

11.18 J. C. Dainty, Ed., *Laser Speckle and Related Phenomena,* Vol. 9, in *Topics in Applied Physics,* Springer-Verlag, New York, 1975.

11.19 H. J. Gerritsen, W. J. Hannan, and E. G. Ramberg, "Elimination of Speckle Noise in Holograms with Redundancy," *Appl. Opt.,* **7,** 2301 (1968).

11.20 D. Gabor, "Laser Speckle and Its Elimination," *IBM J. Res. Develop.,* **14,** 509 (1970).

11.21 H. J. Caulfield, "Speckle Averaging by Spatially Multiplexed Holograms," *Opt. Commun.,* **3,** 323 (1971).

11.22 R. F. Van Ligten, "Speckle Reduction by Simulation of Partially Coherent Object Illumination in Holography," *Appl. Opt.,* **12**, 1255 (1973).

11.23 F. T. S. Yu and E. Y. Wang, "Speckle Reduction in Holography by Means of Random Spatial Sampling," *Appl. Opt.,* **12,** 1656 (1973).

12

Rainbow Holography

The technique of wave front reconstruction, or rather the holographic process, is now over three decades old, having been discovered by Dennis Gabor in 1948 as a possible means of improving the resolving power of the electron microscope (ref. 12.1). It was not until the early 1960s, when Leith and Upatnieks produced the first high-quality holographic image using a strong coherent light source (i.e., the laser) that wider attention was given to holography (refs. 12.2, 12.3). Since then, many applications have been developed for holography.

Although holography has provided a great step toward practical three-dimensional imagery, its acceptance for commercial and educational uses has been slow for a number of reasons. The high cost of the holographic process, low hologram image luminance, and the necessity for special illuminators for quality images are among the primary caues for its rejection. In this chapter a technique of producing hologram images with simple inexpensive white-light sources, high-intensity desk lamps, or even sun light is presented. We show that this type of hologram is capable of producing brighter and more colorful holographic images. Since these types of hologram images are observed through the transmitted light field, they are called white-light transmission holograms (refs. 12.4, 12.5), to distinguish them from Denisyuk's white-light reflection holograms (ref. 12.6). However, this type of hologram is best known as a rainbow hologram because it produces rainbow color hologram images (ref. 12.7).

This chapter discusses techniques for holograms that can be constructed using white-light. We show that color hologram images can be easily generated with the rainbow holographic process. The problems of color blur and resolution limit of the rainbow holograms are calculated. The effect of deep, and high-resolution rainbow holographic images are also described.

12.1 RAINBOW HOLOGRAPHIC PROCESS

From fig. 10.8, it can be seen that real hologram images can be reconstructed with conjugate coherent illumination. Hologram images may also be reconstructed with very small hologram apertures, suffering only a minor degree of resolution loss. In other words, it is possible to reconstruct the entire hologram image when the aperture is reduced to a narrow slit, as shown in fig. 12.1. For convenience of discussion we would call hologram H_1 the master, or primary, hologram.

In rainbow holographic recording, we insert a fresh holographic plate H_2 behind the primary hologram H_1 at a distance near the real hologram images. A convergent reference beam R, derived from the same laser source, is used to illuminate the holographic plate H_2, as shown in fig. 12.1 To obtain a better rainbow hologram image, it is recommended that the holographic plate be located near the hologram image plane. If the holographic plate is properly recorded in the linear region of the T-E curve, the resultant hologram H_2 will be a rainbow hologram.

In order to see the rainbow effect, we first reconstruct the hologram image from the rainbow hologram H_2 by a conjugate divergent coherent source, as shown in fig. 12.2. Because the narrow slit used for the rainbow holographic construction acts as a slit object, a real slit image will be reconstructed downstream from the rainbow hologram H_2, as shown in the figure. If we look through the real slit image, we would expect to see a virtual hologram image behind the holographic plate H_2. If the holographic plate H_2 is inserted behind the hologram image I during the rainbow holographic construction, a real holographic image can be viewed through the slit image by using the reconstruction process shown in fig. 12.2. Since the real slit image is convergently reconstructed, we see a brighter hologram image.

In view of the holographic magnification of eqs. (10.42) and (10.43), we see that the location of the real slit image varies as a function of the reconstruction wavelength of the light source. In other words, for a longer wavelength reconstruction, the slit image appears to be located higher and closer to the hologram aperture H_2, and the slit width is also wider. The same effect applies to the hologram image seen through the slit image, that is, for a longer wavelength

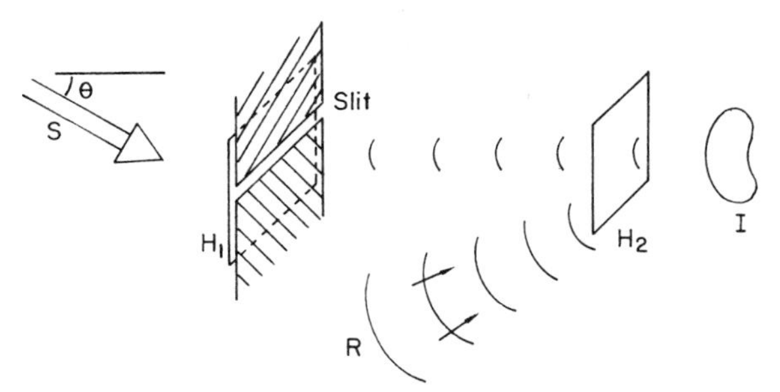

Fig. 12.1 Rainbow holographic recording. S, conjugate coherent illumination; H_1, primary hologram; R, convergent reference beam; H_2, holographic plate; I, real hologram image.

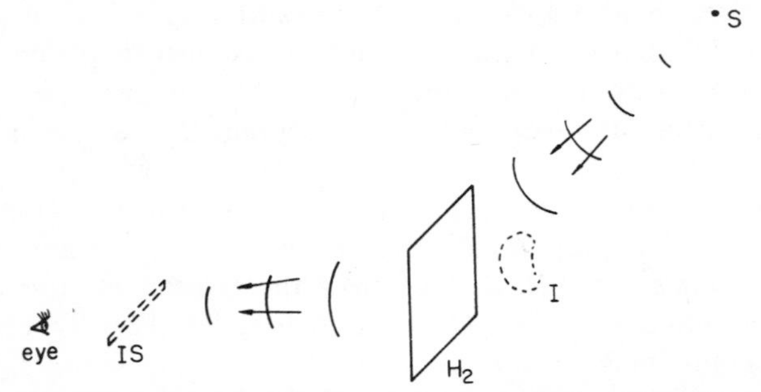

Fig. 12.2 Hologram image reconstruction by a coherent source. *IS*, slit image; H_2, rainbow hologram; *I*, hologram image; *S*, monochromatic point source.

illumination, the holographic image appears to be larger and closer to the holographic plate H_2.

Now, if the hologram H_2 is illuminated by a conjugate divergent white-light source, as shown in fig. 12.3, the hologram slit images, resulting from the different wavelengths of the white-light source, will disperse into rainbow colors in the real slit image space. The holographic image of the object will take on the same rainbow effect behind the hologram. If we transversely view this image through the smeared slit image, the hologram image is observed in a series of rainbow colors one at a time. In other words, if the image is viewed through the red colored smeared slit image, we will see a red colored holographic image. If we peer through the green colored smeared slit image, we will observe a smaller green holographic image.

In concluding this section, we note that, if an observer moves his or her head transversely up and down against the smeared slit image, the "over" or the "under" of the hologram object image will not be observed, in contrast with conventional holography. Vertical parallax is lost with this type of rainbow holography. However, the full horizontal parallax is retained, providing a right-to-left perspective view for binocular stereopsis and motion parallax. Therefore, the sensation of a three-dimensional scene is preserved.

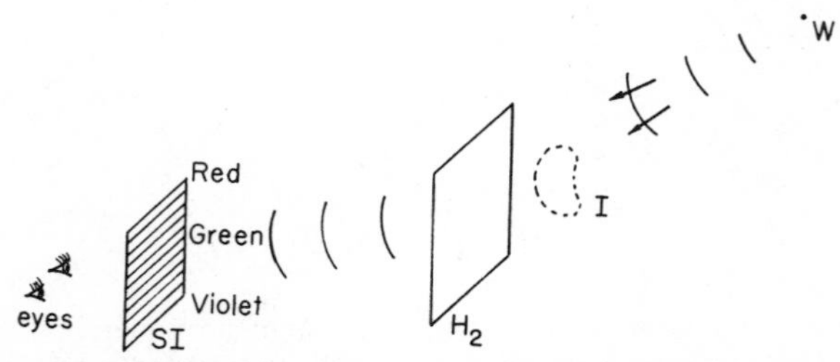

Fig. 12.3 Rainbow hologram reconstruction with white-light source. *SI*, smeared slit image; H_2, rainbow hologram; *I*, hologram image; *W*, divergent white-light source.

12.2 ONE-STEP RAINBOW HOLOGRAMS

In the previous section we have discussed a general concept of the rainbow holographic process developed by Benton (ref. 12.4). This process involves two recording steps. First, a primary hologram is made from a real object with the conventional off-axis holographic technique, and second, the rainbow hologram is recorded from the real hologram image from the master hologram. The relaxation of the coherence requirement for the reconstruction light source arises from the placement of a narrow slit aperture behind the master hologram in the second step of the holographic process. However, a two-step holographic recording process is cumbersome and requires a separate optical setup for each step, a major undertaking for laboratories with limited resources of optical components.

In this section, we illustrate an alternative method that can produce rainbow holograms using a one-step process (ref. 12.8). We note that the one-step process utilizes Rosen's technique of imaging by a lens (ref. 12.9). This new technique offers certain flexibilities in the construction of rainbow holograms, and the optical arrangement is simpler than that for the conventional two-step process.

As noted, the making of a rainbow hologram requires recording a real holographic image of the object through a narrow slit. If the rainbow hologram is illuminated with a monochromatic light source, a real holographic image of the object, limited by the open slit, is produced. However, if the rainbow hologram is illuminated by a white-light source, the holographic image of the slit will disperse the light into a full-view holographic image in rainbow colors. Therefore, the essential goal of rainbow holography is to form an image of the slit aperture between the hologram image of the object and the observer. Figures 12.4 and 12.5 show how this same goal can be achieved in a one-step holographic process, when both the object and the slit are imaged simultaneously by a lens (or a lens system). In this manner, a rainbow hologram can be made without a primary hologram.

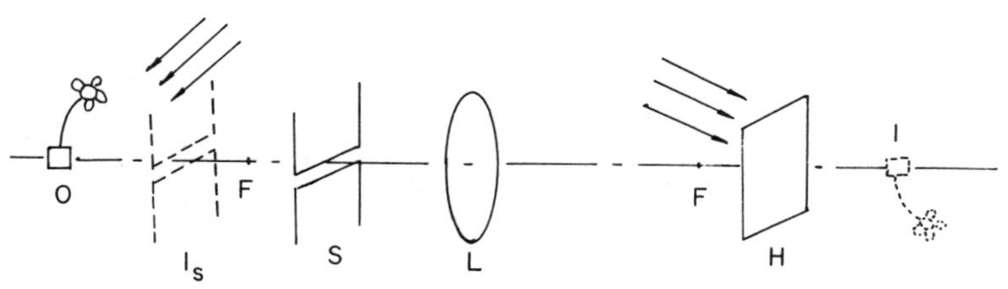

Fig. 12.4 One-step rainbow holographic process for pseudoscopic imaging. O, object; I, image of the object; S, narrow slit; I_s, image of the slit; L, imaging lens; H, recording plate; F, focal point of the lens.

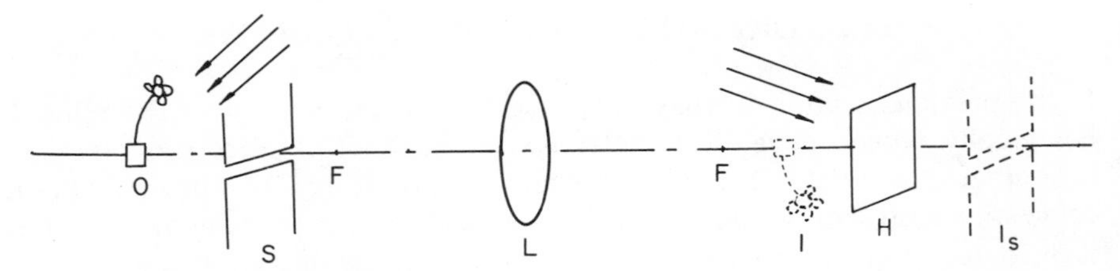

Fig. 12.5 One-step rainbow holographic process for orthoscopic imaging. O, object; I, image of the object; S, narrow slit; I_s, image of the slit; L, imaging lens; H, recording plate; and F, focal point of the lens.

Pseudoscopic Imaging

The optical system of a one-step holographic process is similar to that of a conventional off-axis holographic process, except that an imaging lens and a narrow slit are inserted between the recording plate and the object, as shown in fig. 12.4. A virtual image of the slit is formed between the object and the slit, and a real image of the object is formed behind the recording plate. We see that the depth of the real image in its relation to the slit is inverted by the lens. In readout, we require that the hologram form the slit image downstream from the object image; therefore, the conjugate image of the hologram must be used, and the image is, therefore, pseudoscopic. We note that the three-dimensional effects perceived in viewing a pseudoscopic image can be enhanced if the viewing is done using a pseudoscope (ref. 12.10), a mirror device that causes light that would ordinarily enter the left eye to enter the right eye instead, and vice versa. Thus we see that the one-step arrangement does offer some versatility in the image magnification and good holographic image quality. We note that, as long as the position of the slit allows the passage of light from every point of the object to the entrance of the imaging lens, we could magnify the hologram image to a certain extent and still obtain a full field of view.

Figure 12.6 shows a pseudoscopic holographic image obtained using the rainbow hologram one-step process shown in fig. 12.4. The magnification in this case was near unity. Figure 12.7 shows the same holographic image, but with a magnification of approximately two. Comparing figs. 12.6 and 12.7, we see no evidence of image degradation. However, a relatively high level of speckle noise can be seen in fig. 12.6, which is due primarily to the smaller slit size (about 0.25 mm) used in the construction of the hologram.

Orthoscopic Imaging

A slight modification of the one-step rainbow holographic process results in an orthoscopic hologram image (fig. 12.5). In this case, a real image of the object is formed at the front of the recording plate, and a real image of the slit is formed behind the plate. Thus it is the primary rather than the conjugate image that

Fig. 12.6 A pseudoscopic rainbow holographic image. Magnification $m = 1$.

produces a real image of the slit. However, the field of view is restricted by the size of the lens, which in turn limits the magnification of the hologram image.

Since the image blur is linearly related to the hologram-to-image distance (ref. 12.11), moving the recording plate farther away from the real image creates more severe color blurring in the holographic image. An orthoscopic holographic image, obtained from the one-step process of fig. 12.5, is shown in fig. 12.8. The recording plate is placed about 2.5 cm from the real image.

Fig. 12.7 A pseudoscopic rainbow holographic image. Magnification $m = 2$.

Fig. 12.8 An orthoscopic rainbow holographic image. Magnification $m = \frac{1}{2}$.

One-step aerial imaging by lens rainbow holography is a simple and promising technique. The pseudoscopic imaging process is versatile and capable of providing a relatively sharp magnified hologram image. The orthoscopic imaging process is, however, somewhat more restricted. Nevertheless, the restriction may be alleviated by using a larger imaging lens.

12.3 GENERATION OF COLOR HOLOGRAPHIC IMAGES

In this section we demonstrate a method of constructing true color holographic images using the one-step rainbow holographic technique. This new method is probably the simplest of all the holographic techniques now employed in the generation of true color images.

There are many ways true color holograms may be generated, and each has its own advantages and drawbacks. With the transmission-type holograms, in order to reconstruct a color image, three separate lasers emitting different wavelengths are required. These three lasers must be critically aligned to ensure the exact superposition of the three color images. There is also the problem of spurious images. Techniques to solve these problems generally involve some additional penalties (ref. 12.12). Another method is multiplexing a reflection hologram with three color images (sec. 10.7 and ref. 12.13). The reflection hologram itself acts as a color filter, and its images can be reconstructed with white light. However, the emulsion shrinkage in the fixing of the photographic plate requires a precise swelling process in order to produce faithful color reproductions. Moreover, the diffraction efficiencies of reflection holograms are generally lower than those of the transmission type. We may also generate a color hologram using the Benton type rainbow hologram (ref. 12.14). With this type of color hologram, white light can be used in the reconstruction, and the image is extraordinarily bright. However, the process of constructing a true color rainbow hologram is very cumbersome and time consuming. First, three master holograms have to be constructed with laser lights of different colors. Then the projected real images of these three master holograms are multiplexed onto a fourth hologram sequentially, again with three differently colored laser lights. These three master holograms must be aligned very carefully to assure that their reconstructions are exactly superimposed. Thus we see that these techniques of generating color holographic images are all quite complicated and difficult to implement. The new technique that we describe can produce true color holographic images without requiring any critical alignment of the optical components or swelling of the emulsion. In addition, it allows for white-light reconstruction, and the images of the three primary colors are automatically aligned.

The optical arrangement is illustrated in fig. 12.9. A He-Ne laser is used to provide the red light (6328 Å), and an argon laser is used to provide the green and blue lights (5145 and 4765 Å). The illuminated object is imaged through an achromatic lens to a plane just in front of the hologram. A narrow slit (1.5 mm) is placed between the object and the focal plane of the imaging lens. A collimated reference beam is used in this experiment so that the carrier frequency is the same across the hologram. However, a diverging beam may also be used. The intensities of the three lights are measured independently, and the exposure time for each is calculated. The hologram is first exposed to the red light of the

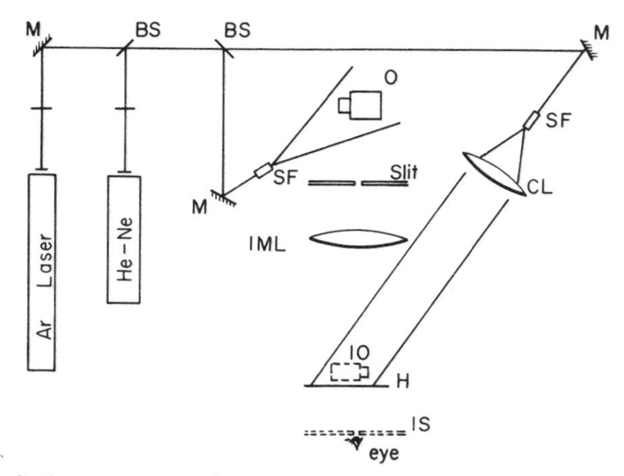

Fig. 12.9 Optical arrangement for the construction of color one-step rainbow hologram: *BS*'s, beam splitters; *M*'s, mirrors; *SF*'s, spatial filters; *CL*, collimating lens; *IML*, achromatic small *f*-number imaging lens; *H*, hologram; *O*, object; *IO*, image of object; *IS*, image of slit and observation plane.

He-Ne laser, then the green light (5145 Å) of the argon laser. The argon laser is then tuned to the blue line (4765 Å), and a third exposure is made.

A Kodak 649F plate is used because of its relatively flat spectral response. After the plate is developed, a rainbow hologram is formed. When the hologram is viewed with a polychromatic point source, a very bright color image is reconstructed. It should be noted that, with this arrangement, the same reference angle is used in the reconstruction, instead of the conjugate angle as with the Benton-type rainbow hologram. Similar to the color rainbow holograms generated with the two-step method using three master holograms, a holographic image with exact color reproduction is observed when the hologram is viewed in the correct plane. If the viewer moves off this plane, different shades of color can still be seen, but the color is different from that of the original object. This is not a serious drawback, since three-dimensionality is preserved along only one axis with rainbow holograms. The greatest disadvantage of this technique is that the field of view is restricted by the lens aperture. In order to achieve a large field of view, an imaging lens with large aperture and small focal length is required, and such a lens tends to be expensive. However, this technique is so much simpler than other techniques for producing color holographic images that it should be a useful addition despite its drawback. In fig. 12.10 we have a black-and-white photograph of a color image generated by the method described. The color image is very bright and sharp, and the color reproduction very distinct and quite faithful.

Fig. 12.10 Black-and-white photograph of a color holographic image reconstructed with a high-intensity desk lamp using the one-step recording technique.

12.4 HIGH-RESOLUTION RAINBOW HOLOGRAPHY

In previous sections a simple one-step rainbow holographic process has been discussed. The one-step technique offers certain flexibilities in the construction of rainbow holograms.

The basic principle of the rainbow holographic process is to form a slit image between the observer and the hologram image. However, the presence of a slit causes a marginal loss of the hologram image resolution in the direction perpendicular to the slit. This marginal resolution loss may limit some applications.

In this section, we illustrate a technique by which this disadvantage may be alleviated (ref. 12.15). Let us illustrate a two-dimensional one-step rainbow holographic recording, as shown in fig. 12.11. Note that a diffuser is required for the rainbow holographic recording. However, the presence of a slit in the holographic construction limits the hologram image resolution. This drawback may be alleviated with the placement of a cylindrical lens behind the object transparency, as shown in fig. 12.12. In addition, the diffuser and the slit aperture are eliminated from this cylindrical lens technique. Thus it is possible to obtain a high-resolution holographic image. In principle, the image resolution is limited only by the imaging lens. Furthermore, this technique also has the advantage of obtaining a higher object beam irradiance, by which it relieves the power requirement of the coherent source in holographic construction. However, the presence of a cylindrical lens introduces hologram image aberration. In principle, this aberration can be removed if the hologram image is reconstructed with the same optical system, as shown in fig. 12.13. A narrow slit is required for the reconstruction process. The presence of a narrow slit also causes a marginal loss of image resolution. However, this marginal resolution loss, as shown in the following experimental demonstration, is less severe than that of the conventional one-step rainbow holographic process. Furthermore, the use of a cylindrical lens in rainbow holography has been introduced by Leith and Chen (ref. 12.16) in deep image rainbow holograms. Later they extended the tech-

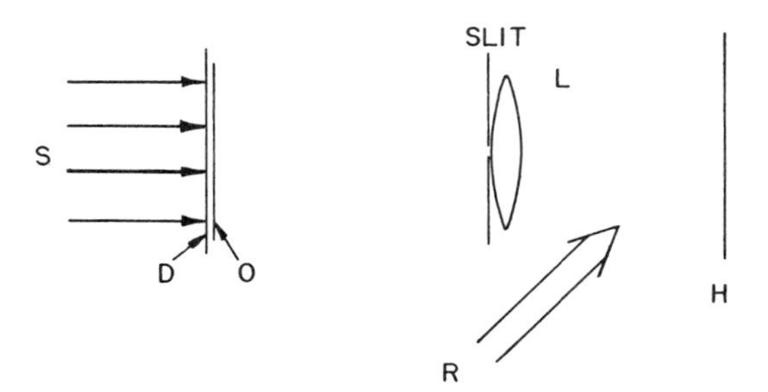

Fig. 12.11 A two-dimensional one-step rainbow holographic construction. *S*, monochromatic plane wave; *D*, diffuser; *O*, object transparency; *L*, imaging lens; *H*, holographic plate; *R*, reference beam.

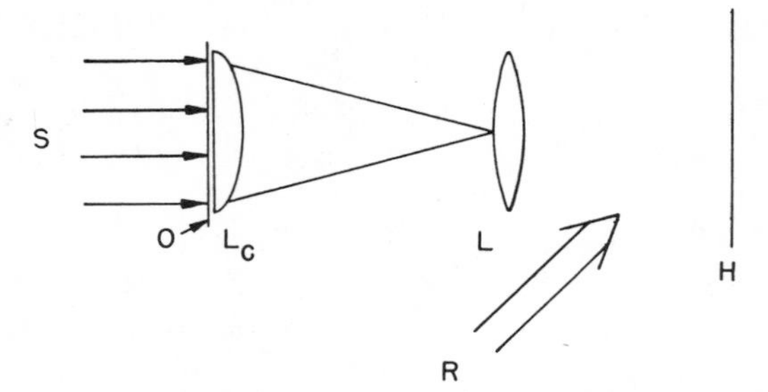

Fig. 12.12 A two-dimensional rainbow holographic construction utilizing a cylindrical lens. S, monochromatic plane wave, O, object transparency; L_c, cylindrical lens; L, imaging lens; H, holographic plate; R, reference beam.

nique to produce a white hologram image (ref. 12.17). In a more recent paper, Chen has also used a cylindrical lens to produce an astigmatic rainbow hologram image (ref. 12.18).

As a means of obtaining a resolution measurement, a USAF resolution target is used as a test object. In our holographic process, Kodak 131 high-speed holographic plates are used throughout the experiments. A series of holographic recordings, with and without the cylindrical lens, under various recording conditions, have been made using the setups diagrammed in figs. 12.11 and 12.12. Images reconstructed with different slit widths are obtained from a white-light source, with the same optical setup as in the construction processes. The hologram images are recorded on black-and-white transparencies for further measurements and evaluations. For convenience of discussion, we have tabulated the holographic construction processes in Table 12.1.

With references to the recorded holographic image transparencies of different construction-reconstruction conditions, the resolution limits are measured as a

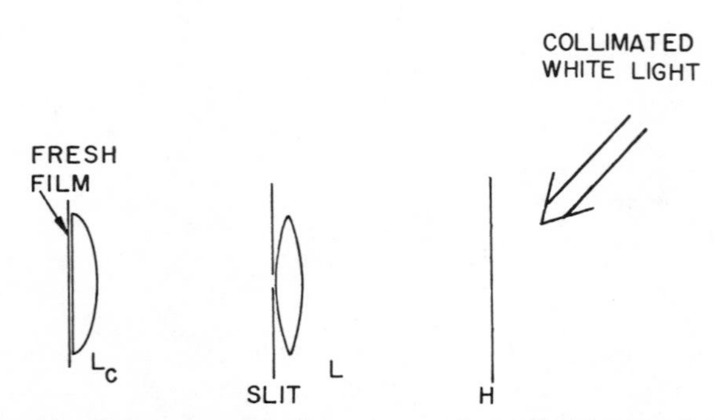

Fig. 12.13 A white-light reconstruction process. L_c, cylindrical lens; L, imaging lens; H, rainbow hologram.

Table 12.1

Test Number	Holographic Construction Conditions
1	With cylindrical lens of fig. 12.12; the image is focused onto the plate
2	With cylindrical lens of fig. 12.12; the plate is 2 cm beyond the image plane
3.1	With cylindrical lens, diffuser, and 1-mm slit; the image is focused onto the plate
3.3	With cylindrical lens, diffuser, and 3-mm slit; the image is focused onto the plate
4.1	With cylindrical lens, diffuser, and 1-mm slit; the plate is 2 cm beyond the image plane
4.3	With cylindrical lens, diffuser, and 3-mm slit; the plate is 2 cm beyond the image plane
5	Without cylindrical lens, as in fig. 12.11; 3-mm slit and the image is focused onto the plate
6	Without cylindrical lens, as in fig. 12.11; 3-mm slit and the plate is 2 cm beyond the image plane

function of reconstruction slit width and are plotted in fig. 12.14. In other words, the resolution measure is obtained from the highest resolution limit of the reconstructed resolution targets. From the results of fig. 12.14, we see that the on-focus rainbow hologram (i.e., test no. 1), obtained with the cylindrical lens and a slit width from 3 to about 6 mm, provides the highest hologram image resolution. The optimum resolution at a reconstruction slit width of 3 mm is about 29 lines/mm which is about a 60 percent improvement on the conventional one-step technique of test no. 5. However, for the off-focus case of test no. 2, the resolution is somewhat lower than the conventional one-step on-focus case of test no. 5, and it is somewhat higher than the one-step off-focus technique of test no. 6. In other words, the resolution limit of the cylindrical lens technique depends upon the location of the object image. Thus the optimum location of the object image is observed to be on the holographic plate.

With reference to fig. 12.14, we see that the variation of the hologram image resolution of the cylindrical lens technique (i.e., test no. 1) is not a monotonic function. The image resolution decreases as the reconstruction slit width decreases. This resolution decrease is primarily due to the diffraction limit of the slit and the speckle noise in the hologram image. Beyond the optimum image resolution limit (i.e., from 3- to 6-mm slit) the resolution decreases monotonically as the reconstruction slit width increases. This is mainly due to the color blurring (refs. 12.19–12.21) of the rainbow hologram image. As for the cases of a diffuse object transparency, in tests 3.3 and 5, the hologram image resolution for the cylindrical lens technique offers a slightly better resolution than that with the conventional one-step technique. We see that the image

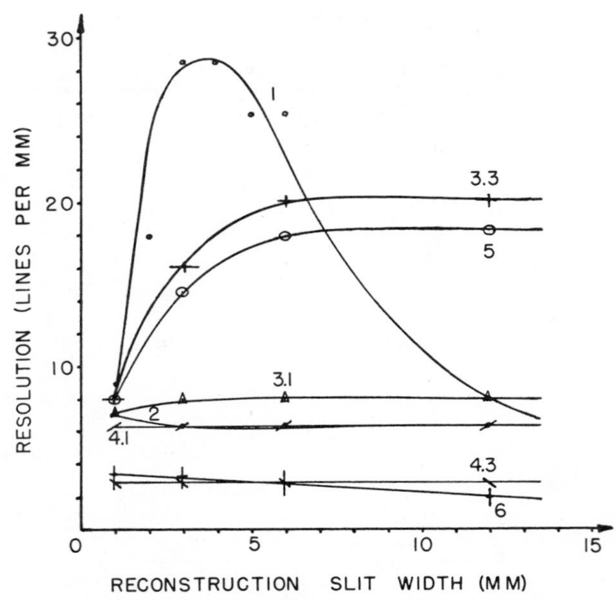

Fig. 12.14 Plots of various hologram image resolutions of Table 12.1 as a function of reconstruction slit width.

resolution for both tests 3.3 and 5 increases monotonically with the reconstruction slit width, and becomes more independent when the reconstruction slit width is larger than the construction slit width.

For color blur and speckle effect, we provide, in fig. 12.15, a series of four photographic results obtained from various holographic conditions. Figure 12.15*a* is a magnified hologram resolution target obtained by the on-focus cylindrical lens technique (i.e., test no. 1) of fig. 12.12, and the hologram image is reconstructed with the same optical setup using a slit size of about 3 mm, as shown in fig. 12.13. The holographic image of fig. 12.15*a* is obtained from an optimum resolution condition of fig. 12.14. Figures 12.15*b* and 12.15*c* show the results obtained with the same cylindrical lens technique (i.e., tests 3.3 and 3.1, respectively), but with the placement of a diffuser in front of the object transparency. Figure 12.15*d* shows the hologram image obtained with the conventional technique (i.e., test no. 5), as described in fig. 12.11, and the image is reconstructed with the same optical system using a 3-mm slit. Finally, from these four experimental photographic results, it is evident that the one using the cylindrical lens and on-focus holographic construction provides a higher hologram image resolution, less speckle noise, and minimum color blur. The cylindrical lens technique also offers a higher object beam irradiance, which may relax the holographic recording conditions.

In conclusion, we would like to point out that a high-resolution rainbow holographic process can be achieved by the simple utilization of a cylindrical

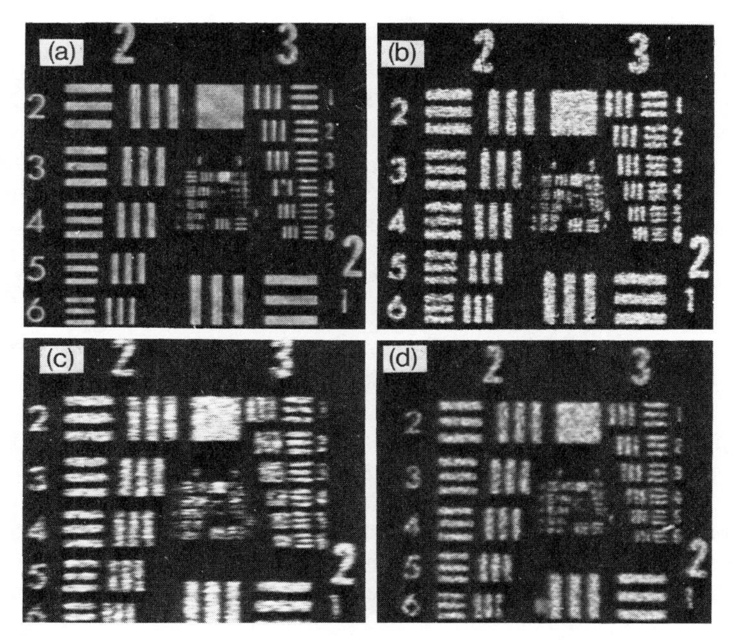

Fig. 12.15 Photographic enlargement of the reconstructed hologram images (magnification 25×): (*a*) Constructed with the cylindrical lens technique of fig. 12.12, and reconstructed with the optical setup of fig. 12.13 using a 3-mm slit. (*b*) Constructed with the cylindrical lens technique of fig. 12.12, but with a diffuser before the object transparency and a 3-mm slit at the front of imaging lens, and reconstructed with the optical setup of fig. 12.13 using a 3-mm slit. (*c*) Constructed with the cylindrical lens technique of fig. 12.12 but with a diffuser before the object transparency and a 1-mm slit at the front of imaging lens, and reconstructed with the optical setup of fig. 12.13 using a 3-mm slit. (*d*) Constructed with the conventional technique of fig. 12.11 and reconstructed with the same optical setup using a 3-mm slit.

lens. We have shown that the cylindrical lens technique offers higher image resolution, less speckle noise, and a lower color blur holographic image. This cylindrical lens technique is particularly suitable for the application of a two-dimensional object transparency. There is another merit of this technique, namely, a higher object beam irradiance, which may relax the power requirement of the coherent light source during the construction process.

12.5 RESOLUTION AND COLOR BLUR OF RAINBOW HOLOGRAMS

We have shown that the making of a rainbow hologram requires recording a real holographic image of the object through a narrow slit. However, the presence of a slit causes a marginal resolution loss and color blur of the holographic image. This marginal resolution loss and the color blur limit some of the rainbow holographic applications. In the previous section we have illustrated a cylindrical lens technique that may alleviate these disadvantages.

The color blur for rainbow holograms has been analyzed by Wyant (ref. 12.11), Tamura (ref. 12.19), and Chen (ref. 12.20). Here we evaluate the image resolution and color blur for a one-step rainbow hologram (ref. 12.21). The results are extended to a cylindrical lens one-step rainbow holographic process. Under certain holographic conditions a higher resolution and smaller degree of color blur can be achieved.

Image Resolution

The image resolution of a one-step rainbow hologram is evaluated in this passage. We approach the analysis with the elementary point spread function and wavelength spread for the rainbow holographic process. For simplicity, we ignore the aberration and the resolution limit of the imaging lens used. Since the resolution limit of the rainbow hologram image is primarily due to the narrow slit aperture, a one-dimensional representation is adequate for the analysis.

With reference to the one-step rainbow holographic process of fig. 12.16, the object beam distribution at the recording plate can be shown (sec. 3.7) as

$$u_1(x; k_1) = C_1 \int_{-W/2}^{W/2} u_0(\xi; k_1) \exp\left\{ck_1\left[s + \frac{(x-\xi)^2}{2s}\right]\right\} d\xi, \qquad (12.1)$$

where C_1 is an appropriate complex constant, $k_1 = 2\pi/\lambda_1$, λ_1 is the wavelength of the coherent light source, W is the slit width,

$$u_0(\xi; k_1) = C_0 \exp\left\{-ik_1\left[(d+s) + \frac{(\xi+h)^2}{2(d+s)}\right]\right\} \qquad (12.2)$$

is the complex light distribution at the slit plane, C_0 is an appropriate complex constant, s is the distance between the slit and the holographic plate, d is the distance between the holographic plate and the point image, and ξ and x are the slit plane and the hologram plane coordinates. For simplicity we express eq. (12.1) in the following form:

$$u_1(x; k_1) = C_1 A(x) B(x), \qquad (12.3)$$

where

$$A(x) = \exp\left(i\frac{k_1 x^2}{2s}\right), \qquad (12.4)$$

$$B(x) = \int_{-W/2}^{W/2} \exp\left\{-ik_1\left[\left(\frac{1}{2g} - \frac{1}{2s}\right)\xi^2 + \left(\frac{h}{g} + \frac{x}{s}\right)\xi\right]\right\} d\xi, \qquad (12.5)$$

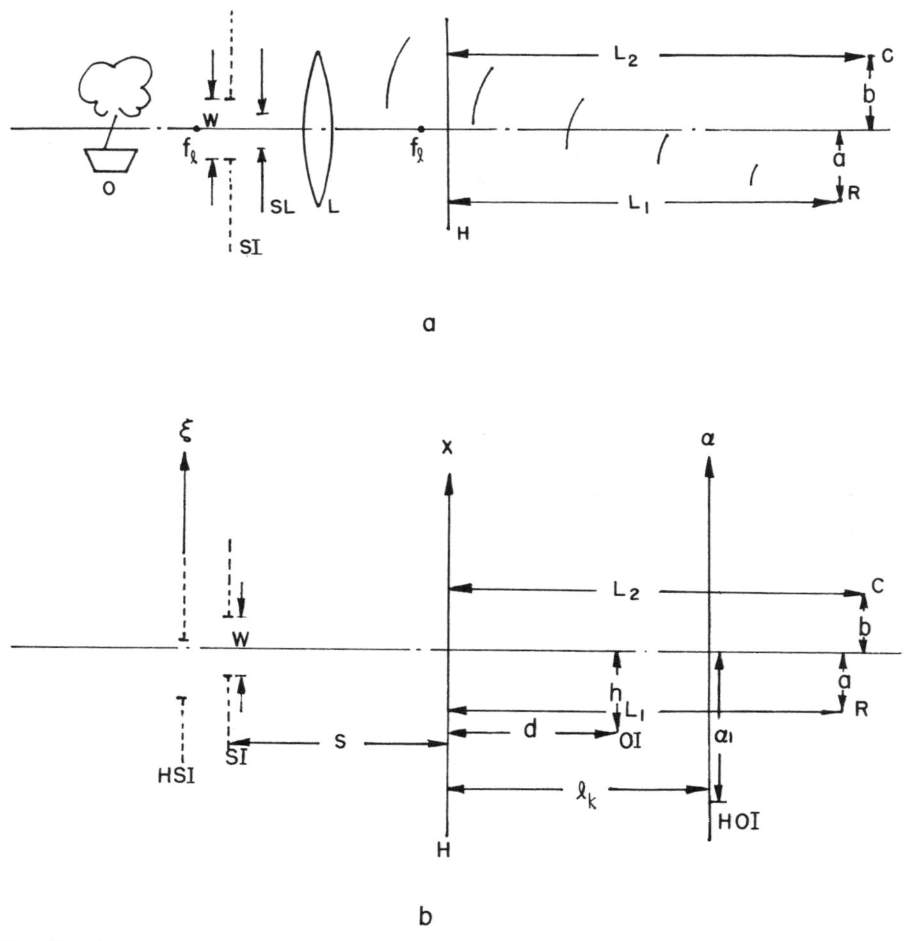

Fig. 12.16 One-step rainbow holographic process. (*a*) Schematic diagram of a one-step rainbow holographic process; (*b*) Geometrical model for calculation. *O*, object; *W*, image slit width; *SI*, slit image; *SL*, slit; f_e, focal length; *L*, imaging lens; *H*, holographic plate; *R*, convergent reference point source; *C*, divergent illuminating point source; *HSI*, hologram image slit; *OI*, object image point; *HOI*, hologram object image.

and $g = d + s$. With reference to fig. 12.16*b*, we also see that the reference beam that arrives at the recording plate is

$$u_2(x; k_1) = C_2 \exp\left\{-ik_1\left[\frac{(x + a)^2}{2L_1}\right]\right\}, \tag{12.6}$$

where C_2 is an appropriate constant. If we assume that the holographic recording is linear, the amplitude transmittance of the recorded hologram is

$$T(x; k_1) = K_0 + K_1 \exp\left\{ik_1\left[\left(\frac{1}{2L_1} + \frac{1}{2s}\right)x^2 + \frac{a}{L_1}x\right]\right\}B(x)$$
$$+ K_2 \exp\left\{-ik_1\left[\left(\frac{1}{2L_1} + \frac{1}{2s}\right)x^2 + \frac{a}{L_1}x\right]\right\}B^*(x), \qquad (12.7)$$

where K_0, K_1, and K_2 are the appropriate complex constants. If this hologram is illuminated by a divergent white-light source, as illustrated in fig. 12.16, then, for every wavelength λ, the complex light arriving at the hologram aperture is

$$u_3(x; k) = C_3 \exp\left\{ik\left[L_2 + \frac{(x - b)^2}{2L_2}\right]\right\}, \qquad (12.8)$$

where $k = 2\pi/\lambda$, λ is the wavelength of the light source, and C_3 is an appropriate complex constant. At a distance l behind the hologram where the image is reconstructed, the complex light field is

$$E(\alpha, k) = \int T(x; k_1)u_3(x; k)h_l(\alpha - x; k)\, dx, \qquad (12.9)$$

where the integral is over the spatial domain of the hologram,

$$h_l(x; k) = \frac{-i}{\lambda l} \exp\left[ik\left(l + \frac{x^2}{2l}\right)\right] \qquad (12.10)$$

is the spatial impulse response, and [eq. (10.42)]

$$l = \frac{\lambda_1 L_1 L_2 d}{\lambda L_2 d - \lambda L_1 L_2 - \lambda_1 L_1 d}. \qquad (12.11)$$

If the reconstructed image term (i.e., the last term) of eq. (12.7) is evaluated, the result is

$$E_r(\alpha; k) = \int_{-W/2}^{W/2} \exp\left\{ik_1\left[\left(\frac{1}{2g} - \frac{1}{2s}\right)\xi^2 + \frac{h}{g}\xi\right]\right\} d\xi$$
$$+ \int_{-\infty}^{\infty} \exp\left\{-ik\left[\frac{\lambda}{\lambda_1}\left(\frac{1}{2d} + \frac{1}{2s}\right)x^2 + \left(\frac{b}{L_2} + \frac{\alpha}{l} + \frac{\lambda a}{\lambda_1 L_1} - \frac{\lambda\xi}{\lambda_1 s}\right)x\right]\right\} dx, \qquad (12.12)$$

where the proportionality constants are ignored. Since the second integral of eq. (12.12) is a Fresnel integral [eq. (3.29)], the following result occurs:

$$E_r(\alpha, k) = \int_{-W/2}^{W/2} \exp\left\{-ik\frac{d}{gl}\left[\alpha + l\left(-\frac{\lambda h}{\lambda_1 d} + \frac{b}{L_2} + \frac{\lambda}{\lambda_1}\frac{a}{L_1}\right)\right]\xi\right\} d\xi$$

$$= C \operatorname{sinc}\left[\frac{\pi}{\lambda}\frac{dW}{gl}(\alpha + \alpha_1)\right], \tag{12.13}$$

where C is a complex constant and

$$\alpha_1 = l\left(-\frac{\lambda h}{\lambda_1 d} + \frac{b}{L_2} + \frac{\lambda a}{\lambda_1 L_1}\right). \tag{12.14}$$

With the use of Rayleigh's criterion (sec. 3.6), the least resolution distance can be shown to be

$$\Delta h_r = \frac{\lambda_1 \lambda L_1 L_2 (d + s)}{(\lambda L_2 d - \lambda L_1 L_2 - \lambda_1 L_1 d)W}. \tag{12.15}$$

If it is an in-line (on-focus) hologram, the least resolution distance becomes

$$\Delta h_r = \frac{\lambda_1 s}{W}, \qquad \text{for } d = 0, \tag{12.16}$$

which is independent of the wavelength of the illuminating light source. On the other hand, if the reference and reconstruction beams are plane waves (i.e., $L_1 = L_2 = a = b = \infty$), then the least resolution distance is

$$\Delta h_r = \frac{\lambda_1 (d + s)}{W}, \tag{12.17}$$

which is also independent of the reconstruction wavelength of the light source, and is proportional to the location of the object image and the distance between the slit image and the holographic plate.

Color Blur

Now let us turn our attention to the color blur of a one-step rainbow holographic process. Note that when a rainbow hologram is illuminated by a conjugate white-light source, there is a finite wavelength spread $\Delta\lambda$ over the reconstructed slit, which causes the color blurring in the hologram image. In other words, the narrower the slit width used, the smaller the degree of color blur. However, the narrower slit width also reduces the hologram image resolution. We seek an optimum rainbow holographic process so that a higher resolution and a smaller degree of color blur can be obtained.

The color blur is defined as the smeared distance of a hologram point image, that is,

$$\Delta H_c \triangleq \frac{\partial \alpha_1}{\partial \lambda} \Delta\lambda, \tag{12.18}$$

where α_1 is the location of a point hologram image on the α axis of the image coordinate as expressed in eq. (12.14), and $\Delta\lambda$ is the wavelength spread over the slit image. We evaluate the complex light distribution at the hologram slit image plane, that is,

$$E_x(\xi; k) = A \int_{-L_{x/2}}^{L_{x/2}} \exp\left\{-ik_1 \frac{x^2}{2s}\right\} \exp\left\{-ik_1\left[\frac{x^2}{2L} + \frac{ax}{L_1}\right]\right\} \cdot \exp\left\{ik\left[\frac{x^2}{2L_2} - \frac{bx}{L_2}\right]\right\} \exp\left\{ik\left[\frac{x^2}{2l_s} - \frac{\xi x}{l_s}\right]\right\} dx, \tag{12.19}$$

where L_x is the width of the hologram aperture, and l_s is the distance between the hologram slit image and the holographic plate. Note that l_s can be written as (chapter 10)

$$l_s = \frac{\lambda_1 s L_1 L_2}{\lambda L_1 L_2 + \lambda s L_2 - \lambda_1 s L_1}. \tag{12.20}$$

By substitution of eq. (12.20) in eq. (12.19),. we have:

$$E_s(\xi; k) = \operatorname{sinc}\left[\frac{\pi}{\lambda_s l_s}(\xi + \xi_1)L_x\right], \tag{12.21}$$

where

$$\xi_1 = l_s\left(\frac{\lambda a}{\lambda_1 L_1} + \frac{b}{L_2}\right) = \frac{s(\lambda L_2 a + \lambda_1 L_1 b)}{\lambda L_1 L_2 + \lambda L_2 - \lambda_1 S L_1} \tag{12.22}$$

is the lateral location of the hologram slit image. Since the width of the hologram slit image is

$$W' = M_s W = \frac{\lambda L_1 L_2 W}{\lambda L_1 L_2 + \lambda s L_2 - \lambda_1 s L_1}, \tag{12.23}$$

where $M_s = (\lambda l_s)/(\lambda_1 s)$ is the lateral magnification, the wavelength spread over W' is

$$\Delta\lambda = \frac{W'}{\partial \xi_1/\partial\lambda} \frac{\lambda(\lambda_1 s L_1 - \lambda L_1 L_2 - \lambda s L_2)}{\lambda_1 s(sa + L_1 b + sb)} W. \tag{12.24}$$

By substitution of eqs. (12.24) and (12.14) into eq. (12.18), we obtain the following color blur expression:

$$\Delta H_c = \frac{\lambda L_1 L_2 d(hL_1 - ad + bL_1 - bd)(\lambda_1 sL_1 - \lambda L_1 L_2 - \lambda sL_2)W}{s(sa + L_1 b + sb)(\lambda L_2 d - \lambda L_1 L_2 - \lambda_1 L_1 d)^2}. \quad (12.25)$$

If both reference and reconstruction beams are plane waves, then the color blur becomes

$$\Delta H_c = \frac{d}{s} W, \quad (12.26)$$

which is the result obtained by Wyant (ref. 12.11), Tamura (ref. 12.19), and Chen (ref. 12.20). From eqs. (12.25) and (12.26), we see that the color blur vanishes for an in-line holographic process (i.e., $d = 0$). Thus in principle, the color blur of the hologram image can be eliminated for an in-line hologram. However, the speckle noise in the hologram image also increases. We now apply our result to the high-resolution rainbow holographic process as described in previous sections. We show that, with the cylindrical lens technique, a higher resolution and a smaller degree of color blur in the hologram image can be obtained.

Let us discuss the cause of a *diffuse* object transparency. First consider the transparency without the cylindical lens. With reference to the parameters defined in fig. 12.17 and with the Gaussian lens equation (ref. 12.22), we have

$$s_0' = \frac{s_0 f_l}{f_l - s_0}, \quad \text{and} \quad d_0' = \frac{d_0 f_l}{d_0 - f_l}, \quad (12.27)$$

where f_l is the focal length of the imaging lens. From the parameters defined in

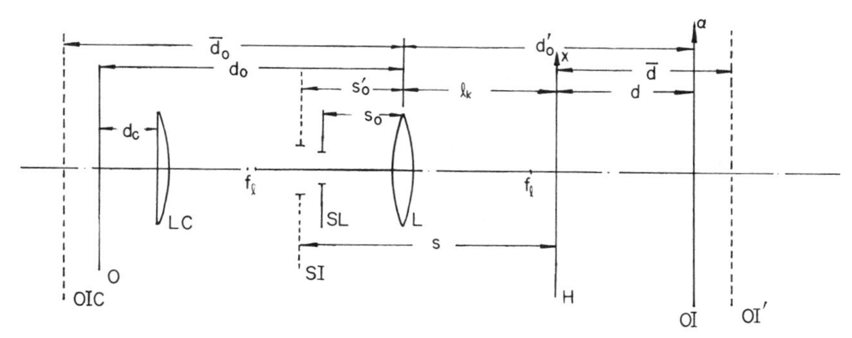

Fig. 12.17 Cylindrical lens technique one-step rainbow holographic process. *OIC*, object image due to cylindrical lens; *O*, object transparency; *LC*, cylindrical lens; f_l, focal point of the imaging lens; *SI*, slit image; *SL*, slit; *L*, imaging lens; *H*, holographic plate; *OI*, object image without cylindrical lens; *OI'*, object image with cylindrical lens.

fig. 12.17, we also see that

$$s = s_0' + l_k = \frac{s_0 f_l}{f_l - s_0} + l_k \tag{12.28}$$

and

$$d = d_0' - l_k = \frac{d_0 f_l}{d_0 - f_l} - l_k. \tag{12.29}$$

For convenient notation we adopt the "bar" sign for the case with the cylindrical lens insertion, as shown in fig. 12.17. Again with the lens equation, we have

$$\bar{d}_0 = \frac{d_c f_c}{f_c - d_c} + f_c = \frac{f_c^2}{f_c - d_c} \tag{12.30}$$

and

$$\bar{d} = \frac{\bar{d}_0 f_l}{\bar{d}_0 - f_l} - l_k, \tag{12.31}$$

where f_c is the focal length of the cylindrical lens. If the reference and reconstruction beams for the cylindrical lens technique of fig. 12.17 are plane waves, then according to the result obtained in eq. (12.17), the least resolution distance is

$$\Delta \bar{h}_r = \frac{\lambda_1(\bar{d} + s)}{W}, \tag{12.32}$$

where the "bar" sign represents the case with the cylindrical lens. In comparison with the one-step technique of eq. (12.17) and with eqs. (12.28)–(12.31), we have the following resolution ratio:

$$\frac{\Delta \bar{h}_r}{\Delta h_r} = \frac{\bar{d} + s}{d + s} = \frac{[\bar{d}_0 f_l(f_l - s_0) + s_0 f_l(\bar{d}_0 - f_l)](d_0 - f_l)}{[d_0 f_l(f_l - s_0) + s_0 f_l(d_0 - f_l)](\bar{d}_0 - f_l)}, \tag{12.33}$$

where $\Delta \bar{h}_r$ is the least resolution distance with the cylindrical lens and Δh_r is the least resolution distance without the cylindrical lens.

From eq. (12.33) we see that, with the appropriate selection of f_c, f_l, and d_0, it is possible to make the least resolution distance of the cylindrical lens technique smaller than the conventional one-step rainbow holographic technique. For example, if we let $s_0 = s_0' = 0$, and $\bar{d}_0 > d_0$, then $\Delta \bar{h}_r < \Delta h_r$. However, if the cylindrical lens is in contact with the object transparency, i.e., $d_0 = \bar{d}_0$, we have $\Delta \bar{h}_r = \Delta h_r$.

Similar to the result of eq. (12.16), the color blur for the cylindrical lens technique is

$$\Delta \bar{H}_c = \frac{\bar{d}}{s} W. \tag{12.34}$$

Therefore, with eqs. (12.29) and (12.31), we have

$$\frac{\Delta \bar{H}_c}{\Delta H_c} = \frac{\bar{d}}{d_c} =$$

$$\frac{\{f_l f_c^2 + s_0 f_l (f_c - d_c) - l_k [f_c^2 + (f_c - d_c)(s_0 - f_l)]\}(f_c + d_c + s_0 - f_l)}{[f_c^2 + (f_c - d_c)(s_0 - f_l)][(f_c + d_c + s_0) f_l - l_k (f_c + d_c + s_0 - f_l)]}. \tag{12.35}$$

If the focal lengths of the imaging lens and cylindrical lens are equal (i.e., $f_l = f_c$), and we let $s_0 = 0$, then eq. (12.35) becomes

$$\frac{\Delta \bar{H}_c}{\Delta H_c} = \frac{f_l^2 - l_k d_c}{f_l^2 + f_l d_c - l_k d_c}. \tag{12.36}$$

Since $f_l^2 < f_l^2 + f_l d_c$, the color blur can be made smaller with the cylindrical lens technique.

We now discuss the case of *gray level* object transparency. From the point spread function of eq. (12.13), the corresponding transfer function is

$$\begin{aligned} T(p) &= \int E_r(\alpha; k) \exp(-ip\alpha)\, d\alpha \\ &= C \operatorname{rect}\left[\frac{\lambda(d + s)lP}{2\pi W d}\right], \end{aligned} \tag{12.37}$$

where p is the spatial frequency in radians per distance, and

$$\operatorname{rect}[\chi] \triangleq \begin{cases} 1, & |\chi| \leqq \frac{1}{2} \\ 0, & \text{otherwise} \end{cases} \tag{12.38}$$

If p_m denotes the spatial frequency limit of the object transparency, then from eq. (12.11) we have

$$p_m = \frac{2\pi d W}{2\lambda(d + s)l} = \frac{2\pi(\lambda L_2 d - \lambda\lambda_1 L_2 - \lambda_1 L_1 d)W}{2\lambda_1 \lambda L_1 L_2 (d + s)}, \tag{12.39}$$

which is proportional to image slit width W of fig. 12.16*b*. We now consider the cylindrical lens technique, as shown in fig. 12.17. If the spatial frequency limit of the object transparency is $\bar{p}_m$, then the corresponding spatial limit at the back focal length of the cylindrical lens becomes

$$\Delta x = \frac{\lambda_1 f_c}{2\pi} \bar{p}_m . \tag{12.40}$$

If we let Δx of eq. (12.40) be equal to W and then substitute into eq. (12.39), we obtain the spatial frequency ratio

$$\frac{\bar{p}_m}{p_m} = \frac{2\lambda L_1 L_2 (d + s)}{f_c(\lambda L_2 d - \lambda L_1 L_2 - \lambda_1 L_1 d)}, \tag{12.41}$$

when $\bar{p}_m$ and p_m are the spatial frequency limits of the object transparency for the cases *with* and *without* the cylindrical lens, respectively. If both reference and reconstruction beams are plane waves, then eq. (12.41) becomes

$$\frac{\bar{p}_m}{p_m} = \frac{2(d + s)}{f_c} = \frac{2(d_0' + s_0')}{f_c}, \tag{12.42}$$

where $d + s = d_0' + s_0'$ is the distance between the slit image and the hologram image, as shown in figs. 12.16 and 12.17. Thus for $d + s > f_c/2$, a higher hologram image resolution can be obtained by the cylindrical lens technique. Furthermore, if the object transparency and the narrow slit are placed in contact with the cylindrical and imaging lenses (i.e., $d_c = 0, f_c = d_0$, and $s_0 = s_0' = 0$), then eq. (12.42) becomes

$$\frac{\bar{p}_m}{p_m} = \frac{2d_0'}{d_0}, \tag{12.43}$$

which is proportional to the hologram image magnification. On the other hand, by direct substitution of eq. (12.40) into eq. (12.39), we have the following relation:

$$\frac{\Delta x}{W} = \frac{f_c(\lambda L_2 d - \lambda L_1 L_2 - \lambda_1 L_1 d)}{2\lambda L_1 L_2 (d + s)}. \tag{12.44}$$

From the color blur equation of eq. (12.25) and the conditions for $d_c = 0$, $s_0 = s_0' = 0$, and $d_0 = f_c$, we obtain the following color blur ratio:

$$\frac{\Delta \bar{H}_c}{\Delta H_c} = \frac{\Delta x}{W} = \frac{d_0}{2d_0'}. \tag{12.45}$$

The analytical results we have obtained are quite compatible with the experimental results obtained in the previous section. For example, in tests no. 3.3 and

5 of table 12.1, the resolutions are about the same for these cases. If we improve the experimental conditions $s_0' = s_0 = 0$ and $f_c = d_0$ for these two cases in eq. (12.30), we have the result $\Delta\bar{h}_r = \Delta h_r$, which is consistent with the experimental result. Moreover, if we consider the conditions of test no. 1, i.e., $s'_0 = 0$, and if we use a unity image magnification, i.e., $d_0' = f_c$, then from eq.(12.38) we would have $\bar{p}_m = 2p_m$; the image resolution with the cylindrical lens of test no. 1 is twice as high as the case without the cylindrical lens of test no. 5. From the experimental results of fig. 12.15, we see that with a 3-mm slit width, the image resolution of test no. 1 is indeed twice as high as that of test no. 5.

As for the color blur, we compare our results with the experimental results from test no. 1 of fig. 12.15*a* and from test no. 5 of fig. 12.15*d*. The color blur is proportional to the size increase of the hologram image in the direction perpendicular to the slit; therefore, from fig. 12.15*a* we calculate the increase in the hologram image to be about 6.5 percent, and, from fig. 12.15*d* the calculated increase is approximately 11 percent. Since our experiments are obtained with unity holographic magnification, it is evident from the color blur relation of eq. (12.41) that $\Delta\bar{H}_c = \Delta H_c/2$. Thus the experimental result is quite consistent with the theoretical result.

12.6 GENERATION OF DEEP IMAGE RAINBOW HOLOGRAMS

The rainbow hologram can be made using a real image (i.e., the object) either from a primary hologram (i.e., two-step technique) or by an imaging lens (i.e., one-step technique). For example, a slit is placed at the first primary hologram, and as a consequence, the resulting rainbow hologram can produce a sharp hologram image under small white-light source illumination. As we have seen in sec. 12.4, the width of the slit affects the quality of the rainbow hologram images. If the slit is too wide, the rainbow hologram image is blurred; if slit is too narrow, the hologram image becomes very speckly. We have shown in the previous section that the optimum condition for hologram image quality is a minimum object-hologram distance. It is evident that, for objects about a half meter from the rainbow hologram plate, the image will either be blurred or speckly or both.

In this section we describe a technique by Leith and Chen (ref. 12.16) that enables rainbow holograms to form images having sharpness and low speckle, even for object-hologram plate distances of the order of a meter.

Let us now consider some pertinent imaging characteristics of the rainbow hologram. As we have seen in previous sections, the object point produces a complex light field limited in the vertical dimension to a width W, which is the consequence of the narrow slit used in making the rainbow hologram, as shown in fig. 12.18. We note that the fringes forming the rainbow hologram are horizontal, since the oblique reference beam is from above. Thus when the rainbow hologram is illuminated by a white-light point source, each spectral wavelength forms the image point at a different location, so that each image

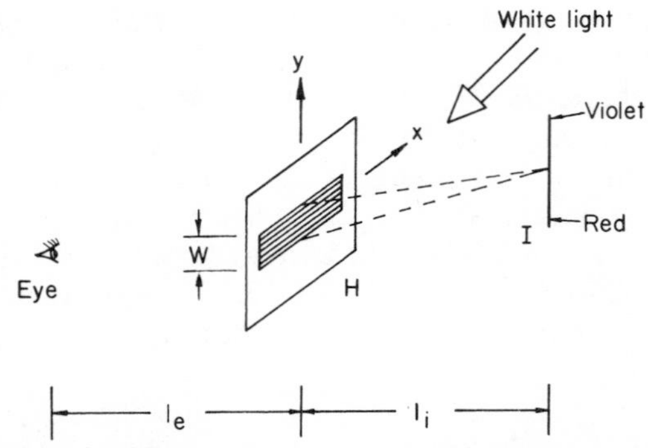

Fig. 12.18 Reconstruction of a point-object rainbow hologram. H, point-object rainbow hologram; I, smeared virtual image.

point will smear in a line of rainbow color at distance l_i behind the hologram, as shown in fig. 12.18. For simplicity of analysis, we assume that the smeared image line is vertical.

Let us now observe the hologram image at a distance l_e from the hologram plate, as depicted in fig. 12.18. Figure 12.19 shows the range of the color smeared image points Δy contributing the diffracted light that enters the observer's eye. With reference to this diagram, the color smeared line segment Δy can be shown to be

$$\Delta y = \frac{W(l_i + l_e)}{l_e} + \frac{\lambda l_i}{W}. \tag{12.46}$$

where λ is the mean wavelength of the diffracted light due to the segment Δy that enters the eye. We note that the first term of this equation represents the smearing factor that was formulated on the basis of geometrical optics, and the second term is an additional smearing factor due to the diffraction phenomenon of the hologram aperture. Since λ is a very small quantity, the above equation can

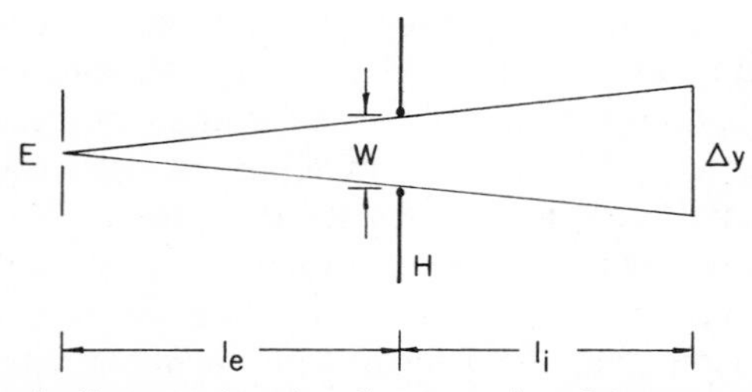

Fig. 12.19 Schematic diagram showing the range Δy of the smeared virtual image from which diffracted light can enter the eye pupil E.

readily be approximated by

$$\Delta y \simeq W\left(1 + \frac{l_i}{l_e}\right). \tag{12.47}$$

From this equation we see that the sharpest hologram image can be produced at the rainbow holographic plate. The smearing that we have described occurs only in the vertical direction, which is due to the high spatial frequency of the rainbow holographic recording. However, in the horizontal direction the image is only slightly smeared, because of the low spatial frequency in that direction. The rainbow hologram image thus formed can be regarded as having astigmatism.

The technique we describe consists of placing a cylindrical lens in front of the slit that covers the primary hologram H_1, as shown in fig. 12.20. This setup causes the y direction of the point image to be focused onto the hologram recording plate H_2, whereas in the x direction the point image is unchanged. Thus the rainbow hologram produces considerable astigmatic effect even under coherent illumination. However, we have achieved our basic objective. In other words, regardless of the width of the slit and the distance that this image (in the x direction) forms from the hologram recording plate H_2, the rainbow hologram image will always be sharp in the vertical, y, direction. Thus in rainbow hologram construction, a width of slit several orders larger than the normal width can be used. As a result of the wider slit, a reduction of speckle noise in the hologram can be obtained and at the same time the image sharpness can be retained.

Since the hologram image in the y direction is focused onto the holographic plate, more blurring is expected on the plane of the x direction, as shown in fig. 12.20. Thus the astigmatism presented to the observer is more severe. In other words, this modified rainbow holographic process gives rise to sharper resolution only in one dimension, when viewing the hologram image in the focal plane of the respective dimensions.

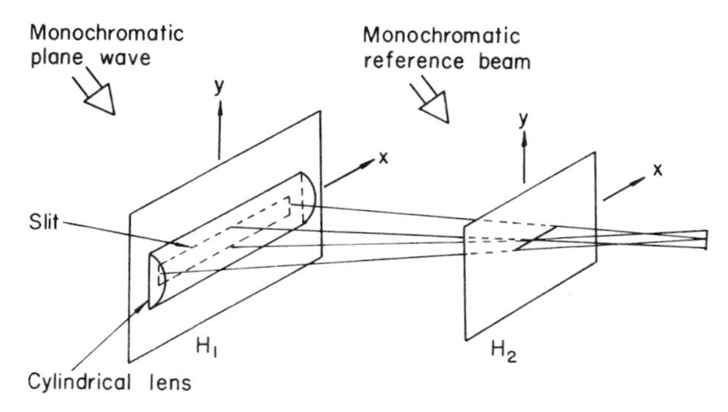

Fig. 12.20 A cylindrical lens is placed over the slit aperture of the primary hologram H_1, so that in the vertical direction y, the image is focused onto the rainbow hologram recording plate H_2.

For experimental demonstration, we first provide the rainbow holographic image (reconstructed by a white-light source) obtained with a normal slit width of about 3 mm, as shown in fig. 12.21*a*. The object (precisely the first letter "0") was located at a distance of about 38 cm from the recording plate. From this figure we see that the hologram image is reasonably sharp but rather speckly. Figure 12.21*b* shows the rainbow hologram image made with a 29-mm slit width but without a cylindrical lens. Although speckle noise has been largely alleviated, the hologram image suffers severe color blurring in the vertical direction. Finally, fig. 12.21*c* shows the rainbow hologram image made with a 29-mm slit width and with a cylindrical lens technique. The cylindrical lens focuses the first letter "0" in the veritcal dimension onto the hologram recording plate. From this figure, we see that the rainbow hologram image is rather sharp and less speckly as compared with fig. 12.21*a* and *b*. Although the experimental results show that using the cylindrical lens technique offers the observer an enormously less speckly rainbow hologram image, we should note that this technique does not increase the depth over which a sharp image can be obtained; rather, it allows distant objects to be sharply imaged at the expense of degraded resolution for near objects, which would otherwise be sharp.

Finally, we note that the cylindrical lens used by Leith and Chen (ref. 12.16) was made by cementing together two pieces of plexiglass, one being bowed so as to touch the other at both ends. A base was attached at the bottom so that a space between the plexiglasses can be filled with water. A suitable focal length may be obtained by changing the refractive index with dissolved sugar.

12.7 MULTIPLEX RAINBOW HOLOGRAM

We now discuss an interesting rainbow holographic stereogram called a multiplex rainbow hologram or cross hologram (refs. 12.24, 12.25), which was invented by Lloyd Cross in 1973. The multiplex rainbow hologram is one of the most significant developments of the last decade in commercial holography. This type of hologram usually appears on a cylindrical piece of film illuminated with a conventional incandescent lamp. Stereoscopic image motion can be observed with this type of hologram, which consists of a series of narrow strips of rainbow holograms, each corresponding to a motion picture frame. This eye-catching stereographic hologram image display has attracted wide interest and attention from a variety of individuals, holographers and laymen alike, In fact, multiplex rainbow holograms have already been used commercially as advertising displays for businesses, exhibit displays for businesses, exhibit displays for trade shows, and novelty displays for variety applications. Cross holography has been featured in television shows, motion pictures, and it has aroused considerable interest in the artistic community. In short, the multiplex rainbow hologram is the result of a hybrid form of holography, which combines cinematography and rainbow holography to create a three-dimensional stereographic hologram image in motion.

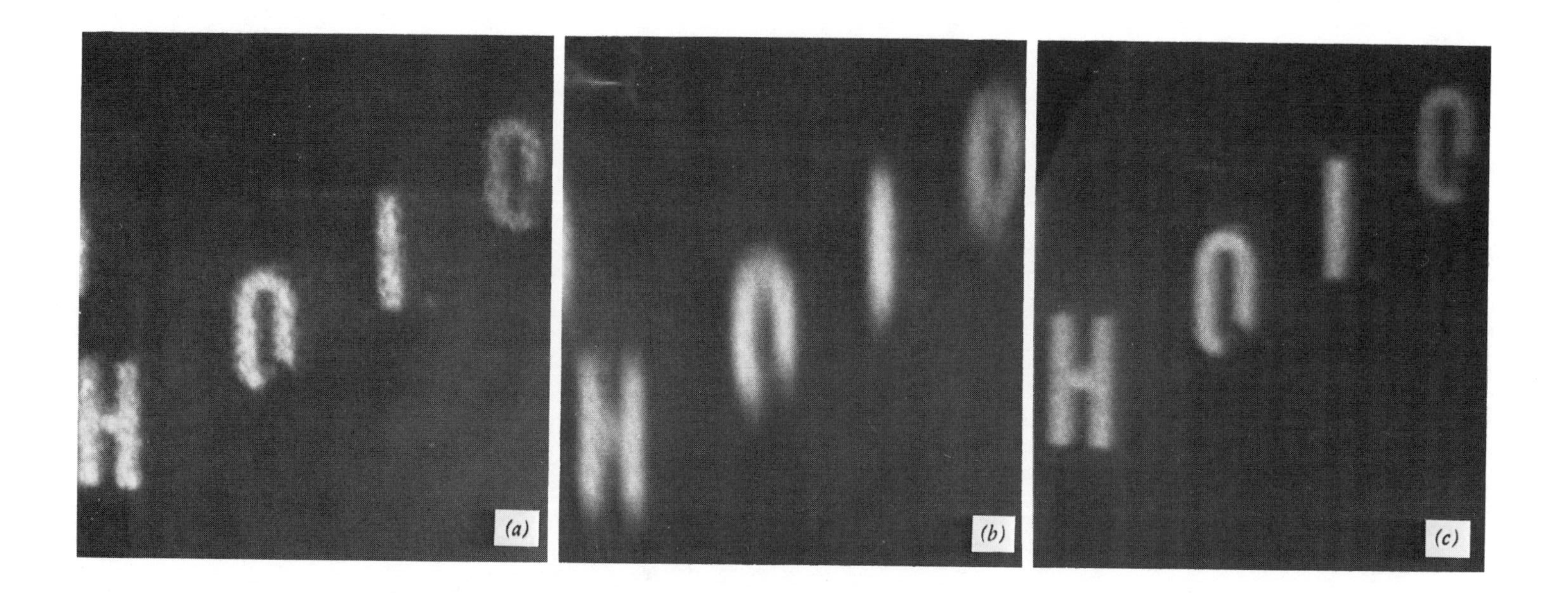

Fig. 12.21 Deep image rainbow hologram reconstruction. (By permission of E. N. Leith.) (*a*) Hologram image obtained with 3-mm slit. (*b*) Hologram image obtained with wide slit (29 mm) but without the cylindrical lens. (*c*) Hologram image obtained with wide slit (29 mm) and with a cylindrical lens.

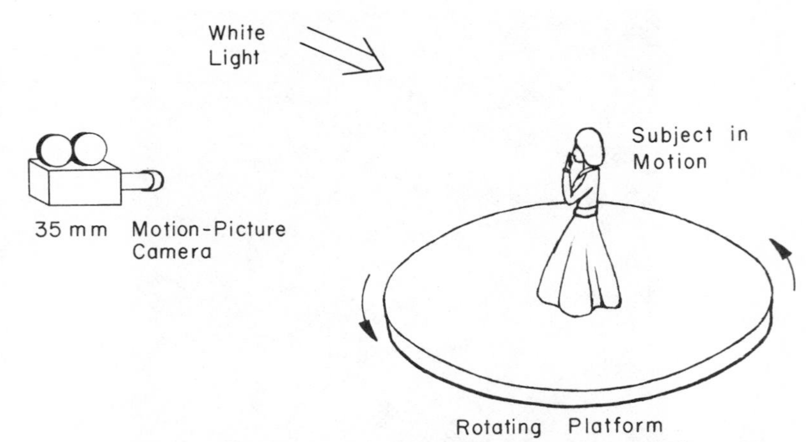

Fig. 12.22 First step of making a multiplex rainbow hologram.

The making of a multiplex rainbow hologram is a two-step process. In general, the first step consists of making 35-mm conventional motion picture film of a subject in motion on a rotating platform, say through 360° rotation. During the filming, this process usually takes about three frames per degree of subject rotation. The resulting movie film thus records the history of the subject motion from a three-dimensional horizontal viewpoint. In other words, a roll of movie film from the front to the back of the subject in motion, as shown in fig. 12.22, is recorded frame-by-frame.

The second step of the process consists of making a spatially multiplexed rainbow hologram, as shown in fig. 12.23. In making the hologram, a movie

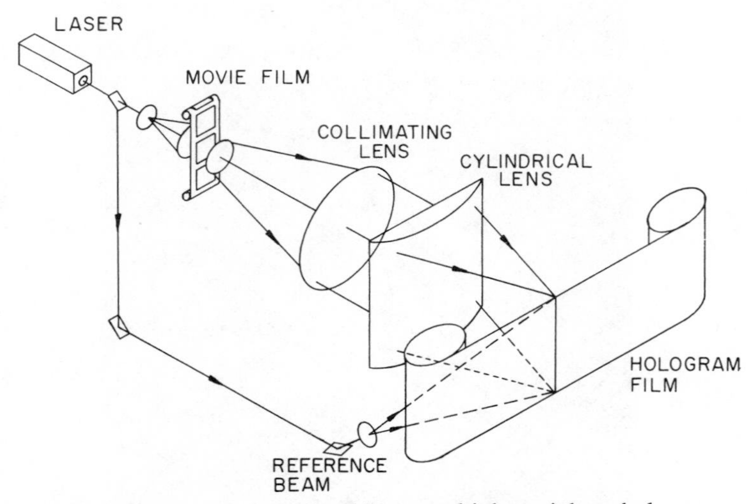

Fig. 12.23 Technique of making a multiplex rainbow hologram.

film transparency is imaged onto the holographic film in the vertical dimension; however, the projection rays are focused by a cylindrical lens to a line on or near the film in the horizontal dimension. A line form of a reference beam is superimposed with the line of the object beam to form a narrow strip of rainbow hologram of each motion picture film frame. The holographic film is advanced one frame width between each exposure and in this manner a spatially multiplexed narrow strip of hologram is recorded on the holographic film. Thus we see that the horizontal exit pupil of the multiplex hologram, which subtends the full horizontal width of the image, is defined by the ray bundle from the cylindrical lens. However, in the vertical dimension, the exit pupil is quite small and, with coherent illumination, only a narrow slice of image ray can be seen. Thus if the hologram is illuminated by a white-light source, a chromatic dispersion of the spectral line due to the high spatial frequency of the holographic grating in the vertical direction can be seen, which increases the vertical exit pupil so that the full vertical of the image in rainbow color can be viewed. The vertical viewing angle is generally small, the color of the hologram image varies from bluish at the upper viewing angle to reddish at lower angle, as shown in fig. 12.24. Figure 12.25 shows three horizontal viewing angles of a multiplex rainbow hologram. From this figure we see that the hologram image in motion can readily be perceived.

In concluding this section, we remark that, despite the promises of the multiplex rainbow hologram, several technical limitations have prevented widespread commercial applications: (1) the vertical parallax of the hologram image is lost; (2) color blurring of the rainbow hologram image occurs; (3) time smear distortion limits the amount of motion; (4) instead of true color rainbow colored

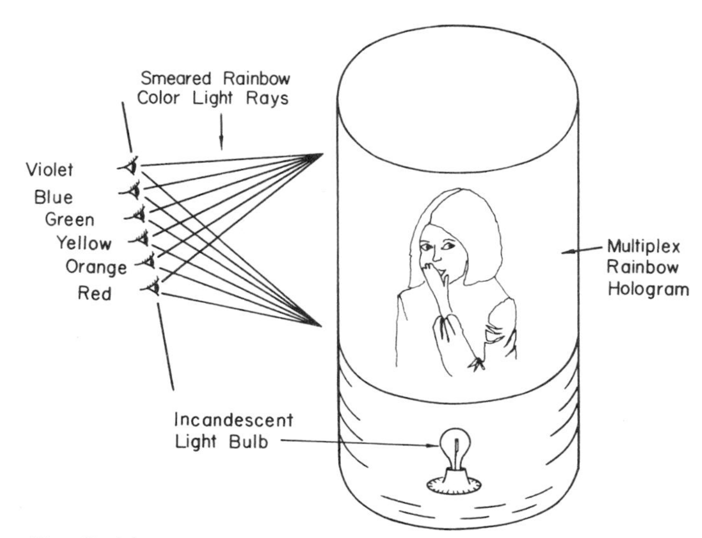

Fig. 12.24 Multiplex rainbow hologram image reconstruction.

Fig. 12.25 Hologram image in motion viewing from three different directions.

images are produced; (5) the depth and fidelity of the hologram image are limited; and so on. Nevertheless, with the further development of this type of multiplex holographic process, some of these limitations may be alleviated.

PROBLEMS

12.1 Let us consider the amplitude transmittance of an off-axis point object hologram H_1 as given in the following:

$$T(x, y; k_1) = K_1 + K_2 \cos\left[k_1\left(\frac{\rho^2}{2R} - x \sin \theta\right)\right],$$

where $k_1 = 2\pi/\lambda_1$. In rainbow holographic recording we superimpose the hologram H_1 with a narrow slit and illuminate the hologram by a conjugate plane wave of the same wavelength, as shown in fig. 12.26. If a recording plate is inserted at a distance l_1 behind the primary hologram H_1 and a convergent reference wave is used for the holographic recording, then:

(a) Draw a system diagram to express the point-object rainbow holographic recording.

(b) If we assume that the hologram image is an ideal point image, then calculate the rainbow holographic recording with respect to point image and the narrow slit.

12.2 If the rainbow hologram of H_2 of prob. 12.1 is illuminated by a divergent

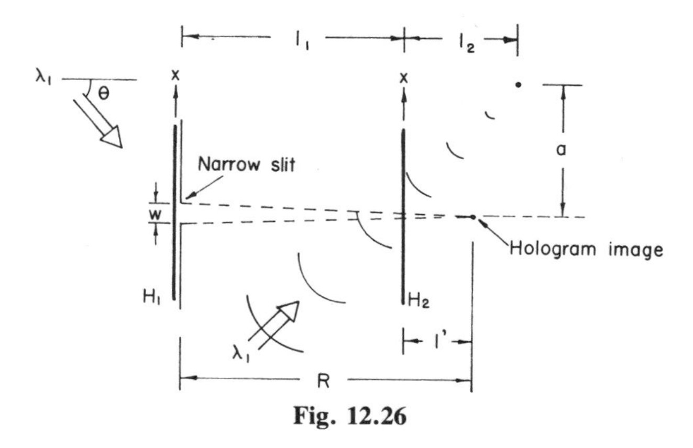

Fig. 12.26

conjugate polychromatic source of wavelengths λ_2 and λ_3, where $\lambda_2 = (1/2)\lambda_1$, and $\lambda_3 = 2\lambda_1$, as shown in fig. 12.27, then:

(a) Sketch a system analog diagram to show the holographic image reconstruction.

(b) Calculate the locations of the slit images and the point images with respect to those wavelengths.

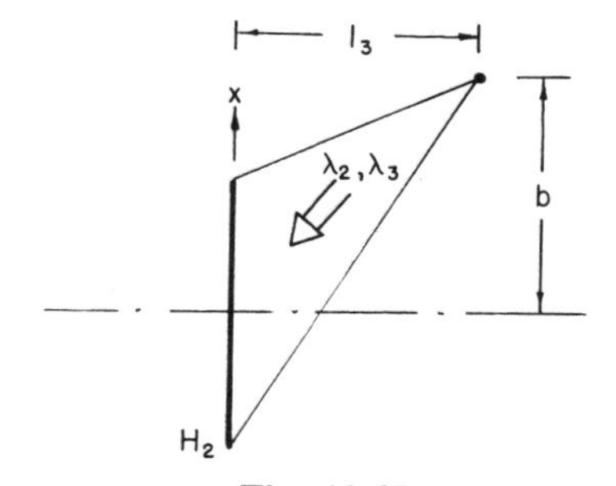

Fig. 12.27

12.3 Given a two-step rainbow holographic process, as shown in fig. 12.28. We assume that the rainbow hologram H_2 is constructed with a red coherent light of wavelength $\lambda = 6328$ Å.

(a) Draw a system analog to represent this rainbow holographic construction.

(b) Ignoring the holographic resolution, calculate the effect of the rainbow hologram recording due to the slit and the image point, respectively.

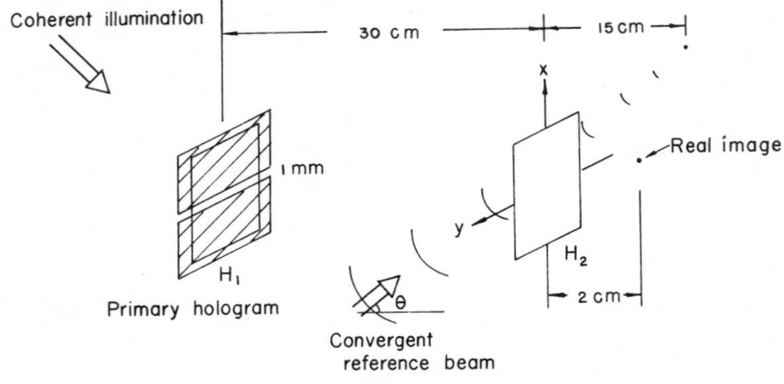

Fig. 12.28

12.4 Let us consider that the rainbow hologram of prob. 12.3 is illuminated by a divergent white-light source, as shown in fig. 12.29. If the rainbow hologram image is viewed by an unaided human eye (the pupil is assumed to be 2 mm wide) over the smeared slit image, then determine the color blur (i.e., smeared image) of the rainbow holographic image.

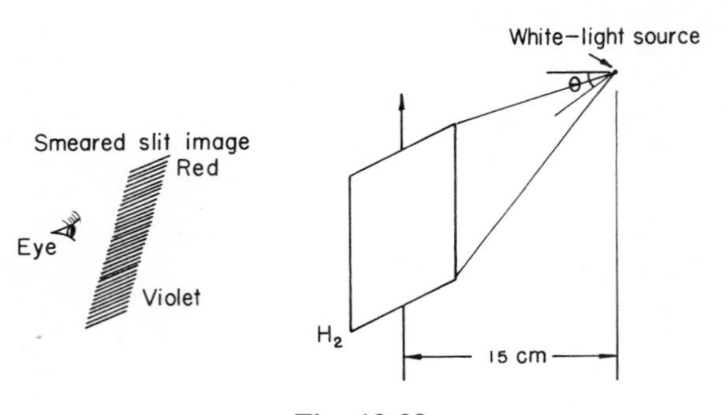

Fig. 12.29

12.5 Given the pseudoscopic one-step rainbow hologram construction of fig. 12.30. We assume that the narrow slit is about $f/4$ away from the imaging lens.

(a) Draw a system analog diagram to evaluate where the narrow slit image will be located.

(b) Evaluate the location of the narrow slit image.

(c) Draw an equivalent schematic diagram to replace the original one with the result obtained in part (b).

(d) Draw a system analog diagram to evaluate the rainbow holographic recording. The impulse responses should be properly defined.

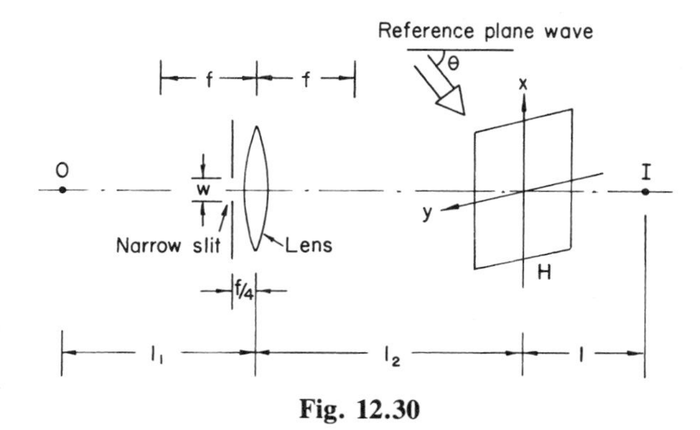

Fig. 12.30

12.6 With reference to the orthoscopic rainbow hologram construction of fig. 12.31:

(a) Determine the slit and object image locations.

(b) What is the lateral magnification of the slit image?

(c) Draw an equivalent diagram of the rainbow holographic recording.

(d) Assuming that the location of the holographic plate intersects the object image, draw a system analog diagram to evaluate the rainbow holographic recording. The impulse responses should be defined.

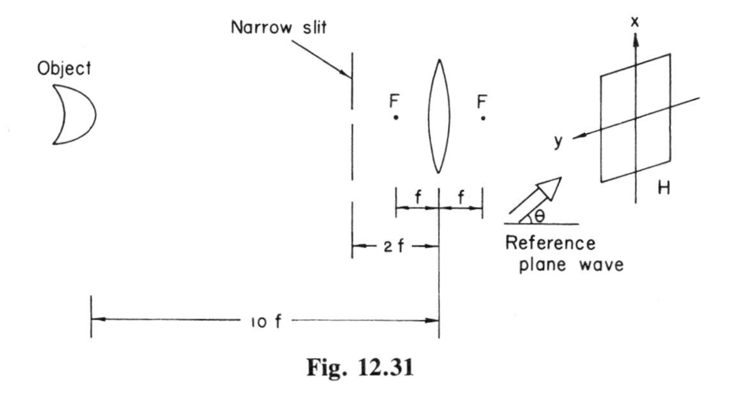

Fig. 12.31

12.7 Given the color rainbow hologram generation shown in fig. 12.32. For simplicity, we assume that the holographic construction is sequentially performed by a red and a green coherent illumination. By simple point concept analysis:

(a) Evaluate this multiwavelength rainbow holographic construction.

(b) If the recorded hologram is reconstructed with a conjugate white-light source, evaluate the location where a true color hologram image can be perceived through the smeared slit images.

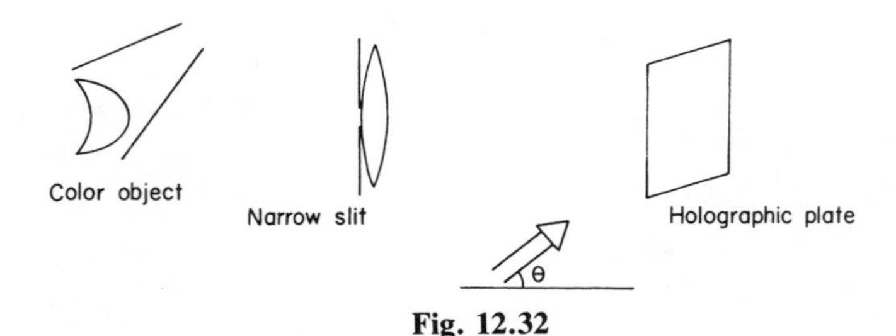

Fig. 12.32

12.8 **(a)** Discuss the requirement for the presence of a narrow slit for rainbow holographic construction.

(b) What are the disadvantages of the presence of a narrow slit during the rainbow hologram construction?

(c) Assuming that the width of the primary hologram is L and the slit width is W, evaluate the amount of spatial frequency reduction of object image during the rainbow holographic construction.

12.9 Let us consider the two-dimensional-object rainbow holographic construction of fig. 12.33. The wavelength of the coherent source is $\lambda_1 = 6328$ Å, and this reference beam is assumed to be an oblique plane wave. If we wish to obtain a minimum resolution limit equal to the color blur of the hologram image, determine the required slit width W of the rainbow holographic process.

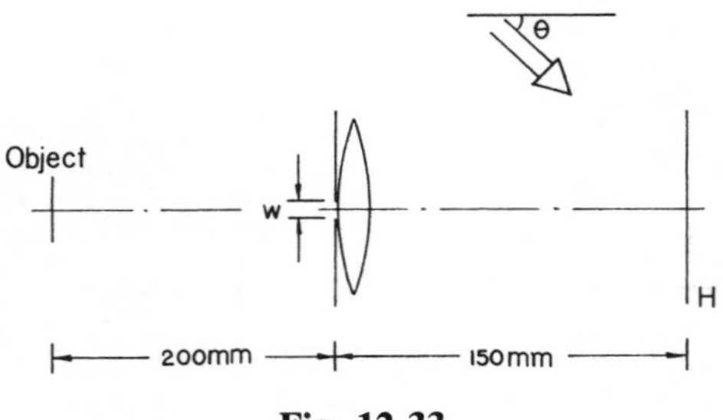

Fig. 12.33

12.10 With reference to the slit width obtained in prob. 12.9, we assume that the object is a two-dimensional transparency.

(a) Calculate the resolution limit of this one-step rainbow holographic process.

(b) If a cylindrical lens, with a focal length of $f_c = 200$ mm, is inserted immediately behind the object transparency, what is the resolution limit of this rainbow holographic process?

12.11 Given the rainbow holographic recording shown in fig. 12.34. We assume that the diameter of the pupil of an unaided human eye is D. If the rainbow hologram is illuminated by a conjugate white-light source, then:

(a) Calculate the image blur due to the wavelength spread.

(b) Evaluate the image blur due to diffraction, which results from both the finite slit size and the eye pupil diameter.

(c) If the color blur due to diffraction is equal to the color blur due to wavelength spread, calculate the required slit width.

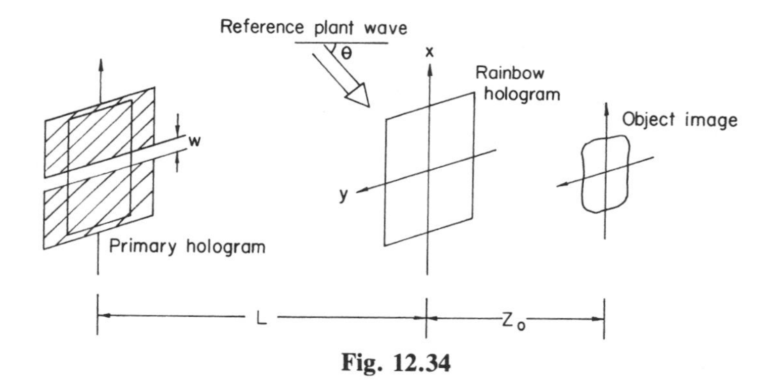

Fig. 12.34

12.12 In order to gain a feeling of magnitudes, we consider a numerical problem. Let us refer to prob. 12.11. If $\lambda = 5000$ Å, $L = 300$ mm, $Z_0 = 50$ mm, and $W = 1$ mm, then:

(a) Calculate the image blur due to wavelength spread and due to diffraction, respectively.

(b) From the numerical results obtained in part (a), discuss the color blur effects due to wavelength spread and diffraction if the slit width is greater or smaller than 1 mm.

REFERENCES

12.1 D. Gabor, "A New Microscope Principle," *Nature,* **161,** 777 (1948).

12.2 E. N. Leith and J. Upatnieks, "Wavefront Reconstruction with Continuous-Tone Objects," *J. Opt. Soc. Am.,* **53,** 1377 (1963).

12.3 E. N. Leith and J. Upatnieks, "Wavefront Reconstruction with Diffused Illumination and Three-Dimensional Objects," *J. Opt. Soc. Am.,* **54,** 1295 (1964).

12.4 S. A. Benton, "Hologram Reconstructions with Extended Light Sources," *J. Opt. Soc. Am.,* **59**, 1545 (1969).

12.5 S. A. Benton, "White-Light Transmission/Reflection Holographic Imaging," in E. Marom, A. A. Friesm, and E. Weiner-Avnear, Eds., *Proceedings of ICO Conference on Application of Holography and Optical Information Processing,* Pergamon Press, New York, 1978, p. 401.

12.6 Y. N. Denisyuk, "Photographic Reconstruction of the Optical Properties of an Object in its Own Scattered Radiation Field," *Sov. Phys.—Doklady,* **7,** 543 (1962).

12.7 E. N. Leith, "White-Light Holograms," *Sci. Am.,* **235,** 80 (1976).

12.8 H. Chen and F. T. S. Yu, "One-Step Rainbow Hologram," *Opt. Lett.,* **2,** 85 (1978).

12.9 L. Rosen, "Holograms of the Aerial Image of a Lens," *Proc. IEEE,* **55,** 79 (1967).

12.10 C. F. Stubenrauch and E. N. Leith, "Use of Holograms in Depth-Cue Experiments," *J. Opt. Soc. Am.,* **61,** 1268 (1971).

12.11 J. C. Wyant, "Image Blur for Rainbow Holograms," *Opt. Lett.,* **1,** 130 (1977).

12.12 R. J. Collier, C. B. Burckhardt, and L. H. Lin, *Optical Holography,* Academic Press, New York, 1971.

12.13 L. H. Lin, K. S. Pennington, G. W. Stroke, and A. E. Labeyric, "Multi-Color Holographic Image Reconstruction with White-Light Illumination," *Bell Syst. Tech. J.,* **45,** 659 (1966).

12.14 P. Hariharan, W. H. Steel, and Z. S. Hegedus, "Multi-color Holographic Imaging with a White-Light Source," *Opt. Lett.,* **1,** 8 (1977).

12.15 F. T. S. Yu, P. H. Ruterbusch, and S. L. Zhuang, "High-Resolution Rainbow Holographic Process," *Opt. Lett.,* **5,** 443 (1980).

12.16 E. N. Leith and H. Chen, "Deep-Image Rainbow Holograms," *Opt. Lett.,* **2,** 82 (1978).

12.17 E. N. Leith, H. Chen, and J. Roth, "White-Light Hologram Technique," *Appl. Opt.,* **17,** 3187 (1979).

12.18 H. Chen, "Astigmatic One-Step Rainbow Hologram Process," *Appl. Opt.,* **18,** 3728 (1979).

12.19 P. N. Tamura, "Multi-color Image From Superposition of Rainbow Holograms," *SPIE,* **126,** 59 (1977).

12.20 H. Chen, "Color Blur of the Rainbow Hologram" *Appl. Opt.,* **17,** 3290 (1978).

12.21 S. L. Zhuang, P. H. Ruterbusch, Y. W. Zhang, and F. T. S. Yu, "Resolution and Color Blur of the One-Step Rainbow Hologram," *Appl. Opt.,* **20,** 872 (1981).

12.22 F. W. Sears, *Optics,* Addison-Wesley Publishing Company, Cambridge, Mass., 1949.

12.23 F. T. S. Yu, A. Tai, and H. Chen, "One-Step Rainbow Holography: Recent Development and Application," *Opt. Eng.,* **19,** 666 (1980).

12.24 L. Cross, "Multiplex Technique for Cylindrical Holographic Stereograms," *SPIE,* **120** (1977).

12.25 L. Huff, "Color Holographic Stereograms," *Opt. Eng.,* **19,** 691 (1980).

13

Applications of Holography

Leith and Upatnieks' revival and improvement of Gabor's holography have led to a great number of scientific applications. These have not been limited to optical wavelengths; applications are found in the microwave and radio-frequency regions (ref. 13.1), as well as in acoustics (refs. 13.2, 13.3). It would require an exhaustive effort to cover each of the many applications. This is not the main objective of this chapter. We do, however, devote some time to discussing a few of the most interesting applications in the optical region.

13.1 MICROSCOPIC WAVE FRONT RECONSTRUCTION

The invention of holography was motivated by the hope for improvements in electron microscopy. Gabor's original work (refs. 13.4–13.6), of course, had this orientation, as did that of El-Sum (ref. 13.7).

Holographic microscopy may—at least in principle—have four advantages. First, *holographic magnifications* may be accomplished by constructing the wave front with a very short wavelength and then reconstructing the hologram image by means of a longer wavelength. Second, the lateral resolution of the image reconstruction is limited by the size of the hologram aperture; the larger the hologram aperture, the better the resolution. Thus a high resolution image may be attainable for a large field of view. Third, wave front aberrations in holography may be corrected to some degree by suitable recording and reconstructing procedures. Thus a nearly aberration-free hologram image may be achieved. Fourth, images of high longitudinal resolution may be achieved. Thus holographic microscopy may offer enormous depth of focus.

Several practical problems in the application of microscopic wave front reconstruction prevent this new technique from being a serious threat to conventional microscopy at present. Figures 13.1 and 13.2 (from ref. 13.8) are examples of the results to be expected from the technique.

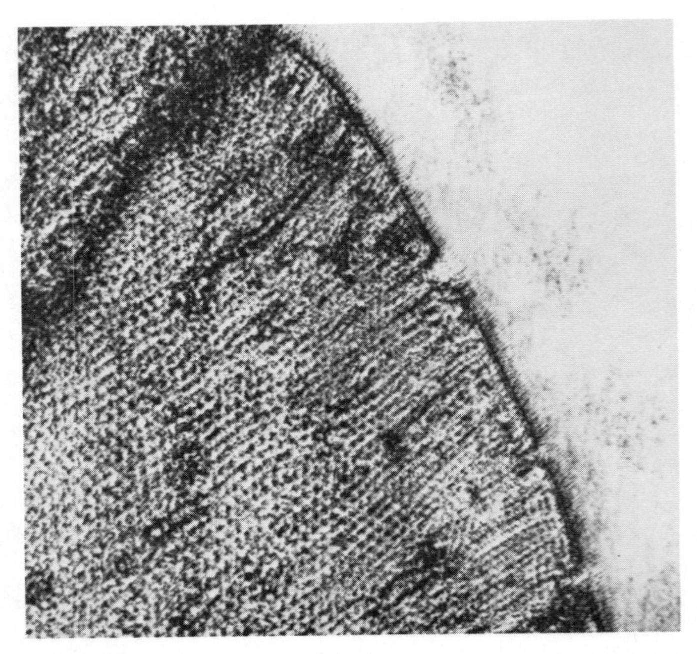

Fig. 13.1 Holographic microscopy of a fly's wing. The lateral magnification is about 60×. (By permission of E. N. Leith.)

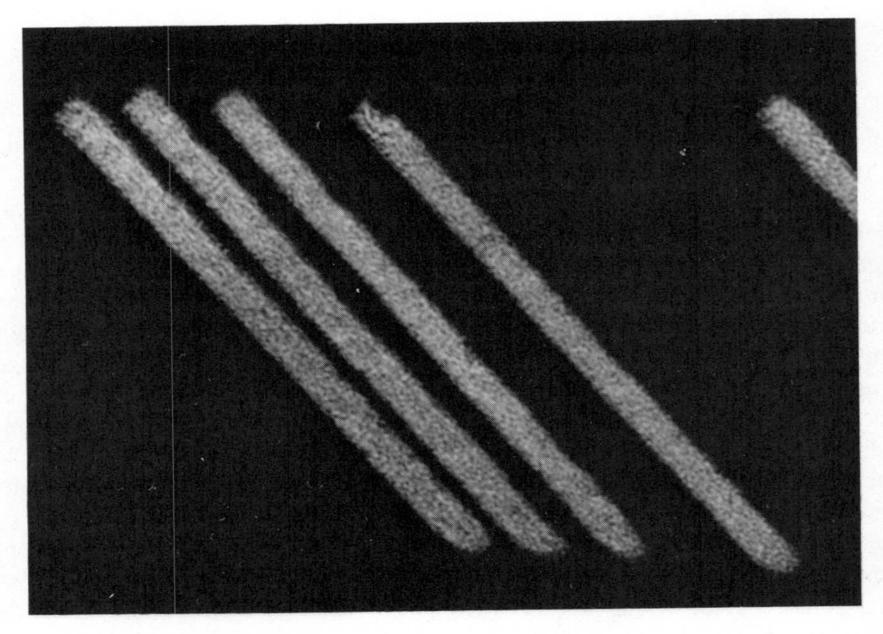

Fig. 13.2 Holographic microscopy of a test chart. The lateral magnification is about 120×. The spacing between lines is about 10 microns. (By permission of E. N. Leith.)

13.2 MULTIEXPOSURE HOLOGRAPHIC INTERFEROMETRY

One of the most interesting and important of the applications of holography may be to interferometry. Holograms preserve the amplitude and phase distributions of the recording field. The same hologram can also be the record of two or more different sets of wave fronts (multiexposure holography). Therefore it is possible to produce a superposition of two or more hologram images, between which an interference image will be reconstructed. This powerful *holographic interferometry* is based on the principle of coherent additions of the complex wave fields (ref. 13.9).

To analyze the process, let us assume that a photographic plate has been holographically and sequentially exposed to N independent exposures. If we denote $I_n(x, y)$ as the nth irradiance of the recording, then the total exposure is

$$E(x, y) = \sum_{n=1}^{N} t_n I_n(x, y), \tag{13.1}$$

where t_n is the nth exposure time. The irradiance $I_n(x, y)$ may be written as

$$I_n(x, y) = |u_n(x, y)|^2 + |v|^2 + u_n(x, y)v^* + u_n^*(x, y)v, \tag{13.2}$$

where $u_n(x, y)$ and v are the complex light fields due to the object and reference waves, respectively.

If this multiexposure recording is properly biased in the linear region of the T-E curve of the photographic emulsion, then the transmittance of the recorded hologram is given by

$$T(x, y) = K_1 \sum_{n=1}^{N} |u_n(x, y)|^2 + K_2 + K_3 \sum_{n=1}^{N} u_n(x, y)v^* + K_4 \sum_{n=1}^{N} u_n^*(x, y)v, \tag{13.3}$$

where the K's are the appropriate positive constants.

If the hologram of eq. (13.3) is illuminated by a monochromatic plane wave v, then a sequence of virtual images will be reconstructed. The resulting complex light field behind the hologram aperture may be written as

$$E_{\text{v}}(x, y) = |v|^2 K_3 \sum_{n=1}^{N} u_n(x, y), \tag{13.4}$$

where the subscript v denotes the virtual image diffraction. It can be seen from eq. (13.4) that the object light fields are superimposed on each other; thus they will mutually interfere.

It is obvious that, if the hologram is illuminated by a conjugate monochromatic plane wave v^*, then a sequence of real hologram images will be constructed. The complex light field due to these real images is

$$E_r(x, y) = |v^2|K_4 \sum_{n=1}^{N} u_n^*(x, y), \tag{13.5}$$

where the subscript r refers to the real image. These object light fields will also mutually interfere.

One of the most interesting applications of multiexposure holography is due to Heflinger et al. (ref. 13.10). Examples of double-exposure holographic interferograms are given in figs. 13.3 and 13.4. The interferogram of fig. 13.3 was made by using a Q-switched pulsed ruby laser. The first pulse (i.e., the first exposure) records only the hologram of the diffuse background, and the second pulse records the hologram in which a bullet is traveling in argon gas in front of the background. The shock wave generated by the bullet causes changes in the refractive index of the surrounding gas. Thus the two hologram images will mutually interfere. This interference pattern in turn describes the shock wave

Fig. 13.3 A double-exposure holographic interferogram of the shock wave of a 22-caliber bullet moving through argon gas at 1060 meters/second. The hologram was made with Q-switched ruby laser. (By permission of L. O. Heflinger.)

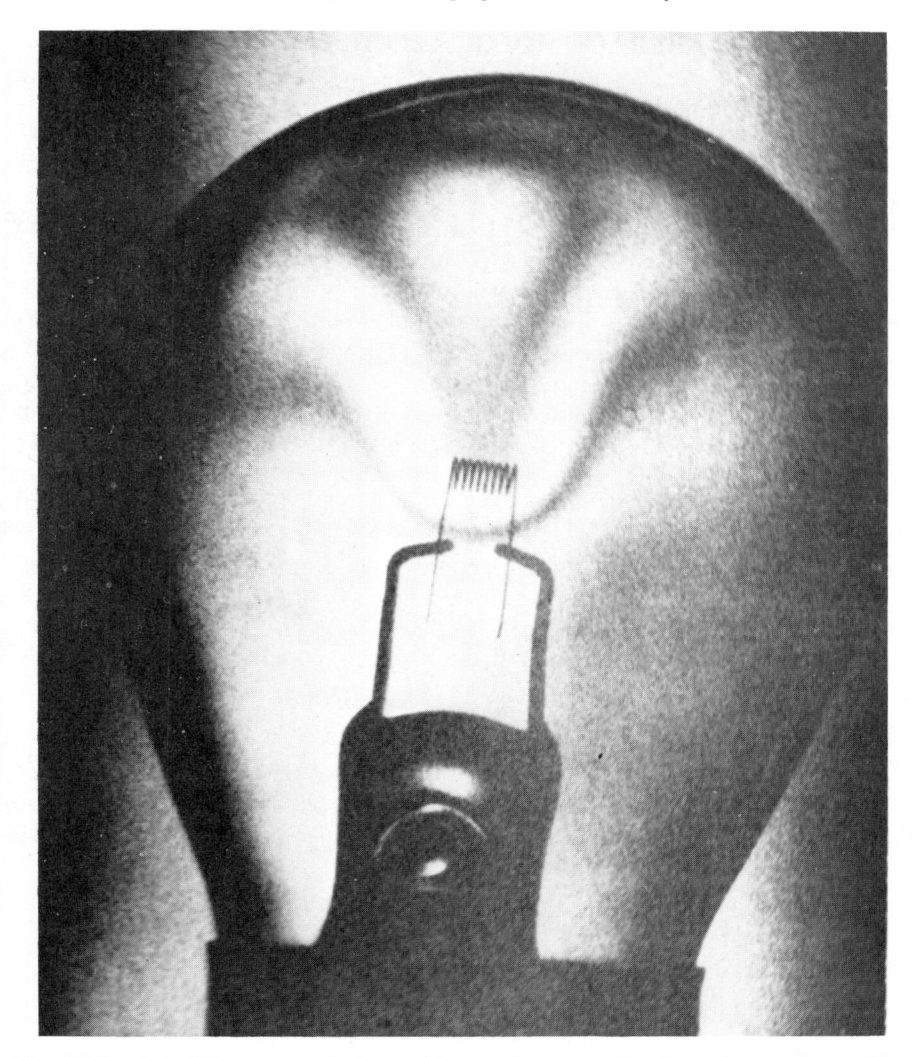

Fig. 13.4 A double-exposure holographic interferogram of an incandescent lamp. The hologram was made with a Q-switched ruby laser. (By permission of L. O. Heflinger.)

behavior generated by the bullet. The resulting interference fringes form a three-dimensional pattern. The interferogram of fig. 13.4 was made in a similar fashion; one exposure was made with a cold filament and the other exposure was made with a heated filament. The distribution of the gas in the bulb is not the same for each case; thus the holographic image reconstructions will mutually interfere.

13.3 TIME-AVERAGE HOLOGRAPHIC INTERFEROMETRY

The concept of multiexposure interferometry may be extended to continuous-exposure interferometry, which we will call *time-average holographic interferometry*. The principle of this technique is identical to that of multiexposure holography, except that the exposure is a continuous rather than discrete variable. That is, a large number of infinitesimal holographic recordings of the same object take place within a given exposure time. Each of these infinitesimal subholograms will produce its own holographic images, and an interference pattern will result. The earliest application of time-average holographic interferometry is the Powell and Stetson study of a vibrating object (ref. 13.11).

The analysis of this technique will now be given. Let a planar object be situated in the $\boldsymbol{\xi}$ coordinate system, in an arrangement similar to that shown in fig. 10.7. If the object is vibrating in the $\xi\eta$ plane, the object function satisfies

$$O[\xi(t), \eta(t), \zeta'(t)] = O[\xi_0 + \xi'(t), \eta_0 + \eta'(t), \zeta'(t)] \qquad (13.6)$$

where ξ_0 and η_0 are some average coordinates of the object function, and $\xi'(t)$, $\eta'(t)$ and $\zeta'(t)$ are the time-dependent coordinate variables about $(\xi_0, \eta_0, 0)$.

By the Fresnel–Kirchhoff theory, the resulting complex light field on the recording medium may be written as

$$u(\boldsymbol{\rho}; k, t) = \iint_S O[\boldsymbol{\xi}(t); k] E_l^{+}[\boldsymbol{\rho} - \boldsymbol{\xi}(t); k]\, d\xi(t)\, d\eta(t), \qquad (13.7)$$

where S denotes the surface integral, and where

$$E_l^{+}[\boldsymbol{\xi}(t); k] = -\frac{i}{\lambda l} \exp\left\{ ik\left[l - \zeta(t) + \frac{\xi^2(t) + \eta^2(t)}{2l} \right] \right\}$$

is the spatial impulse response.

The transmission function of the time-average hologram is

$$T(\boldsymbol{\rho}; k) = K \int_0^{\Delta t} [|u(\boldsymbol{\rho}; k, t)|^2 + |v|^2 + u(\boldsymbol{\rho}; k, t)v^* + u^*(\boldsymbol{\rho}; k, t)v]\, dt, \qquad (13.8)$$

where Δt is exposure time and K is an appropriate positive constant. It may be seen from eq. (13.8) that the reconstructed wave front is indeed a summation of the wave fronts produced by the object at each of its positions. Since each wave front and its associated subhologram are coherent with all the others, the summed holographic image will contain interference fringes.

Of special interest is the case of vibration along the optical axis only; i.e., $\xi(t) = \eta(t) = 0$, $\zeta(t) \neq 0$. The complex light field from the object of eq. (13.7) is then

$$u(\boldsymbol{\rho}; k, t) = -\frac{i}{\lambda l} \exp(ikl) \iint_S O[\xi_0, \eta_0, \zeta'(t)] \exp[-ik\zeta'(t)] \times \exp\left\{\frac{k}{2l}[(x - \xi_0)^2 + (y - \eta_0)^2]\right\} d\xi\, d\eta. \qquad (13.9)$$

If we illuminate the hologram of eq. (13.8) by a monochromatic wave front, the resultant virtual hologram image diffraction will be

$$E(\boldsymbol{\xi}; k) = C \int_0^{\Delta t} O[\boldsymbol{\xi}_0, \eta_0 \zeta'(t)] \exp[-ik\zeta'(t)]\, dt. \qquad (13.10)$$

where C is a complex constant.

If the displacement $\zeta'(t)$ is small, eq. (13.10) may be approximated by

$$E(\boldsymbol{\xi}; k) = CO(\xi_0, \eta_0) \int_0^{\Delta t} \exp[-ik\zeta'(t)]\, dt. \qquad (13.11)$$

As an example, if the vertical motion is a sinusoidal vibration,

$$\zeta(t) = a(\xi_0, \eta_0) \cos \omega t, \qquad \text{for } t \ll \Delta t, \qquad (13.12)$$

where $a(\xi_0, \eta_0)$ is the amplitude function over the planar object, then the solution of eq. (13.11) is

$$E(\boldsymbol{\xi}; k) = CO(\xi_0, \eta_0) J_0[ka(\xi_0, \eta_0)], \qquad (13.13)$$

where J_0 is the zero-order Bessel function.

That is to say, the resultant hologram image is the object function weighted by a zero-order Bessel function, with a primary maximum value at $a(\xi_0, \eta_0) = 0$. The irradiance at each point of the resultant hologram image depends on the object amplitude function $a(\xi_0, \eta_0)$.

Examples of time-average holographic interferograms by Powell and Stetson are given in fig. 13.5. These interferograms are those of a vibrating diaphragm. The different order modes of vibration can be easily seen; by counting the number of fringes from the edge of the diaphragm, we may determine the vibration amplitude at any point.

13.4 CONTOUR GENERATION

Contour generation by wave front reconstruction processes was first reported by Hildebrand and Haines (refs. 13.12, 13.13). The two methods of contour gener-

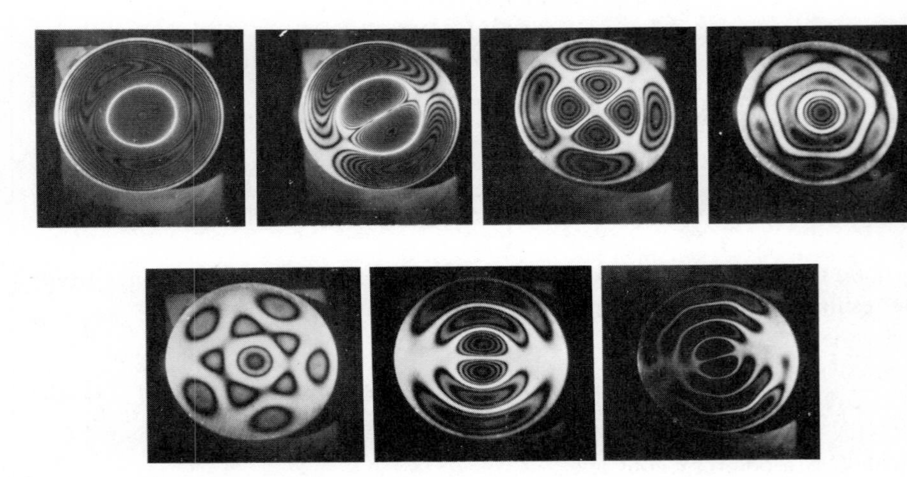

Fig. 13.5 Time-average holographic interferograms of a vibrating diaphragm in different modes of vibration. (By permission of R. L. Powell.)

ation are known as multisource generation and multiwavelength generation. Multisource generation is accomplished by two mutually coherent but separate point sources. Wave front recording in this method can be obtained either by simultaneous illumination of the object (by these two coherent sources), or by double-exposure sequential illuminations of the object. For either type of illumination, interference fringes will appear on the surface of the object. For example, in the multisource arrangement of fig. 13.6 point sources are arranged in such a way that the interference fringes follow hyperbolic curves of constant path-length difference. If this illuminated object is recorded on a plate oriented parallel to the fringe pattern of the object, surface variations of the object will give rise to contours in the hologram image.

However, the multisource technique has a practical disadvantage. It requires that the directions of the illuminations be precise. This constraint means that

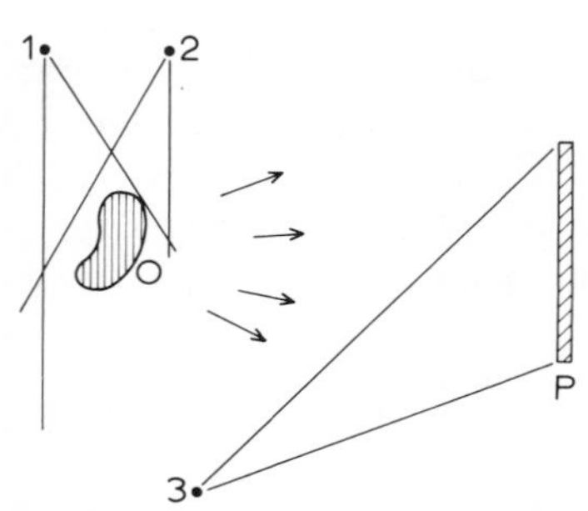

Fig. 13.6 Recording geometry for multisource contour generation. 1,2, point source of illumination; 3, reference source; O, object; P, photographic plate.

some parts of an irregularly shaped object may not be illuminated at all. This disadvantage can be easily removed by means of the multiwavelength technique. If the source emits only two wavelengths, then the wave front construction will consist of two superposed patterns. If this two-wavelength hologram is illuminated by a monochromatic source, then two hologram images, with slightly different locations and magnifications, will be reconstructed in superposition. Thus interference takes place, and a bright and dark fringe pattern will be generated on the surface of the resultant image.

Figure 13.7 gives the reconstructed images of a U.S. quarter. The image reconstruction of fig. 13.7*a* was produced by a single-wavelength hologram, while the image shown in fig. 13.7*b* was produced by a two-wavelength hologram. The wavelengths used for recording were two spectral lines produced by an argon laser. These two lines were separated by 65 Å and the elevations contoured in fig. 13.7*b* are spaced at about 0.02 mm.

Contour generation may also be accomplished by changing the refractive index of the region surrounding an object (fig. 13.8). A first exposure is made

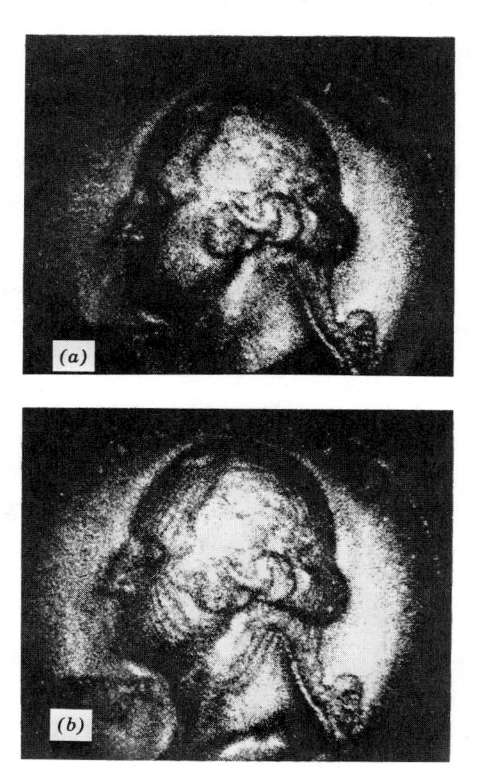

Fig. 13.7 Contour generation by the two-wavelength technique. (*a*) Single-wavelength image without fringes. (*b*) Two-wavelength image with fringes. (By permission of B. P. Hildebrand.)

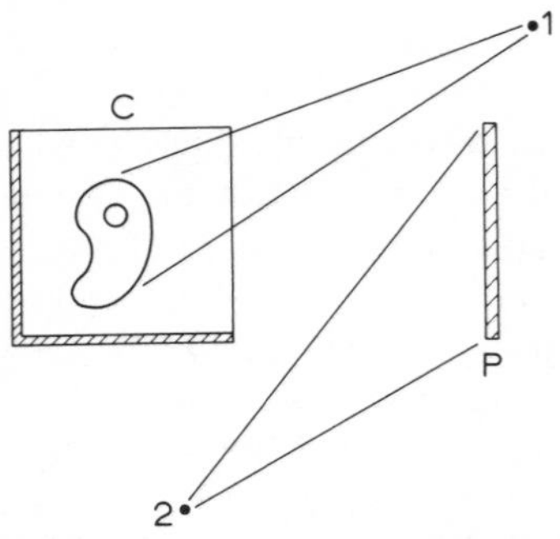

Fig. 13.8 Contour generation by means of changes in the refractive index of the fluid surrounding the object. *C*, transparent container; *O*, object; *P*, photographic plate; 1, illuminating source; 2, reference source.

with only the object inside a transparent container. A second exposure takes place when the container, with the unmoved object, is filled with some appropriate liquid or gas. Thus the optical path lengths are changed, and superposition gives rise to interference fringes on the reconstructed image.

13.5 IMAGING THROUGH A RANDOMLY TURBULENT MEDIUM

Coherent optical systems may be applied to the problem of imaging through randomly turbulent media. It is well known that the random refractive index of the turbulence interferes with light propagation. Thus conventional imaging techniques may suffer some major distortions. However, imaging by means of holographic techniques are often able to offset some of the degradation caused by the turbulence. Several holographic imaging techniques through a randomly aberrating medium were proposed by Leith and Upatnieks (ref. 13.14), Kogelnik (ref. 13.15), and Goodman et al. (ref. 13.16).

In this section we briefly illustrate two of these techniques. First, let us consider the geometry shown in fig. 13.9. In the presence of the turbulent medium, the spatial impulse response is modified to give

$$E_{lN}^{+}(\boldsymbol{\xi}, t, k) = E_{l}^{+}(\boldsymbol{\xi}; k)N(\boldsymbol{\xi}, t; k), \tag{13.14}$$

where $E_l^+(\xi; k)$ is the spatial impulse response in the absence of turbulence, $\boldsymbol{\xi}(\xi, \eta, \zeta)$ is the coordinate system, and $N(\boldsymbol{\xi}, t; k)$ is the time-dependent complex random function of the turbulence. $N(\boldsymbol{\xi}, t; k)$ may be written as

$$N(\boldsymbol{\xi}, t; k) = |N(\boldsymbol{\xi}, t; k)| \exp[i\psi(\boldsymbol{\xi}, t; k)], \tag{13.15}$$

where $\psi(\boldsymbol{\xi}, t; k)$ is the random phase function. It is clear that $0 \leq |N(\boldsymbol{\xi}, t; k)| \leq 1$, because the turbulence is assumed to be in a passive medium.

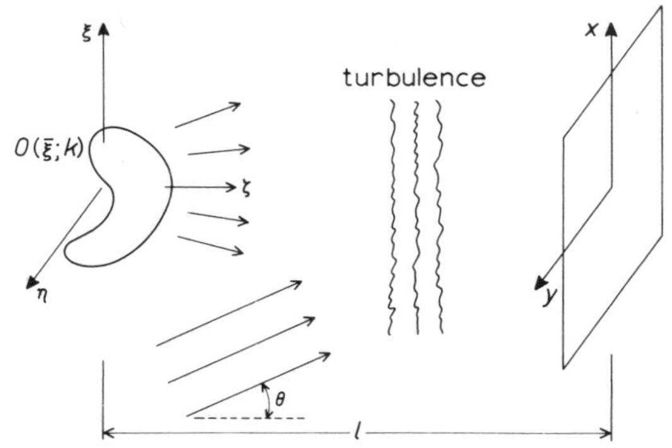

Fig. 13.9 Holographic imaging through a randomly turbulent medium.

The complex light amplitude distribution on the photographic plate is determined, as usual, by means of the Fresnel–Kirchhoff theory:

$$u(\boldsymbol{\rho}, t; k) = \iint_S O(\boldsymbol{\xi}; k) E_{lN}^{+}(\boldsymbol{\rho} - \boldsymbol{\xi}, t; k)\, d\xi\, d\eta, \tag{13.16}$$

where $O(\boldsymbol{\xi}; k)$ represents the object function, S denotes the surface integration, and $\boldsymbol{\rho}(x, y)$ is the chosen coordinate system.

We would like to illustrate, however, a special case. Let there be a thin layer of turbulence located very close to the surface of the recording medium. The complex light field distributions on the recording surface due to the object and to the reference light field, respectively, may be approximated by

$$u(\boldsymbol{\rho}, t; k) \simeq |N(\boldsymbol{\rho}, t; k)| \exp[i\psi(\boldsymbol{\rho}, t; k)] u_0(\boldsymbol{\rho}; k) \tag{13.17}$$

and

$$v(\boldsymbol{\rho}, t; k) \simeq |N(\boldsymbol{\rho}, t; k)| \exp\{i[kx \sin\theta + \psi(\boldsymbol{\rho}, t; k)]\}, \tag{13.18}$$

where

$$u_0(\boldsymbol{\rho}; k) = |u_0(x, y)| \exp[i\phi(x, y)] = \iint_S O(\xi; k) E_l^{+}(\boldsymbol{\rho} - \boldsymbol{\xi}; k)\, d\xi\, d\eta$$

is the object light field distribution without the effect of the turbulence. The irradiance of the wave front recording is therefore

$$\begin{aligned} I(\boldsymbol{\rho}, t; k) &= [u(\boldsymbol{\rho}, t; k) + v(\boldsymbol{\rho}, t; k)][u(\boldsymbol{\rho}, t; k) + v(\boldsymbol{\rho}, t; k)]^* \\ &= |N(\boldsymbol{\rho}, t; k)|^2\{1 + |u_0(x, y)|^2 + 2|u(x, y)| \cos[kx \sin\theta + \phi(x, y)]\}. \end{aligned} \tag{13.19}$$

The transmittance of the recorded hologram may be shown to be

$$T(\boldsymbol{\rho}; k) = K \int_0^{\Delta t} I(\boldsymbol{\rho}, t; k)\, dt = K\{1 + |u_0(x, y)|^2 + 2|u(x, y)| \cos[kx \sin\theta + \phi(x, y)] \int_0^{\Delta t} |N(\boldsymbol{\rho}, t; k)|^2\, dt\}, \quad (13.20)$$

where K is a proportionality constant and Δt is the exposure time.

If the time integral of eq. (13.20) is relatively *uniform* over the recording aperture, then the transmittance function may be written as

$$T(\boldsymbol{\rho}; k) = K\{1 + |u_0(x, y)|^2 + 2|u(x, y)| \cos[kx \sin\theta + \phi(x, y)]\}\, |N|^2\, \Delta t. \quad (13.21)$$

Thus for this special case, the random turbulence does not affect the image.

Typical results of holography through a randomly turbulent medium are shown in fig. 13.10. The hologram was made when a stationary phase-distorting medium (an ordinary shower glass) was introduced near the recording aperture. For comparison, fig. 13.11 shows the result obtained by replacing the recording

Fig. 13.10 Holographic imaging through a randomly phase-distorting medium.

Fig. 13.11 Conventional imaging through a randomly phase-distorting medium.

medium by a conventional positive lens. We see that the image formed by the conventional technique suffers major distortion, while the holographic image has suffered only minor degradation.

If a layer of stationary phase-distorting medium is inserted between the object and the recording medium in the manner shown in fig. 13.12, then the resultant

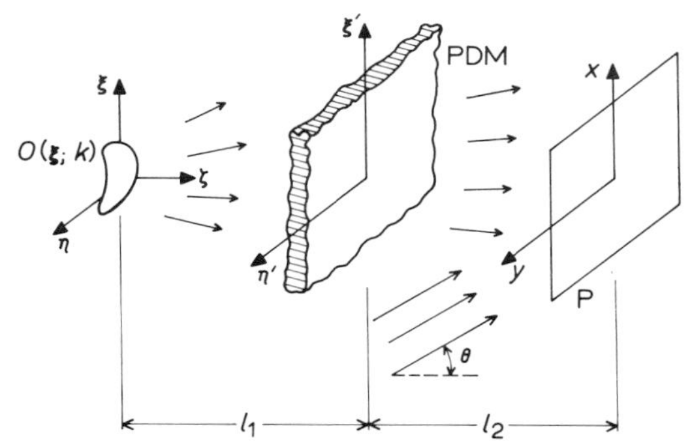

Fig. 13.12 Holographic imaging through a phase-distorting medium. O, object function; PDM, phase-distorting medium; P, photographic plate. Note the position of the plane reference wave.

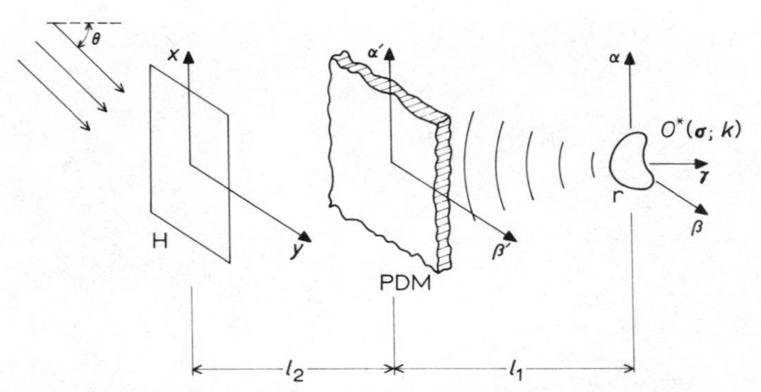

Fig. 13.13 Image reconstruction by means of identical phase compensation. H, hologram; r, real image. Plane monochromatic illumination, negative θ. The phase-distorting medium PDM is the same used in the construction shown in fig. 13.12.

complex light field scattered from the distorting medium is

$$u(\boldsymbol{\rho}; k) = \iint_{S_2} u_0(\xi', \eta') \exp[-i\psi(\xi', \eta')] E^{+}_{l_2}(\boldsymbol{\rho} - \boldsymbol{\xi}'; k)\, d\xi'\, d\eta', \tag{13.22}$$

where

$$u_0(\xi', \eta') = \iint_{S_1} O(\boldsymbol{\xi}; k) E^{+}_{l_1}(\boldsymbol{\xi}' - \boldsymbol{\xi}, k)\, d\xi\, d\eta$$

is the object light field distribution on the $\boldsymbol{\xi}'$ coordinate plane, $\psi(\xi', \eta')$ is the stationary random phase function, S_1 and S_2 denote the surface integrations over the $\boldsymbol{\xi}$ and $\boldsymbol{\xi}'$ coordinate planes, and $E^{+}_{l_1}$ and $E^{+}_{l_2}$ are the respective spatial impulse responses from the $\boldsymbol{\xi}$ to $\boldsymbol{\xi}'$ coordinate systems and from the $\boldsymbol{\xi}'$ to $\boldsymbol{\rho}$ coordinate systems.

If a reference plane wave is applied *behind* the distorting medium, as shown in the same figure, then the transmittance of the recorded hologram will be

$$T(\boldsymbol{\rho}; k) = K[|u|^2 + |v|^2 + uv^* + u^*v], \tag{13.23}$$

where K is the proportionality constant and $v = \exp(ikx \sin\theta)$ is the plane reference wave.

If this hologram described by eq. (13.23) is illuminated by an oblique monochromatic plane wave of negative θ, as shown in fig. 13.13, then the real image diffraction at a distance l_2 behind the hologram aperture can be shown to be proportional to the conjugate wave field $u^*(\alpha', \beta')$. If the same random phase-distorting medium is inserted at the $\boldsymbol{\sigma}'$ coordinate system, then at a distance l_1

behind the phase-distorting plate the complex light field can be shown to be

$$E(\boldsymbol{\sigma}; k) = K_1 O^*(\boldsymbol{\sigma}; k), \tag{13.24}$$

which is the conjugate of the object function. The phase distortion has been completely eliminated, and a distortion-free image is formed. However, extremely accurate positioning of the phase-distorting medium is required during the holographic reconstruction process; otherwise the reconstructed image may not be completely free from distortion.

13.6 COHERENT OPTICAL TARGET RECOGNITION THROUGH A RANDOMLY TURBULENT MEDIUM

It is well recognized that conventional imaging is unsuccessful when a randomly phase-distorting medium closely overlays the observation system. However, in the last section, we have seen that holographic imaging is a possible solution, since identical phase distortions occurring in the signal and reference fields will be cancelled during wave front recording. In practice, this means that a bright area or point must be positioned near the object being imaged. However, in this section (based upon refs. 13.17, 13.18), we consider what degree of recognition can be attained even though it may not be possible to generate such a reference field.

The complex light field scattered from the target through a turbulent medium under the "far-field" geometry of fig. 13.14 may be written as the convolution

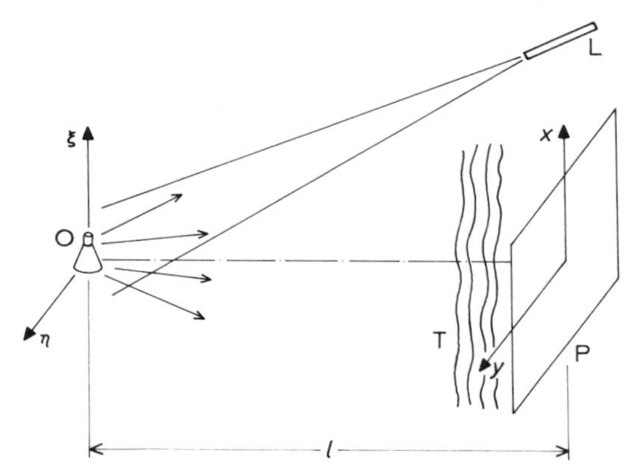

Fig. 13.14 Geometry of far-field object-spectrum recording. O, object; T, turbulence; L, laser; P, photographic plate.

integral

$$u(\boldsymbol{\rho}, t; k) = \iint_S O(\boldsymbol{\xi}; k) E_{lN}^{+}(\boldsymbol{\rho} - \boldsymbol{\xi}, t; k)\, d\xi\, d\eta, \tag{13.25}$$

where $O(\boldsymbol{\xi}; k)$ is the target function, E_{lN}^{+} is defined by eq. (13.14), and S denotes surface integration.

If it is assumed that the separation l is very large, then eq. (13.25) may be approximated by

$$u(\boldsymbol{\rho}, t; k) \simeq -\frac{i}{\lambda l} \exp\left[ik\left(l + \frac{\rho^2}{2l}\right)\right] \iint_S O(\boldsymbol{\xi}; k) N(\boldsymbol{\rho} - \boldsymbol{\xi}, t; k) \times \exp\left[-i\frac{k}{l}(x\xi + y\eta)\right] d\xi\, d\eta, \tag{13.26}$$

which is the Fourier transform of the convolution of the object function and the complex random function N.

If the turbulence is close to the surface of the recording aperture, then eq. (13.26) may be further approximated by

$$u(\boldsymbol{\rho}, t; k) \simeq -\frac{i}{\lambda l} \exp\left[ik\left(l + \frac{\rho^2}{2l}\right)\right] N(\boldsymbol{\rho}, t; k) \mathcal{F}[O(\boldsymbol{\xi}; k)]. \tag{13.27}$$

Thus the transmittance of the recorded photographic plate may be written as

$$T(\boldsymbol{\rho}; k) = K|\mathcal{F}[O(\boldsymbol{\xi}; k)]|^2 \int_0^{\Delta t} |N(\boldsymbol{\rho}, t; k)|^2\, dt, \tag{13.28}$$

where K is a proportionality constant, and Δt is the exposure time.

If the ensemble time average of eq. (13.28) is uniform over the recording aperture, then the above equation may be reduced to

$$T(\boldsymbol{\rho}; k) = K|N|^2 \Delta t\, |\mathcal{F}[O(\boldsymbol{\xi}; k)]|^2, \tag{13.29}$$

which is proportional to the power spectrum of the target function. If the turbulence is a phase-only distorting medium, then eq. (13.29) may be written as

$$T(\boldsymbol{\rho}; k) = K\, \Delta t\, |\mathcal{F}[O(\boldsymbol{\xi}; k)]|^2, \tag{13.30}$$

which, it can be seen, is completely free of phase disturbance.

Since the target (object) is far field, the power spectrum of the object function spreads all over the recording surface. In general, it is difficult to recognize the shape of the object by means of the recorded power spectrum. However, it may be possible to discover the shape of the object by the corresponding autocorrelation. To do this, we illuminate the recorded photographic plate by a monochromatic plane wave, as shown in fig. 13.15. Then the autocorrelation function of the recorded object may be obtained at the back focal plane of the transform lens, such that

$$R(\boldsymbol{\sigma}; k) = \iint T(\boldsymbol{\rho}; k) \exp\left[-i(x\alpha + y\beta)\right] dx\, dy, \qquad (13.31)$$

where $\boldsymbol{\sigma}(\alpha, \beta)$ is the coordinate system at the back focal plane.

To illustrate the effect of a close-lying phase distortion on an object's autocorrelation function relative to its effect on the conventional image, the two techniques were compared using a 6328 Å He-Ne laser for illumination. The far field condition of fig. 13.14 was simulated by means of a transform lens. A shower glass near the front of the recording aperture produced the close-lying phase distortion. Conventional imaging was obtained by placing a positive lens behind the shower glass. The images with and without the intervening phase distortion are shown in figs. 13.16*a* and 13.16*b*, respectively.

For the autocorrelation technique, the power spectrum was recorded directly on a photographic plate. The developed transparency was then Fourier transformed by a lens, and the autocorrelation function was obtained at the back focal plane. The autocorrelation photographs with and without the intervening phase distortion are shown in figs. 13.17*a* and 13.17*b*. Comparison of figs. 13.16*b* and 13.17*b* shows that the autocorrelation method is significantly less sensitive to phase distortion than is the conventional imaging technique. Similar successful tests have been made on several other object shapes, and no difficulty was encountered in associating any of the distorted autocorrelation functions with their respective undistorted autocorrelation and original object functions.

We see that successful elimination of phase distortion may be accomplished by detecting the corresponding power spectrum of the object rather than the object function. Although there exists no unique relation between the object

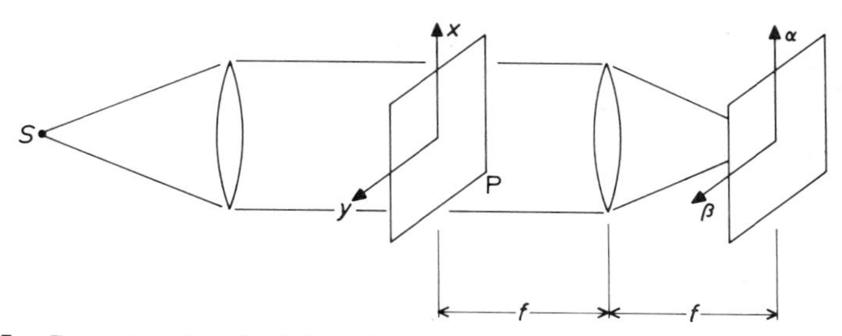

Fig. 13.15 Geometry for obtaining the autocorrelation function. *S*, monochromatic point source; P, recorded film.

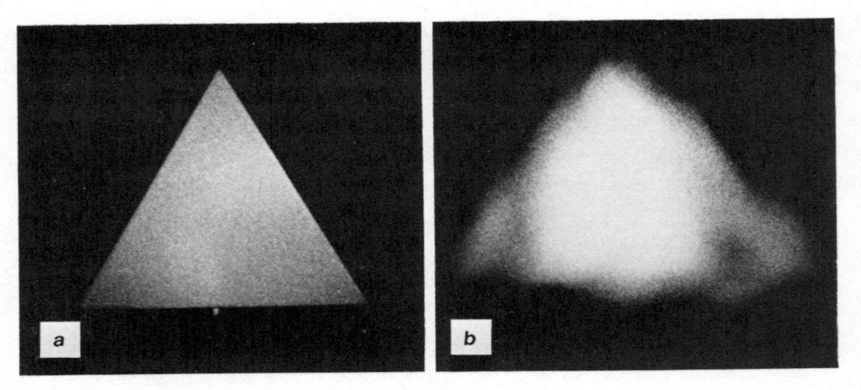

Fig. 13.16 Conventional imaging. (*a*) Without phase-distorting medium. (*b*) With phase-distorting medium.

function and the corresponding power spectrum, for a small number of distinguishable objects it may be possible to recognize the individual objects by means of their corresponding autocorrelation functions. Of course, this scheme eliminates the original phase information. However, for strong turbulence, the phase distortion hinders far more than the phase information helps.

13.7 COLOR ENCODING OF HOLOGRAPHIC FRINGE PATTERN WITH WHITE-LIGHT PROCESSING

Interferometry is one of the most important applications of holography (ref. 13.19). Since most holographic interferometric image reconstruction processes require coherent light, these conventional techniques possess two major drawbacks: the holographic fringe pattern reconstruction requires a coherent light source and a large amount of speckle noise is unavoidable.

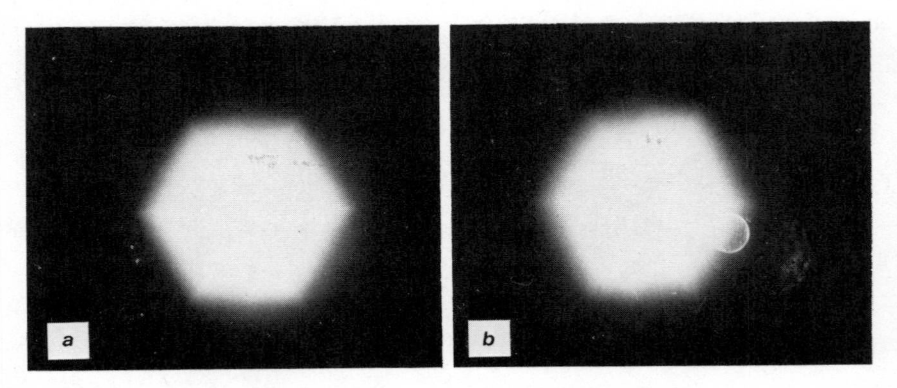

Fig. 13.17 Autocorrelation technique. (*a*) Without phase-distorting medium. (*b*) With phase-distorting medium.

In chapter 7 we have described a white-light processing technique that can be used to process an image in complex amplitude and with coherent optical noise suppression. Further applications of the white-light processing technique can be found in the literature. In chapter 8, we have demonstrated that the white-light processing technique is suitable for color image processing and that it is generally simple and economical to operate. In this section, we utilize the same white-light processing concept to color encode multiplexed holographic fringe patterns (ref. 13.20). An on-focus holographic construction process of Rosen (ref. 13.21) is used to generate the multiplexed holographic interferogram. This technique avoids the marginal resolution loss that originates from the narrow slit rainbow holographic process. The holographic fringe patterns resulting from different physical effects are displayed and color encoded simultaneously. This technique eliminates the problem associated with the multiwavelength and multislit rainbow holographic processes (ref. 13.22) and allows complex amplitude processing to be performed in the Fourier transform plane.

Figure 13.18 illustrates the technique for construction of an on-focus multiplexed holographic interferogram. A set of N interferograms are recorded on a single holographic plate and each interferogram is recorded with reference beams at different angular positions. The amplitude transmittance of the multiplexed interferograms is approximated by the following equation (see chapter 10):

$$T(x, y) = K_1 + K_2 \sum_{n=1}^{N} a_n(x, y) \cos [\phi_n(x, y) - k_0 x_n \sin \theta_n], \quad (13.32)$$

where K_1 and K_2 are arbitrary constants, $k_0 = 2\pi/\lambda_0$ is a wave number for the

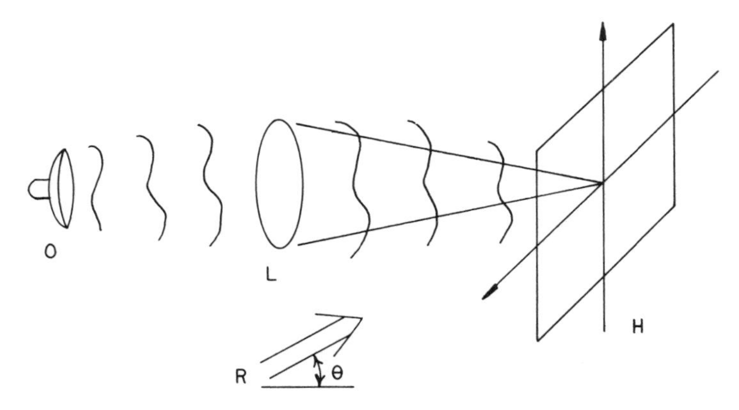

Fig. 13.18 An on-focus, time-average, multiplexed holographic interferometric construction. O, vibrational object; L, imaging lens; H, holographic plate; R, reference angle; θ, oblique angle.

coherent light source, and θ_n is the oblique angle between the reference and object beams. The quantities

$$s_n(x, y) = a_n(x, y) \exp [i\phi_n(x, y)], \tag{13.33}$$

are the complex optical waveforms of the holographic interferograms, and the x_n's are the spatial carrier frequency axes.

Figure 13.19 illustrates the white-light optical processor with the multiplexed hologram positioned in the input plane P_1. The Fourier transform in the back focal plane P_2 of the achromatic lens is given by the expression

$$E(\alpha, \beta; \lambda) = C_1 \delta(\alpha, \beta; \lambda) + C_2 S_n\left[\alpha_n - \frac{\lambda f}{2\pi} p_0, \beta_n; \lambda\right] + C_3 S_n\left[\alpha_n + \frac{\lambda f}{2\pi} p_0, \beta_n; \lambda\right], \tag{13.34}$$

where C_1, C_2, and C_3 are the appropriate complex constants, $p_0 = k_0 \sin \theta$ is the spatial carrier frequency, f is the transform lens focal length, (α_n, β_n) are the coordinates of the spatial frequency plane P_2 with respect to the input plane coordinates (x_n, y_n), and $S_n(\alpha_n, \beta_n; \lambda)$ is the corresponding Fourier transform of the multiplexed input signals $s_n(x_n, y_n; \lambda)$. The quantities $S_n[\alpha_n - (\lambda f/2\pi)p_0$,

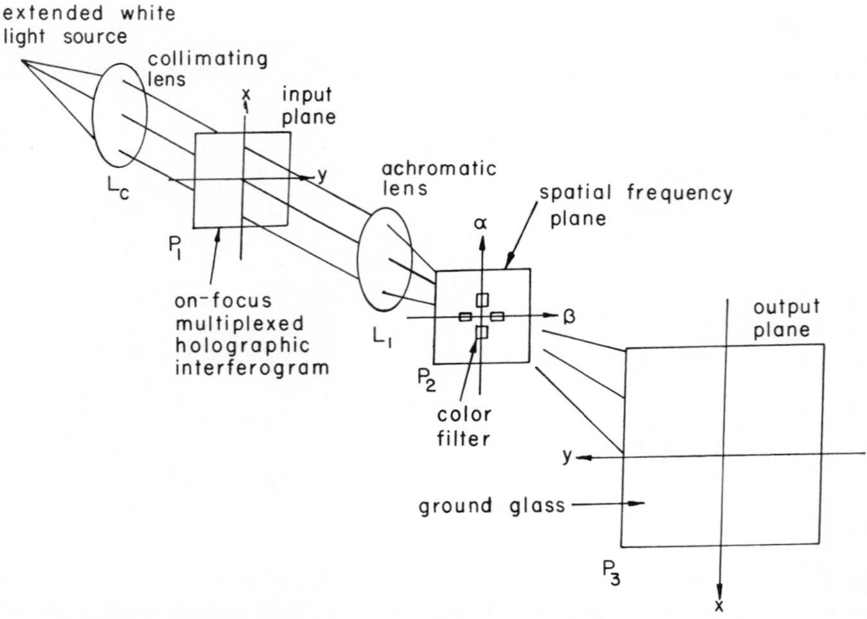

Fig. 13.19 White-light optical color encoding of holographic fringe patterns.

$\beta_n]$ are linearly translated along the α_n axis with respect to the wavelength λ of the light source. If we consider the whole spectral band of the light source, each Fourier spectrum will be smeared into a rainbow color band along the α_n axis.

Individual color band pass filters are placed over each of the smeared Fourier spectra, $S_n[\alpha_n - (\lambda f/2\pi)p_0, \beta_n]$ in the spatial frequency plane. The color-filtered Fourier spectra are incoherently added at the output image plane P_3 to form the multicolor encoded holographic fringe pattern. The color encoded intensity distribution is approximated by the following expression:

$$I(x, y) \simeq \sum_{n=1}^{N} \Delta\lambda_n |S_n(x, y; \lambda_n)^2, \tag{13.35}$$

where $\Delta\lambda_n$ and λ_n are narrow spectral bandwidths and the mean wavelengths of the color filters. The above results illustrate that the white-light processing technique is capable of performing color encoding of the holographic fringe data and offers a direct viewing capability for a closed circuit television camera or optical viewing system.

The initial experimental demonstration consists of an aluminum rod positioned by a gimbal mount. A double-exposure on-focus hologram is made with the rod end surface perpendicular to the sensitivity direction. The first exposure is recorded on the holographic plate before rotating the object about a vertical axis. A second exposure is recorded after the rotation. Both exposures are made with a single reference beam R, as shown in fig. 13.18. Another double-exposure hologram is made on the same holographic plate with a different axis of rotation and different reference beam. We can call the two reference beams R_1 and R_2; they are approximately the same angular distance from the object beam, but their relative orientations are 90° apart. The pair of double-exposure interferograms are consequently multiplexed on the same holographic plate, but they have different principal axes of spatial frequency carriers.

The photographically developed multiplexed hologram is reconstructed using the white-light optical processor illustrated in fig. 13.19. Two sets of color smeared Fourier spectra are produced approximately 90° apart in the spatial frequency plane P_2. Red and green color filters are placed over the respective smeared Fourier spectra, which are located on the principal spatial frequency carrier axes as shown in fig. 13.20. The angles for reference beams R_1 and R_2 (fig. 13.18 and preceding paragraph) should be sufficiently small to enable the images to pass through the achromatic lens in fig. 13.19. The color encoded holographic interferograms are reconstructed in plane P_3.

Figure 13.21 contains several black-and-white photographs of color encoded holographic interferograms of the cylindrical rod surface. Figures 13.21*a* and 13.21*b* illustrate a pair of nonparallel rotational motion fringes with one set color encoded red and the other green. These individual fringe patterns are achieved by blocking one pair of the fringe spectra in the Fourier transform plane and placing a color filter over the other corresponding set. Figure 13.21*c* illustrates

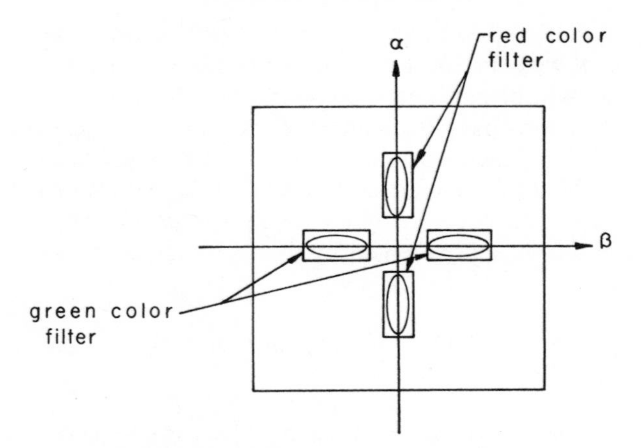

Fig. 13.20 Color filtering at the spatial frequency plane.

a combination of the two color encoded fringe patterns. The reconstructed image in the output plane contains predominately four colors. A red fringe corresponds to the addition of a dark fringe on one pattern and a bright fringe that is color encoded red on a second fringe pattern. A similar situation occurs for the green fringe areas. A dark fringe corresponds to the addition of a pair of dark fringes and a yellow fringe to a pair of light fringes with equal intensities. The addition of light fringes with unequal intensities gives rise to transition colors such as the orange fringes in fig. 13.21*c*.

Another example utilizes time-average holograms of a vibrating acoustic speaker. A small 8-ohm, 0.2-W speaker is driven by 2 kHz sinusoidal waveform. A pair of multiplexed time-average interferograms are recorded for input voltages of 2 V p-p. A pair of color encoded fringe patterns are seen at the output plane of the white-light optical processor.

Figure 13.22*a* contains the broad fringe pattern that is color encoded green and corresponds to a 2-V p-p voltage input. The red, finer fringe pattern in fig. 13.22*b* corresponds to a 4-V p-p force input. Figure 13.22*c* contains the combined color encoded multiplexed time-average fringe patterns. Since the individual fringe patterns have nearly identical intensity functions, the superposition of two bright fringes corresponds to a yellow hue. The brightest fringes on a time-average interferogram are nodal regions where the vibration amplitude is zero. This phenomenon is described mathematically by an intensity distribution that is proportional to J_0^2, where J_0 is a zero-order Bessel function. The less intense yellow regions correspond to a superimposed pair of higher order bright fringes. The red and green fringes similarly correspond to a superposition of one dark and one light fringe.

The white-light processing technique allows a direct comparison between different sets of holographic fringe patterns. Since the holographic image reconstruction is accomplished by an extended white-light source, the degree of coherent artifact noise is substantially less than that of the conventional tech-

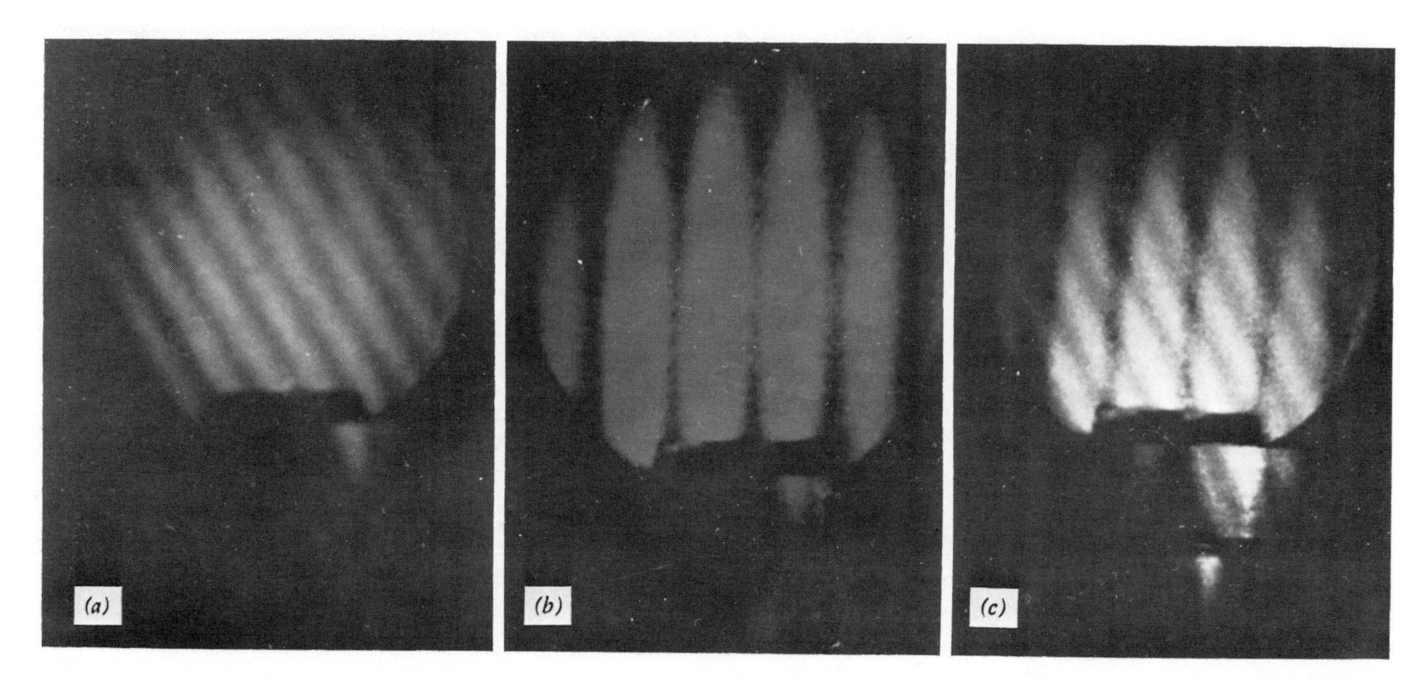

Fig. 13.21 Black-and-white photographs of color encoded double-exposure interferograms. (*a*) A black-and-white picture of a green encoded interferometer due to a rotational motion. (*b*) A black-and-white picture of a red color encoded interferogram due to a rotational motion about a different axis and recorded using a different reference beam. (*c*) A black-and-white picture of a multiplexed color encoded interferogram of (*a*) and (*b*). In color the vertical dark fringes are in green, the slanted fringes are in red, and the bright fringes are in yellow.

niques, which use coherent light. The multiplex holographic techniques described are simple, versatile, and economical. This approach to holographic fringe analysis shows great promise for comparing two or more fringe patterns to identify nodal or zero motion regions on the vibrating test object. The color encoding of intensity variations makes the task of identifying time-average fringe orders much easier and more pleasing to the eye.

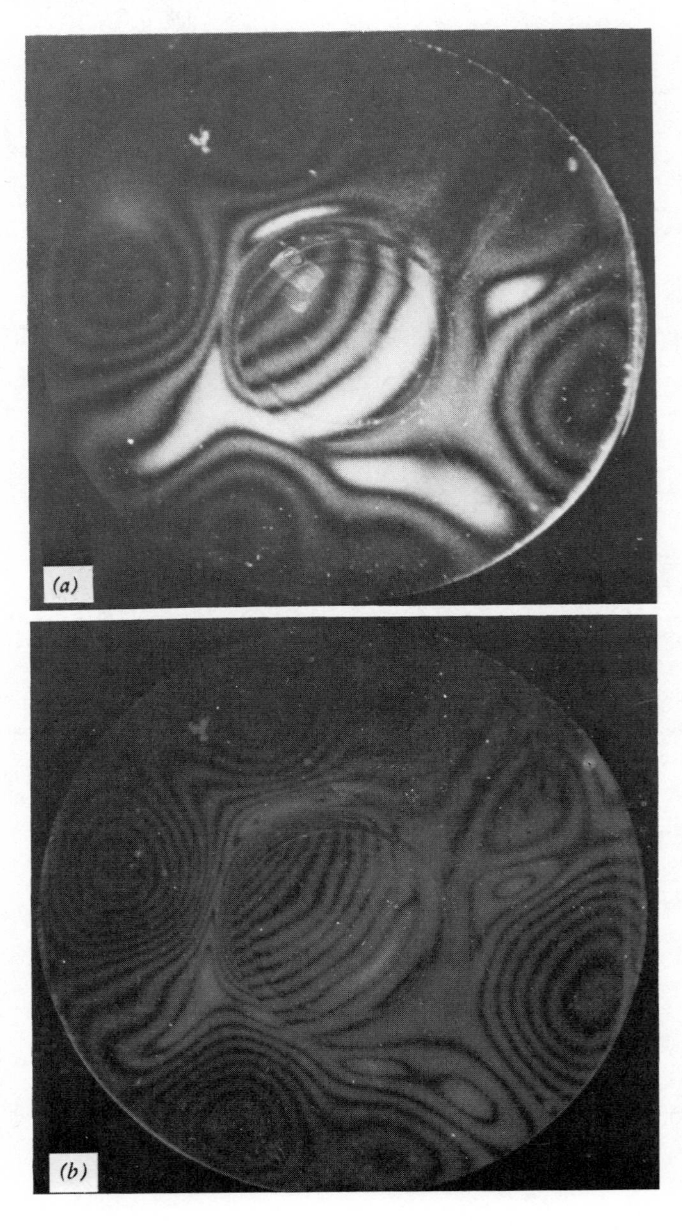

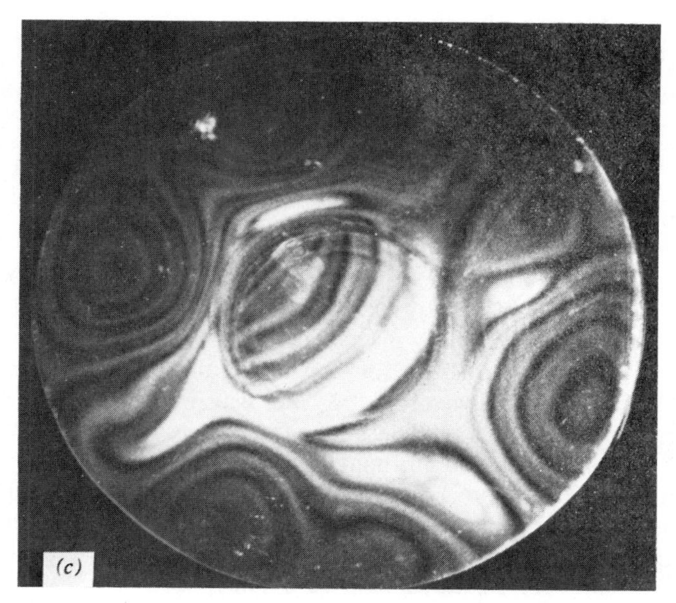

Fig. 13.22 Black-and-white pictures of color encoded time-average interferograms of a vibrating speaker. (*a*) A black-and-white picture of a green color encoded interferogram of a speaker oscillating at 2 kHz with an input voltage of 2 V p-p. (*b*) A black-and-white picture of a red color encoded interferogram of the same speaker oscillating at 2 kHz with an input voltage of 4 V p-p. (*c*) A black-and-white picture of a multiplexed color encoded interferogram of (*a*) and (*b*). In color, the dark fringes are in red, the gray fringes are in green, and the bright areas are in yellow.

13.8 REAL HOLOGRAPHIC IMAGE CONTOURING

In certain contouring applications, it is often necessary to perform the contouring at a distance from the object, that is, without actual contact with the object. Such techniques are very important in the measurement of delicate biological specimens where the surface may be so vulnerable it could easily be damaged or contaminated by physical contact with contouring gauges, and, if the surface is too soft, it may be difficult to even define the surface. Another important application may be in the preservation of ancient artifacts. In measuring the fragile archaeological finds, a minimum amount of handling is preferred, to avoid further damage to precious objects that are already disintegrating. Prior to holography, such remote contouring was done by methods such as stereophotography and photoelectric detection of displacement from focus. However, these methods require intensive manual operation; to obtain any data with reasonable accuracy requires a prohibitive amount of time, and such techniques are subject to human error. Because of its capability for reconstructing the wave front of an object faithfully, holography together with conventional interferometry have suggested several techniques for contouring applications. However, most techniques center on the visual hologram image. The most commonly used technique is perhaps the two-wavelength method described in sec. 13.4. However, this technique only provides a visual contour of the object; thus, the data may not be read automatically. With a different procedure, which we discuss in the following passage, the real hologram image projection offers a new technique for automatic contouring.

If a conventional hologram is illuminated by a conjugate reference beam, a real hologram image can be easily projected. The precision of holographic contouring depends upon the degree of accuracy of the hologram image. The holographic aberrations are fully discussed in chap. 10. It has been concluded that, in order to obtain an aberration-free hologram image, the simplest procedure is the use of plane waves of the same wavelength for construction and reconstruction. The uneven thickness variation of the recording emulsion will also cause complicated hologram image distortion, as described in sec. 11.4. Nevertheless, these distortions can be substantially reduced by the application of an index-matching liquid gate. Thus with all these precautions, a high-precision hologram image can be obtained. The precision of the holographic image is important in most holographic contouring applications.

To detect the surface boundary of a real hologram image, we can use a photodetecting probe to scan across the image boundary. If the size of the real hologram image is small compared with the hologram aperture, detection of the image surface boundary will be no major difficulty. The location of the highest output voltage from the photodetector represents the surface boundary. However, if the size of the hologram image is larger than that of the hologram aperture, it beomes very difficult to define the surface boundary, since we are dealing with a diverging reconstruction. Nevertheless, the surface boundary may be detected with an encoding technique.

One of the earliest encoding techniques, to the author's knowledge, is the one attributed to Stetson (ref. 13.23). He encodes the object by using white paint with black toner. However, this encoding process offers very limited applications, since it will deface the object. Obviously, it cannot be used for certain biological specimens and ancient artifacts. Furthermore, with this encoding technique the output signal is often difficult to define. In this section a different encoding technique is introduced by which precise real hologram image contouring may be obtained (ref. 13.24).

There are many ways of producing interference modulation (i.e., encoding) of a hologram image; however, we restrict our discussion to two of the simplest methods, namely, double exposures and fringe encoding. We first discuss the double-exposure technique; that is, the modulation fringes that are produced on the image surface boundary by sequential double holographic exposures. The construction of the double-exposure hologram is accomplished by a slightly different displacement illumination of the object. The corresponding irradiances of the sequential holographic exposures are

$$I_1(x, y) = v + |u_1(x, y)| + 2v|u_1(x, y)| \cos[kx \sin\theta + \phi_1(x, y)] \tag{13.36}$$

and

$$I_2(x, y) = v + |u_2(x, y)| + 2v|u_2(x, y)| \cos[kx \sin\theta + \phi_2(x, y)], \tag{13.37}$$

where $u_1(x, y) = |u_1(x, y)| \exp[i\phi_1(x, y)]$ and $u_2(x, y) = |u_2(x, y)| \exp[i\phi_2(x, y)]$ are the respective sequential object beams distributed on the recording medium, $v = \exp[ikx \sin\theta]$ is the oblique reference plane wave, k is the wave number, and (x, y) is the chosen coordinate system for the recording medium. The sequential object beams can be determined by the Fresnel–Kirchhoff theory, i.e.,

$$u_1(x, y) = \iint O(\xi, \eta, \zeta) E_1^+(\boldsymbol{\rho} - \boldsymbol{\xi}; k)\, d\xi\, d\eta \tag{13.38}$$

and

$$u_2(x, y) = \iint O(\xi, \eta, \zeta) \exp[-ik\xi \sin\psi] E_1^+(\boldsymbol{\rho} - \boldsymbol{\xi}; k)\, d\xi\, d\eta, \tag{13.39}$$

where $E_1^+(\boldsymbol{\xi}; k)$ is the spatial impulse response, $\boldsymbol{\xi}(\xi, \eta, \zeta)$ and $\boldsymbol{\rho}(x, y, z)$ are the chosen coordinates for the object and recording medium, $O(\xi, \eta, \zeta)$ is the object function, and ψ is the angular orientation of the illuminating beam.

Assume that this double-exposure holographic recording is linear; then the hologram transparency is proportional to the sum of $I_1 + I_2$. If this hologram is illuminated by an oblique conjugate beam, two real hologram images will be superimposedly reconstructed. Thus the composite real image irradiance is

$$I_r(\alpha, \beta, \gamma) = K[1 + \cos(k\alpha \sin\psi)]|O(\alpha, \beta, \gamma)|^2, \tag{13.40}$$

where K is a proportionality constant. From the above equation, we see that a surface interference modulation is encoded. Because the fringe spacing is determined by the displacement angle ψ, it can be controlled. However, this double-exposure encoding technique has one major drawback: the two exposures should be made under certain optimum recording conditions so that the visibility of the interference modulation is maximum. Furthermore, the small angle of orientation of the object illuminations is not always easy to maintain.

Perhaps the simplest and most effective way of producing the surface modulating fringes is by a two-point source encoding technique, as shown in fig. 13.23. The fringe encoding is derived from the two-point source illumination,

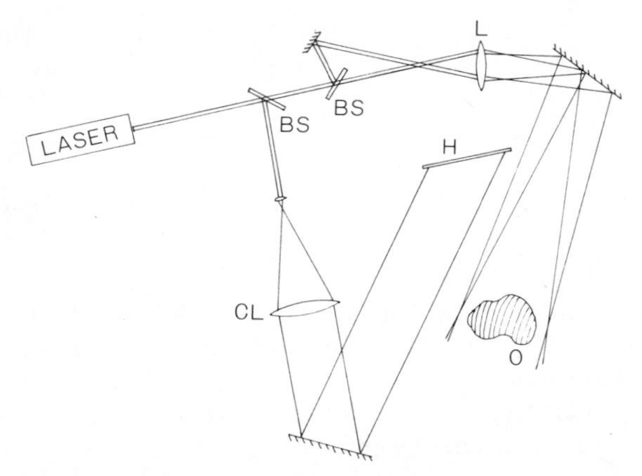

Fig. 13.23 Holographic recording of a modulated object. BS, beam splitter; L, small focal length lens; CL, collimated lens; H, hologram; O, object.

during the holographic recording. Thus the complex light field distributed on the recording medium from the object is

$$u(x, y) = \iint O(\xi, \eta, \zeta)[1 + \exp(ip\xi)]E_1^{+}(\boldsymbol{\rho} - \boldsymbol{\xi}; k)\, d\xi\, d\eta, \tag{13.41}$$

where p is the spatial frequency of the fringes. Again, if the holographic recording is linear and the hologram is illuminated by a conjugate plane wave of the same wavelength, the encoded real hologram image irradiance is

$$I_r(\alpha, \beta, \gamma) = K(1 + \cos p\alpha)\,|O(\alpha, \beta, \gamma)|^2. \tag{13.42}$$

Thus we see that the image surface boundary is modulated by a sinusoidal fringe pattern in the α direction. It should be emphasized that the fringe spacing on the hologram image can easily be adjusted by a simple technique (ref. 13.25). The resolution of the real image contouring depends upon the spatial frequency of the modulated fringes. Thus a higher spatial frequency of the modulation represents a higher contouring resolution. However, there is a practical limit, which depends upon the coarseness of the objective surface, the speckle effect, and the film-grain noise of the recording medium.

In contouring the modulated hologram image, the width of the scanning aperture of the photodetector should be smaller than one-half of the fringe spacing. In order to increase the output signal voltage, we can use a narrow rectangular slit. The photodetector scans across the real hologram image in such a manner that the narrow slit is parallel to the modulated fringes, as shown in fig. 13.24. With a repeated scanning process across the image surface boundary, high-accuracy contouring may be achieved.

For illustration, assume that a modulated real hologram image is situated in

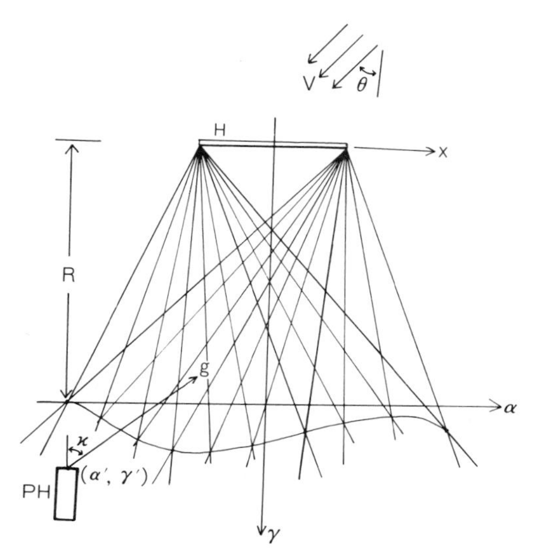

Fig. 13.24 Detection of modulated real hologram image. H, hologram; PH, photodetector.

space. Let a photodetector scan across the surface boundary with an obliquity angle κ. Thus the output signal voltage of the detector is

$$V(t) = K \iint_R I(\sigma)\, dR, \tag{13.43}$$

where $V(t)$ is a function of the position coordinate σ, $I(\sigma)$ is the position irradiance of the light field, R is the open aperture of the slit, and t is the time variable.

Equation (13.43) can be approximated by the following simplified form:

$$V(t) \simeq K_1 + K_2 \exp\{-[\gamma(t) - \gamma_0]^2\} \iint_R I(\sigma)\, dR, \tag{13.44}$$

where K_1 and K_2 are the appropriate constants, and γ_0 is the location of the hologram image surface boundary. Thus for a narrow rectangular slit with dimension $w \times h$, the output signal voltage is

$$V(t) \simeq K_1 + K_2 wh \exp[-(\gamma' - \gamma_0 + gt \cos \kappa)^2]\{1 + \cos[p(\alpha' + gt \sin \kappa)]\}, \tag{13.45}$$

where g is the velocity of the scan. Thus we see that an envelope modulated signal is obtained, as shown in fig. 13.25. Maximum modulation represents the location of surface boundary. From eq. (13.45) we see that, since the scanning

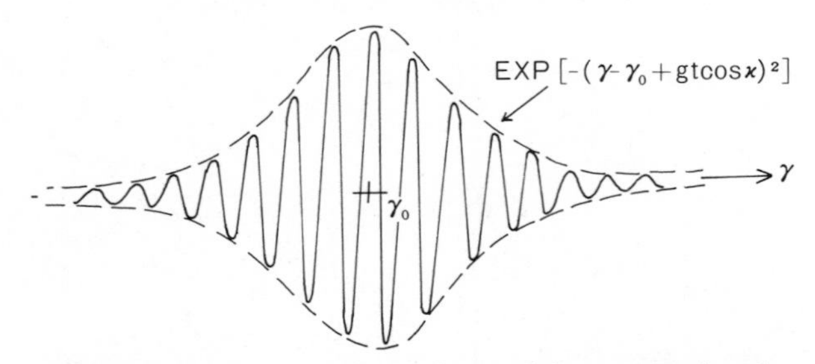

Fig. 13.25 A typical wave front of the output signal.

angle κ is smaller, and the detecting envelope is narrower, the cycles of modulation are fewer. Thus in practice, an optimum scanning angle must be found. However, the ultimate limitation of this real image contouring technique lies primarily in the coarseness of the object surface, the speckle effect of the hologram image, and in the granularity of the recording medium. Despite these limitations, high-accuracy contouring has been reported by Gara et al. (ref. 13.25).

The significance of real hologram image contouring is that, unlike the virtual image, the image may be optically touched by a photodetector. Thus the precision of surface contouring can be made very high. There is, however, a practical limitation, which depends upon the coarseness of the surface, the granularity of the film, as well as the holographic speckles. Nevertheless, the major advantage of this technique is that the contouring takes place on the real hologram image rather than on the object, and the contouring may be performed automatically. With the incorporation of a multichannel photodetector and a digital computer, a major time-saving and a high-accuracy data printout can be achieved.

REFERENCES

13.1 G. L. Tyler, "The Bistatic, Continuous-Wave Radar Method for the Study of Planetary Surfaces," *J. Geophys. Res.*, **71,** 1559 (1966).

13.2 R. K. Mueller and N.K. Sheridan, "Sound Holograms and Optical Reconstruction," *Appl. Phys. Lett.*, **9,** 328 (1966).

13.3 A. F. Metherell et al., "Introduction to Acoustical Holography," *J. Acoust. Soc. Am.*, **42,** 733 (1967).

13.4 D. Gabor, "A New Microscope Principle," *Nature,* **161,** 777 (1948).

13.5 D. Gabor, "Microscopy by Reconstructed Wavefronts," *Proc. Roy. Soc.*, ser. A, **197,** 454 (1949).

13.6 D. Gabor, "Microscopy by Reconstructed Wavefronts, II," *Proc. Phys. Soc.*, ser. B, **64,** 449 (1951)

13.7 H. M. A. El-Sum, "Reconstructed Wavefront Microscopy," doctoral dissertation, Stanford University, 1952 (available from University Microfilms, Ann Arbor, Mich.).

13.8 E. M. Leith and J. Upatnieks, "Microscopy by Wavefront Reconstruction," *J. Opt. Soc. Am.*, **55,** 569 (1965).

13.9 D. Gabor et al., "Optical Image Synthesis (Complex Amplitude Addition and Subtraction) by Holographic Fourier Transformation," *Phys. Lett.*, **18,** 116 (1965).

13.10 L. O. Heflinger, R. F. Wuerker, and R. E. Brooks, "Holographic Interferometry," *J. Appl. Phys.*, **37,** 642 (1966).

13.11 R. L. Powell and K. A. Stetson, "Interferometric Vibration Analysis by Wavefront Reconstruction," *J. Opt. Soc. Am.*, **55,** 1593 (1965).

13.12 B. P. Hildebrand and K. A. Haines, "The Generation of Three-Dimensional Contour Maps by Wavefront Reconstruction," *Phy. Lett.*, **21,** 422 (1966).

13.13 B. P. Hildebrand and K. A. Haines, "Multiple-Wavelength and Multiple-Source Holography Applied to Contour Generation," *J. Opt. Soc. Am.*, **57,** 155 (1967).

13.14 E.N. Leith and J. Upatnieks, "Holographic Imagery Through Diffusing Media," *J. Opt. Soc. Am.*, **56,** 523 (1966).

13.15 H. Kogelnik, "Holographic Image Projection Through Inhomogeneous Media," *Bell Syst. Tech. J.*, **44,** 2451 (1965).

13.16 J. W. Goodman et al., "Wavefront-Reconstruction Imaging Through Random Media," *Appl. Phys. Lett.*, **8,** 311 (1966).

13.17 F. T. S. Yu and H. W. Rose, "Coherent Optical Target Recognition Through a Randomly Turbulent Medium," *J. Opt. Soc. Am.*, **59,** 474 (1969).

13.18 H. W. Rose, T. L. Williamson, and F. T. S. Yu, "Coherent Optical Target Recognition Through a Phase Distorting Medium," *Appl. Opt.*, **10,** 515 (1971).

13.19 C. M. Vest, *Holographic Interferometry,* John Wiley, New York, 1979.

13.20 G. Gerhart, P. H. Ruterbusch and F. T. S. Yu, "Color Encoding of Holographic Interferometric Fringe Patterns with White-Light Processing," *Appl. Opt.*, **20,** 3085 (1981).

13.21 L. Rosen, "Holograms of the Aerial Image of a Lens," *Proc. IEEE,* **55,** 85 (1967).

13.22 F. T. S. Yu, A. Tai, and H. Chen, "One-Step Rainbow Holography: Recent Development and Application," *Opt. Eng.*, **19,** 666 (1980).

13.23 K. A. Stetson, "Holographic Surface Contouring by Limited Depth of Focus," *Appl. Opt.*, **7,** 987 (1968).

13.24 F. T. S. Yu and A. Tai, "Modulation-Detection Holographic Image Contouring," *Japanese J. Appl. Phys.*, **14,** 213 (1975).

13.25 A. D. Gara, R. F. Majkowski, and T. T. Stapleton, "Holographic System for Automatic Surface Mapping," *Appl. Opt.*, **12,** 2172 (1973).

14

Applications of Rainbow Holography

One major drawback of conventional holographic processes is the requirement for coherent illumination for the image reconstruction. The rainbow (white-light transmission) holographic process of Benton (refs. 14.1, 14.2) is a major improvement in holographic display. Like reflection holograms (refs. 14.3, 14.4), rainbow holograms can be viewed with a white-light source, but the reconstructed rainbow image is exceptionally bright (sec. 2.1). However, the two-step rainbow holographic process has provided few useful scientific applications, the principal reason being the complex nature of the two-step process. Before the rainbow holograms can be made, a primary hologram must be constructed. Thus if we are to apply this two-step rainbow holographic process to a double-exposure interferogram, a conventional double-exposure hologram must first be made to serve as the master hologram. From the standpoint of interferometric applications, the master interferogram already provides all the information that we are seeking. The purpose of the second step is merely to transform the display into a rainbow hologram. Therefore, for most scientific holographic applications, the additional step in the holographic process is generally not worth the extra effort.

If the rainbow holographic process can be reduced to only a single step, it will become advantageous to perform interferometry in the form of a rainbow hologram. First, it is always easier and less costly to view the hologram with a simple white-light source, especially in industrial and commercial applications. The incoherent illumination would allow the hologram to be viewed any time and any place without the use of a laser. Second, speckle effects are much less severe with rainbow holograms under white-light illumination. It is well known that the speckle effect is a serious problem both in the viewing and recording of holographic images. The speckles limit the resolution of the image by making small details difficult to define. To make matters worse, in many types of holographic interferometry, the object image and the interference fringes form in different planes. In order to record a sharp image of both the object and the fringes, it is necessary to use a small camera aperture setting to provide the necessary depth of field. This technique results in undesirably large speckles. On the other hand, the speckle effect in rainbow holograms is generally much less severe, offering better fringe visibility.

In chap. 12 we have demonstrated the feasibility of generating rainbow holograms in one step. This procedure is an easy to perform as conventional holography. In this chapter some of the rainbow holographic applications that can be implemented with this new technique (ref. 14.5) are demonstrated. A unique process of multiwavelength and multislit interferometry using the one-step technique are introduced. With this technique the effects of different physical conditions can be studied and compared. This technique provides excellent discrimination between the different sets of interferometric fringe patterns by displaying them individually or together in different colors.

There is, however, one disadvantage with the one-step process using either of the arrangements we have described in chap. 12. The field of view is restricted by the aperture of the imaging lens so that when the viewer moves too far off the optical axis, the image disappears. In order to achieve a large field of view, an imaging lens with a large aperture and small focal length is required. Thus for general display purposes, Benton's two-step technique may still prove to be more practical. However, the one-step process is just about ideal for many applications of holographic interferometry. The reason is that, for many interferometry applications, it is generally not necessary to have a large field of view. In fact, with many applications, the observation should be confined to the optical axis if the interference fringes are to provide accurate quantitative measurements.

In order to perform holographic interferometry, either of the arrrangements we describe can be used. Both techniques are utilized in this chapter. For most applications, it is generally not a serious detriment to have a pseudoscopic image reconstruction. In fact, the pseudoscopic arrangement is generally the preferred technique because of its advantages of larger field of view and ease of setting up.

14.1 RAINBOW HOLOGRAPHIC INTERFEROMETRY

Holographic interferometry is one of the most interesting and important applications of holography (ref. 14.6). However, in most cases, coherent illumination is required for the reconstruction process. Experience indicates that conventional holographic interferometry has the following shortcomings: first, a coherent source is required for reconstruction and second, significant speckle noise will be present.

We have proposed a one-step rainbow holographic interferometric process in which these two shortcomings of conventional holographic interferometry may be alleviated (ref. 14.7). As noted, making a rainbow hologram requires recording a real holographic image of an object through a narrow slit. Consequently, if the rainbow hologram is illuminated by a monochromatic light source, a real image is formed, which is limited in its positions of viewability. However, if the rainbow hologram is illuminated by a white-light source, the hologram image of the slit will disperse the light into a full view holographic image in rainbow colors. Thus the basic reconstruction process of a rainbow hologram is to reproduce a real image of the slit between the hologram image of the object and the observer. We have shown in chap. 12 that the same goal can be achieved by a one-step rainbow holographic recording process.

The optical system for one-step rainbow holographic interferometry is similar to that of conventional holographic interferometry, except that an imaging lens and a narrow slit are inserted between the object and the recording plate, as shown in fig. 12.4. Thus a virtual image of the slit is formed between the object and the slit, but a real image of the object is formed behind the recording plate. The depth of the real image in its relation to the slit is inverted by the lens. In readout, we require that the slit image be formed downstream from the object image. Therefore, the conjugate image of the hologram must be used, and the image that is viewed by the observer through the slit image is pseudoscopic. By a slight modification of fig. 12.4, an orthoscopic holographic image can be observed, as shown in fig. 12.5. However, in holographic interferometry the required measurements can be carried out equally well whether the image is orthoscopic or pseudoscopic.

A double-exposure rainbow holographic interferogram is made using an aluminum strip under stress for an object. A first exposure is made when the object is under stress, and a second exposure is made when the mechanical stress is increased slightly. Figure 14.1*a* shows the virtual image, formed by white-light illumination. In order to compare this result with that obtained by the conventional technique, the experiment is repeated exactly as before without use of the slit. The readout is done with coherent illumination, resulting in the image shown in fig. 14.1*b*. Comparison of figs. 14.1*a* and 14.1*b* shows the rainbow holographic interferogram to provide higher contrast fringes, and less speckle noise. Also, the rainbow image is far brighter even though the illumination (mW/cm^2) on the hologram is lower.

In the application to time-average holographic interferometry, a rainbow hologram is made by the same optical setup (fig. 12.4) using a loudspeaker about 5 cm in diameter as a vibrating object. The speaker is driven by a sinuosidal signal of about 1350 Hz. Figure 14.2*a* shows the result of a time-average rainbow holographic interferogram. In order to compare this with the conventional time-averaging technique, a hologram is made of the same speaker with the same optical setup but without the slit. The image formed with coherent illumination is shown in fig. 14.2*b*. Comparison of figs. 14.2*a* and 14.2*b* shows again that the rainbow method results in higher contrast fringes and less speckle; in particular, the grating structure at the center portion of the speaker for the conventional case is completely washed out by the speckle noise, but it is much clearer in the rainbow case. Thus we expect a rainbow holographic interferogram to offer somewhat better resolution than a conventional holographic interferogram.

The one-step rainbow process can be easily applied to most cases of holographic interferometry, yielding the advantages of white-light readout, brighter images, less speckle noise, and better resolution. Although orthoscopic image interferograms can be easily obtained, pseudoscopic images will usually serve just as well. The two-step rainbow process can produce the results shown here; however, it is more cumbersome and requires a two-step coherent recording process. Therefore, a higher level of speckle noise in the holographic interferogram is unavoidable.

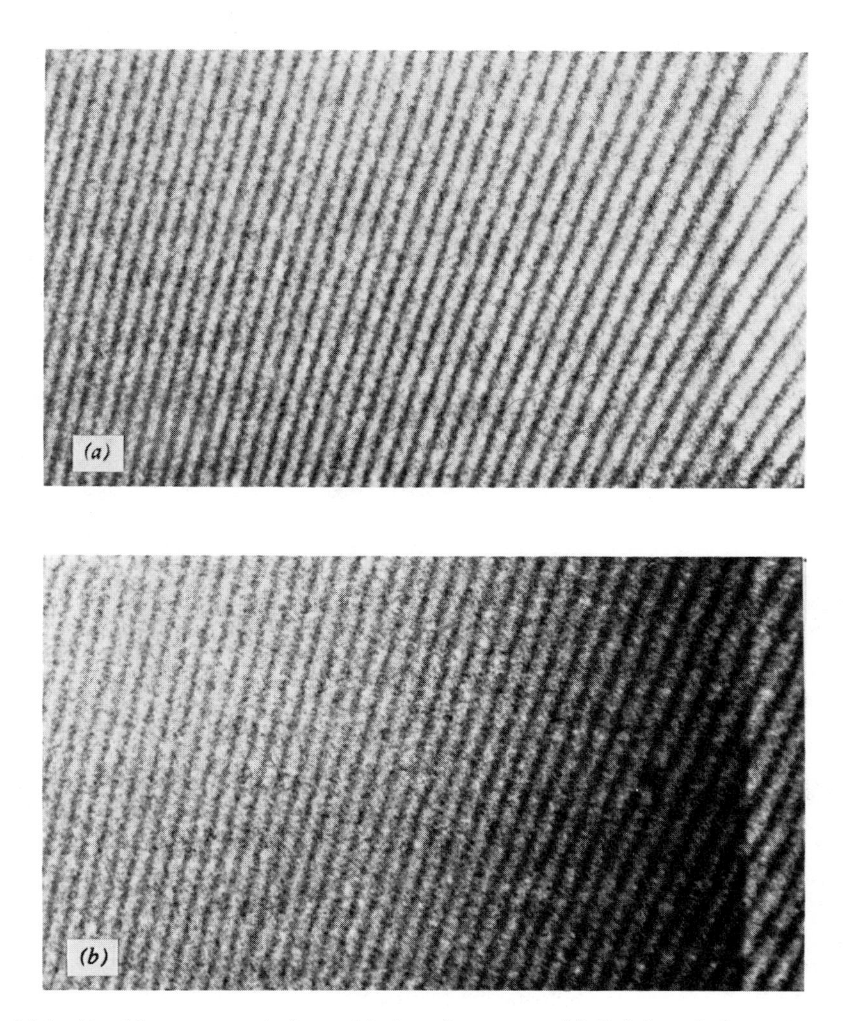

Fig. 14.1 Double-exposure holographic interferograms. (*a*) Rainbow holographic interferogram. (*b*) Conventional holographic interferogram.

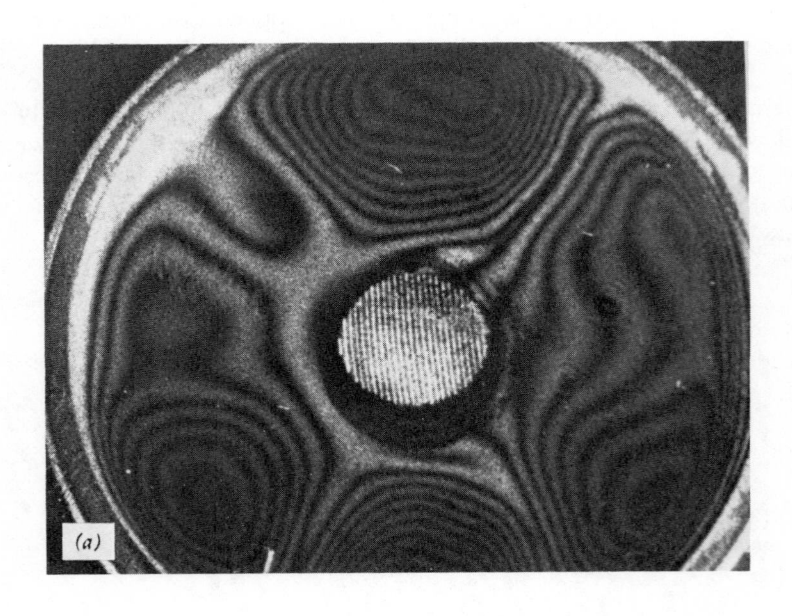

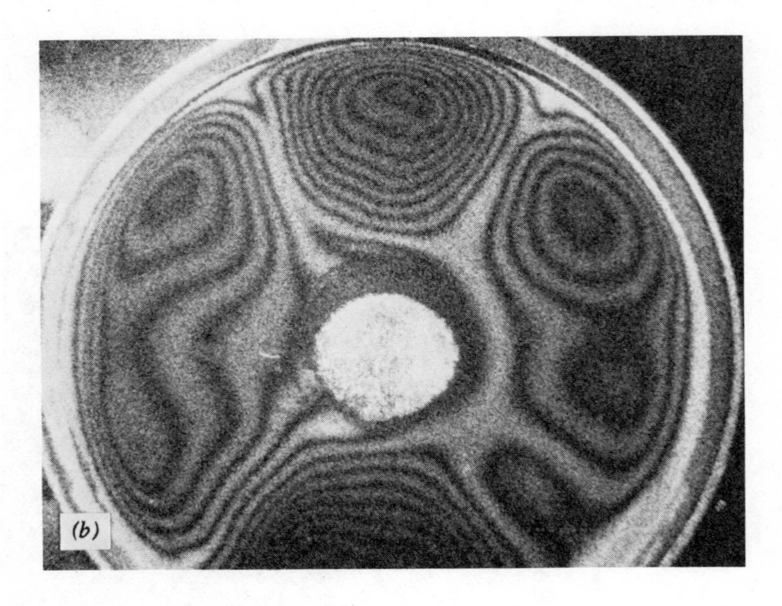

Fig. 14.2 Time-average holographic interferogram. (*a*) Rainbow holographic interferogram. (*b*) Conventional holographic interferograms.

14.2 MULTIWAVELENGTH RAINBOW HOLOGRAPHIC INTERFEROMETRY

In Sec. 12.3 we have demonstrated that a true color image can be generated by a one-step rainbow holographic process. By the same technique, multiwavelength rainbow holographic interferometry can be performed (refs. 14.5, 14.8). Interesting and unique results can be obtained by this technique. The advantage of this multiwavelength technique is that it allows the comparison of different physical effects (e.g., stress and heat) by displaying the interference fringes due to the different effects individually or together in different colors. The arrangement used in multiwavelength interferometry is shown in fig. 14.3. The holographic recording is basically the same as the arrangement that we have described in the previous section, except that two lasers emitting at different wavelengths are utilized. To demonstrate the effect, we affix a plate of aluminum onto a gimbal mount. Using only one of the laser beams (Ar^+ 5145 Å), an exposure is made of the object. We then tilt the mount slightly vertically and a second exposure is made using the same laser source. Another exposure is made with the other laser source (He-Ne 6328 Å). The mount is then tilted slightly about the 45° axis and a fourth exposure is made using the second laser source. Thus we have in the holographic plate two interferometric holographic images, one recorded in green color corresponding to the vertical tilt and the other recorded in red color corresponding to the 45° tilt. The two holograms multiplexed in the plate have different principal spatial frequencies corresponding to $\sin\theta/\lambda_1$ and $\sin\theta/\lambda_2$, where λ_1 and λ_2 are the wavelengths utilized, and θ is the reference angle. After the holographic plate is developed, the hologram is reconstructed with a collimated white-light source. Since the spatial frequencies of the multiplexed holograms are different, two transversely displaced slits are reconstructed, each with the appropriate color disperion as illustrated in fig.

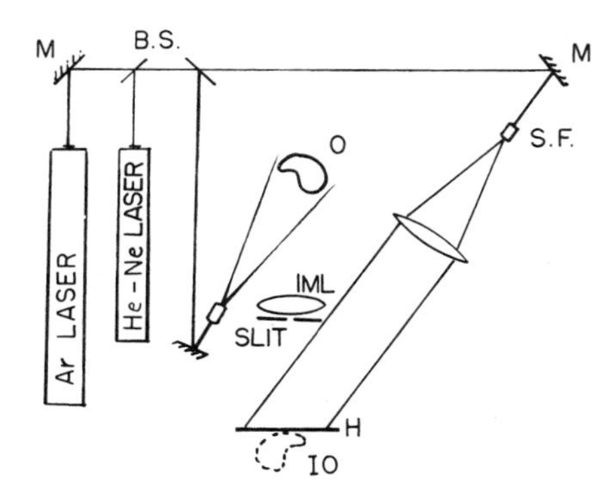

Fig. 14.3 Optical arrangement for multiwavelength holographic interferometry.

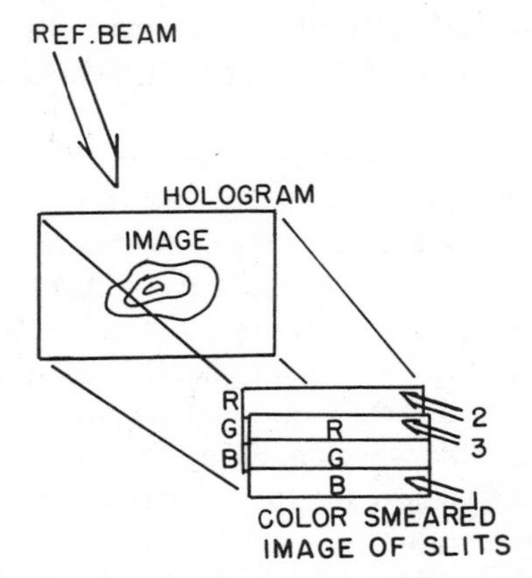

Fig. 14.4 Reconstructed image of slits and various positions of observation.

14.4. When the viewer observes the image at one edge of the color smeared slit image at position 1, he or she can see only the image with the inference fringes corresponding to the 45° tilt in bluish-green color. A black-and-white picture of the reconstructed image viewed at this position is shown in fig. 14.5*a*. If the viewer moves to position 2 at the other end of the color smeared slit image, he or she can see only the image and fringe pattern corresponding to the vertical tilt in red color, as shown in fig. 14.5*b*. When observing from position 3 in the middle of the color smeared slit image, both holographic interferograms will appear together exactly superimposed, as shown in fig. 14.5*c*, with the fringes corresponding to the vertical tilt in green and the fringes corresponding to the 45° tilt in red. This technique allows direct comparison between the two sets of fringes, and it provides excellent discrimination between the two sets of fringes, since they are displayed in different colors.

This multiwavelength technique can also be utilized for vibration analysis in time-average holographic interferograms. To demonstrate this application, we use a small 5-cm speaker as the object. The arrangement shown in fig. 14.3 is used again. When exposed to the green light of an argon laser, the speaker is driven by a 6-kHz sinusoidal signal. Then a 7-kHz signal is used when exposing the plate to the red light of the He-Ne laser. After the plate is developed, the holographic images are reconstructed with a collimated white light. At position 1, the viewer can see only the holographic images of the speaker with the mode structure corresponding to the 7-kHz oscillation in bluish-green color, as shown in fig. 14.6*a*. At position 2, the viewer can see the mode structure corresponding to 6-kHz oscillation in red, as shown in fig. 14.6*b*. At position 3, the viewer can see both mode structures simultaneously, with the 6-kHz structure appearing in green and the 7-kHz structure in red. A black-and-white picture of the superimposed color images is shown in fig. 14.6*c*.

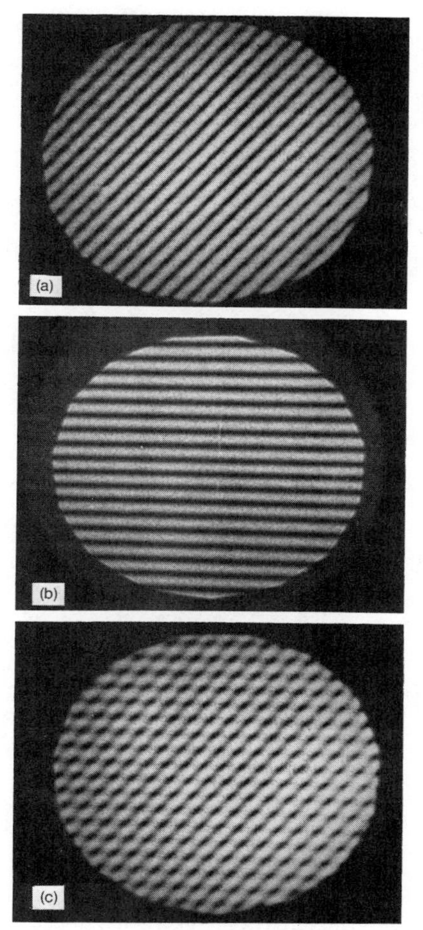

Fig. 14.5 Black-and-white pictures of reconstructed images observed at different positions. (*a*) Fringe pattern in bluish-green due to 45° tilt observed at position 1. (*b*) Fringe pattern in red due to vertical tilt observed at position 2. (*c*) Superimposed fringe patterns of 45° tilt in red and vertical tilt in green observed at position 3.

14.3 MULTISLIT RAINBOW HOLOGRAPHIC INTERFEROMETRY

In the previous section we have shown that interference fringes due to different physical effects can be displayed individually or together in different colors by a multiwavelength technique. However, the one-step holographic technique suffers one serious drawback when applied to holographic interferometry: the presence of the slit causes a loss of parallax along the axis perpendicular to the direction of the slit. For some interferometric applications, such as time-average

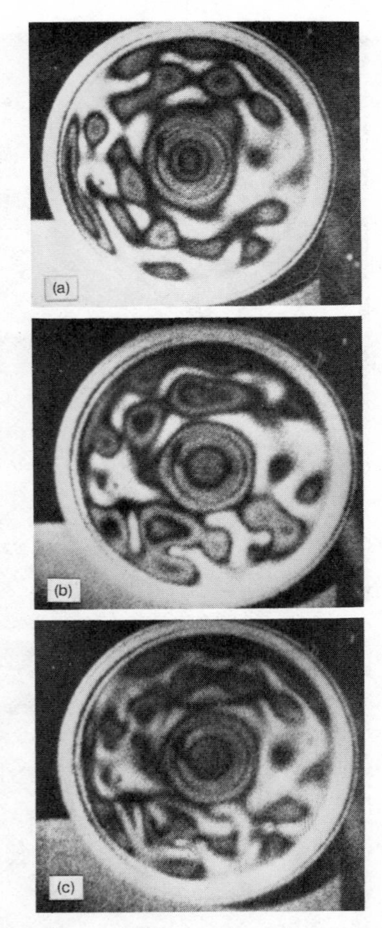

Fig. 14.6 Black-and-white pictures of reconstructed images of time-average hologram. (*a*) Mode structure in bluish-green under 7-kHz excitation observed at position 1. (*b*) Mode structure in red under 6-kHz excitation observed at position 2. (*c*) Superimposed mode structures of 7-kHz in red and 6-kHz in green observed at position 3.

and contouring, the loss of parallax along one direction has no adverse effect. However, for other applications, such as displacement measurements, the loss of parallax along one direction greatly reduces the accuracy of the interferometric measurements. In general, fringe measurements from at least two perspectives are required along each axis. The one-step rainbow holographic technique allows measurements from only a single perspective along one of the axes. To remedy this deficiency, we introduce in this section a variation of the one-step technique using a double-slit arrangement (refs. 14.5, 14.9).

The double-slit holographic recording arrangement is illustrated in fig. 14.7. Two parallel slits are placed at the top and bottom of the imaging lens. The

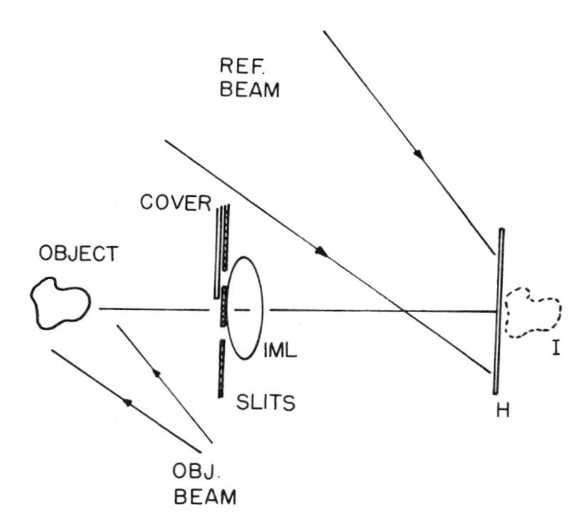

Fig. 14.7 Multislit one-step rainbow holographic recording arrangement.

object image is projected near the holographic plate. A converging reference beam is used to illuminate the hologram from above. Four separate exposures are made. With the object at rest in the original position, a first exposure is made with one of the slits covered up. Then a second exposure is made with the other slit covered. The object is then subjected to whatever physical effect is desired. The third and fourth exposures are then made with each of the slits alternatively covered. After the plate is developed, two sets of double-exposure interferograms are produced, each taken from a different slit position. If the real holographic images are reconstructed with a diverging white-light point source, two color smeared slit images are formed, as shown in fig. 14.8. When the hologram is viewed at position 1, the interference fringes as seen through the top slit position are observed. From position 3, the fringes are through the bottom slit position. When viewing the hologram at position 2, the interference fringes corresponding to the two slit positions are seen, superimposed and displayed in

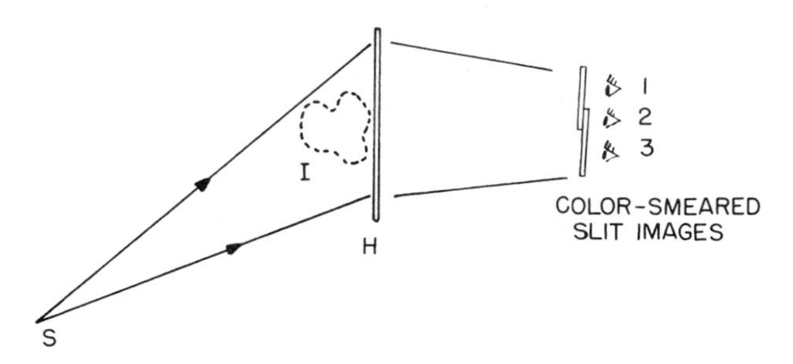

Fig. 14.8 Holographic reconstruction with a white-light source: S, white-light point source; H, hologram; I, reconstructed image; 1, 2, and 3, viewing positions.

different colors. Using this arrangement, fringe measurements can be made from more than one perspective along both axes. In addition, the result obtained with the multislit method is visually very similar to that produced by the multiwavelength technique. Both techniques produce two sets of holograms with different average spatial frequencies. With the multislit arrangement, the offset angle is varied, while with the multiwavelength technique, the wavelength is changed. In figs. 14.9*a* and 14.9*b*, we show the interferometric images obtained with the double-slit technique; we note that the fringe location changes with the perspective.

We have presented a variation of the one-step rainbow holographic technique that allows fringe measurements from more than one perspective along both axes. The new arrangement effectively remedies the most serious drawback of the one-step technique as applied to holographic interferometry. Similar to the multiwavelength technique, the multislit arrangement makes use of spatial frequency multiplexing to store and display two sets of fringe information. In the

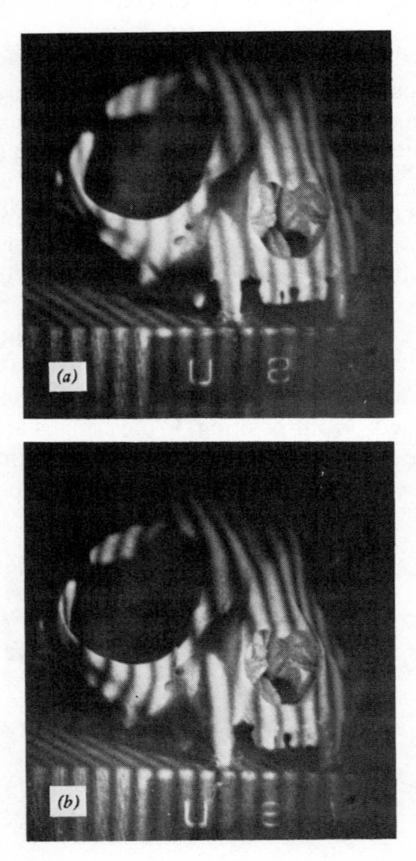

Fig. 14.9 Interferometric images obtained with double-slit holographic interferometry. (*a*) and (*b*) Taken from different slit positions.

multiwavelength technique, fringe patterns due to different physical effects are recorded, while in the multislit technique, data concerning the fringe patterns from two different perspectives are kept.

14.4 CONTOUR GENERATION BY ONE-STEP RAINBOW HOLOGRAPHIC TECHNIQUE

Let us apply the one-step rainbow process to holographic contouring.

There are two basic holographic contouring techniques. One makes use of liquids or gases with different refractive indexes and the other makes use of two different wavelengths from a coherent light source. We shall select the two-wavelength technique for our demonstration, since in our judgment it is a better test for the one-step process. For simplicity, we utilize the orthoscopic arrangement for this application, as shown in fig. 14.10. The arrangement is almost identical to that introduced by Varner (ref. 14.10). The only difference is the insertion of a slit aperture. The object is illuminated normally with a beam splitter and its image is projected to a plane near the holographic plate. The diffracted plane wave from the grating is used as the reference so that the spatial frequencies of the holograms generated with different wavelengths are independent of the wavelengths. That is, $\sin \theta_1/\lambda_1 = \sin \theta_2/\lambda_2$, where λ_1 and λ_2 are the wavelengths used in the generation of the holograms and θ_1 and θ_2 are the corresponding reference angles. Two exposures are made using two different wavelengths. Since the spatial frequencies of the two holograms in the photographic plate are the same, the two reconstructed images are exactly superimposed transversely. However, the images are displaced longitudinally along the optical axis. They interfere to produce contour fringes with spacing corresponding to a longitudinal distance of $D = \lambda_1\lambda_2/(2|\lambda_2 - \lambda_1|)$. In our experimental demonstration, we make use of the 5017 and 4965 Å lines of an argon

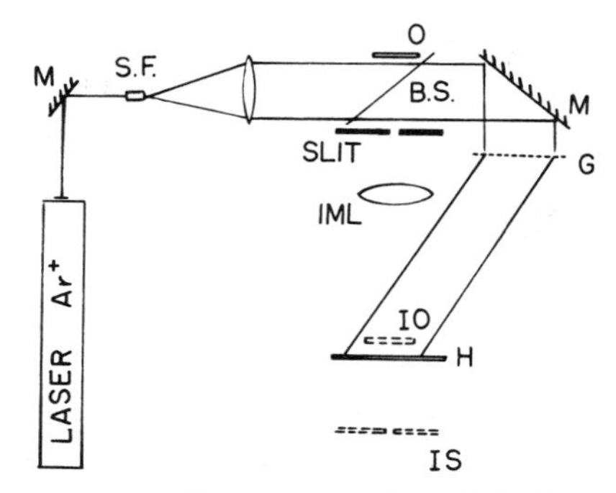

Fig. 14.10 Optical arrangement for two-wavelength holographic contouring: B.S., beam spitter; G, grating; IML, imaging lens; O, object; IO, image of object; IS, image of slit; H, hologram.

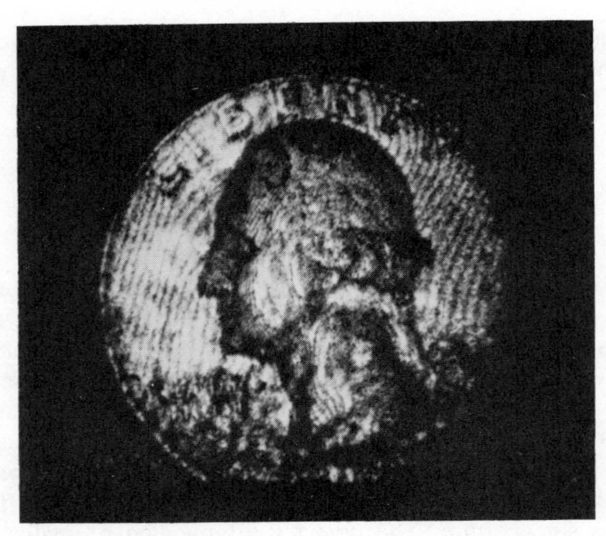

Fig. 14.11 Contoured image of a coin reconstructed with a white-light source.

laser, corresponding to a contour spacing of 24 μm. In fig. 14.11, a rainbow holographic contour image of a coin reconstructed with a tungsten arc lamp is shown.

14.5 ARCHIVAL STORAGE OF COLOR FILMS BY RAINBOW HOLOGRAPHIC PROCESS

A major problem of the film industry, which has long been unresolved, is that of the archival storage of color films. Organic dyes used in color films usually become unstable under prolonged storage, resulting in a gradual fading of the color film. Although several techniques currently exist to preserve the color image, they all possess certain disadvantages. One commonly used technique is the preservation of the color image on three different rolls of black-and-white film by the repetitive recording of the three primary color images. Reproduction of the original color image requires the use of three primary color projectors advancing synchronously and with fine precision. The major drawbacks to this technique are that the storage volume of the film is tripled, and that the technique is generally very elaborate and expensive to operate.

An alternative method is using a color Fourier hologram technique (ref. 14.11). However, this method also possesses several drawbacks. For example, the technique requires three critically aligned coherent sources for construction and reconstruction, and coherent artifact noise is unavoidable.

In a recent paper (ref. 14.12), we have demonstrated that the archival storage of color films might be achieved by a one-step rainbow holographic process. Although this technique possesses several advantages over the conventional techniques, it still suffers from a number of drawbacks. First, the presence of a narrow slit in the rainbow holographic recording causes a marginal loss of resolution in the direction perpendicular to the slit. Second, there is an inherent color blurring present in the rainbow hologram image (sec. 12.5). Third, because a coherent source is required for rainbow hologram construction, coherent noise in the hologram image cannot be avoided. However, we have shown that a high-resolution rainbow holographic process can be achieved by a cylindrical lens technique (sec. 12.4). We have shown that a higher image resolution, less speckle noise, and a lower color blur image can be obtained.

In this section, we apply a one-step rainbow holographic technique, utilizing a cylindrical lens, to the archival storage of color films (ref. 14.13). This technique eliminates the use of a diffuser and a narrow slit during the holographic construction, and thereby makes it possible to obtain a high-resolution holographic image. We show several results, as applied to color transparencies, for the archival storage of color films.

The optical arrangement for the construction of color rainbow holograms utilizing a cylindrical lens is illustrated in fig. 14.12. This system is similar to the conventional one-step rainbow holographic process (sec. 12.2), except that the diffuser and narrow slit are removed. Instead, a cylindrical lens is inserted immediately behind the object transparency, which focuses the object image onto the imaging lens. A He-Ne laser emitting a wavelength of 6328 Å, and two argon lasers with wavelengths of 5145 and 4880 Å are utilized for the holo-

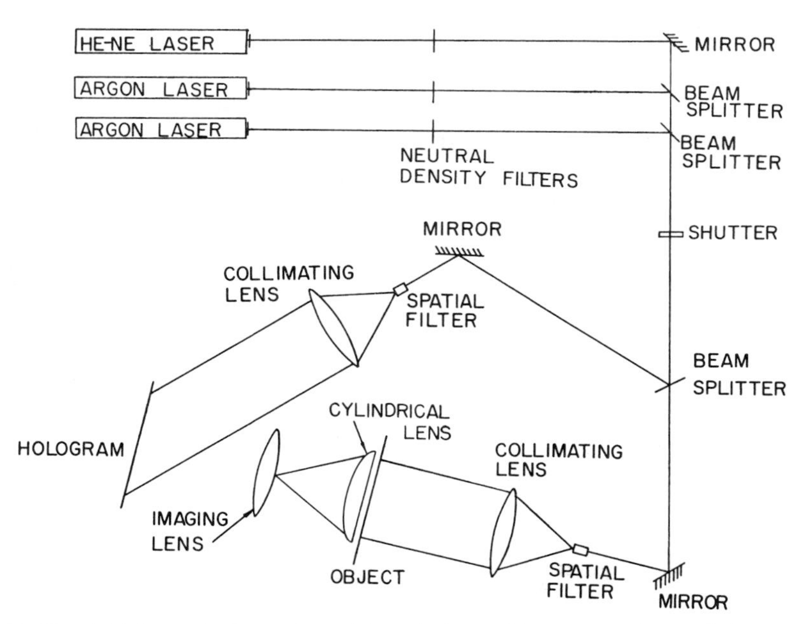

Fig. 14.12 A high-resolution color rainbow holographic construction utilizing a cylindrical lens technique.

graphic construction. Each laser beam is attenuated individually with neutral density filters to compensate for the differences in the light intensities as applied to the spectral response of the recording medium. For simplicity, a collimated reference beam is used in the holographic construction. We have noted, in sec. 12.4, that the optimum position for the holographic recording plate is on the image plane. With the intensitites of the three color lasers adjusted to the appropriate photographic response, a single holographic exposure can be accomplished. After the exposed holographic plate is developed, a color rainbow hologram has been constructed.

Reconstruction is done using a collimated white-light illumination at a conjugate angle, as shown in fig. 14.13. The optimum holographic reconstruction is obtained when the rainbow hologram image is formed using an optical setup identically symmetric to the optical setup used in the holographic construction. When the rainbow hologram is illuminated by a conjugate collimated white-light, each of the rainbow subholograms formed by the different wavelengths will be diffracted into continuous spectral bands of images. However, by placing a narrow slit (about 3 to 6 mm; see sec. 12.4) at the back focal length of the imaging lens, only the appropriate wavelengths are allowed to propagate through the narrow slit, thus allowing a true color hologram image to be reconstructed at the output plane. One advantage to using the identically symmetric optical setup for hologram image reconstruction is that light diffracted by the hologram simply retraces the light path of the holographic construction, minimizing the hologram image aberration.

The laser wavelenth selection for construction of color holograms has been discussed by Collier, et al. (ref. 14.14), based on the sensations of color that can be produced by three selected laser lines and the power requirement for each laser to produce an equal energy white for human observers. Because the human eye has a low luminous efficiency in the blue region of the visible spectrum, a large power requirement for the blue wavelength is found in practice. However, the objective of archival storage of color films is to reproduce the color image

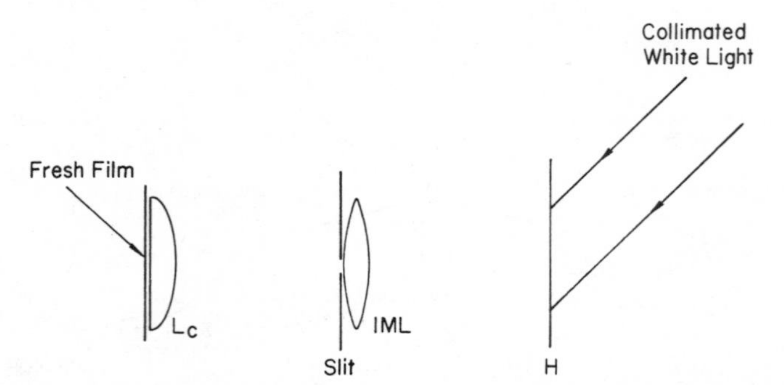

Fig. 14.13 Color rainbow holographic reconstruction with a collimated white-light. H, rainbow hologram; IML, imaging lens; L_c, cylindrical lens.

onto another color film. Thus the power requirement should be adjusted to the spectral response of the color film instead of to the human eye.

To obtain a color balance with the rainbow holographic process, a multicolor test pattern is used. Different tests for this one-step rainbow holographic process, using Kodak 649F plates, are performed. A 75-W xenon arc lamp is used for the reconstruction process shown in fig. 14.13. The color hologram images are recorded on Ektachrome and Kodachrome daylight films. Figure 14.14 shows a black-and-white photograph of a color holographic image obtained with the white-light reconstruction. The upper region of the test-pattern hologram image is green, the lower left corner is red, the lower right is blue, the center region is white, and the other regions between the primaries are the secondary colors.

The artifact noise appearing in the reconstructed test pattern is primarily due to the uncleanness and the quality of the optical components and equipment used.

Once the color balance requirement of the rainbow holographic recording is achieved, the application to the archival storage of color films can be made. Figures 14.15*a* and 14.15*b* show a set of black-and-white photographs of the original color transparency and a rainbow hologram image obtained by this one-step technique. The coherent artifact noise that appeared in fig. 14.15*b* is primarily due to the uncleanness of the optical system, which in principle can be reduced. From this figure we notice the varying shades of color in the background, due primarily to the inhomogeneity of the illuminating and reference beams. This variation in background color can be corrected with a better setup of the holographic system. As compared with the original color transparency of fig. 14.15*a*, there is, however, a degree of color misregistration with the rainbow holographic technique. This color misregistration can be minimized

Fig. 14.14 A black-and-white photograph of a color rainbow hologram image of a color test pattern. In the color image the upper region is green, the lower left corner is red, the lower right corner is blue, and the center region is white. The coherent artifact noise is due to the uncleanness and quality of the optical components and equipment used.

Fig. 14.15 (*a*) A black-and-white photograph of the original color transparency. (*b*) A black-and-white photograph of a color hologram image obtained by the rainbow holographic process. Note the resolution of the lettering on the tires of the hologram image.

with a finer adjustment of the optical parameters. The original object transparency is chosen not only for the test of color reproduction but also as a means of testing the hologram image resolution, In comparison of figs. 14.15*a* and 14.15*b*, we see that the lettering on the tires of the color hologram image is exceptionally good.

Finally, we would like to point out that it is possible to produce a high-quality color holographic image through the one-step rainbow holographic process utilizing a cylindrical lens technique; however, there are still several problems that remain to be solved. For example, the coherent artifact noise and the color blur in the rainbow holographic image still cannot be totally eliminated, and the operating procedure to achieve a true color hologram image is still relatively elaborate and expensive. Therefore, the practical application of this rainbow holographic technique to the archival storage of color films is still questionable.

14.6 SPATIAL FILTERED PSEUDOCOLOR HOLOGRAPHIC IMAGING

Virtually all research in optical image processing to date has been limited to intensity-modulated black-and-white images. The human eye, however, perceives color as well as brightness, and, therefore, the processing techniques should eventually include color coded images as well. In addition, human observation can generally differentiate the variation in colors better than gray levels. Thus an encoded color image can provide an easier discrimination for visual observation.

In this section we discuss a technique for a coherent optical pseudocolor encoding that utilizes spatial filtering with a one-step rainbow holographic process (refs. 14.15, 14.16). This pseudocolor encoding technique is achieved by sequential spatial filtering with coherent sources of different colors. The filtered image is stored in a multiplexed rainbow hologram. Once the multiplexed rainbow hologram is made, the pseudocolor of the hologram image can be viewed through the hologram slit image by simple white-light illumination. The color of the hologram image is derived from the white-light source instead of from color dyes. Thus the use of spatial filters provides a simple and versatile encoding technique, for which a wide range of possible pseudocolor encoding operations can be performed. In addition, the encoding system also provides the capability of image filtering in the spatial frequency plane.

In the viewing of the reconstruction process, the pseudocolor holographic image provides easy color discrimination and also produces an image that is easily accommodated in an electronic system. For example, a color television camera can easily pick up the pseudocolor hologram image and encode it for transmission or storage.

This pseudocolor encoding technique is also capable of producing a wide range of pseudocolor for the coded images. By varying the viewing angles, we can selectively observe combinations of the coded images in different colors. This unique feature cannot be implemented by the traditional (optical or computer) encoding technique, without additional processing.

Consider a holographic pseudocolor encoding imaging system using the multiwavelength technique, as shown in fig. 14.16. We see that the pseudocolor encoding system is a one-step rainbow holographic recording in tandem with a multiwavelength coherent optical spatial filtering system. If we place a black-and-white object transparency $s(x, y)$ at the input plane p_1, then at the back focal plane of the transform lens L_1 the complex light distribution is

$$S(p, q) = C \exp\left[\frac{f}{2k}\left(\frac{1 - v}{v}\right)(p^2 + q^2)\right] \cdot \iint s(x, y) \exp[-i(px + qy)]\, dx\, dy, \qquad (14.1)$$

where C is a complex constant, $k = 2\pi/\lambda$ is the wave number, λ is the wave-

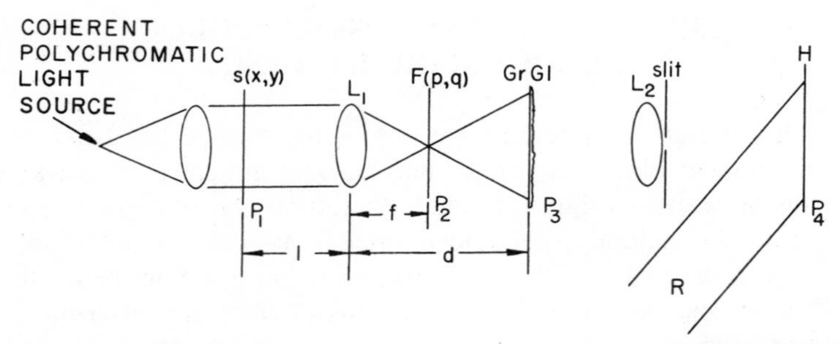

Fig. 14.16 Optical arrangement of spatial filtering pseudocolor holographic imaging. BS, beam splitter; M, mirror; SF, spatial filter; $s(x, y)$, input signal transparency; L_1, transform lens; f, focal length; GrGl, ground glass; L_2, imaging lens; R, reference beam; H, holographic plate.

length of the coherent source, f is the focal length of the transform lens, $v = f/l$, l is the distance between the input plane P_1 and the transform lens, and (p, q) is the spatial frequency coordinate system. Except for a quadratic phase factor, $S(p, q)$ is the Fourier transform of the input object transparency. Thus the encoding system is capable of performing complex spatial filtering in the spatial frequency plane P_2. At a distance $d = fl/(l - f)$ behind the transform lens L_1, an inverted image $s(mx, my)$ is formed at the diffuser plane P_3, where $m = d/l$ is the lateral magnification. The image irradiance behind the diffuser is, therefore,

$$I(x, y) = K|s(mx, my)|^2, \tag{14.2}$$

where K is a proportionality constant. For simplicity, let $m = 1$.

In pseudocolor encoding we insert a complex spatial filter $H(p, q)$ in the spatial frequency plane P_3; the image irradiance distributed on the diffuser is

$$I(x, y) = K|s(x, y)| * |h(x, y)|^2, \tag{14.3}$$

where $*$ denotes the convolution operation and $h(x, y)$ is the impulse response of the complex filter $H(p, q)$. If the rainbow holographic recording is sequentially recorded with different spatial filters for different wavelengths of the light source, we obtain an encoded pseudocolor hologram. For example, if the first exposure is made with the red light together with a spatially encoded filter $F(p, q)$, and the second and third exposures are then made with the green and blue lights with spatial filters $G(p, q)$ and $H(p, q)$, respectively, then a multiplexed multiwavelength rainbow hologram is recorded. If this rainbow hologram is illuminated by a conjugate collimated white light, the composite hologram images of $s(x, y) * f(x, y)$ in red, $s(x, y) * g(x, y)$ in green, and $s(x, y) * h(x, y)$ in blue can be viewed through the hologram image slits, as shown in fig.

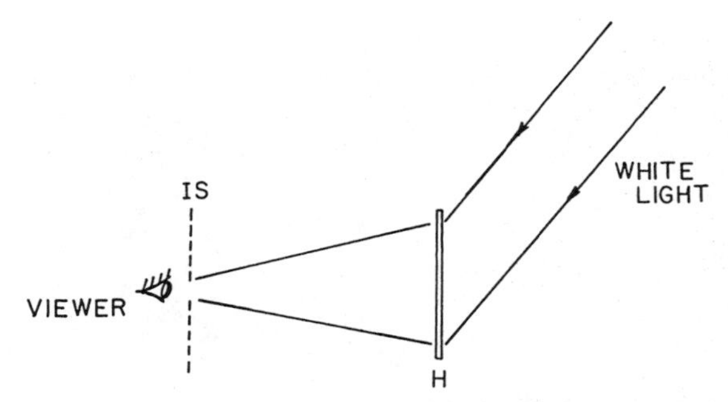

Fig. 14.17 Direct viewing of pseudocolor hologram image. H, hologram; IS, image of slit.

14.17, where $f(x, y)$, $g(x, y)$, and $h(x, y)$, are the impulse responses of the spatially encoded filters $F(p, q)$, $G(p, q)$, and $H(p, q)$, respectively, and $*$ denotes the convolution operation.

Thus a reconstructed pseudocolor hologram image can be observed by placing the eye at the slit image plane. If we view the pseudocolor hologram image at the exact location of the original slit, the pseudocolor will appear in the original encoded colors. However, if we move transversely off the location of the original slit, the pseudocolor of the hologram image will appear in different shades of colors. Moreover, if we observe further off the original slit location, one or two of the filter encoded hologram images may disappear from observation. This loss of the subhologram images at some transversal observation may be useful in the pseudocolor holographic imaging, since it allows us to observe the different pseudocolor coded hologram images all at once, or different combinations of two hologram images, or even a single encoded image.

In the experimental demonstrations, we use the red of a He-Ne laser (6328 Å) and the green of the argon laser (5145 Å) as light sources. A slit width of about 1.5 mm is used and the diffused image is focused about 1 cm in front of the holographic plate. A resolution target is used in our first experiment. The first exposure is made with the He-Ne laser as the light source and a high-pass filter at the spatial frequency plane P_2. A second exposure is then made with the argon laser together with a low-pass filter at P_2. After the holographic plate is developed, the hologram image is reconstructed with a collimated white light, illuminating at the conjugate reference angle. When viewing the hologram image at the position of the original slit, we can observe a true color coded hologram image, as shown in the black-and-white picture of fig. 14.18. If we observe at a position slightly off that of the original slit, a different color encoded hologram image can be seen. If we move farther off transversely in one direction, the high-passed hologram image disappears, leaving only the low-passed image. However, if we move in the opposite direction, only the high-passed hologram image can be seen. In a second experiment, we apply the same spatial filterings to an aerial radar image transparency, with red for the low-pass filter and green

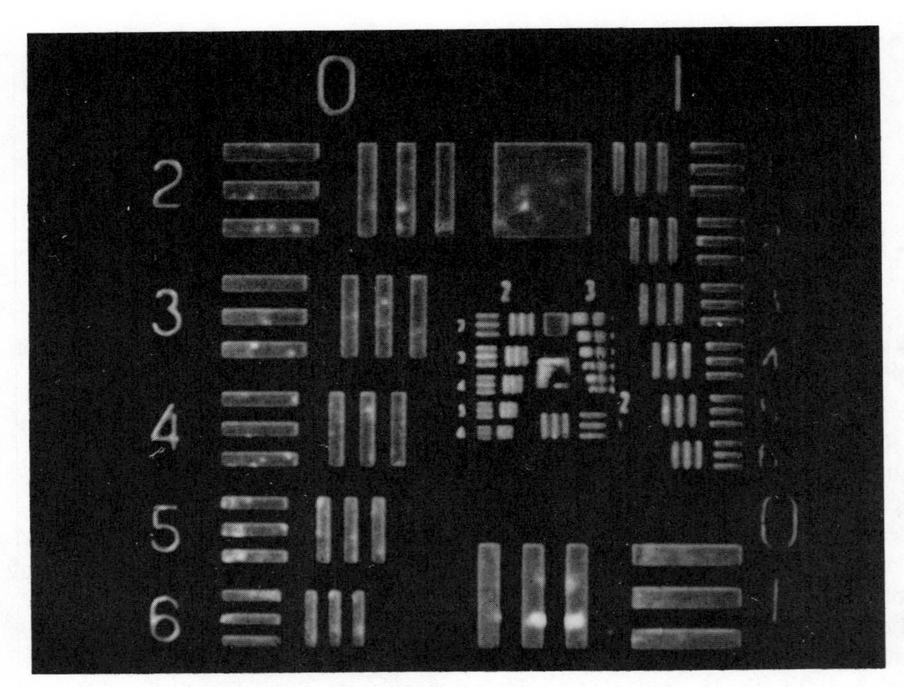

Fig. 14.18 A black-and-white pseudocolor encoded hologram image. The edges of the resolution target represent the high spatial frequency components, which are displayed in white (red). The inner regions represents the low spatial frequency components, which are displayed in black (green).

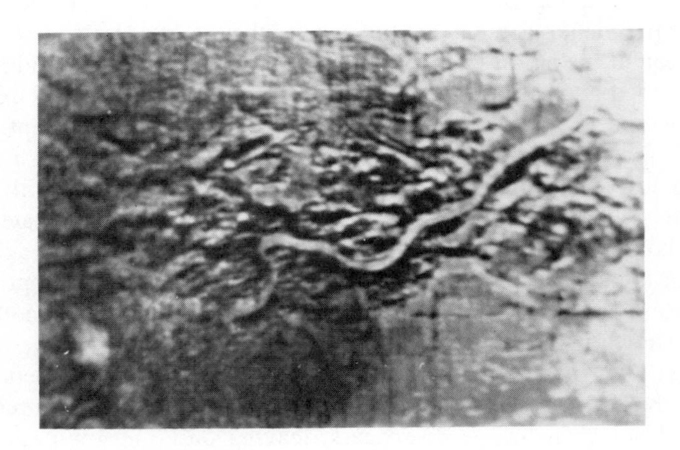

Fig. 14.19 A black-and-white pseudocolor hologram image of an aerial radar image; white (red) represents even terrain and black (green) represents uneven terrain.

for the high-pass filter. The resulting pseudocolor encoding of a black-and-white pseudocolor hologram image is shown in fig. 14.19. The white (red) coded image represents even terrain and the black (green) represents uneven terrain of the radar image.

14.7 MULTISLIT PSEUDOCOLOR ENCODING RAINBOW HOLOGRAM

In sec. 14.3 we have introduced a multislit technique as applied to rainbow holographic interferometry. The basic advantage of the multislit technique is the utilization of a single coherent light source. We now demonstrate that spatial filtered pseudocolor holographic imaging can also be obtained by a multislit technique (ref. 14.17).

In multislit pseudocolor encoding, a set of prepositioned narrow parallel slits is inserted for sequential spatial filtered encoding, as shown in fig. 14.20. For simplicity of demonstration, a double-slit spatial filter pseudocolor encoder is used. A high-pass spatial filter is used for encoding in the spatial filter is used for encoding in the spatial frequency plane P_2. Two exposures are made, with and without the spatial filter, alternating which slit is open. In this manner a spatial filtered multiplexed rainbow hologram is recorded. If this recorded hologram is illuminated by a conjugate white-light source, two color hologram images may be seen through the color smeared slit images. At a given viewing angle, the color of the hologram image is predetermined by the wavelength λ of the source, the reference beam angle θ, the separation ΔL of the slits, and the distance D between the imaging lens and output plane.

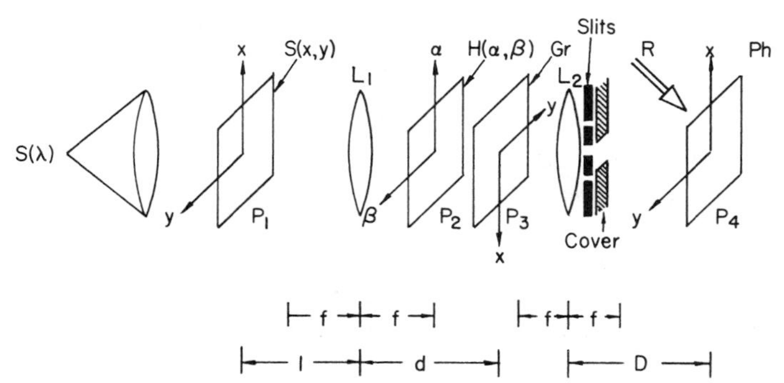

Fig. 14.20 Multislit spatial filtered pseudocolor holographic recording $s(x, y)$, input signal transparency; L_1, transform lens; f, focal length; Gr, ground glass; L_2, imaging lens; R, reference beam; Ph, holographic plate.

The nominal fringe spacing d in the hologram can be determined by the following expression:

$$d = \frac{\lambda}{\sin\left[\theta + \tan^{-1}\left(\frac{\Delta L}{D}\right)\right]}. \tag{14.4}$$

For small $\left(\frac{\Delta L}{D}\right)$,

$$\tan^{-1}\left(\frac{\Delta L}{D}\right) = \left(\frac{\Delta L}{D}\right)$$

and the fringe spacing can be approximated by

$$d \simeq \frac{\lambda}{\sin\theta + \frac{\Delta L}{D}\cos\theta}. \tag{14.5}$$

In the reconstruction process, the hologram image that was derived from each of the smeared slit images can be described by

$$d \sin\theta = \lambda + \Delta\lambda, \tag{14.6}$$

where $\Delta\lambda$ is the incremental deviation in wavelength. From eqs. (14.5) and (14.6) the separation of the slits that causes the generation of color hologram images as a function of the wavelength λ and wavelength deviation $\Delta\lambda$ can be written as

$$\Delta L = \frac{-\Delta L}{\lambda + \Delta\lambda} D \tan\theta. \tag{14.7}$$

Thus to encode the input image transparency with appropriate wavelengths (i.e., λ and $\lambda + \Delta\lambda$), we can place the slit apertures in the appropriate predetermined locations. The same result can also be obtained by changing the reference beam angle during the recording process. However, changing the position of a slit is easier to accomplish in practice than changing the reference beam angle.

The experimental result of a pseudocolor encoded holographic image in a black-and-white picture of a resolution target is seen in fig. 14.21. The hologram image of the high spatial frequency components is displayed in green together with the image of the low spatial frequency components in red. If this hologram is viewed through the smeared hologram slit images, different sets of pseudocolor hologram images can be seen. For example, at one viewing angle the edge

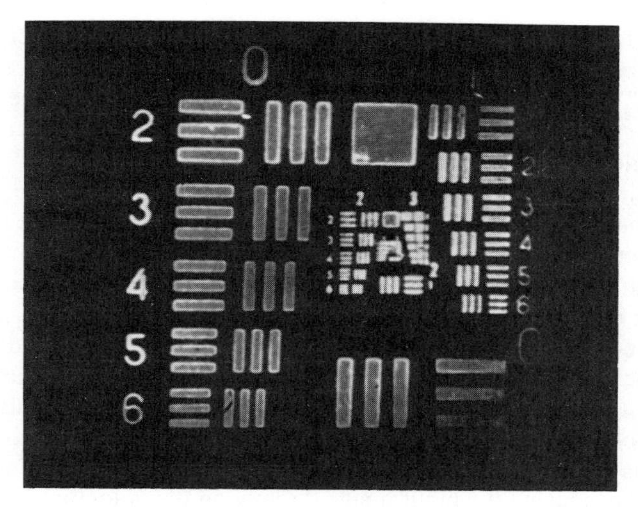

Fig. 14.21 Black-and-white picture of a multicolor hologram image of a resolution target. High spatial frequency components (edges of the rectangles) appear white (green), while the low spatial frequency components appear black (red).

contour will be seen in blue while the background (low spatial frequency image) will be seen in green.

Figure 14.22 is a black-and-white picture of the reconstructed three-color hologram image of a word "OPTICS." For this illustration, the encoding takes place at the diffuse image plane rather than the spatial frequency plane as in the

Fig. 14.22 Black-and-white picture of the multicolor hologram image of the word "OPTICS." "O" and "T" are in blue, "P" and "C" in green, and "I" and "S" in red at one viewing angle.

previous example. The multiplexed rainbow hologram is constructed by using a different pair of letters and a different slit position for each of the three exposures. The slit positions are chosen to yield a hologram image with the letters "O" and "T" in blue, "P" and "C" in green, and "I" and "S" in red at one viewing angle. This example shows the relative ease of controlling the pseudocolor encoding process.

In addition to the color blur problem in the rainbow holographic process, the multislit pseudocolor encoding technique also suffers from the problem of color displacement. With reference to Tamura's result (ref. 14.18), he shows that the lateral color displacement (vertical direction) differs from the longitudinal displacement (z-direction) by the factor $\tan^3 \theta$. If θ is 20°, the vertical displacement is about 5 percent of the longitudinal displacement. Thus the longitudinal color displacement is expected to be more severe than the vertical color displacement. However, for large transparent objects, the misregistration due to lateral displacement is negligible. In addition, longitudinal color displacement can be minimized by focusing the input image directly on the holographic plate during the encoding process, but at the cost of increasing the holographic speckle effect.

CONCLUSION

In the reconstruction of rainbow holographic images, a point source such as a tungsten or a xenon arc lamp should be used in order to obtain the best resolution. The requirement of the source size is dependent upon the distance between the image of the object and the hologram. The source size requirement is less stringent if the image is focused right on the holographic plate. Under such a condition even a fairly extended source size, such as a microscope illuminator or high intensity desk lamp, may be used for illumination with fairly acceptable image resolution.

However, by imaging the object right on the holographic plate, the image redundancy is lost and the image quality is much more sensitive to the imperfections in the reference beam. Since the reconstructed image appears on the same plane as the hologram, the gross interference pattern on the hologram may also affect the image quality. Thus there is a trade-off between source size requirement and image quality. We have found that a distance of 1 to 2 cm between the object image and the hologram generally produces a satisfactory compromise. Another factor that can affect the image quality is the slit size. Generally, the amount of color blur decreases with the decrease in slit size. However, the intensity of the object wave would also be proportionally decreased. This problem can be partially solved by utilizing photographic plates with high sensitivity such as Kodak 131 plates. Moreover, making the slit size small decreases the resolution and increases the speckle size in the reconstructed image. Thus once again, there is a trade-off between color blur and speckle size, and a compromise is required.

For both double-exposure interferometry and the two-wavelength contouring technique, the interference fringes are determined only by the wavelengths of the light source in the holographic recording process. The fringes are not dependent on the wavelengths of the illuminating light in the reconstruction of the holographic images. When the hologram is viewed at the image plane of the reconstructed slit, the image with the interference fringes appears in a single color. The color changes as the viewer moves up or down along the direction of the color smear, but the image and the fringe pattern remain unchanged. If the viewer moves back away from the image plane of the slit, the image appears in a rainbow of colors, but the image and the fringe pattern remain unchanged.

There are certain limits to the application of this white-light holographic technique, and it cannot replace entirely the conventional form of holography. For example, it may not be possible to apply this technique for real-time holographic interferometry. Also, in some holographic interferometry applications, the parallax between the fringes and the object can provide substantial information. This information is lost along one direction with the proposed technique due to the presence of the narrow slit. However, this limitation may be overcome by the multislit technique. We stress that this new technique can be applied to a significant number of interferometric applications and its white-light reconstruction capability offers an important advantage, especially in industrial applications.

In demonstrating multiwavelength and multislit interferometry, we make use of only two wavelengths and two slits. Strictly speaking, a mixture of multiwavelengths and multislits can be used to obtain an even more interesting result. However, there are problems and penalties involved. First, as we multiplex more sets of images into one plate, the diffraction efficiency for each reconstructed image decreases by $1/n^2$, where n is the number of multiplexed holograms. Second, presenting too much information at once sometimes makes the results confusing. In other words, the frequency separation between each color becomes less as the number of wavelengths or slits is increased. This makes discrimination between the different images more difficult.

In the applications of this one-step rainbow holographic process, we have shown that color images can be easily generated by the one-step technique. This one-step technique is simpler than all the other color holographic techniques existing to date. We have also shown the feasibility of the one-step technique as applied to the archival storage of color films, as well as several advantages that the one-step archival process offers. In practice, however, some degree of color blur and coherent artifact noise still cannot be avoided. The operating procedure to achieve a true color hologram image is still generally elaborate and expensive. Thus the practical problem of applying this rainbow holographic technique to the archival storage of color films is still unsolved.

We have also introduced techniques for pseudocolor holographic imaging using the one-step technique. The techniques provide a wide range of spatial encodings, for which a vast variety of pseudocolor holographic images can be obtained. The possible applications of the pseudocolor encoding technique may range from the pseudocolor coding of aerial photography for remote sensing to the construction of multicolor advertising displays. An important property of this pseudocolor encoding method is that no color filters or color dyes are used in the encoding process. Thus, more natural pseudocolor hologram images can be observed.

REFERENCES

14.1 S. A. Benton, "Hologram Reconstructions with Extended Light Sources," *J. Opt. Soc. Am.*, **59,** 1545 (1969).

14.2 S. A. Benton, "White-Light Transmission/Reflection Holographic Imaging," in E. Marom and A. A. Friesem, Eds., *Proceedings of ICO Conference on Applications of Holography and Optical Information Processing,* Pergamon Press, New York, 1977, pp. 401–409.

14.3 E. N. Leith, "White-Light Holograms," *Sci Am.*, **235,** 80 (1976).

14.4 H. Chen and F. T. S. Yu, "One-Step Rainbow Hologram," *Opt. Lett.*, **2,** 85 (1978).

14.5 F. T. S. Yu, A. Tai, and H. Chen, "One-Step Rainbow Holography: Recent Development and Application," *Opt. Eng.*, **19,** 666 (1980).

14.6 C. M. Vest, *Holographic Interferometry,* John Wiley, New York, 1979.

14.7 F. T. S. Yu and H. Chen, "Rainbow Holographic Interferometry," *Opt. Commun.*, **25,** 173 (1978).

14.8 F. T. S. Yu, A. Tai, and H. Chen, "Multiwavelength Rainbow Holographic Interferometry," *Appl. Opt.*, **18,** 212 (1979).

14.9 A. Tai, F. T. S. Yu, and H. Chen, "Multislit One-Step Rainbow Holographic Interferometry," *Appl. Opt.*, **18,** 6 (1979).

14.10 J. R. Varner, "Simplified Multiple-Frequency Holographic Contouring," *Appl. Opt.*, **10,** 212 (1971).

14.11 C. S. Ih, "Multicolor Imagery from Holograms by Spatial Filtering," *Appl. Opt.*, **14,** 438 (1975).

14.12 F. T. S. Yu, A. Tai, and H. Chen, "Archival Storage of Color Films by Rainbow Holographic Technique," *Opt. Commun.*, **27,** 307 (1978).

14.13 P. H. Ruterbusch, S. L. Zhuang, and F. T. S. Yu, "Progress Report on Archival Storage of Color Films by a One-Step Rainbow Holographic Process," *Opt. Commun.*, **37,** 335 (1981).

14.14 R. J. Collier, C. B. Burckhardt, and L. H. Lin, *Optical Holography,* Academic Press, New York, 1971.

14.15 F. T. S. Yu, "New Technique of Pseudocolor Encoding of Holographic Imaging," *Opt. Lett.*, **3,** 57 (1978).

14.16 F. T. S. Yu, A. Tai, and H. Chen, "Spatial Filtered Pseudocolor Holographic Imaging," *J. Opt.*, **9,** 269 (1978).

14.17 F. T. S. Yu, T. H. Chao, and M. S. Dymek, "Multislit Spatial Filtered Pseudocolor Holographic Imaging," *Opt. Commun.*, **32,** 225 (1980).

14.18 P. N. Tamura, "Multi-color Image From Superposition of Rainbow Holograms," *SPIE,* **126,** 59 (1977).

Index